Virus Evolution

Current Research and Future Directions

Edited by:

Scott C. Weaver[1], Mark Denison[2], Marilyn Roossinck[3] and Marco Vignuzzi[4]

[1]Institute for Human Infections and Immunity and Departments of Pathology, Microbiology & Immunology, Galveston, TX, USA
[2]Division Infectious Diseases, Vanderbilt University School Medicine, Nashville, TN, USA
[3]Center for Infectious Disease Dynamics, The Pennsylvania State University, University Park, PA , USA
[4]Viral Populations and Pathogenesis Group, Pasteur Institut, Paris, France

Caister Academic Press

ISBN: 978-1-910190-23-4 (paperback)
ISBN: 978-1-910190-24-1 (ebook)

Ebooks

Cover design adapted from Figure 3.3

Contents

Current books of interest

- Omics in Plant Disease Resistance 2016
- Acidophiles: Life in Extremely Acidic Environments 2016
- Climate Change and Microbial Ecology 2016
- Biofilms in Bioremediation: Current Research and Emerging Technologies 2016
- Microalgae: Current Research and Applications 2016
- Gas Plasma Sterilization in Microbiology 2016
- Virus Evolution: Current Research and Future Directions 2016
- Arboviruses: Molecular Biology, Evolution and Control 2016
- Shigella: Molecular and Cellular Biology 2016
- Aquatic Biofilms: Ecology, Water Quality and Wastewater Treatment 2016
- Alphaviruses: Current Biology 2016
- Thermophilic Microorganisms 2015
- Flow Cytometry in Microbiology: Technology and Applications 2015
- Probiotics and Prebiotics: Current Research and Future Trends 2015
- Epigenetics: Current Research and Emerging Trends 2015
- Corynebacterium glutamicum 2015
- Advanced Vaccine Research Methods for the Decade of Vaccines 2015
- Antifungals 2015
- Bacteria-Plant Interactions: Advanced Research and Future Trends 2015
- Aeromonas 2015
- Antibiotics: Current Innovations and Future Trends 2015
- Leishmania: Current Biology and Control 2015
- Acanthamoeba: Biology and Pathogenesis (2nd edition) 2015
- Microarrays: Current Technology, Innovations and Applications 2014
- Metagenomics of the Microbial Nitrogen Cycle 2014
- Pathogenic Neisseria 2014
- Proteomics: Targeted Technology, Innovations and Applications 2014
- Biofuels: From Microbes to Molecules 2014
- Human Pathogenic Fungi: 2014
- Applied RNAi: From Fundamental Research to Therapeutic Applications 2014
- Halophiles: Genetics and Genomes 2014
- Molecular Diagnostics: Current Research and Applications 2014
- Phage Therapy: Current Research and Applications 2014
- Bioinformatics and Data Analysis in Microbiology 2014
- The Cell Biology of Cyanobacteria 2014

Related books

Arboviruses
Molecular Biology, Evolution and Control
Edited by: N Vasilakis, DJ Gubler
c. 209 pp, April 2016
Book: ISBN 978-1-910190-21-0
£159/$319
Ebook: ISBN 978-1-910190-22-7
£159/$319

Topics range from taxonomy and genome organization, to arbovirus diagnostics and control. Essential reading.

Alphaviruses
Current Biology
Edited by: S Mahalingam, L Herrero, B Herring
x + 184 pp, January 2016
Book: ISBN 978-1-910190-15-9 £159/$319
Ebook: ISBN 978-1-910190-16-6
£159/$319

As the first to focus exclusively on alphaviruses, this book serves as an invaluable resource for researchers and clinicians working with alphaviruses and related viruses. Essential reading.

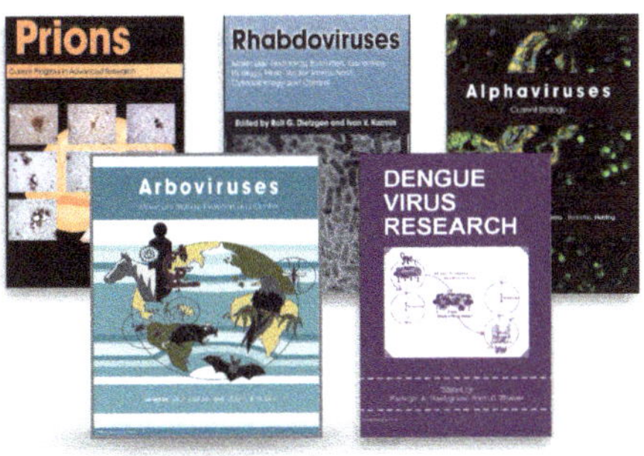

Advanced Vaccine Research Methods for the Decade of Vaccines
Edited by: F Bagnoli, R Rappuoli
xiv + 456 pp, April 2015
Book: ISBN 978-1-910190-03-6 £180/$360
Ebook: ISBN 978-1-910190-04-3
£180/$360

A thorough and up-to-date review of vaccinology research in age of omics technologies. Essential reading for everyone working in vaccine research and development.

Applied RNAi
From Fundamental Research to Therapeutic Applications
Edited by: P Arbuthnot, MS Weinberg
x + 252 pp, June 2014
Book: ISBN 978-1-908230-43-0 £159/$319
Ebook: ISBN 978-1-908230-67-6
£159/$319

"Essential reading" (Biotechnol Agron Soc Environ); "recommended" (Fungal Diversity); "an excellent resource" (Doodys)

Phage Therapy
Current Research and Applications
Edited by: J Borysowski, R Międzybrodzki, A Górski
xvi + 378 pp, April 2014
Book: ISBN 978-1-908230-40-9 £180/$360
Ebook: ISBN 978-1-908230-74-4
£180/$360

"comprehensive overview" (BioSpektrum); "timely and comprehensive" CID

1

Viral Mutation Rates

Rafael Sanjuán

Instituto Cavanilles de Biodiversidad y Biología Evolutiva, 46980 Paterna, Valencia, Spain
rafael.sanjuan@uv.es

Abstract

Spontaneous mutations resulting from uncorrected replication errors, edition of the genetic material, or spontaneous nucleic acid damage are the primary source of genetic diversity, and central to the evolutionary process. Viruses show ample variation in their rates of spontaneous mutation. RNA viruses show extremely high rates whereas DNA viruses mutate more slowly, yet still considerably faster than their hosts. This chapter addresses the molecular mechanisms underlying the production of spontaneous mutations in viruses, such as polymerase infidelity, lack of proofreading, avoidance of post-replicative repair, or host-mediated nucleoside deamination. The evolution of viral mutation rates is also addressed, focusing on the association between mutation rates and other basic genomic properties, on how viruses can modify replication fidelity in response to selection, and on the ability of viruses to optimize mutation rates for maximal adaptability. Finally, some implications for pathogenesis, the emergence of drug resistance, and vaccination are discussed.

Overview

The mutation rate is defined as the probability that a change in genetic information is passed on to the next generation. Mutation rates thus refer to molecular processes such as replication errors, base editions or nucleic acid damage which fail to be corrected by proofreading mechanisms of post-replicative repair. By convention, a viral generation is defined by the completion of an infection cycle at the cellular level, and thus mutation rates are commonly expressed in per-cell units. Alternatively, the mutation rate can be given per round of genome copying, although this definition does not take into account the fact that some mutations appear outside replication. Importantly, the mutation rate is different from the process of allele fixation at the viral population level. The latter is determined not only by the mutation rate, but also by virological, demographical, ecological, and population-genetic factors, including the duration of the infection, latency, transmission route, tissue tropism, selective pressures, etc. and thus should be better referred to as the rate of molecular evolution (Duffy *et al.*, 2008).

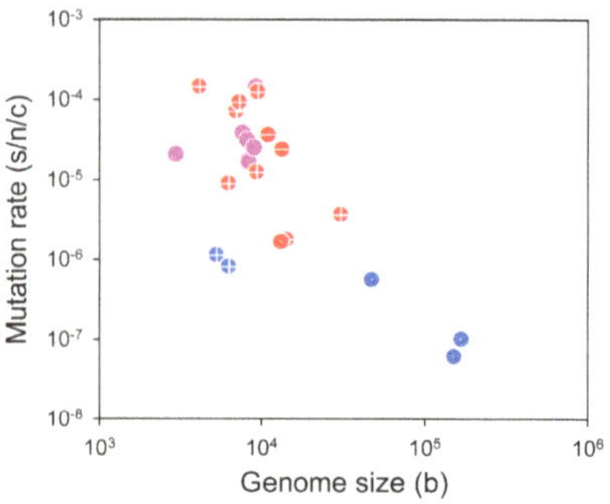

Figure 1.1. Variability in viral mutation rates. Mutation rates as substitutions per nucleotide per cell (s/n/c) are shown as a function of genome size (bases, b). RNA viruses are indicated in red (plus and minus strand viruses depicted with + and – symbols), reverse-transcribing viruses in pink, and DNA viruses in blue (single strand with crossed circles, and double strand with dotted circles). Data were taken from Sanjuán *et al.*, (2010) and Bradwell *et al.*, (2013).

Viruses are the biological systems showing the highest variability in mutation rates, from 10^{-8} to 10^{-4} mutations per nucleotide per cell (m/n/c) (Sanjuán *et al.*, 2010; Figure 1.1). For comparison, eubacterial mutation rates vary approximately by 50-fold and those of eukaryotes (from yeast to humans) by 100-fold (Lynch, 2010). The evolutionary properties of different types of viruses vary according to their mutation rates. Typically, RNA viruses mutate and thus evolve faster than DNA viruses, although there may be some exceptions to this rule (Duffy and Holmes, 2008). Some RNA viruses and viroids show the highest mutation rates described in any biological entity, with 1.4×10^{-4} m/n/c in bacteriophage Qβ (Bradwell *et al.*, 2013) and one mutation every 400 bases in chrysanthemum chlorotic mottle viroid (Gago *et al.*, 2009). Thus, in RNA viruses, each single-cell burst releases dozens of new mutations, giving rise to extremely variable and fast-evolving populations often referred to as molecular quasispecies (Domingo *et al.*, 1978; Holland *et al.*, 1982).

Viral mutation rates have important biomedical implications. For instance, in patients infected with HIV-1, any possible single-base substitution conferring drug resistance is produced daily, making the simultaneous administration of multiple drugs necessary in order to control the infection (Perelson, 2002). The viral mutation rate is also an important parameter in the assessment of vaccination strategies (Davenport *et al.*, 2008), and determines the likelihood that live attenuated vaccines revert to virulence (Vignuzzi *et al.*, 2008). Fast mutation can also compromise the ability of the host immune system to clear the infection and, as such, the mutation rate can be considered as a virulence factor (Pfeiffer and Kirkegaard, 2005; Vignuzzi *et al.*, 2006). The viral mutation rate also influences the risk of disease emergence at the epidemiological level (Holmes, 2009; Pepin *et al.*, 2010). Finally, high mutation rates make evolutionary molecular clocks 'tick' faster in RNA viruses than in any other natural evolving system, making it possible, for instance, to identify the source of recent outbreaks (Bhattacharya, 2014).

Quantitation of viral mutation rates

The fidelities of RNA virus polymerases, such as for instance the HIV-1 reverse transcriptase (Menéndez-Arias, 2009), poliovirus 3D protein (Liu *et al.*, 2013), or hepatitis C virus (HCV) NS5B protein (Powdrill *et al.*, 2011), have been determined *in vitro* by analyzing the pre-steady-state kinetics of base misincorporation for different base combinations, as well by performing replication assays in which the template carries phenotypic reporters for subsequent mutation scoring (Menéndez-Arias, 2009). Although these *in-vitro* studies provide invaluable information, the mutation rates inferred in the test tube tend to differ substantially from those measured in cell cultures or *in vivo*. In biochemical assays, saturating amounts of essential compounds such as nucleotides or divalent cations are often used, potentially modifying polymerase fidelity and the type of mutations produced. More importantly, viral mutation rates are also determined by host factors which are typically not accounted for by these studies.

A variety of methods have been developed to measure viral mutation rates under highly controlled, cell culture conditions. A simple approach consists of inoculating cells with a quasi-genetically homogeneous virus (e.g. an isolated plaque, or an infectious clone) and quantifying the genetic diversity produced in the short term (one ore few transfers), usually by sequencing. The mutation rate can then be estimated by dividing the observed number of mutations by the number of infection cycles elapsed. However, this approach fails to sample many deleterious mutations, since their frequency is rapidly lowered by selection, although selection bias can be corrected if the distribution of mutational fitness effects is known (Sanjuán *et al.,* 2010). In retroviruses, culturing systems have been developed that allow propagating the virus in the absence of selection, thus avoiding this problem. Another common source of bias in these studies is the introduction of artifactual mutations during RT-PCR and sequencing. This problem is particularly acute when using next-generation sequencing (NGS) because per-read error rates are much higher than the actual frequency of the spontaneous mutations to be detected. Recent technological advances such as the CircSeq (Acevedo and Andino, 2014) and the Duplex sequencing (Kennedy *et al.,* 2014) methods increase the accuracy of NGS by orders of magnitude and will permit the analysis of rare variants with unprecedented depth. Finally, a more classical method for measuring viral mutation rates is the Luria-Delbrück fluctuation test (Luria and Delbrück, 1943; Zheng, 2005). This requires seeding a large number of parallel cultures with a small initial number of viruses each. After viral growth has taken place in a standard, permissive medium, spontaneous mutants are phenotypically scored by plating the virus in the presence of a selective agent, such a monoclonal antibody, an antiviral, or a non-permissive cell type. The distribution of the number of mutants per assayed culture can then be used for estimating the mutation rate.

In contrast to the variety of solutions devised for measuring viral mutation rates in cell cultures, performing such measurements *in vivo* is a complex task because observed mutation frequencies are affected by a number of uncontrolled factors, including selection against deleterious mutations, the unknown number of rounds of copying, population bottlenecks, and so on. Transgenic plants have been engineered to redundantly express viral genes and thus relieve these genes from selection by trans-complementation (Malpica *et al.,* 2002; Tromas and Elena, 2010). Another strategy consists of focusing exclusively on mutations that fully suppress viral infectivity (lethal mutations). Since, by definition, these mutants do not contribute to the next viral generation, their population frequency should stay constant and equal to the mutation rate. This method has allowed estimation of the *in vivo* mutation rate of a viroid (Gago *et al.,* 2009) and HCV (Cuevas *et al.,* 2009b; Ribeiro *et al.,* 2012). However, lethal mutations are extremely rare, and their observed frequencies can be strongly affected by RT-PCR and sequencing errors. Finally, for extremely well characterized viruses such as HCV, it is possible to estimate the number of rounds of copying elapsed during acute infection, thus making it possible to calculate mutation rates from sequence data in a way analogous to cell culture studies (Ribeiro *et al.*, 2012).

Molecular determinants of viral mutation rates

Intrinsic fidelity of viral polymerases

Viral mutation rates are primarily determined by the intrinsic fidelity of their polymerases. Variants with different degrees of fidelity have been isolated in picornaviruses, alphaviruses, and retroviruses, although those which do not severely compromise viral fitness tend to have mild effects on mutation rate (Arias *et al.*, 2008; Coffey *et al.*, 2011; Graci *et al.*, 2012; Menéndez-Arias, 2009; Pfeiffer and Kirkegaard, 2003). It has also been established that intrinsic fidelity can be determined by residues located outside the catalytic domain (Gong and Peersen, 2010; Korneeva and Cameron, 2007). For instance, reorientation of the triphosphate moiety of the incoming nucleotide is a key fidelity checkpoint for the poliovirus 3D protein (Arnold *et al.*, 2005). Finally, intrinsic polymerase fidelity also depends on the properties of the template. Template-primer misalignment at homopolymeric runs can cause frameshift mutations or lead to base substitutions through a template-primer dislocation mechanism (Kunkel, 1985), and replication pausing sites appear to be particularly prone to mutation (Ji *et al.*, 1994). Also, RNA secondary structures have been shown to promote template switching, a primary source of recombinant variants (Pathak and Temin, 1992; Galetto *et al.*, 2004; Simon-Loriere *et al.*, 2009). However, the presence/absence of 3' exonuclease proofreading activity in viral polymerases is probably the major determinant of RNA virus mutation rates. All RNA virus polymerases studied thus far, except those of coronaviruses (Denison *et al.*, 2011; Ulferts and Ziebuhr, 2011), lack 3' exonuclease activity (Roberts *et al.*, 1988; Steinhauer *et al.*, 1992), thus explaining the higher mutation rates of RNA viruses compared to DNA viruses. Although DNA virus-encoded polymerases are generally believed to exhibit high fidelity, there appear to be some exceptions. For instance, polymerase X encoded by the African swine fever virus (ASFV) exhibits an exceptionally high

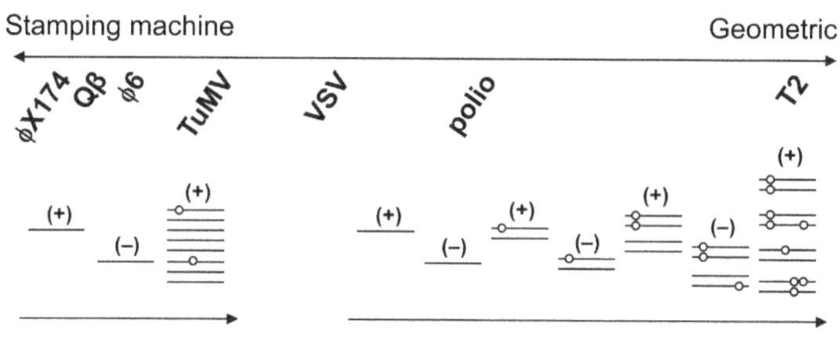

Figure 1.2. Effect of replication mode on per-cell mutation rates. Left: stamping machine replication mode, in which a single antigenome (–) is used to synthesize all progeny genomes within a cell (+). Right: semi-conservative (binary), geometric replication, in which the number of progeny genomes increases by twofold in each replication cycle. Arrows indicate the progression of viral replication within a cell. Top: approximate location of several viruses in the stamping machine-geometric continuum, as determined by Chao *et al.*, (2002), Sanjuán *et al.*, (2010), Martínez *et al.*, (2011), Garcia-Villada and Drake (2012), and Schulte *et al.*, (2015).

error rate, potentially contributing to the genetic diversity of this virus (de Villiers *et al.,* 2010; Lamarche *et al.,* 2006).

Replication mode

Viruses fundamentally differ from cells in that they can adopt diverse replication modes, from which two conceptually different modes can be distinguished (Figure 1.2). First, in *stamping machine* replication, a single template is used to produce all progeny strands within a given cell, therefore implying that there is only one round of copying per cell. This theoretical scenario, however, cannot be realized because synthesis of a reverse-complementary intermediate is always required. In practice, a virus is said to replicate via the stamping machine model if the parental genome is copied only once to a reverse-complementary genome, which is then used as template for synthesizing all progeny genomes. Second, in the semi-conservative, cell-like replication mode, progeny strands immediately become available as templates for the next round of copying, and the number of strands increases by two-fold in each cycle. This implies that there has to be multiple rounds of copying per cell in order to produce a sufficient number of viral progeny and, therefore, that mutations can be geometrically amplified within cells. As such, even if the intrinsic fidelity of a polymerase remains unchanged, the mean and variance of the number of mutations produced per infected cell depends on the mode of replication. Indeed, semi-conservative and stamping machine replication are just two extremes of a continuum of possible replication modes. For instance, multiple progeny strands can be synthesized from the initial parental genome, but these can then be used for subsequent rounds of replication in the same cell.

Whereas large double-stranded DNA viruses use semi-conservative replication, it has been suggested that stamping machine replication has been favored in RNA viruses because it reduces the accumulation of deleterious mutations in the population (Sardanyés *et al.,* 2009; Sardanyés and Elena, 2011; Thébaud *et al.,* 2010). However, no systematic empirical assessment of RNA virus replication modes has been undertaken. In some cases, the replication mode is largely constrained by the molecular details of replication. For instance, microviruses such as bacteriophage φX174 use the rolling circle mechanism, which enforces a stamping machine-like component (Gillam *et al.,* 1984; Hutchison, III and Sinsheimer, 1966). An easy-to-quantify parameter that provides information about the replication mode is the relative concentration of direct and reverse-complementary strands (i.e. plus/minus). Under fully semi-conservative replication, these two concentrations should be equal, whereas marked asymmetries suggest that replication is closer to the stamping machine model. Also, the Luria-Delbrück fluctuation test provides mutation rates per round of copying, whereas other methods provide them in per-cell units. Therefore, the ratio of the rates estimated using these different units provides an idea of the number of rounds of copying that the virus undergoes in each cell (Sanjuán *et al.,* 2010). The results obtained to date using these various approaches tend to confirm that the stamping machine model prevails in several RNA viruses, including bacteriophages Φ6 (Chao *et al.,* 2002) and Qβ (Garcia-Villada and Drake, 2012), as well as turnip mosaic virus (TuMV) (Martínez *et al.,* 2011). However, some level of geometric amplification appears to take place in TuMV and vesicular stomatitis virus (VSV) (Sanjuán *et al.,* 2010), and a recent study suggests that poliovirus undergoes multiple rounds of copying per cell (Schulte et al. 2015).

Host factors modulating RNA virus mutation rates

Cell culture and *in-vivo* studies have established that viral mutation rates are influenced by the host. A major source of mutations in HIV-1 is constituted by apolipoprotein B mRNA-editing catalytic polypeptide-like enzymes (APOBEC3), particularly APOBEC 3G, 3F and 3H (Harris *et al.,* 2003; Lecossier *et al.,* 2003; Mangeat *et al.,* 2003). The activity of these cytidine deaminases has a very well defined sequence context dependence which results in editing of the second base of some CC and TC dinucleotides in the retroviral cDNA (Armitage *et al.,* 2008). This leads to large numbers of G-to-A base transitions in the genomic RNA. Expression levels of APOBEC3G in patients may determine the rate at which hypermutated HIV-1 genomes accumulate throughout the course of an infection, and may influence disease progression (Land *et al.,* 2008). Another important host-encoded protein that can edit viral genomes is the double-stranded RNA-dependent adenosine deaminase (ADAR), which changes adenosines to inosines. This enzyme was first shown to produce A-to-G hypermutation in measles virus (Cattaneo *et al.,* 1988), and subsequent studies have suggested that it can also hypermutate other paramyxoviruses (van den Hoogen *et al.,* 2014), influenza virus (Susèpne *et al.,* 2011), and rhabdoviruses (Carpenter *et al.,* 2009).

In addition to these editing enzymes, there are other host factors that can potentially modulate viral mutation rates such as the intra-cellular balance and concentration of dNTPs (Bebenek *et al.,* 1992; Diamond *et al.,* 2004; Julias and Pathak, 1998). Some cellular metabolites can induce nucleic acid damage, such as ethanol-derived reactive oxygen species (ROS), which were found to have a mutagenic effect on HCV (Seronello *et al.,* 2011). Also, the type of mutations observed in cucumber mosaic virus (CMV) was found to depend on the plant species (Schneider and Roossinck, 2001; Pita *et al.,* 2007). These differences may be caused by among-host differences in viral mutation rates, or in other evolutionary processes such as selection strength or bottlenecks associated with cell-to-cell movement. In support of the former possibility, and in particular of the role of viral replication fidelity, differences in the ability of CMV to produce genetic diversity were found to be associated with among-host changes in the activity of two virus-encoded polymerases (Pita and Roossinck, 2013).

Despite the purported importance of host-encoded factors, a recent study found that the HIV-1 mutation rate did not vary in T lymphoblasts, glioblastoma and human embryonic kidney cells, although there were differences in the mutation spectrum among these host cells, which might be associated with APOBEC activity levels (Holtz and Mansky, 2013). Similarly, the mutation rate of VSV was found to be constant across different cell types, including primary and immortalized cells, and cells derived from various tissues (Combe and Sanjuán, 2014). However, this study revealed differences between mammalian and insect cells, and similar results were obtained for chikungunya virus (CHIKV) (Rozen-Gagnon *et al.,* 2014). These are two arthropod-borne viruses, and it is generally accepted that this type of transmission imposes evolutionary constraints to viruses because selective pressures can differ extensively between the mammalian host and the insect vector (Jenkins *et al.,* 2002; Woelk and Holmes, 2002). However, the above recent findings suggest that differences in the rate of spontaneous mutation may also contribute to the relatively slow evolution of arthropod-borne viruses.

DNA viruses and post-replicative repair

The classical view that DNA viruses evolve slowly has been challenged during the last decade, after the discovery of some fast-evolving DNA viruses. For instance, the molecular evolution rates of canine parvovirus (Shackelton *et al.,* 2005), human B19 erythrovirus (Shackelton and Holmes, 2006), and polyomavirus (Shackelton *et al.,* 2006) fall within the range accepted for many RNA viruses (Duffy *et al.,* 2008). Indeed, DNA viruses show ample variation in per-base mutation rates (Sanjuán *et al.,* 2010). Probably the main determinant of DNA virus mutation rates is access to post-replicative repair, since this system can filter out more than 99% of spontaneous mutations (Fijalkowska *et al.,* 2012). A well-studied case is the single-stranded DNA bacteriophage φX174, which uses the host DNA III holoenzyme for replication. The polymerase has similar intrinsic fidelity on phage and bacterial DNA (Fersht, 1979; Fersht and Knill-Jones, 1981), yet the mutation rate of the phage is three orders of magnitude higher than that of the host (Cuevas *et al.,* 2009a; Raney *et al.,* 2004). In *E.coli*, strand-specific bidirectional methyl-directed mismatch repair (MMR) is performed by the Dam/MutHLS system (Jiricny, 2013), which discriminates the parental and daughter strands by the presence of a methyl group in the adenosine of a GATC sequence motif located on either side of the mismatch. MutH then cleaves the non-methylated daughter strand, which is degraded and re-synthesized (Fukui, 2010; Li, 2008; Marti *et al.,* 2002; Modrich and Lahue, 1996; Schofield and Hsieh, 2003). However, the genome of bacteriophage φX174 contains no GATC sites, impeding methylation and thus avoiding MMR (Cuevas *et al.,* 2011).

The interaction with DNA viruses and repair pathways is still poorly understood in eukaryotes. Several vertebrate DNA viruses are known to alter the DNA damage response (DDR), which detects DNA lesions and initiates cell cycle arrest and repair. Adenoviruses, herpesviruses, polyomaviruses, and papillomaviruses degrade DDR components and promote DNA damage in the host, and recruitment of DDR components into viral replication centers is also a very common process (Luftig, 2014). However, how this impacts viral mutation rates is still unknown. Indeed, there is very little information about mutation rates in eukaryotic DNA viruses, with essentially one published estimate for herpes simplex virus (HSV-1) (Drake and Hwang, 2005). This scarceness of data is in part due to the fact that classical Sanger sequencing does not provide sufficiently high throughput data to infer the rates of spontaneous mutation in DNA viruses, whereas use of NGS technologies has been limited by the high per-read errors.

Evolution of viral mutation rates

Genomic correlates of viral mutation rate

As noticed more than 20 years ago, mutation rates have coevolved with genome sizes in DNA-based microorganisms, including viruses, bacteria, and unicellular eukaryotes (Drake, 1991; Drake *et al.,* 1998). As a result, the mutation rate per genome has stayed remarkably invariant across these widely different taxa, averaging 0.003 per round of copying. However, the evolutionary forces and the mechanistic basis underlying this observation remain poorly understood. Recent research has addressed whether a similar rule applies to RNA viruses. This has been relatively difficult to assess, because RNA virus genome sizes vary only by tenfold as opposed to a 1000-fold variation in DNA viruses and microorganisms

(Figure 1.1). Nevertheless, several lines of evidence support that mutation rates and genome sizes have also coevolved in RNA viruses. First, coronaviruses show the biggest genomes among RNA viruses (on the order of 30 kb), and are the only RNA viruses encoding a 3′exonuclease proofreading activity (Denison *et al.*, 2011; Ulferts and Ziebuhr, 2011). Ablation of this activity increases the mutation frequency dramatically, thus demonstrating its significant contribution to replication fidelity (Eckerle *et al.*, 2010). Second, genome sizes show a weak, yet significantly negative, correlation with molecular evolution rates in RNA viruses (Sanjuán, 2012). Third, it has been shown that bacteriophage Qβ, which has one of the smallest genomes among RNA viruses, displays a particularly high rate of spontaneous mutation (Bradwell *et al.*, 2013). Several hypotheses may explain why mutation rates and genome sizes should correlate negatively. As originally posited by quasispecies theory, there is an upper limit to the number of mutations a genome can tolerate, a limit known as the error threshold (Eigen *et al.*, 1988; Eigen, 2002; Manrubia *et al.*, 2010). If RNA viruses replicate close to this limit, any increase in genome size should be accompanied by a proportional reduction in the rate of spontaneous mutations. Although the error threshold explanation should not apply to the more slowly mutating DNA viruses, the genetic load of a population still depends directly on the genomic mutation rate. Also, it is possible that RNA virus mutation rates sit at the optimum allowing for maximal adaptability and that, as predicted by population genetics theory, this optimal value depends inversely on genome size (Orr, 2000; Sanjuán, 2012).

Differences in mutation rate are also observed between viruses with single- and double-stranded genomes (Figure 1.1). Genome polarity influences many aspects of RNA virus biology. In positive-stranded viruses, the genomic RNA acts directly as mRNA; whereas in double-stranded RNA viruses (and also in negative-stranded viruses) a previous transcription step is required. In the latter, the genomic RNA is highly protected with nucleoproteins to avoid double-stranded RNA-sensing proteins, which would trigger innate antiviral responses. Although no systematic mutation rate differences between positive- and negative-stranded RNA viruses have been demonstrated, the only available estimate for a double-stranded RNA virus being that of bacteriophage φ6 (Chao *et al.*, 2002), suggests a lower rate compared to single-stranded RNA viruses. This view is indirectly supported by phylogenetic studies, which indicate that double-stranded RNA viruses tend to evolve more slowly than single-stranded RNA viruses (Sanjuán, 2012). The divide between single- and double-stranded viruses is much more evident for DNA viruses, mutation rates being more than tenfold higher in the former (Sanjuán *et al.*, 2010).

Evolutionary processes shaping viral mutation rates

Since spontaneous mutations are the ultimate source of genetic variation, they are required for adaptation. However, mutations are more often deleterious than beneficial and thus tend to increase the genetic load of the population. Based on the principle that natural selection operates in the short term, theory predicts that mutation rates should tend to evolve towards the lowest possible values to minimize this load (Kimura, 1967; Lynch, 2010). However, this is clearly not the case for most viruses, particularly RNA viruses. Several possible explanations have been put forward to explain this. First, the ability to replicate RNA faithfully may be simply biochemically constrained. However, the presence of 3′

exonuclease proofreading activity in coronaviruses contradicts this hypothesis. Second, although mechanistically possible, replication fidelity mechanisms may be too costly for most viruses. Supporting this view, it has been shown that replication rate and fidelity are negatively correlated in VSV and HIV-1 (Furió *et al.,* 2005; Furió *et al.,* 2007). However, in theory the cost of fidelity should also apply to other organisms for which fast replication is important, thus failing to explain why RNA virus mutation rates are orders of magnitude higher than those of DNA viruses. Third, the genetic load depends directly on the fraction of spontaneous mutations that are deleterious. One could hypothesize that this fraction is lower in viruses than in bacteria or eukaryotes, thus reducing the selection pressure for minimizing mutation rates in the former. However, direct measurements of selection coefficients show exactly the opposite pattern, with 20-40% of single-point mutations being effectively lethal in a variety of RNA and single-stranded DNA viruses (Sanjuán, 2010). Fourth, no selection-driven change in mutation rate reduction is expected if the effect of selection on mutation rate-modifier alleles is smaller than that of random genetic drift, a limit known as the *drift barrier* (Lynch, 2011; Sung *et al.,* 2012). However, this barrier should be encountered only when the mutation rate has attained a sufficiently low value and the population size is small, a situation that thus does not seem to fit to most viruses. Fifth, models and experimental data using bacteria have shown that mutator alleles can reach high frequencies in populations subject to strong positive or diversifying selection (LeClerc *et al.,* 1996; Pal *et al.,* 2007; Sniegowski *et al.,* 1997; Taddei *et al.,* 1997). However, these mutators are typically short-lived, particularly in the presence of recombination, thus failing to explain the stably high mutation rates of many viruses. Finally, there are other possible factors determining the evolution of mutation rates such as the topology of the fitness landscape (Clune *et al.,* 2008) or population structure (Jiang *et al.,* 2010). Current evolutionary theory thus provides a variety of explanations for the high mutation rates of many viruses, but none is fully satisfactory.

Response to selection

Direct evidence for the ability to evolutionarily adjust mutation rates comes from a large number of studies in which viruses have been treated with mutagenic nucleoside analogues. In response to these treatments, RNA viruses can evolve changes in key residues of the viral polymerase which control replication fidelity or the affinity for the drug, thus compensating for mutagenesis induction (Agudo *et al.,* 2010; Arias *et al.,* 2008; Coffey *et al.,* 2011; Graci *et al.,* 2012; Pfeiffer and Kirkegaard, 2003; Sierra *et al.,* 2007). Earlier work showed that, in the large DNA bacteriophage T4, a series of polymerase variants can also efficiently suppress the effect of chemical mutagens (Drake *et al.,* 1969; Drake and Greening, 1970). The selective pressure exerted by these treatments and thus, the response to selection, differ depending on viruses and drugs. For instance, 5-azacytidine (AZC) tends to increase the frequency of U→C, A→G, and C→G substitutions in bacteriophage Qβ (Cabanillas *et al.,* 2014); whereas only the third type of change is induced by AZC in lympho-choriomeningitis virus (LCMV) (Grande-Pérez *et al.,* 2002) and in HIV-1 (Dapp *et al.,* 2009). Also, whether nucleoside analogues select for amino acid replacements in the polymerase active site, in other domains of the protein, or even in other proteins will depend on the repertoire of mutations available for each virus and their associated fitness costs.

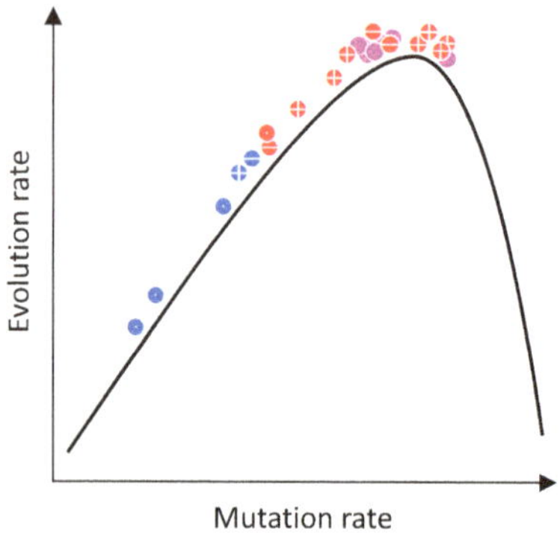

Figure 1.3. Relationship between mutation rate and evolution rate. The neutral theory of molecular evolution posits a linear relationship between these two rates; however, as shown in Sanjuán (2012), evolution rates increase less than linearly and stagnate with increasing mutation rates, suggesting that background selection against deleterious mutations retards the fixation of neutral or beneficial mutations. This creates an evolutionarily optimal mutation rate, which appears to have been reached for RNA viruses. Beyond this optimum, evolution rates would decay sharply due to an excessive genetic load, as also shown experimentally for a number of viruses (see text). Different types of viruses are indicated with the same symbols as in Figure 1.1.

Related to this, viruses which do not encode their own polymerase are particularly interesting for studying the type of changes selected by chemical mutagenesis. Serial passaging of bacteriophage φX174 in the presence of the base analogue 5-fluorouracil (FU) selected for an amino acid replacement in the periplasmic N-terminal domain lysis protein E, which reduced lysis efficiency, allowing the virus to prolong the cell infection cycle and thereby increase its per-cell burst size (Domingo-Calap *et al.,* 2012; Pereira-Gómez and Sanjuán, 2014). A possible explanation for why this change confers resistance to FU has to do with the replication mode of the virus. As detailed above, under stamping machine replication, the number of rounds of copying per cell is fixed. In turn, the number of such cycles (and thus rounds of copying) required for infecting a certain number of host cells is determined by the burst size of the virus. For instance, if each cell releases 100 infectious viral particles, two infection cycles will be required for infecting 10,000 cells starting from a single viral particle, but four cycles will be needed if the burst size is only 10. In the latter case then, the virus would need twice as many rounds of copying to infect all cells, meaning that spontaneous mutations will have more opportunities to accumulate. Therefore, by changing the properties of its lysis protein, bacteriophage φX174 probably reduces the number of rounds of copying required to infect a given number of cells and thereby alleviates the mutagenic effect of the drug.

Optimality of viral mutation rates

The above experiments have solidly established that slight elevations of the mutation rate have a negative impact on viral fitness and adaptability, suggesting that the spontaneous mutation rates of RNA viruses are close to highest tolerable level. On the other hand, RNA viruses carrying increased-fidelity polymerase variants tend to have reduced fitness and virulence in animal models (Gnadig *et al.,* 2012; Pfeiffer and Kirkegaard, 2005; Vignuzzi *et al.,* 2006). In combination, these observations suggest that RNA virus mutation rates are close to their evolutionarily optimum. An additional source of evidence supporting this view comes from phylodynamic studies using dated field isolates (Sanjuán, 2012). In a purely neutral evolutionary scenario, the rate of molecular evolution should increase linearly with mutation rate (Kimura, 1983). This prediction was confirmed for slowly-mutating DNA viruses. However, evolution rates increase less than linearly and stagnate with increasing mutation rates (Figure 1.3). This pattern was explained by a model in which the genetic load associated with transient deleterious mutations retards the fixation of neutral or beneficial alleles. According to this, further increases of the mutation rate would have a negative impact on viral evolvability.

Since the mutation rates of DNA viruses are considerably lower than those of RNA viruses, yet their lifestyles are similar, one may conclude that DNA virus mutation rates are suboptimal in terms of adaptability. Indeed, chemical mutagenesis has not the same detrimental effect in DNA as in RNA viruses. For instance, in a study conducted with bacteriophage T7, a massive number of mutations were detected by NGS in viruses serially passaged under mutagenic conditions, but this increase did not appear to negatively impact viral adaptability (Springman *et al.,* 2009). In another study, the adaptability and molecular evolution rates of several RNA and DNA bacteriophage species were compared using experimental evolution (Domingo-Calap and Sanjuán, 2011). During the initial steps of adaptation, higher mutation rates allowed RNA bacteriophages to adapt up to five times faster than their DNA counterparts, and the former fixed mutations in the population three times faster. However, this difference vanished once viruses reached a moderately high level of adaptation. Therefore, DNA viruses evolve and adapt more slowly than RNA viruses, but the difference is not as large as expected from their average mutation rates. One possible explanation is that DNA viruses do not show a constant mutation rate along their genomes. Supporting this view, in the *Bordetella* phage BPP-1, a site-specific, reverse transcription-based system is used to selectively introduce large numbers of mutations in a tail fiber gene required for host ligand recognition (Liu *et al.,* 2002). It has also been shown that vaccinia virus deploys so-called genetic accordions to transiently elevate the gene copy number of a *locus* involved in blocking the cellular antiviral response (Elde *et al.,* 2012).

Importance of viral mutation rates for disease

Virulence

The characterization of single amino acid replacements that modify the intrinsic fidelity of RNA virus polymerases has allowed a direct assessment of the effect of viral mutation rates on the ability to cause disease. Serial passaging of poliovirus in the presence of the nucleoside analogue ribavirin led to the selection of substitution G64S in the viral polymerase, and subsequent analysis of genetic diversity by sequencing indicated that the G64S substitution increased polymerase fidelity by approximately threefold, while replication kinetics was apparently retained (Pfeiffer and Kirkegaard, 2003). It was found that the loss of genetic diversity associated with the G64S replacement compromised the ability of the virus to evolve drug resistance or to escape antibody neutralization in cell cultures over the short term and that, furthermore, the G64S substitution produced a moderate (Pfeiffer and Kirkegaard, 2005) to strong (Vignuzzi *et al.,* 2006) attenuation of the virus in a murine infection model. These results led to the suggestion that the maintenance of a high mutation rate is required for the virus to evade the innate immune response, or to adapt to different microenvironments *in vivo.* In the case of poliovirus, virulence is mainly determined by the ability to reach and productively infect the nervous system, a capacity that was lost in the G64S variant (Vignuzzi *et al.,* 2006). Similar results were later obtained with an increased-fidelity variant of CHIKV (Coffey *et al.,* 2011). As for poliovirus, this alphavirus variant exhibited reduced genetic diversity and impaired ability to infect and disseminate in the natural mosquito host. Later on, another variant of the poliovirus polymerase (H273R) with reduced replication fidelity was identified (Korboukh *et al.,* 2014). Interestingly, this variant was also found to be attenuated *in vivo*, as judged by the ability to cause paralysis in mice. Again, reduced virulence correlated with impaired ability of the H273R virus to productively infect certain organs. Overall, these studies establish a connection between the viral mutation rate and virulence.

As mentioned above, DNA virus mutation rates are substantially lower than those of RNA viruses, yet DNA viruses can also reach high virulence levels, as exemplified by the smallpox virus or ASFV. As discussed above, the ability of DNA viruses to reach high per-base mutation rates is impeded by genetic load. However, DNA viruses also need to respond to immune pressure and adapt to host tissues, in the same ways RNA viruses do. Undoubtedly, the large genomes of some DNA viruses allow them to encode a battery of genes that antagonize the innate immune response (Bahar *et al.,* 2011), as opposed to RNA viruses. Still, phylogenetic analyses reveal a fast rate of evolution in DNA virus genes involved in the host-pathogen arms race (McLysaght *et al.,* 2003). Specific elevation of the mutation rate at these *loci* using mechanisms such as genetic accordions probably contributes to this fast evolution (Elde *et al.,* 2012). Therefore, despite their high average replication fidelity, DNA viruses manage to create the genetic diversity required to defeat host defenses.

Vaccination

Many RNA virus proteins have the ability to block innate immunity (Versteeg and Garcia-Sastre, 2010). As stated above, the small genomes of RNA viruses limit their ability to deploy complex batteries of immune-evasion proteins similar to those found in large DNA viruses. However, alternative strategies have evolved. For instance, HIV-1 achieves neutralizing antibody evasion by concealing surface epitopes located in the envelope protein gp120 under a dense glycan shield (Reitter *et al.,* 1998; Wei *et al.,* 2003). Other human viruses, on the other hand, such as HCV have not seemingly adopted this same strategy, or at least not so extensively. Probably the most common mechanism allowing RNA viruses to keep ahead of host adaptive immunity is the fast production of amino acid changes at key antigenic sites. This produces a rapid *antigenic drift*, as has been extensively documented for seasonal influenza (Smith *et al.,* 2004). Since the rate of evolution in antigenic sites rapidly erases cross-protection, influenza vaccines need to be updated annually. Adaptation to host cellular immunity by the rapid fixation of escape mutations in CTL epitopes has also been evidenced in HIV-1 (Kawashima *et al.,* 2009) and HCV (Gaudieri *et al.,* 2006). In these two viruses, considerable within-host variability at antigenic sites and a huge global diversity pose serious obstacles to the development of vaccines (Korber *et al.,* 2009). Another problem associated with the extremely high mutation rates of RNA viruses is reversion to virulence of live attenuated vaccines. In poliovirus, individuals vaccinated with the standard Sabin strain occasionally shed vaccine-derived viruses which have the potential to cause pathogenic infections in non-vaccinated individuals (Georgescu *et al.,* 1997). Work on polymerase fidelity variants suggests that RNA virus mutation rates could be engineered to create new, safer live vaccines. For instance, the G64S and H273R fidelity variants of poliovirus have been shown to elicit a protective immune response (Vignuzzi *et al.*, 2008; Korboukh *et al.*, 2014). High-fidelity variants seem to be better vaccine candidates than low-fidelity ones, because they are less likely to fix new virulence mutations or undergo reversion of attenuating mutations. Introducing changes that reduce the rate of recombination would also be desirable to minimize the risk of appearance of recombinants between the vaccine strain and circulating wild-type viruses.

Drug resistance and antiviral strategies

Antiviral drugs inhibiting essential steps of the viral infection cycle can be extremely effective in the short term. However, drug-resistant viruses tend to evolve rapidly, leading to treatment failure. Owing to the high mutation rates of RNA viruses, all possible single nucleotide substitutions and many double mutants are produced daily in each infected patient. Therefore, resistance mutants often pre-exist in the viral population as minority variants before treatment starts. Use of NGS now allows the detection of these low-frequency polymorphisms and is thus a useful tool for assisting decision making in the clinical setting (Nelson and Hughes, 2015). A famous example of drug resistance is AZT, the first FDA-approved treatment against AIDS. Although transiently effective, resistance mutations located in the HIV-1 reverse transcriptase inevitably appear at the highest tolerated doses. However, in combination with other antiretroviral drugs, AZT contributes to efficiently halting viral replication and dramatically reduces the risk of drug resistance. These lessons should also be valid for other fast-mutating RNA viruses, such as HCV. Characterization of the rate and spectrum of mutations can also help predicting drug resistance. For instance, the HCV polymerase NS5B has a highly biased mutation spectrum, producing some types of substitutions thousands of times faster than others (Powdrill *et al.*, 2011). Drug resistance is usually conferred by one or few amino acid substitutions, which thus involve only a few nucleotide changes. If these substitutions are amongst the rarest in the mutational spectrum, the emergence of drug resistant variants will be delayed. This, together with the fitness costs associated with resistance mutations, determine the so-called *genetic barrier* to drug resistance (Svarovskaia *et al.*, 2014).

Beyond the evolutionary optimal mutation rate, viral fitness and adaptability decays dramatically due to the accumulation of deleterious mutations which directly reduce fitness and interfere with the fixation of beneficial mutations. The viral population may eventually cross the error threshold, leading to the irreversible loss of genetic information. Based on this, it was suggested that RNA viruses could be combated using specific mutagens (Anderson *et al.*, 2004; Bull *et al.*, 2007; Domingo, 2006). The idea that RNA viruses may be particularly sensitive to lethal mutagenesis was spurred by early experiments demonstrating that slight increases in mutation frequencies led to dramatic fitness losses in poliovirus and VSV (Holland *et al.*, 1990). Lethal mutagenesis later proved effective under cell culture conditions and in animal models against a variety of RNA viruses including enteroviruses (Crotty *et al.*, 2001; Graci *et al.*, 2007), aphtoviruses (Sierra *et al.*, 2000), hantaviruses (Chung *et al.*, 2007), arenaviruses (Grande-Pérez *et al.*, 2005), and lentiviruses (Dapp *et al.*, 2009; Loeb *et al.*, 1999). Indeed, some classical antivirals such as ribavirin or amiloride have been later shown to be mutagens for RNA viruses (Crotty *et al.*, 2001; Levi *et al.*, 2010), and analysis of

HCV sequences from patients treated with interferon/ribavirin also suggested a mutagenic effect *in vivo* (Cuevas *et al.,* 2009b). A clinical trial tested the feasibility of lethal mutagenesis in HIV-1, but was unsuccessful (trial NCT00129194) (Mullins *et al.,* 2011). An added problem is that, contrary to initial expectations, lethal mutagenesis also selects for resistances, including the emergence of higher-fidelity variants.

References

Acevedo, A. and Andino, R. (2014). Library preparation for highly accurate population sequencing of RNA viruses. Nat. Protoc. *9*, 1760-1769.

Agudo, R., Ferrer-Orta, C., Arias, A., de la Higuera, I., Perales, C., Pérez-Luque, R., Verdaguer, N., and Domingo, E. (2010). A multi-step process of viral adaptation to a mutagenic nucleoside analogue by modulation of transition types leads to extinction-escape. PLoS Pathog. *6*, e1001072.

Anderson, J. P., Daifuku, R., and Loeb, L. A. (2004). Viral error catastrophe by mutagenic nucleosides. Annu. Rev. Microbiol. *58*, 183-205.

Arias, A., Arnold, J. J., Sierra, M., Smidansky, E. D., Domingo, E., and Cameron, C. E. (2008). Determinants of RNA-dependent RNA polymerase (in)fidelity revealed by kinetic analysis of the polymerase encoded by a foot-and-mouth disease virus mutant with reduced sensitivity to ribavirin. J. Virol. *82*, 12346-12355.

Armitage, A. E., Katzourakis, A., de Oliviera, T., Welch, J. J., Belshaw, R., Bishop, K. N., Kramer, B., McMichael, A. J., Rambaut, A., and Iversen, A. K. (2008). Conserved footprints of APOBEC3G on Hypermutated human immunodeficiency virus type 1 and human endogenous retrovirus HERV-K(HML2) sequences. J. Virol. *82*, 8743-8761.

Arnold, J. J., Vignuzzi, M., Stone, J. K., Andino, R., and Cameron, C. E. (2005). Remote site control of an active site fidelity checkpoint in a viral RNA-dependent RNA polymerase. J. Biol. Chem. *280*, 25706-25716.

Bahar, M. W., Graham, S. C., Chen, R. A., Cooray, S., Smith, G. L., Stuart, D. I., and Grimes, J. M. (2011). How vaccinia virus has evolved to subvert the host immune response. J. Struct. Biol. *175*, 127-134.

Bebenek, K., Roberts, J. D., and Kunkel, T. A. (1992). The effects of dNTP pool imbalances on frameshift fidelity during DNA replication. J. Biol. Chem. *267*, 3589-3596.

Bhattacharya, S. (2014). Science in court: Disease detectives. Nature *506*, 424-426.

Bradwell, K., Combe, M., Domingo-Calap, P., and Sanjuán, R. (2013). Correlation between mutation rate and genome size in riboviruses: mutation rate of bacteriophage Qβ. Genetics *195*, 243-251.

Bull, J. J., Sanjuán, R., and Wilke, C. O. (2007). Theory of lethal mutagenesis for viruses. J. Virol. *81*, 2930-2939.

Cabanillas, L., Sanjuán, R., and Lázaro, E. (2014). Changes in protein domains outside the catalytic site of the bacteriophage Qbeta replicase reduce the mutagenic effect of 5-azacytidine. J. Virol. *88*, 10480-10487.

Carpenter, J. A., Keegan, L. P., Wilfert, L., O'Connell, M. A., and Jiggins, F. M. (2009). Evidence for ADAR-induced hypermutation of the *Drosophila* sigma virus (Rhabdoviridae). BMC Genet. *10*, 75-10.

Cattaneo, R., Schmid, A., Eschle, D., Baczko, K., ter Meulen, V., and Billeter, M. A. (1988). Biased hypermutation and other genetic changes in defective measles viruses in human brain infections. Cell *55*, 255-265.

Chao, L., Rang, C. U., and Wong, L. E. (2002). Distribution of spontaneous mutants and inferences about the replication mode of the RNA bacteriophage φ6. J. Virol. *76*, 3276-3281.

Chung, D. H., Sun, Y., Parker, W. B., Arterburn, J. B., Bartolucci, A., and Jonsson, C. B. (2007). Ribavirin reveals a lethal threshold of allowable mutation frequency for Hantaan virus. J. Virol. *81*, 11722-11729.

Clune, J., Misevic, D., Ofria, C., Lenski, R. E., Elena, S. F., and Sanjuán, R. (2008). Natural selection fails to optimize mutation rates for long-term adaptation on rugged fitness landscapes. PLoS Comput. Biol. *4*, e1000187.

Coffey, L. L., Beeharry, Y., Borderia, A. V., Blanc, H., and Vignuzzi, M. (2011). Arbovirus high fidelity variant loses fitness in mosquitoes and mice. Proc. Natl. Acad. Sci. USA *108*, 16038-16043.

Combe, M. and Sanjuán, R. (2014). Variation in RNA virus mutation rates across host cells. PLoS Pathog. *10*, e1003855.

Crotty, S., Cameron, C. E., and Andino, R. (2001). RNA virus error catastrophe: direct molecular test by using ribavirin. Proc. Natl. Acad. Sci. USA *98*, 6895-6900.

Cuevas, J. M., Duffy, S., and Sanjuán, R. (2009a). Point mutation rate of bacteriophage ΦX174. Genetics *183*, 747-749.

Cuevas, J. M., Gonzalez-Candelas, F., Moya, A., and Sanjuán, R. (2009b). The effect of ribavirin on the mutation rate and spectrum of Hepatitis C virus *in vivo*. J. Virol. *83*, 5760-5764.

Cuevas, J. M., Pereira-Gómez, M., and Sanjuán, R. (2011). Mutation rate of bacteriophage PhiX174 modified through changes in GATC sequence context. Infect. Genet. Evol. *11*, 1820-1822.

Dapp, M. J., Clouser, C. L., Patterson, S., and Mansky, L. M. (2009). 5-Azacytidine can induce lethal mutagenesis in human immunodeficiency virus type 1. J. Virol. *83*, 11950-11958.

Davenport, M. P., Loh, L., Petravic, J., and Kent, S. J. (2008). Rates of HIV immune escape and reversion: implications for vaccination. Trends Microbiol. *16*, 561-566.

de Villiers, E. P., Gallardo, C., Arias, M., da Silva, M., Upton, C., Martin, R., and Bishop, R. P. (2010). Phylogenomic analysis of 11 complete African swine fever virus genome sequences. Virology *400*, 128-136.

Denison, M. R., Graham, R. L., Donaldson, E. F., Eckerle, L. D., and Baric, R. S. (2011). Coronaviruses: an RNA proofreading machine regulates replication fidelity and diversity. RNA Biol. *8*, 270-279.

Diamond, T. L., Roshal, M., Jamburuthugoda, V. K., Reynolds, H. M., Merriam, A. R., Lee, K. Y., Balakrishnan, M., Bambara, R. A., Planelles, V., Dewhurst, S., and Kim, B. (2004). Macrophage tropism of HIV-1 depends on efficient cellular dNTP utilization by reverse transcriptase. J. Biol Chem. *279*, 51545-51553.

Domingo, E (2006). Quasispecies: concept and implications for virology. Springer.

Domingo, E., Sabo, D., Taniguchi, T., and Weissmann, C. (1978). Nucleotide sequence heterogeneity of an RNA phage population. Cell *13*, 735-744.

Domingo-Calap, P., Pereira-Gómez, M., and Sanjuán, R. (2012). Nucleoside analogue mutagenesis of a single-stranded DNA virus: evolution and resistance. J. Virol. *86*, 9640-9646.

Domingo-Calap, P. and Sanjuán, R. (2011). Experimental evolution of RNA versus DNA viruses. Evolution *65*, 2987-2994.

Drake, J. W. (1991). A constant rate of spontaneous mutation in DNA-based microbes. Proc. Natl. Acad. Sci. USA *88*, 7160-7164.

Drake, J. W., Allen, E. F., Forsberg, S. A., Preparata, R. M., and Greening, E. O. (1969). Genetic control of mutation rates in bacteriophageT4. Nature *221*, 1128-1132.

Drake, J. W., Charlesworth, B., Charlesworth, D., and Crow, J. F. (1998). Rates of spontaneous mutation. Genetics *148*, 1667-1686.

Drake, J. W. and Greening, E. O. (1970). Suppression of chemical mutagenesis in bacteriophage T4 by genetically modified DNA polymerases. Proc. Natl. Acad. Sci. USA *66*, 823-829.

Drake, J. W. and Hwang, C. B. (2005). On the mutation rate of herpes simplex virus type 1. Genetics *170*, 969-970.

Duffy, S. and Holmes, E. C. (2008). Phylogenetic evidence for rapid rates of molecular evolution in the single-stranded DNA begomovirus tomato yellow leaf curl virus. J. Virol. *82*, 957-965.
Duffy, S., Shackelton, L. A., and Holmes, E. C. (2008). Rates of evolutionary change in viruses: patterns and determinants. Nat. Rev. Genet. *9*, 267-276.
Eckerle, L. D., Becker, M. M., Halpin, R. A., Li, K., Venter, E., Lu, X., Scherbakova, S., Graham, R. L., Baric, R. S., Stockwell, T. B., Spiro, D. J., and Denison, M. R. (2010). Infidelity of SARS-CoV Nsp14-exonuclease mutant virus replication is revealed by complete genome sequencing. PLoS Pathog. *6*, e1000896.
Eigen, M. (2002). Error catastrophe and antiviral strategy. Proc. Natl. Acad. Sci. USA *99*, 13374-13376.
Eigen, M., McCaskill, J., and Schuster, P. (1988). Molecular quasi-species. J. Phys. Chem. *92*, 6881-6891.
Elde, N. C., Child, S. J., Eickbush, M. T., Kitzman, J. O., Rogers, K. S., Shendure, J., Geballe, A. P., and Malik, H. S. (2012). Poxviruses deploy genomic accordions to adapt rapidly against host antiviral defenses. Cell *150*, 831-841.
Fersht, A. R. (1979). Fidelity of replication of phage phi X174 DNA by DNA polymerase III holoenzyme: spontaneous mutation by misincorporation. Proc. Natl. Acad. Sci. USA *76*, 4946-4950.
Fersht, A. R. and Knill-Jones, J. W. (1981). DNA polymerase accuracy and spontaneous mutation rates: frequencies of purine.purine, purine.pyrimidine, and pyrimidine.pyrimidine mismatches during DNA replication. Proc. Natl. Acad. Sci. USA *78*, 4251-4255.
Fijalkowska, I. J., Schaaper, R. M., and Jonczyk, P. (2012). DNA replication fidelity in *Escherichia coli*: a multi-DNA polymerase affair. FEMS Microbiol. Rev. *36*, 1105-1121.
Fukui, K. (2010). DNA mismatch repair in eukaryotes and bacteria. J. Nucleic Acids. *2010. pii*, 260512.
Furió, V., Moya, A., and Sanjuán, R. (2005). The cost of replication fidelity in an RNA virus. Proc. Natl. Acad. Sci. USA *102*, 10233-10237.
Furió, V., Moya, A., and Sanjuán, R. (2007). The cost of replication fidelity in human immunodeficiency virus type 1. Proc. Biol. Sci. *274*, 225-230.
Gago, S., Elena, S. F., Flores, R., and Sanjuán, R. (2009). Extremely high mutation rate of a hammerhead viroid. Science *323*, 1308.
Galetto, R., Moumen, A., Giacomoni, V., Veron, M., Charneau, P., and Negroni, M. (2004). The structure of HIV-1 genomic RNA in the gp120 gene determines a recombination hot spot *in vivo*. J Biol. Chem. *279*, 36625-36632.
Garcia-Villada, L. and Drake, J. W. (2012). The three faces of riboviral spontaneous mutation: spectrum, mode of genome replication, and mutation rate. PLoS Genet. *8*, e1002832.
Gaudieri, S., Rauch, A., Park, L. P., Freitas, E., Herrmann, S., Jeffrey, G., Cheng, W., Pfafferott, K., Naidoo, K., Chapman, R., Battegay, M., Weber, R., Telenti, A., Furrer, H., James, I., Lucas, M., and Mallal, S. A. (2006). Evidence of viral adaptation to HLA class I-restricted immune pressure in chronic hepatitis C virus infection. J. Virol. *80*, 11094-11104.
Georgescu, M. M., Balanant, J., MacAdam, A., Otelea, D., Combiescu, M., Combiescu, A. A., Crainic, R., and Delpeyroux, F. (1997). Evolution of the Sabin type 1 poliovirus in humans: characterization of strains isolated from patients with vaccine-associated paralytic poliomyelitis. J. Virol. *71*, 7758-7768.
Gillam, S., Astell, C. R., Jahnke, P., Hutchison, C. A., III, and Smith, M. (1984). Construction and properties of a ribosome-binding site mutation in gene E of phiX174 bacteriophage. J. Virol. *52*, 892-896.
Gnadig, N. F., Beaucourt, S., Campagnola, G., Borderia, A. V., Sanz-Ramos, M., Gong, P., Blanc, H., Peersen, O. B., and Vignuzzi, M. (2012). Coxsackievirus B3 mutator strains are attenuated *in vivo*. Proc. Natl. Acad. Sci. USA *109*, E2294-E2303.

Gong, P. and Peersen, O. B. (2010). Structural basis for active site closure by the poliovirus RNA-dependent RNA polymerase. Proc. Natl. Acad. Sci. USA *107*, 22505-22510.

Graci, J. D., Gnadig, N. F., Galarraga, J. E., Castro, C., Vignuzzi, M., and Cameron, C. E. (2012). Mutational robustness of an RNA virus influences sensitivity to lethal mutagenesis. J. Virol. *86*, 2869-2873.

Graci, J. D., Harki, D. A., Korneeva, V. S., Edathil, J. P., Too, K., Franco, D., Smidansky, E. D., Paul, A. V., Peterson, B. R., Brown, D. M., Loakes, D., and Cameron, C. E. (2007). Lethal mutagenesis of poliovirus mediated by a mutagenic pyrimidine analogue. J Virol. *81*, 11256-11266.

Grande-Pérez, A., Lázaro, E., Lowenstein, P., Domingo, E., and Manrubia, S. C. (2005). Suppression of viral infectivity through lethal defection. Proc. Natl. Acad. Sci. USA *102*, 4448-4452.

Grande-Pérez, A., Sierra, S., Castro, M. G., Domingo, E., and Lowenstein, P. R. (2002). Molecular indetermination in the transition to error catastrophe: systematic elimination of lymphocytic choriomeningitis virus through mutagenesis does not correlate linearly with large increases in mutant spectrum complexity. Proc. Natl. Acad. Sci. USA *99*, 12938-12943.

Harris, R. S., Bishop, K. N., Sheehy, A. M., Craig, H. M., Petersen-Mahrt, S. K., Watt, I. N., Neuberger, M. S., and Malim, M. H. (2003). DNA deamination mediates innate immunity to retroviral infection. Cell *113*, 803-809.

Holland, J. J., Domingo, E., de la Torre, J. C., and Steinhauer, D. A. (1990). Mutation frequencies at defined single codon sites in vesicular stomatitis virus and poliovirus can be increased only slightly by chemical mutagenesis. J. Virol. *64*, 3960-3962.

Holland, J. J., Spindler, K., Horodyski, F., Grabau, E., Nichol, S., and VandePol, S. (1982). Rapid evolution of RNA genomes. Science *215*, 1577-1585.

Holmes, E C (2009). The evolution and emergence of RNA viruses. Oxford University Press.

Holtz, C. M. and Mansky, L. M. (2013). Variation of HIV-1 mutation spectra among cell types. J. Virol. *87*, 5296-5299.

Hutchison, C. A., III and Sinsheimer, R. L. (1966). The process of infection with bacteriophage phi-X174. X. Mutations in a phi-X Lysis gene. J. Mol. Biol. *18*, 429-447.

Jenkins, G. M., Rambaut, A., Pybus, O. G., and Holmes, E. C. (2002). Rates of molecular evolution in RNA viruses: a quantitative phylogenetic analysis. J. Mol. Evol. *54*, 156-165.

Ji, J., Hoffmann, J. S., and Loeb, L. (1994). Mutagenicity and pausing of HIV reverse transcriptase during HIV plus-strand DNA synthesis. Nucleic Acids Res. *22*, 47-52.

Jiang, X., Mu, B., Huang, Z., Zhang, M., Wang, X., and Tao, S. (2010). Impacts of mutation effects and population size on mutation rate in asexual populations: a simulation study. BMC Evol. Biol. *10*, 298.

Jiricny, J. (2013). Postreplicative mismatch repair. Cold Spring Harb. Perspect. Biol. *5*, a012633.

Julias, J. G. and Pathak, V. K. (1998). Deoxyribonucleoside triphosphate pool imbalances *in vivo* are associated with an increased retroviral mutation rate. J. Virol. *72*, 7941-7949.

Kawashima, Y., Pfafferott, K., Frater, J., Matthews, P., Payne, R., Addo, M., Gatanaga, H., Fujiwara, M., Hachiya, A., Koizumi, H., Kuse, N., Oka, S., Duda, A., Prendergast, A., Crawford, H., Leslie, A., Brumme, Z., Brumme, C., Allen, T., Brander, C., Kaslow, R., Tang, J., Hunter, E., Allen, S., Mulenga, J., Branch, S., Roach, T., John, M., Mallal, S., Ogwu, A., Shapiro, R., Prado, J. G., Fidler, S., Weber, J., Pybus, O. G., Klenerman, P., Ndung'u, T., Phillips, R., Heckerman, D., Harrigan, P. R., Walker, B. D., Takiguchi, M., and Goulder, P. (2009). Adaptation of HIV-1 to human leukocyte antigen class I. Nature *458*, 641-645.

Kennedy, S. R., Schmitt, M. W., Fox, E. J., Kohrn, B. F., Salk, J. J., Ahn, E. H., Prindle, M. J., Kuong, K. J., Shen, J. C., Risques, R. A., and Loeb, L. A. (2014). Detecting ultralow-frequency mutations by Duplex Sequencing. Nat. Protoc. *9*, 2586-2606.
Kimura, M. (1967). On the evolutionary adjustment of spontaneous mutation rates. Genet. Res. , Camb. *9*, 23-34.
Kimura, M (1983). The neutral theory of molecular evolution. Cambridge University Press.
Korber, B. T., Letvin, N. L., and Haynes, B. F. (2009). T-cell vaccine strategies for human immunodeficiency virus, the virus with a thousand faces. J. Virol. *83*, 8300-8314.
Korboukh, V. K., Lee, C. A., Acevedo, A., Vignuzzi, M., Xiao, Y., Arnold, J. J., Hemperly, S., Graci, J. D., August, A., Andino, R., and Cameron, C. E. (2014). RNA virus population diversity, an optimum for maximal fitness and virulence. J. Biol. Chem. *289*, 29531-29544.
Korneeva, V. S. and Cameron, C. E. (2007). Structure-function relationships of the viral RNA-dependent RNA polymerase: fidelity, replication speed, and initiation mechanism determined by a residue in the ribose-binding pocket. J. Biol. Chem. *282*, 16135-16145.
Kunkel, T. A. (1985). The mutational specificity of DNA polymerase-beta during *in vitro* DNA synthesis. Production of frameshift, base substitution, and deletion mutations. J. Biol. Chem. *260*, 5787-5796.
Lamarche, B. J., Kumar, S., and Tsai, M. D. (2006). ASFV DNA polymerse X is extremely error-prone under diverse assay conditions and within multiple DNA sequence contexts. Biochemistry *45*, 14826-14833.
Land, A. M., Ball, T. B., Luo, M., Pilon, R., Sandstrom, P., Embree, J. E., Wachihi, C., Kimani, J., and Plummer, F. A. (2008). Human immunodeficiency virus (HIV) type 1 proviral hypermutation correlates with CD4 count in HIV-infected women from Kenya. J. Virol. *82*, 8172-8182.
LeClerc, J. E., Li, B., Payne, W. L., and Cebula, T. A. (1996). High mutation frequencies among Escherichia coli and Salmonella pathognes. Science *274*, 1208-1211.
Lecossier, D., Bouchonnet, F., Clavel, F., and Hance, A. J. (2003). Hypermutation of HIV-1 DNA in the absence of the Vif protein. Science *300*, 1112.
Levi, L. I., Gnadig, N. F., Beaucourt, S., McPherson, M. J., Baron, B., Arnold, J. J., and Vignuzzi, M. (2010). Fidelity variants of RNA dependent RNA polymerases uncover an indirect, mutagenic activity of amiloride compounds. PLoS Pathog. *6*, e1001163.
Li, G. M. (2008). Mechanisms and functions of DNA mismatch repair. Cell Res. *18*, 85-98.
Liu, M., Deora, R., Doulatov, S. R., Gingery, M., Eiserling, F. A., Preston, A., Maskell, D. J., Simons, R. W., Cotter, P. A., Parkhill, J., and Miller, J. F. (2002). Reverse transcriptase-mediated tropism switching in Bordetella bacteriophage. Science *295*, 2091-2094.
Liu, X., Yang, X., Lee, C. A., Moustafa, I. M., Smidansky, E. D., Lum, D., Arnold, J. J., Cameron, C. E., and Boehr, D. D. (2013). Vaccine-derived mutation in motif D of poliovirus RNA-dependent RNA polymerase lowers nucleotide incorporation fidelity. J. Biol. Chem. *288*, 32753-32765.
Loeb, L. A., Essigmann, J. M., Kazazi, F., Zhang, J., Rose, K. D., and Mullins, J. I. (1999). Lethal mutagenesis of HIV with mutagenic nucleoside analogs. Proc. Natl. Acad. Sci. USA *96*, 1492-1497.
Luftig, M. A. (2014). Viruses and the DNA damage response: activation and antagonism. Annu. Rev. Virol. *1*, 605-625.
Luria, S. E. and Delbrück, M. (1943). Mutations of bacteria from virus sensitivity to virus resistance. Genetics *28*, 491-511.
Lynch, M. (2010). Evolution of the mutation rate. Trends Genet. *26*, 345-352.
Lynch, M. (2011). The lower bound to the evolution of mutation rates. Genome Biol. Evol. *3*, 1107-1118.

Malpica, J. M., Fraile, A., Moreno, I., Obies, C. I., Drake, J. W., and Garcia-Arenal, F. (2002). The rate and character of spontaneous mutation in an RNA virus. Genetics *162*, 1505-1511.
Mangeat, B., Turelli, P., Caron, G., Friedli, M., Perrin, L., and Trono, D. (2003). Broad antiretroviral defence by human APOBEC3G through lethal editing of nascent reverse transcripts. Nature *424*, 99-103.
Manrubia, S. C., Domingo, E., and Lázaro, E. (2010). Pathways to extinction: beyond the error threshold. Philos. Trans. R. Soc Lond B Biol. Sci. *365*, 1943-1952.
Marti, T. M., Kunz, C., and Fleck, O. (2002). DNA mismatch repair and mutation avoidance pathways. J. Cell Physiol. *191*, 28-41.
Martínez, F., Sardanyés, J., Elena, S. F., and Daròs, J. A. (2011). Dynamics of a plant RNA virus intracellular accumulation: stamping machine vs. geometric replication. Genetics *188*, 637-646.
McLysaght, A., Baldi, P. F., and Gaut, B. S. (2003). Extensive gene gain associated with adaptive evolution of poxviruses. Proc. Natl. Acad. Sci. USA *100*, 15655-15660.
Menéndez-Arias, L. (2009). Mutation rates and intrinsic fidelity of retroviral reverse transcriptases. Viruses *1*, 1137-1165.
Modrich, P. and Lahue, R. (1996). Mismatch repair in replication fidelity, genetic recombination, and cancer biology. Annu. Rev. Biochem. *65*, 101-133.
Mullins, J. I., Heath, L., Hughes, J. P., Kicha, J., Styrchak, S., Wong, K. G., Rao, U., Hansen, A., Harris, K. S., Laurent, J. P., Li, D., Simpson, J. H., Essigmann, J. M., Loeb, L. A., and Parkins, J. (2011). Mutation of HIV-1 genomes in a clinical population treated with the mutagenic nucleoside KP1461. PLoS ONE *6*, e15135.
Nelson, C. W. and Hughes, A. L. (2015). Within-host nucleotide diversity of virus populations: Insights from next-generation sequencing. Infect. Genet. Evol. *30C*, 1-7.
Orr, H. A. (2000). The rate of adaptation in asexuals. Genetics *155*, 961-968.
Pal, C., Macia, M. D., Oliver, A., Schachar, I., and Buckling, A. (2007). Coevolution with viruses drives the evolution of bacterial mutation rates. Nature *450*, 1079-1081.
Pathak, V. K. and Temin, H. M. (1992). 5-Azacytidine and RNA secondary structure increase the retrovirus mutation rate. J. Virol. *66*, 3093-3100.
Pepin, K. M., Lass, S., Pulliam, J. R., Read, A. F., and Lloyd-Smith, J. O. (2010). Identifying genetic markers of adaptation for surveillance of viral host jumps. Nat. Rev. Microbiol. *8*, 802-813.
Pereira-Gómez, M. and Sanjuán, R. (2014). Delayed lysis confers resistance to the nucleoside analogue 5-fluorouracil and alleviates mutation accumulation in the single-stranded DNA bacteriophage φX174. J. Virol. *88*, 5042-5049.
Perelson, A. S. (2002). Modelling viral and immune system dynamics. Nat. Rev. Immunol. *2*, 28-36.
Pfeiffer, J. K. and Kirkegaard, K. (2003). A single mutation in poliovirus RNA-dependent RNA polymerase confers resistance to mutagenic nucleotide analogs via increased fidelity. Proc. Natl. Acad. Sci. USA *100*, 7289-7294.
Pfeiffer, J. K. and Kirkegaard, K. (2005). Increased fidelity reduces poliovirus fitness and virulence under selective pressure in mice. PLoS Pathog. *1*, e11.
Pita, J. S., de Miranda, J. R., Schneider, W. L., and Roossinck, M. J. (2007). Environment determines fidelity for an RNA virus replicase. J Virol. *81*, 9072-9077.
Pita, J. S. and Roossinck, M. J. (2013). Mapping viral functional domains for genetic diversity in plants. J. Virol. *87*, 790-797.
Powdrill, M. H., Tchesnokov, E. P., Kozak, R. A., Russell, R. S., Martin, R., Svarovskaia, E. S., Mo, H., Kouyos, R. D., and Gotte, M. (2011). Contribution of a mutational bias in hepatitis C virus replication to the genetic barrier in the development of drug resistance. Proc. Natl. Acad. Sci. USA *108*, 20509-20513.
Raney, J. L., Delongchamp, R. R., and Valentine, C. R. (2004). Spontaneous mutant frequency and mutation spectrum for gene A of phiX174 grown in E. coli. Environ. Mol. Mutagen. *44*, 119-127.
Reitter, J. N., Means, R. E., and Desrosiers, R. C. (1998). A role for carbohydrates in immune evasion in AIDS. Nat. Med. *4*, 679-684.

Ribeiro, R. M., Li, H., Wang, S., Stoddard, M. B., Learn, G. H., Korber, B. T., Bhattacharya, T., Guedj, J., Parrish, E. H., Hahn, B. H., Shaw, G. M., and Perelson, A. S. (2012). Quantifying the Diversification of Hepatitis C Virus (HCV) during Primary Infection: Estimates of the *in vivo* mutation rate. PLoS Pathog. *8*, e1002881.

Roberts, J. D., Bebenek, K., and Kunkel, T. A. (1988). The accuracy of reverse transcriptase from HIV-1. Science *242*, 1171-1173.

Rozen-Gagnon, K., Stapleford, K. A., Mongelli, V., Blanc, H., Failloux, A. B., Saleh, M. C., and Vignuzzi, M. (2014). Alphavirus mutator variants present host-specific defects and attenuation in mammalian and insect models. PLoS Pathog. *10*, e1003877.

Sanjuán, R. (2010). Mutational fitness effects in RNA and ssDNA viruses: common patterns revealed by site-directed mutagenesis studies. Phil. Trans. R. Soc. Lond. *365*, 1975-1982.

Sanjuán, R. (2012). From molecular genetics to phylodynamics: evolutionary relevance of mutation rates across viruses. PLoS Pathog. *8*, e1002685.

Sanjuán, R., Nebot, M. R., Chirico, N., Mansky, L. M., and Belshaw, R. (2010). Viral mutation rates. J. Virol. *84*, 9733-9748.

Sardanyés, J. and Elena, S. F. (2011). Quasispecies spatial models for RNA viruses with different replication modes and infection strategies. PLoS ONE *6*, e24884.

Sardanyés, J., Solé, R. V., and Elena, S. F. (2009). Replication mode and landscape topology differentially affect RNA virus mutational load and robustness. J. Virol. *83*, 12579-12589.

Schneider, W. L. and Roossinck, M. J. (2001). Genetic diversity in RNA virus quasispecies is controlled by host-virus interactions. J. Virol. *75*, 6566-6571.

Schofield, M. J. and Hsieh, P. (2003). DNA mismatch repair: molecular mechanisms and biological function. Annu. Rev. Microbiol. *57*, 579-608.

Schulte, M. B., Draghi, J. A., Plotkin, J. B. and Andino, R. (2015). Experimentally guided models reveal replication principles that shape the mutation distribution of RNA viruses. eLife 4: e03753.

Seronello, S., Montanez, J., Presleigh, K., Barlow, M., Park, S. B., and Choi, J. (2011). Ethanol and reactive species increase basal sequence heterogeneity of hepatitis C virus and produce variants with reduced susceptibility to antivirals. PLoS ONE *6*, e27436.

Shackelton, L. A. and Holmes, E. C. (2006). Phylogenetic evidence for the rapid evolution of human B19 erythrovirus. J. Virol. *80*, 3666-3669.

Shackelton, L. A., Parrish, C. R., Truyen, U., and Holmes, E. C. (2005). High rate of viral evolution associated with the emergence of carnivore parvovirus. Proc Natl. Acad. Sci. USA *102*, 379-384.

Shackelton, L. A., Rambaut, A., Pybus, O. G., and Holmes, E. C. (2006). JC virus evolution and its association with human populations. J. Virol. *80*, 9928-9933.

Sierra, M., Airaksinen, A., González-López, C., Agudo, R., Arias, A., and Domingo, E. (2007). Foot-and-mouth disease virus mutant with decreased sensitivity to ribavirin: implications for error catastrophe. J. Virol. *81*, 2012-2024.

Sierra, S., Dávila, M., Lowenstein, P. R., and Domingo, E. (2000). Response of foot-and-mouth disease virus to increased mutagenesis: influence of viral load and fitness in loss of infectivity. J. Virol. *74*, 8316-8323.

Simon-Loriere, E., Galetto, R., Hamoudi, M., Archer, J., Lefeuvre, P., Martin, D. P., Robertson, D. L., and Negroni, M. (2009). Molecular mechanisms of recombination restriction in the envelope gene of the human immunodeficiency virus. PLoS Pathog. *5*, e1000418.

Smith, D. J., Lapedes, A. S., de Jong, J. C., Bestebroer, T. M., Rimmelzwaan, G. F., Osterhaus, A. D., and Fouchier, R. A. (2004). Mapping the antigenic and genetic evolution of influenza virus. Science *305*, 371-376.

Sniegowski, P. D., Gerrish, P. J., and Lenski, R. E. (1997). Evolution of high mutation rates in experimental populations of E. coli. Nature *387*, 703-705.

Springman, R., Keller, T., Molineux, I., and Bull, J. J. (2009). Evolution at a high imposed mutation rate: adaptation obscures the load in phage T7. Genetics *184*, 221-232.

Steinhauer, D. A., Domingo, E., and Holland, J. J. (1992). Lack of evidence for proofreading mechanisms associated with an RNA virus polymerase. Gene *122*, 281-288.

Sung, W., Ackerman, M. S., Miller, S. F., Doak, T. G., and Lynch, M. (2012). Drift-barrier hypothesis and mutation-rate evolution. Proc. Natl. Acad. Sci. USA *109*, 18488-18492.

Suspène, R., Petit, V., Puyraimond-Zemmour, D., Aynaud, M. M., Henry, M., Guetard, D., Rusniok, C., Wain-Hobson, S., and Vartanian, J. P. (2011). Double-stranded RNA adenosine deaminase ADAR-1-induced hypermutated genomes among inactivated seasonal influenza and live attenuated measles virus vaccines. J. Virol. *85*, 2458-2462.

Svarovskaia, E. S., Dvory-Sobol, H., Parkin, N., Hebner, C., Gontcharova, V., Martin, R., Ouyang, W., Han, B., Xu, S., Ku, K., Chiu, S., Gane, E., Jacobson, I. M., Nelson, D. R., Lawitz, E., Wyles, D. L., Bekele, N., Brainard, D., Symonds, W. T., McHutchison, J. G., Miller, M. D., and Mo, H. (2014). Infrequent development of resistance in genotype 1-6 hepatitis C virus-infected subjects treated with sofosbuvir in phase 2 and 3 clinical trials. Clin. Infect. Dis. *59*, 1666-1674.

Taddei, F., Radman, M., Maynard-Smith, J., Toupance, B., Gouyon, P. H., and Godelle, B. (1997). Role of mutator alleles in adaptive evolution. Nature *387*, 700-702.

Thébaud, G., Chadoeuf, J., Morelli, M. J., McCauley, J. W., and Haydon, D. T. (2010). The relationship between mutation frequency and replication strategy in positive-sense single-stranded RNA viruses. Proc. Biol. Sci. *277*, 809-817.

Tromas, N. and Elena, S. F. (2010). The rate and spectrum of spontaneous mutations in a plant RNA virus. Genetics *185*, 989.

Ulferts, R. and Ziebuhr, J. (2011). Nidovirus ribonucleases: Structures and functions in viral replication. RNA Biol. *8*, 295-304.

van den Hoogen, B. G., van, B. S., de, R. J., van, N. S., Smith, D. J., Laksono, B., Gultyaev, A., Osterhaus, A. D., and Fouchier, R. A. (2014). Excessive production and extreme editing of human metapneumovirus defective interfering RNA is associated with type I IFN induction. J. Gen. Virol. *95*, 1625-1633.

Versteeg, G. A. and Garcia-Sastre, A. (2010). Viral tricks to grid-lock the type I interferon system. Curr. Opin. Microbiol. *13*, 508-516.

Vignuzzi, M., Stone, J. K., Arnold, J. J., Cameron, C. E., and Andino, R. (2006). Quasispecies diversity determines pathogenesis through cooperative interactions in a viral population. Nature *439*, 344-348.

Vignuzzi, M., Wendt, E., and Andino, R. (2008). Engineering attenuated virus vaccines by controlling replication fidelity. Nat. Med. *14*, 154-161.

Wei, X., Decker, J. M., Wang, S., Hui, H., Kappes, J. C., Wu, X., Salazar-Gonzalez, J. F., Salazar, M. G., Kilby, J. M., Saag, M. S., Komarova, N. L., Nowak, M. A., Hahn, B. H., Kwong, P. D., and Shaw, G. M. (2003). Antibody neutralization and escape by HIV-1. Nature *422*, 307-312.

Woelk, C. H. and Holmes, E. C. (2002). Reduced positive selection in vector-borne RNA viruses. Mol. Biol. Evol. *19*, 2333-2336.

Zheng, Q. (2005). New algorithms for Luria-Delbruck fluctuation analysis. Math. Biosci. *196*, 198-214.

2

Viral Informatics: Tools for Understanding the Evolution of Biology's Most Varied Genomes

Siobain Duffy

Department of Ecology, Evolution and Natural Resources, New Brunswick, NJ 08901, USA
duffy@aesop.rutgers.edu

Abstract

Our understanding of viral molecular evolution is being advanced by a combination of massive amounts of sequencing data and statistical algorithms that can extract information from them. Classic and modern approaches to phylogenetics, population genetics and other evolutionary analyses are reviewed. The sometimes conflicting language and desires of computational scientists and virologists as they approach the same issue are discussed. It has never been easier to accurately model viral molecular evolution, except that there are few suggestions or guidelines for how to conduct those analyses, and thus there is potential for garbage-in, garbage-out results that obfuscate the study of viral evolution.

Introduction

Most studies of viral evolution are studies of molecular evolution. Overwhelmingly, these are of nucleotide sequence evolution, where the genomic regions analyzed may not be genes with known function, are usually without solved protein structures, and where mutations cannot be easily connected to phenotypes. In fact, we are increasingly studying the evolution of predicted open reading frames that haven't necessarily been shown to be expressed during viral infection by experimentation, and of viruses that have never been cultured nor had an observable phenotype in a lab setting. While all of biology is undergoing a transition towards more bioinformatic research (Stevens, 2013), virology is perhaps the most affected organismal discipline (Seto, 2010). Viruses are harder to observe and quantify than cellular organisms, and sequencing-based, culture-independent techniques have dramatically expanded the number and kinds of viruses under study, even more so than among Bacteria and Archea (Greene and

Reid, 2014). Viral evolution, therefore, has become a field dominated by viral genomics, with diverse computational methods helping to make sense of it all.

Any chapter on viral informatics cannot be truly comprehensive: not only are new statistical methods being developed annually that refine or create new ways to study viral evolution, there are too many individually hosted web applications and downloadable tools that could be brought to bear on viral datasets to even list here. This diversity is a blessing and a curse for researchers, as tools taking different approaches to old problems can build better mouse traps, but these improvements may never rise in popularity and displace weaker options. Most of these new programs are developed at academic institutions and made freely available, which means that researchers can try many of them to find what works best for their datasets and which interfaces they prefer. Since some of these free tools can also be freely run on servers in the cloud, viral molecular evolution research can be conducted by any researcher with an internet connection, in both developed and developing countries.

This review aims to provide an introduction to the diversity of databases and software tools currently available, and a primer on the rationale behind common approaches viral evolutionary biologists take when analyzing nucleotide sequence data. Many of the methods discussed here were developed for studying mammalian evolution and refined by researchers working with viral data sets, typically Human immunodeficiency virus and Influenza virus, over time. Protein evolution gets short shrift here in favor of discussing genomic evolution at length. While some of the methods for studying genomic and protein evolution are similar (*e.g.*, alignment, phylogeny), computational tools for understanding protein structural evolution are unique (Carrillo-Tripp *et al.*, 2009; Cerny *et al.*, 2014). Tools for understanding viral protein evolution have been reviewed elsewhere (Mohabatkar *et al.*, 2012).

Where the data are

Most virologists starting an evolutionary informatics project *de novo*, or looking for sequences to accompany those they have generated themselves, head to one of the major databases that archives biological sequences. The largest genetic sequence database is maintained and mirrored through the repositories of the International Nucleotide Sequence Database Collaboration: the DNA Data Bank of Japan (http://www.ddbj.nig.ac.jp/), the European Nucleotide Archive (http://www.ebi.ac.uk/ena/) and GenBank (http://www.ncbi.nlm.nih.gov/genbank/). For researchers conducting BLAST searches (Altschul *et al.*, 1990) these databases will provide lists of closely related sequences to include in analyses. Protein databases, including structures, are also maintained by the respective regional entities (*e.g.*, European Bioinformatics Institute, the US National Center for Biotechnology Information), and can be well integrated with the nucleotide files. Of particular interest to virologists is the NCBI viral genome resource (Table 2.1) which provides links to the Reference Sequences of all viruses – a single genome from each approved viral species with a complete genome in GenBank. While this resource is always expanding, it has become a compact, standardized set of sequences to use if one were interested in asking a question across all sampled and sequenced viruses. There are also links on this viral genome resource to viral sequences organized by genomic material: single-stranded DNA, double-stranded

Table 2.1. Databases of viral genomic sequences

Virus-specific sequence databases	
NCBI viral genome resource	http://www.ncbi.nlm.nih.gov/genomes/GenomesHome.cgi
NCBI viral variation resource	http://www.ncbi.nlm.nih.gov/genome/viruses/variation/
Virus Pathogen Resource	http://www.viprbrc.org/brc/.gov/genbank/
Viral Bioinformatic Resource Center	http://virology.ca/
LANL HIV genome database	http://www.hiv.lanl.gov/
PBI dsDNA phages (of certain hosts) database	http://phagedb.org
Nucleic Acid Research's updated list of viral genome databases	http://www.oxfordjournals.org/our_journals/nar/database/subcat/5/18
Viral metagenome portals	
CAMERA	http://camera.crbs.ucsd.edu/ddc/
VIROME	http://virome.dbi.udel.edu/
METAVIR	http://metavir-meb.univ-bpclermont.fr/

DNA, single-stranded RNA (further divided into positive and negative sense ssRNA genomes), double-stranded RNA and retro-transcribing viruses. (links to the viral entries in TaxBrowser, http://www.ncbi.nlm.nih.gov/Taxonomy/Browser/wwwtax.cgi?name=Viruses). These categories roughly reflect the Baltimore classification by genome type (Baltimore, 1971), but narrower classification levels as approved by the International Committee for the Taxonomy of viruses (ICTV) are also delineated: orders, families, genera and species. Sequences not yet classified as species are also listed in separate sections within the TaxBrowser framework. For researchers looking to include all sequences from a given species or set of related species, TaxBrowser is an invaluable resource.

For some specific viruses, these universal databases are not necessarily the best place to gather sequence data, as other databases exist for more specialized purposes. Within GenBank, there are specific portals for a few highly sequenced human viruses (Influenza virus (Bao *et al.*, 2008), Dengue virus (Resch *et al.*, 2009), West Nile virus, Middle East respiratory syndrome virus, and Ebolavirus, Table 2.1). These subdatabases contain additional metadata for the viral sequences that are part of the non-redundant nucleotide database, and ensure common standards of sequence formatting and annotation. Many curated databases of viral genomes mirror the content of GenBank, supplemented with enhanced metadata. For instance, another resource for those interested in mammalian viruses would be the Virus Pathogen Resource, or for large dsDNA viruses, its Canadian counterpart, the Viral Bioinformatics Resource Center (Table 2.1).

However, viral databases not run by NCBI do not necessarily overlap 100% with sequences indexed in GenBank. Notable examples include the Los Alamos National Lab's HIV genome database, the World Health Oraganization's repositories for viral strain typing and epidemiology (*e.g.*, measles virus, http://www.who-measles.org/) and the Pittsburgh Bacteriophage Institute's databases for double-stranded DNA phages of *Arthrobacter*, *Bacillus*, *Mycobaterium* and *Streptomyces* (Table 2.1). A long, but incomplete, list of viral genomic databases is published annually by the journal Nucleic Acids Research (Fernandez-Suarez *et al.*, 2014). While there are advantages to running focused databases for communities studying particular pathogens, comparative virology and viral evolution research is harmed by sequestering data away from GenBank. Depositing sequences in comprehensive databases is a more significant contribution to the field than simply making them publicly available elsewhere.

Curated, species-identifiable databases meet the needs of many researchers but are not the only place to look for viral sequences. The best efforts of these traditional sequence repositories have not risen to the challenge that is the burgeoning amount of metagenomic reads and contigs that are produced annually by biologists. Viral sequences occur in all kinds of sequencing projects, from eukaryotic genome sequencing (*e.g.* Yoon *et al.*, 2011) to prokaryote metagenomes (Edwards and Rohwer, 2005), so it is hard to exhaustively search all potential places viral sequences may have been archived. It is challenging to even compare against sequences isolated from the exclusively viral fraction of environments (or nearly exclusively, as contamination with cellular genes occurs, and some stressed prokaryotes can be smaller than the 0.22µm filter size often used to remove cellular organisms, Fancello *et al.*, 2012; Kuhn *et al.*, 2014). In 2014 the environmental virology community was dealt a blow by funding issues closing one of the most popular tools for querying previously sequenced metagenomes, CAMERA. Thankfully its rich collection of organized metagenomes will remain online at its data distribution center (Table 2.1). Interested users can download these large resources, or query some of them through alternative portals such as VIROME (Wommack *et al.*, 2012) or METAVIR (Roux *et al.*, 2014). More specific tools for working with viral metagenomes are reviewed elsewhere (Fancello *et al.*, 2012).

Alignment

Not all evolutionary analyses require that sequences be aligned with one another prior to analysis, but most do require this meaningful step. Aligning genomic sequences is the single biggest assumption of most bioinformatic analyses, more important than model choice or tree-building methodologies (Phillips, 2006). Consequently, a commensurate amount of research effort has been dedicated to the development of alignment algorithms, and there are many choices and software packages that implement them. Over time and improved algorithms, multiple sequence alignment (MSA) has become the default for most analyses. Ideally, MSA involves simultaneously aligning all sequences dynamically to find an optimal alignment minimizing genetic distances between the sequences (given the parameters set by the user). However, this is computationally intractable for even fairly small datasets (NP-hard, Just, 2001). Instead, several conceptually different approximations to true MSA have been successfully implemented, including progressive alignment [*e.g.*, ClustalW, (Larkin *et al.*, 2007) and improvements thereon, such as MUSCLE (Edgar, 2004), MAFFT (Katoh and Standley, 2013), T-Coffee, (Di Tommaso *et al.*, 2011)], and those using hidden Markov models (*e.g.*, ClustalΩ, Sievers *et al.*, 2011). There continues to be room for improvement in MSA design to more confidently align larger datasets in less time. New tools are benchmarked against the many existing methods, especially the popular Clustal

Table 2.2. Stand-alone programs for sequence alignment and places to archive alignments

Alignment viewing software	
Aliview	http://www.ormbunkar.se/aliview/
BioEdit	http://www.mbio.ncsu.edu/bioedit/bioedit.html
Se-Al	http://tree.bio.ed.ac.uk/software/seal/
Data archiving	
Dryad	http://datadryad.org
TreeBASE	http://treebase.org/treebase-web/

algorithms (http://www.clustal.org/), but there is no consensus on how to test the methods that would help highlight which methods work best in which contexts, such as what would be best for fast-evolving viruses (Iantorno *et al.,* 2013). Further, most MSA programs require the user to find appropriate parameters or risk being biased by the default assumptions. Defaults are often optimized for cellular datasets and may not necessarily be optimized for viral genes.

Reproducibility is an essential part of science, but the meaning of reproducibility is becoming narrower with regards to sequence alignment (Anisimova *et al.,* 2010). Rather than wanting to be able to generate the same results in a paper using the authors' alignments and specified parameters of analysis, readers are often more interested in being able to generate the same results starting from the same, unaligned sequences in public databases. This leaves the alignment processes solely to MSA algorithms that can be run with the authors' specified parameters. While this is usually necessary for large datasets and for all but the smallest genomes, manual verification of alignments is important to assure that the alignment parameters were appropriate for each kind of dataset.

Luckily, many virologists focus on single gene analyses or viruses with small genomes, and it is usually still possible to hand-check entire alignments before proceeding to subsequent analyses. Indeed, the authors of many alignment algorithms explicitly endorse subsequent manual checking to correct errors in alignment (Bradley *et al.,* 2009), and a recent survey found that more than 75% of papers with phylogenetic trees mentioned manual modification of alignments in their methods (Morrison, 2009). Viral datasets in particular benefit from careful checking of the placement of insertions and deletions, as some viruses frequently experience indel mutations (Domingo *et al.,* 2012). Unfortunately these manual changes are the irreproducible step in sequence analysis, and scanning the alignments by eye does affect the reproducibility of pipelined analyses (Blouin *et al.,* 2009). To compensate, it is not sufficient to merely list accession numbers in a publication, or to describe the kinds of optimality criteria used for manual alignment in the methods section, but these manually modified alignments must be made available for interested readers should they wish to reproduce or modify the published results. TreeBASE ((Piel *et al.,* 2002) and Dryad Digital Repository ((Vision, 2010) are data archives that both accept alignments and are well-integrated, although submitting to TreeBASE requires that the alignment had been

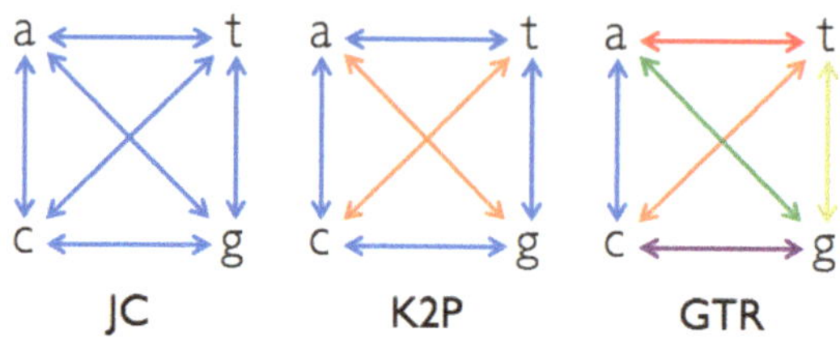

Figure 2.1. Graphic representation of three nucleotide substitution models. The JC model requires all rates of substitution to be the same, signified by the same color arrows. The K2P has two rates: one for transitions (on the diagonals) and another for the transversions. The GTR allows all substitution pairs to have a different rate. All of these are two-headed arrows, indicating that both substitution directions on the arrow have the same rate.

used to generate a published phylogenetic tree (Table 2.2). Other alternative locations to deposit alignments are as supplementary material in publications, on personal webpages and through institutional repositories, but they will be of maximum utility in public databases where they can be indexed and searched efficiently.

Alignments can be viewed (necessary for manual checking) in most DNA sequencing software packages, which are dominated by commercial products like Sequencher (http://www.genecodes.com). Free, open source alternatives exist such as seqtrace (Stucky, 2012), integrated software packages offer this functionality (see Integration, below) and numerous stand-alone alignment viewing programs exist (Table 2.2). Like much molecular evolution freeware, many of these are no longer updated or supported (*e.g.*, Se-Al and BioEdit), but others routinely rise up to take their place (*e.g.*, Aliview).

Whether or not the alignment was created by eye or by analysis pipeline, researchers often do not want to include all of an alignment in downstream analysis. This is because, by default, all columns of an alignment are of equal importance, but computer algorithms and researchers are not equally confident about the quality of all columns of the alignment. Regions with many gaps may not be reflecting true homology, which could then misinform downstream analyses. These regions can be identified by eye, or with software such as Gblocks (Talavera and Castresana, 2007).

Nucleotide Substitution Models

Sometimes during alignment, and for most phylogenetic inference, researchers need to model how often one nucleotide changes into another. The naive assumption, modeled in 1969 by Jukes and Cantor (JC, Jukes and Cantor, 1969), that an adenine has an equal chance of turning into a cytosine, guanine or thymine, is biologically unrealistic (Figure 2.1). Transitions occur, and are tolerated more often than transversions (Garcia-Villada and Drake, 2012), especially in RNA viruses where guanine can bond with either pyrimidine base, cytosine or uracil. This means that at a minimum, sequence evolution should be modeled by the Kimura two parameter (K2P, Kimura, 1980) model, which favors transitions. Many programs, not just those explicitly reconstructing phylogenetic trees, allow a user to choose the K2P instead of the default JC, such as those involved in recombination detection. More complex models, up to the general time-reversible (GTR) model, which allows each substitution pair (*e.g.*, a→g and g→a) to occur at its own rate, are widely implemented. The first addition to these nucleotide substitution matrices corrected for unequal base frequencies – in case a dataset did not contain roughly 25% of each of the 4 nucleotides (Felsenstein, 1981). Finally, the likelihood of a base experiencing mutation is often modeled. This is typically taken into account two ways: estimating the proportion of sites that do not experience mutation (proportion of invariant sites, I) and by fitting a gamma distribution (abbreviated Γ) to model how often a site experiences mutation, since some sites will have experienced more substitutions than others.

These corrections are important not only to accurately model the distributions of substitutions in a data set but to model the biological reality that many mutations have occurred that are not seen in the dataset because the same sites have been mutated multiple times. As two sequences diverge in time, so too will their genetic distance, with the complication that nucleotides are limited to only four character states (a, c, g, t/u) and sometimes mutations will make the sequences more identical instead of more divergent. Ultimately, the sequences hit a point where divergence fails to substantially increase despite substantial periods of time, which is termed saturation. This means there is a sweet spot of divergence where nucleotide sequences are useful for assessing evolutionary patterns and phylogenetic relationships that ends when the sequences are saturated, and protein sequences would be more useful (Goldman, 1998).

An alternative approach for protein-coding nucleotide sequences is to use a codon position model, which explicitly allows for synonymous third position changes to occur more frequently. Codon models embrace another biologically relevant bias – that synonymous mutations are more often tolerated than nonsynonymous mutations (Chare and Holmes, 2004). They are much more complex than nucleotide substitution matrices, and a time-reversible codon model can have more than 1800 independent parameters compared to GTR's six rates (Anisimova, 2012; Zaheri *et al.,* 2014). These models have become essential for tests of selection pressure that rely on relative rates of nonsynonymous and synonymous mutations. Codon models are also affected by saturation, but still match or outperform protein-based analyses on the same datasets (Seo and Kishino, 2008). However, these models aren't appropriate for some viral datasets because of overlapping reading frames (Chirico *et al.,* 2010); there are currently no modifications to these models to appropriately model nucleotides that simultaneously serve in two codon positions (Anisimova and Liberles, 2012). Additionally, nucleotides in viral genomes frequently serve more than just a protein-coding function, as secondary structures are often essential for RNA virus protein expression (*e.g.*, IRES, Kieft, 2008) and encapsidation (Schneemann, 2006).

Model choice is often conducted by software programs that compare the fit of various models. Some analysis websites have their own implementations (*e.g.*, www.datamonkey.org, Los Alamos National Labs), and there have been several stand-alone programs, but the most comprehensive model testing program available is jModeltest2 (Darriba *et al.,* 2012). jModeltest2 tests all 1624 candidate models that are derivatives of the GTR to trees created from an alignment (typically neighbor-joining trees) and compares their fit. All of these model-fitting programs penalize more complex models for having additional parameters and produce a very good fitting result that isn't necessarily the most complex model. None of these programs compares the fit of codon-based models to those that assume a regular nucleotide substitution matrix and vice versa; codon-based models are separately fit to data within programs that use them. However, there is a middle ground between a simple nucleotide substitution matrix and a full codon model. This compromise considers each of the three codon positions in a protein-coding gene dataset separately and models each with a potentially different nucleotide substitution matrix (*e.g.*, as implemented in Partition Finder (Lanfear *et al.*, 2012) and BEAST (Bouckaert *et al.*, 2014). Codon position models have been shown to fit RNA virus datasets very well (Shapiro *et al.*, 2006a). Another approach is to consider each site in an alignment separately – not just as relative codon positions – and group nucleotide sites along an alignment according to how each would be most accurately modeled (Wu *et al.*, 2013). No one is advocating reconstructing phylogenies with each individual site in an alignment having its own substitution model, but the techniques exist to create complex nucleotide substitution models that are uniquely matched for each alignment.

Interestingly, all of these complex improvements to nucleotide substitution models still assume time reversibility of substitution pairs. While the GTR is flexible and appropriate for many datasets, it fails to take into account mutational biases that may be highly significant in viruses, such as the c→t transitions that occur far more frequently in single-stranded DNA viruses than the reverse substitution (Chithambaram *et al.*, 2014; Duffy and Holmes, 2008; Rokyta *et al.*, 2005). This underscores an important point: most of the computational tools available for understanding sequence evolution were not designed for viral datasets. Explicitly, more complex models than the GTR were abandoned because mammalian data sets did not seem to need additional precision (Yang, 1994), and they are infrequently implemented in modern programs. For instance, MrBayes was able to implement them when first released (Huelsenbeck and Ronquist, 2001) but subsequently turned this feature off because it was not widely used. In many cases this is not a problem, or not a very large problem. Many dsDNA viruses mutate at similar rates to mammalian cells (Sanjuán *et al.*, 2010) and so these viruses may have a similar degree of polymorphism, saturate at similar speeds and their degree of divergence may mimic the datasets on which these tools were honed. On the other hand, virus-specific models of protein evolution have been shown to improve phylogenetic accuracy for retro-transcribing elements (Dimmic *et al.*, 2002) and Influenza virus (Dang *et al.*, 2010). There are many assumptions that go into phylogenetic inference, and as the viral evolution community strives for increasing accuracy, models developed on and for viral datasets may improve viral phylogenetic reconstruction.

Recently a new kind of model was derived for Influenza nucleoprotein evolution from experimental mutagenesis and selection experiments, based entirely on empirically determined mutation rates and selection coefficients (Bloom, 2014). This was the best model to reconstruct the phylogeny of twentieth and twenty-first century Influenza A nucleoprotein sequences, but its optimization for one protein likely limits its application to other datasets. However, this is a potentially promising approach for creating provably accurate matrices for additional virus genes as more of these data become available, or for important pathogens for which funding would be available to support the wet lab research needed to create these matrices. This is one, arduous way to ensure that the specific biases that influence a virus' evolution are accurately modeled.

Phylogenetic inference

A phylogeny is a hypothetical evolutionary relationship among taxa – a best guess of how and how closely isolates are related to one another. Many evolutionary questions can be answered with accurate phylogenetic trees. Viral molecular epidemiology often employs phylogenies to determine from where and when a pathogen arrived (Duffy and Holmes, 2007). Phylogenetic methods (including computationally trivial neighbor joining trees) are often used to identify species and known subtypes (Lam *et al.,* 2010), or to find groupings that could be considered subtypes in the future (Pierce and Stepien, 2012). Many viral evolution studies culminate in a phylogenetic analysis, and the majority of the conclusions of the papers are based on statistical support for groupings on the tree.

There are numerous methods for building trees, and older software packages implement multiple methods: MEGA (Tamura *et al.,* 2013), PAUP* (Swofford, 2003), PHYLIP (Felsenstein, 1989). Fifteen or twenty years ago, there were significant arguments about the relative merits of different algorithms, and to convince readers one needed to use many methods to build trees and look for consistency among them (*e.g.*, Bowen *et al.,* 2000). Over the last decade maximum parsimony and minimum evolution tree building methods have become less popular in favor of maximum likelihood (ML) methods, including Bayesian inference. While the aforementioned classic programs can construct trees using ML estimation, additional programs strictly based on these approaches have proliferated [Garli (Zwickl, 2006), PhyML (Guindon *et al.,* 2010), RaxML (Stamatakis, 2014), Table 2.3], as have implementations of phylogenies in statistical programs like R (R Development Core Team, 2012). The diversity of programs is due the different ways each approaches the fundamental problem of likelihood analyses: that it is computationally impossible to create every possible tree topology, assign a likelihood, and pick the best one. Therefore, each algorithm uses different heuristics and approaches to finding the most likely tree or trees. For trees with many taxa (>500 sequences) Garli and RaxML perform very well (Stamatakis, 2006), while the other algorithms have not been optimized for these situations. All of these programs are available both for download and available on external servers, such as those run by the Cyberinfrastructure for Phylogenetic Research (CIPRES), the Swiss Institute of Bioinformatics, and the South of France bioinformatics platform (Table 2.3).

Both ML and Bayesian programs evaluate the match between the aligned data and trees, given the nucleotide substitution models and parameters specified in the analysis, but they do this in complementary ways. ML algorithms evaluate how likely it is that the aligned data were a product of a candidate tree topology: which tree makes the data the most likely result, given the models specified (Page and Holmes, 1998). Bayesian inference takes the reverse tack, optimizing the posterior probability of finding a tree given the data and the model as prior distributions (Huelsenbeck *et al.,* 2001). These two approaches also search for the best trees differently. ML programs will find a pretty good tree and then modify it by detaching

Table 2.3. Some programs that build and use phylogenetic trees, and portals to access them

ML phylogenetic inference	
Garli	http://www.bio.utexas.edu/faculty/antisense/garli/garli.html
PAUP*	http://paup.csit.fsu.edu/
PhyML	http://www.atgc-montpellier.fr/phyml/
RaxML	http://sco.h-its.org/exelixis/web/software/raxml/index.html
Bayesian phylogenetic inference	
BEAST	http://beast2.org/
MrBayes	http://mrbayes.sourceforge.net/
RevBayes	http://revbayes.github.io/
Web portals	
CIPRES	http://www.phylo.org/
South of France bioinformatics	http://www.atgc-montpellier.fr/
Swiss Institute of Bioinformatics	http://www.expasy.org/
Further analysis using phylogenies	
BaTS	http://evolve.zoo.ox.ac.uk/Evolve/BaTS.html
HyPhy	http://hyphy.org/w/index.php/Main_Page
Mesquite	http://mesquiteproject.wikispaces.com/
PAML	http://abacus.gene.ucl.ac.uk/software/paml.html
TreeMap	http://sydney.edu.au/engineering/it/~mcharles/software/treemap/treemap3.html
Integrated Packages	
CLC Workbench	http://www.clcbio.com/
DAMBE	http://dambe.bio.uottawa.ca/dambe.asp
Geneious	http://www.geneious.com/
MEGA	http://www.megasoftware.net/
SNAP Workbench	http://carbonelab.org/workbench

and reattaching branches to see if it can find a better one. Some of these perturbation algorithms are heuristic themselves, like nearest neighbor interchange and tree-bisection reconnection, but branch and bound is exhaustive (Page and Holmes, 1998). Bayesian phylogenetic programs employ Markov chain Monte Carlo methods to search tree space. The program starts with a random tree, which is stochastically modified in each iteration, according to the prior distributions. The MCMC algorithm usually quickly finds much more likely trees and spends the vast majority of the analysis around these highly likely trees, which is why the first ~10% of the search is discarded as "burn in" so as to not allow the initial poor trees to influence the overall results. Bayesian inference should be run for a sufficiently large number of chains until the results stably converge. Tracer (Rambaut *et al.,* 2014) is an excellent program for viewing the results of both BEAST and MrBayes outputs, and has a number of ways to check that the analyses have sampled a wide range of likely trees and have converged. The output of Bayesian phylogenetic inference is often a maximum posterior probability (MPP) tree.

Phylogeneticists calculate the confidence they have in the best trees obtained by likelihood and Bayesian approaches differently as well. ML trees must be subjected to additional testing, typically bootstrap analysis, to determine how much the alignment supports each grouping in the tree topology (Page and Holmes, 1998). Bootstrapping involves resampling with replacement the sites in an alignment to generate alternative alignments (pseudoreplicates), which are then used to build trees. Usually this is done >>100 times (1000 is common) and the percent of the pseudoreplicate trees where a clade was reconstructed is that clade's bootstrap value. These are intuitive scores, and readers know that a clade reconstructed in 95% of the resamplings is well-supported, and one that was supported only 60% of the time might not be worth basing conclusions on. Bayesian trees are supported post-hoc only, by determining what proportion of the trees in the analysis (excluding burn-in) supports the clades in the MPP tree. These posterior probabilities supporting clades are often very high – 0.9 or more – and cannot be considered equivalent to bootstrap percentages. While clades with very high posterior probabilities of 0.97 or 0.99 are very likely to be well-supported, a clade with posterior probability of 0.85 should not be afforded the same credibility as one with a bootstrap score of 85% (Cummings *et al.,* 2003).

Further phylogenetic analyses

Phylogenies are more useful for evolutionary analyses than just what can be gleaned from their topology and clade support. Even in the early days of computational phylogenetics programs there was a division between programs primarily for building trees and those for testing hypotheses using trees. PAML (Yang, 2007), for instance, remains a widely used tool in molecular evolution research even though it was never recommended for tree building and users have to import trees built in other programs. Using phylogenetic information as part of downstream comparative analyses is often essential when using many sequences from a single species or population. As statistical tests often assume that data are independent from one another, the phylogenetic dependence of related sequences must be accounted and controlled for. For instance, phylogenetic comparative methods have dramatically decreased the false positive rate for detecting epistasis in viral genes (Poon *et al.,* 2007; Shapiro *et al.,* 2006b). Perhaps the most common phylogenetically informed analysis involves testing the strength and location of natural selection, but this is covered in a separate section (see Natural Selection, below). This section focuses on three other kinds of analyses that are only possible with a well-supported phylogenetic tree or set of trees: robust estimations of rates and dates, phylodynamic reconstructions of demographic histories and tests of trait association.

Rates and dates

BEAST distinguished itself from other phylogenetic and statistical genetic software packages early on with its focus on time-rooted trees and its inclusion of a relaxed molecular clock model (Drummond and Rambaut, 2007). It quickly gained a reputation as the go-to program to estimate viral nucleotide substitution rates, and to calculate divergence dates. Typically these analyses are conducted using tip-dated sequences: sequences with established times of isolation and sequencing, specified in increments from days to years. The genetic distance between sequences combined with separation in time allow the program to calibrate evolutionary rates, which can vary on each branch of the phylogeny. Hundreds of viral substitution rates have been calculated using these methods in BEAST (Duffy *et al.,* 2008; Hicks and Duffy, 2014). Alternatively sequences can be analyzed without dates but with a rate specified by the researcher, or the researcher can specify some divergence times and allow these specified nodes to calibrate the clock. BEAST methods have been employed to help understand the trajectory and timing of viral emergences, including the 2009 H1N1 pandemic (Smith *et al.,* 2009) and the 2014 Ebola outbreak (Dudas and Rambaut, 2014). Dating particular branching events on a tree, or knowing when the last common ancestor of a group of sequences existed (TMCRA, time to most recent common ancestor) is often useful in determining when viruses shifted into new hosts or to date migration events (De Bruyn *et al.,* 2012). BEAST was used to help exonerate Bulgarian doctors accused of infecting Libyan orphans with HIV because analyses showed that it was certain that the HIV outbreak had started in the orphanage before the doctors came to Libya (de Oliveira *et al.,* 2006).

The timescale of the dataset used for analysis influences the results, as data sampled over a short time will have a higher ratio of transitory deleterious mutations compared to substitutions that have or will fix in the population (Ho *et al.,* 2007). This can mean there isn't significant temporal signal in the alignment, and randomized date controls can reveal this (Firth *et al.,* 2010). The timescale of the questions being asked by the researcher also is important – tip-dated sequences sampled over the last 25 years cannot accurately calibrate and inform evolutionary events that happened 2500 years ago. Indeed, the timescale-dependency of these analyses is becoming increasingly clear and researchers should exercise extreme caution when comparing analyses over very different timescales (Duchene *et al.,* 2014).

While BEAST dominates tip-dated viral phylogenetic research, there are many other programs that implement strict and relaxed molecular clock methods, including PAML. The assumption of a strict molecular clock is still often useful for a first approximation, for instance, when testing the temporal signal within an alignment prior to BEAST analysis with Path-o-gen (http://tree.bio.ed.ac.uk/software/pathogen/).

Phylodynamics

The evolutionary history of viruses is shaped by genomic properties, ecological pressures and epidemiological realities, causing their diversity and demography to be united by the concept of phylodynamics (Grenfell *et al.,* 2004). Bayesian coalescent inference as implemented in BEAST can use the phylogenetic relationships among sequences to explore the population genetics of this population in the past (Pybus and Rambaut, 2009). One of the popular features of this approach is reconstructing how large the effective population size of a population has been, and whether it has gone through changes over time. For viruses the calculated parameter isn't exactly effective population size, but the effective number of infections, which is more reflective of how many viruses were transmitting between hosts through the past (Frost and Volz, 2010). Sometimes it will still be referred to in the literature, however, as effective population size, or N_e (Rasmussen *et al.,* 2014a). The coalescent time for various clades within the highly likely phylogenies can be assessed, and the pattern in timing analyzed to estimate effective number of infections over time (Minin *et al.,* 2008). This can reveal population bottlenecks (Cattoli *et al.,* 2011) or increases in viral diversity (Bertolotti *et al.,* 2007), which can be connected to human interventions like antiviral drugs and infection control measures (Castro-Nallar *et al.,* 2012) or disease outbreaks (Siebenga *et al.,* 2010).

Associating phenotypic traits

The ability to connect ecological and phenotypic traits with clades and divergence events is a very popular motivation for those using phylogenetic methods. For virologists, this has come in a desire to test the co-speciation of viruses with their hosts (Ramsden *et al.,* 2009), to look for host-shifting events (Kuzmin *et al.,* 2012) or to look for quantitiative support for migration events (Lemey *et al.,* 2009). However, trait association models are general, and can be used with any phenotypic character or ecological condition. They are implemented in software packages, notably Mesquite (Maddison and Maddison, 2014) and can be implemented in R (caper, http://cran.r-project.org/web/packages/caper/index.html). However, among viral evolutionary biologists, BEAST is currently ascendant. Current versions of BEAST obviate the need for additional software like BaTS (Parker *et al.,* 2008).

Host-virus co-speciation questions are still a complex matter, especially when there can be uncertainty about both the host and the virus phylogenies. TreeMap (Charleston and Robertson, 2002) is a program that has been used by virologists to test the null hypothesis of co-speciation, and it is still being maintained and updated (Table 2.3).

Population genetics

The five forces that can alter population genetics are mutation, recombination, drift, selection and migration (Hartl and Clark, 1997). All of these forces shape the viral datasets that researchers analyze, though it can be hard to disentangle the strength of these forces. There are many classical methods still in use today that can be found in aging software packages such as DnaSP (Rozas *et al.,* 2003), in continually updated suites like Arlequin (Excoffier and Lischer, 2010), and more phylogenetically informed approaches are included in more modern packages [BEAST 2 (Bouckaert *et al.,* 2014), HyPhy (Kosakovsky Pond *et al.,* 2005)].

Mutation

Whether or not mutation is in itself an evolutionary force or merely the medium upon which other forces act is a matter of semantics and debate (Nei, 2013), but it is undeniably one of the factors that must be accounted for when studying viral datasets. Often mutation is assumed to be the only driver of sequence diversity (Domingo *et al.,* 2012), but in practice it is usually considered responsible for the sequence variation which cannot be unambiguously assigned to recombination (*e.g.*, Lima *et al.*, 2013), and thus its contribution to standing genetic diversity may be overestimated. Molecular variation within a population can be measured with a number of statistics, ranging from haplotype frequency, which is rarely informative since almost all sequences in a viral population are at least different by a single mutation, to measures designed to accommodate the high diversity of viral populations, such as partitioning analysis of quasispecies (Baccam *et al.,* 2001; Ojosnegros *et al.,* 2008). Most of these measures look at all pairwise genetic distances in order to calculate their values, and are usually calculated on data that have already been multiply aligned (*i.e.* have introduced gaps to improve the alignment). Nucleotide diversity (π, Nei and Li, 1979), which is the complement of average percent nucleotide identity – the average proportion of dissimilar sites between two sequences in the population – may be the most popular of these metrics (Gorbachev, 2012), including with deep sequencing data (Cornman *et al.,* 2014; Shao *et al.,* 2014). However, all of these diversity measures are underutilized, as many papers that purport to look at viral diversity do so without any quantification. Despite widespread acceptance that many viral populations are very diverse, we lack benchmarks to compare population diversities in all but a small number of human-infecting viruses, such as Influenza virus and Dengue virus (Parameswaran *et al.,* 2012; Wilker *et al.,* 2013).

Recombination

Many viruses have high rates of recombination (Batorsky *et al.,* 2011; Froissart *et al.,* 2005; Lefeuvre *et al.,* 2009), which can create complicated networks of evolutionary relationships better modeled as a web than a phylogenetic tree. Because recombination violates one of the assumptions of phylogenentic reconstruction programs (and of most coalescent models (McVean and Cardin, 2005), many virologists are interested in detecting recombination in datasets for the purpose of removing recombinant taxa to have the most appropriate dataset possible for phylogenetic reconstruction. Others study recombination patterns – the breakpoint locations and the interconnected network of viral genomes. Both of these purposes are served by a number of recombination detection algorithms that are encoded in both stand alone programs [SimPlot (Lole *et al.,* 1999), RDP (Martin *et al.,* 2010), VisRD (Strimmer *et al.,* 2003), Table 2.4] and as part of analysis packages such as HyPhy (Kosakovsky Pond *et al.,* 2006).

RDP has been particularly popular in virology for two reasons: it combines results from many recombination detection algorithms (at least 10, but users need not run all of them) into a graphical user interface and it was created and is maintained by viral evolutionary biologists. This means important features for working with some viral datasets, such as detecting recombination in whole circular genomes that are artificially linearized for alignment, can be activated by one click in the program. Because recombination is such a significant force in viral datasets, several virologists were authors on early papers on algorithms and programs to assess recombination [Claude Fauquet (Padidam *et al.,* 1999), Adrian Gibbs (Gibbs *et al.,* 2000), Peter Simmonds (Simmonds and Smith, 1999)]. Advances in recombination detection continue to driven by viral datasets (*e.g.* jpHMM, Schultz *et al.,* 2009),

Table 2.4. Programs for recombination analysis

Recombination detection	
jpHMM	http://jphmm.gobics.de/
LAMARC	http://evolution.genetics.washington.edu/lamarc/index.html
LDHat	http://ldhat.sourceforge.net/
RDP	http://web.cbio.uct.ac.za/~darren/rdp.html
SimPlot	http://sray.med.som.jhmi.edu/SCRoftware/simplot/
VisRD	http://www2.cmp.uea.ac.uk/~vlm/visrd/
Network graphs	
Network	http://www.fluxus-engineering.com/sharenet.htm
SplitsTree	http://www.splitstree.org/
TCS	http://w3.ualg.pt/~rcastil/SOFTWARE_WINDOWS/TCS1.21/docs/TCS1.21.html
Ancestral recombination graphs	
ARGUS	http://gobics.de/software
ARGweaver	http://mdrasmus.github.io/argweaver/
BEAGLE	http://www.eecs.berkeley.edu/~yss/software.html
GraphML	http://revbayes.github.io/

though many improvements come from bioinformaticians outside of virology as well.

Methods to detect recombination fall into a few categories, the most popular of which are those that look for different phylogenetic relationships for different portions of an alignment [the majority of current methods, such as the eponymous RDP (Martin and Rybicki, 2000) and quartet scanning in VisRD], those that look for changes in strength of pairwise alignment [*e.g.*, GENECONV (Padidam *et al.*, 1999), SimPlot] and those that use population genetic data to estimate recombination rates [*e.g.* LAMARC (Kuhner, 2006), LDHat (Auton and McVean, 2007)]. Researchers have their individual favorites, but most viral evolution papers employ multiple methods, if only for the ease of doing so in RDP. These methods can also be used to assess reassortment (pseudorecombination) in segmented viruses if genomic segments are concatenated (Stainton *et al.*, 2012).

Detecting recombination is not enough for virologists who want to envison the relationships among their sequences on a more accurate network rather than in tree form. These can be created in any program that creates statistical parsimony haplotype networks [Network, TCS (Clement *et al.*, 2000), Table 2.4]. More popular is creating either a neighbor-net (analogous to a neighbor-joining tree) or a split network using SplitsTree (Huson and Bryant, 2006), which has been the leading program at statistical analysis of phylogenetic networks. As more biologists realize the necessity of not ignoring recombination in phylogenetic reconstruction (Arenas *et al.*, 2010), more methods for working with evolutionary networks are being developed. Another way of visualizing phylogenetic relationships is through ancestral recombination graphs. These differ from full network representations by being rooted, and often tree-like except for the additional connections necessary to reflect hybridization/horizontal gene transfer events. These graphs are optimized to have as few recombination events as possible. Again, some programs to calculate these were generated to solve problems in viral evolution [ARGUS, (Bulla *et al.*, 2010) but alternatives have been developed and can be applied to viral datasets [BEAGLE (Lyngsø *et al.*, 2005), ARGweaver (Rasmussen *et al.*, 2014b), Table 2.4].

It is worth mentioning a distinction between how bioinformaticians and virologists talk about recombination. Phylogenetic reconstruction is hampered when two sequences in the same tree recombine and produce hybrid offspring genomes that are also included in the analysis, as this means that all nodes do not have one and only one parent anymore (Nakleh, 2011). This creates reticulation, and a network other than a strictly branching graph results. When genomes recombine with taxa that are not in the tree it potentially creates a long branch in order to model the incoming diversity as a series of mutational events, but does not make more than one node a parent of a child node (the other parental sequence is not included in the analysis). While long branches can affect phylogenetic reconstruction (Bergsten, 2005), and violate some of the assumptions of coalescent theory, these are easier problem to overcome than reticulation, which is against the conceptual framework of most phylogenetics programs. Virologists, on the other hand, are often concerned with easy-to-detect interspecific recombination events even if they are not including both parental sequences in a phylogenetic analysis. While these hybrids arise from the mechanism of recombination, and would validly be considered 'recombination' in virology journals, from a population genetics perspective they can be considered gene flow from one species to another species, which is a kind of migration.

Migration

The movement of viruses among populations is typically studied on two scales: assessing if genetic variation is structured by geography or if migration is an homogenizing force, and inferring biogeographic movements through phylogenetic relationships (Holsinger and Weir, 2009; Slatkin and Maddison, 1989). In the case of the former, classical methods are used to describe the connectivity of viral populations. Wright's fixation index for haploids can describe how homogenous viral populations are, and help estimate migration rates among them (Holsinger and Weir, 2009; Lapidot *et al.*, 2014). Migration events can de detected by geographically separated taxa grouping closely in phylogentic trees (*e.g.*, Duffy and Holmes, 2007), and time-structured sequences can be used to look at when migration events occurred along a phylogenetic tree using the phenotypic trait association models discussed above (Bahl *et al.*, 2011; Lefeuvre *et al.*, 2010).

Genetic drift

Drift as an evolutionary force underlies many of the analyses in molecular evolution, but often it is the null hypothesis against which natural selection is detected. By and large, viral datasets are found to not be evolving neutrally. For instance, they overwhelmingly reject neutrality when testing with Tajima's D (Tajima, 1989). However, exploration of drift typically ends with this rejection. It is difficult to determine from genomic data what nucleotide changes in viruses are truly neutral instead of transient deleterious mutations (Sanjuán, 2012) or even recurrent mutations. Synonymous sites in codons are often under selection (see Codon Usage Bias, below) and even non-coding regions in viruses have been shown to include adaptive mutations (Remold *et al.,* 2008). Both neutral and non-neutral sites can both hitchhike with adaptations during viral evolution, further complicating the study of neutral processes from genomic data alone (Zanini and Neher, 2013a).

Natural selection

Most molecular evolution research is focused on adaptive processes and detecting the impact of natural selection on nucleotide sequences. This is part of the overall adaptationist history of evolutionary biology, but also reflects important motivations for why people study the evolution of viruses. Often researchers are trying to understand emerging viruses, or how viruses adapt to novel environments or overcome pharmaceutical challenges (Seoighe *et al.,* 2007). A better understanding of viral adaptive processes can also lead to improved forecasting of viral evolution, which is useful for vaccine design and for public health planning (Luksza and Lassig, 2014).

There are statistics that can look at a polymorphism across a whole sequence (or a sliding window of sequence) and determine if there are deviations from neutral evolution. Tajima's D was mentioned above, and the McDonald-Kreitman test has proven robust at detecting selection on viral datasets (Bhatt *et al.,* 2010). Other tests of neutrality are reviewed elsewhere (Fu, 2001). However, the bulk of detecting selection in protein-coding nucleotide sequences is based on d_N and d_S, the rates of nonsynonymous and synonymous substitutions, often expressed as the ratio d_N/d_S (Nei and Gojobori, 1986). The logic of this is clear – mutations will arise in genomes, but selection should purge most nonsynonymous changes to proteins and tolerate most synonymous changes, causing low d_N/d_S values for most genes, and for most codons within genes. Diversifying positive selection, when nonsynonymous changes arise and are maintained in the population, can be identified by large d_N/d_S ratios. This kind of selection is frequently observed in datasets of viral epitopes that interact with the mammalian adaptive immune system. What is most important is to be different from what the immune system already recognizes, causing many, diverse nonsynonymous mutations to be favored (Suzuki, 2006). Directional positive selection, when only one nonsynonymous change arises and is selected over time, is harder to detect using these methods, but also produces elevated d_N/d_S ratios. For instance, the changes fixed by canine parvovirus as it was emerging in dogs were associated with a single branch of a phylogenetic tree with d_N/d_S=0.803 (Shackelton *et al.,* 2005). Truly neutrally evolving regions should have a d_N/d_S of 1.

Since the late 1990s these tests have been implemented with codon models, first as codeml in PAML, and then many other programs, but notably HyPhy and its web instantiation (http://www.datamonkey.org/). While they were originally developed to look at orthologous and paralogous mammalian genes, they have become essential for understanding viral adaptive evolution (Nielsen and Yang, 1998). Often these tools are applied to sequences from a single viral species, from a single population, which violates some of the assumptions of the method

(Kryazhimskiy and Plotkin, 2008). Also, viral sequences themselves challenge some of the underlying assumptions of the analysis: synonymous sites may be more constrained than in cellular data sets (see Codon Usage Bias, below) and some viral genomic architectures include overlapping reading frames and/or have functional secondary structures that impact fitness so the viral d_S may be an imperfect baseline. However flawed some of the underlying assumptions are, these methods have been used to identify critical sites under both positive and purifying selection in viruses (Kosakovsky Pond *et al.,* 2012). Researchers have even attempted mapping epitopes of viral proteins using the strength of positive selection as a guide (REEvolution, http://bioinfo.ib.usp.br/revolver/). d_N/d_S values that show no evidence for purifying selection on a gene have also proven useful for identifying spuriously predicted ORFs in viral sequences (Wang *et al.,* 2011).

Within HyPhy, researchers can use several methods to detect whether any codons in their alignment are experiencing larger amounts of selection than others, from the most conservative Single Likelihood Ancestor Counting (SLAC) to complex models involving fixed effect likelihoods that can detect whether a small fraction of branches is experiencing selection that the remainder of the tree is not (Kosakovsky Pond *et al.,* 2012). SLAC is more efficient and able to be used on larger datasets than the other models in HyPhy (fixed effect likelihood, relative effect likelihood, model of episodic directional selection).

All of the methods to detect selection discussed at length here are only appropriate for protein-coding genes, and those without overlapping reading frames. Methods of selection for overlapping reading frames is an active area of development (Mir and Schober, 2014; Wei and Zhang, 2015) with obvious applications to viruses (Sabath *et al.,* 2008). Additionally, strides are being made towards detecting natural selection on noncoding regions (Kern and Haussler, 2010), but these have been applied in only in a limited way to viruses (Wong and Nielsen, 2004), in part because viral genomes are largely protein-coding (Koonin, 2009).

Codon usage bias

A significant amount of time and attention has been devoted to examining how well viruses match the codon biases of the hosts in which they replicate (Bahir *et al.,* 2009). There are many ways to quantify the degree of codon usage bias (CUB) in an organism (Roth *et al.,* 2012), but often the subjectively interpreted effective number of codons is used (Wright, 1990). Another popular metric is the relative synonymous codon usage (RSCU, Sharp *et al.,* 1986), which normalizes each amino acid's degenerate codon use to 1, the equal use of all codons. Viral CUB can be compared to host CUB, producing a single value in a Codon Adaptation Index (Sharp and Li, 1987) or by regression of individual codon usage by rRSCU (Cardinale *et al.,* 2013; Chithambaram *et al.,* 2014). While some of these approaches are available in stand alone programs (*e.g.*, CAIcal http://genomes.urv.cat/CAIcal/), the most recommended implementation would be in the DAMBE software package (Xia, 2013), as it contains the most refined forms of each statistic for assessing codon bias.

While CUB has been instrumental in determining what genes have likely been horizontally transferred into prokaryotic genomes (Karlin, 2001; Moszer *et al.,*

1999), the utility of viral CUB has not been similarly established. Many viruses poorly match the codon bias of their hosts (Jenkins and Holmes, 2003), some viruses have very broad ranges of hosts with divergent CUB (Sau *et al.*, 2007), and large dsDNA viruses often bring their own tRNA genes to alter the host cell's supply (Michely *et al.*, 2013). Consequently, codon bias is thought to be driven more by viral mutation processes than by selection for fast translation by host machinery, and has been used as a window onto the effects of genetic drift in viruses (Holmes, 2008). However, synonymous codons in viral genomes can be under positive (Kryazhimskiy *et al.*, 2008) and negative selection (Zanini and Neher, 2013b), including to not match host CUB to control translation speed for accurate protein folding (Ding *et al.*, 2014). Finally the role of constraints such as genomic secondary structure (Gu *et al.*, 2010) has not yet been sufficiently considered as a shaper of codon usage in viruses (Cardinale *et al.*, 2013).

Viral systematics

One of the most fundamental issues facing biologists is how to classify the diversity of life (Dunn, 2008). This has been an issue of particular concern for virologists in the last two decades because sequencing technology has enabled the discovery of viral genomes at a much more rapid pace than their phenotypes can be studied. For the first time, ICTV definitions on how viruses should be named and their species delineated are moving from ELISA-based definitions to genome sequence similarity (King *et al.*, 2012). Species names, and potentially even higher taxonomic levels, can now be proposed from purely genomic data, without substantiation through wet lab work (Gorbalenya, 2014). Additionally, computational approaches are being used to revisit existing taxonomic designations; measures of protein divergence have been explored to delineate groupings in RNA viruses (Lauber and Gorbalenya, 2012). Thus, we now study viral systematics – and patterns of macroevolution – largely through informatics as well.

As percent nucleotide identity is being widely used for identification of novel viral strains and species, tools to automate this process are being developed. The Species Demarcation Tool (http://web.cbio.uct.ac.za/~brejnev/) was developed by geminivirologists to deal with their increasingly specious family of viruses (Muhire *et al.*, 2013). Percent nucleotide identity approaches rely on pairwise alignment, but the tools can be applied to already aligned datasets as well. Perhaps due to the frequent gene flow among viral species, few are suggesting that named viral groupings be monophyletic, and phylogenetics does not yet have a large role in viral taxonomy.

Within a group of less well-characterized sequences, operational taxonomic units (OTUs) can be assigned by separating sequences by percent nucleotide identity thresholds that are user-specified. While they are more often used in prokaryotic metagenomics analyses, programs like QIIME (http://qiime.org/) can both determine numbers of OTUs and compare presence of each OTU in different metagenomic samples. This approach works well with amplicon data from a mixed sample, such as a metagenome using primers specific for the capsid gene of a viral group of interest (such as the gp23 gene of T4-like phages (Comeau and Krisch, 2008) or the capsid gene of ssDNA phages (Hopkins *et al.*, 2014) or when there is a reference database supplied, such as the phage proteomic tree

(Caporaso *et al.,* 2011). This is a common first step in determining the species richness or Shannon's diversity of viral metagenomes (Emerson *et al.,* 2013).

Integration

Ideally, many of us would like one tool – a single portal – in which to search databases, conduct phylogenetic and population genetic tests and be able to make figures describing these results. Both free and commercially available packages attempt to provide this kind of functionality. Several of the curated viral resources (*e.g.*, NCBI, VIPR) have integrated tools for some analyses, providing some one-stop shopping for narrowly focused analyses. With data obtained on one's own, MEGA is one widely available suite that houses many analysis tools. Since it is free and runs on the three major operating systems, it can be an attractive choice for educational settings. However, many others abound, with varying strengths and weaknesses in analytical tools [*e.g.*, DAMBE, Mesquite, SNAP workbench (Price and Carbone, 2005), Table 2.4]. Additionally, there are more customizable environments for analysis, which rely on users comfortable with command line interfaces, as in R. Excitingly, there are multiple attempts to bridge the gap between the modularity of current command line methods with graphic interfaces, such as the developing RevBayes and the established Galaxy (http://galaxyproject.org/, (Giardine *et al.,* 2005). Galaxy has greater integration with sequence databases, and there are clear tutorials to help new users. However, the current gold standard in integrated database and tools for studying viral evolution research may be Geneious, and as its ".com" URL implies, it is not free (Table 2.4). Other commercial packages (*e.g.*, CLC Workbench) also exist. While no single portal provides all the extant tools needed to study all aspects of viral molecular evolution, we are much closer than a decade ago to that being the case, and an integrated package may be the best and easiest choice for many researchers.

Garbage in, garbage out

The peril of computational analysis is that almost always a user will get an answer. Misaligned sequences fed into a phylogenetics program will produce a tree, and perhaps one with strong statistical support. But this tree will not reflect how genomes are actually evolving because the alignment underpinning the analysis wasn't accurately representing homologous sites. Similarly, BEAST analyses will deduce rates of nucleotide substitution even if the dataset weren't sampled over a long range of time, and therefore doesn't have sufficient temporal signal to calibrate any molecular clock (Duffy and Holmes, 2009). It is not infrequent that one will run across a paper in the literature guilty of one of these kinds of errors – published informatic analyses without the authors understanding the limitations of the programs, the way parameters influence results, and what the biological meaning of the output is. Some of these errors are obvious to readers, or readers familiar with the methods, but errors in alignment are obscured unless the alignment is publicly available, or until others try to replicate the analysis without success, and publish alternative conclusions.

More subtle problems can arise in viral bioinformatic studies that are harder to detect. Because most of the tools used were recently developed, not all of the limitations of the methods are known or what kinds of control analyses are most appropriate to run. For instance, one common assumption in bioinformatic analyses is that more data is always better because it can average out individual sequence errors, such as those introduced during PCR amplification (Smith and Peay, 2014). Just as with alignments, where it can improve accuracy to remove sections that are of questionable homology, using a smaller but well-vetted dataset can improve the accuracy of BEAST phylogenomic analyses. It was recently shown that only one misdated sequence in a 65 taxon data set was enough to cause non-overlapping 95% credibility intervals between the accurate and misdated results (Hicks and Duffy, 2012). What is standard practice when using informatic tools can change rapidly, and subscribing to user groups or participating in community discussions when available can be an important way to keep up with the most accurate methods of using different programs.

Final thoughts

Compared to other sections of this book, this chapter will likely not wear well over time. This is a snapshot of viral molecular evolution methods in 2014, and an inexhaustive picture at that (Sharma *et al.,* 2015). While all of these will still be viable analytical tools for the next decade or beyond, the best tools for the job are continuously improving and many of the current programs will be supplanted. For instance, alignment-free methods may become more popular in virology (Gao and Luo, 2012; Lauber *et al.,* 2013). Minimally, URLs will change, programs will be orphaned, funding lapses will close webservers, and tutorials will become out of date. Sites that implement many methods like CIPRES may have the best chance of surviving because they are used by many kinds of biological researchers and they are not single-program sites that are privately maintained by a lab that might not pay attention to the resource after a grant expires or a motivated student graduates. The virology community has good reason to support efforts for data archiving and gateways for computationally intensive analyses both within and among our nations. Most of the viral phylogenetic analyses that are published can be run through these sites on high performance computers, and these portals can provide the computational resources for many labs' needs.

Finally, all of the approaches to studying viral evolution discussed in this chapter are limited by being bioinformatic studies. While sequence motifs can be identified by strong statistical support, without empirical, wet-lab testing, the functional significance of those motifs for viruses cannot necessarily be inferred. It's a modern day spandrels of San Marcos situation (Gould and Lewontin, 1979), except we know that when strongly conserved sequences are experimentally mutated sometimes the results violate expectation: many highly divergent sequences can be functional (Dorit and Riley, 2002). There will also always be some questions for which sequence data and computational analysis are not sufficient – including studies of phenotypic masking, or maternal effects that could impact viral fitness and evolution (Duffy *et al.,* 2006; Holland *et al.,* 1989). Sequence data is cheaper to generate than most other kinds of virological data, and materials for data analysis are orders of magnitude cheaper than wet-lab experimentation. But as we find the majority of viral genomes by their sequences alone, there are many lifetimes of ecological and molecular characterization that could – and often should – be done to help us understand how these viruses function and what role they serve ecologically. Our understanding of viral evolution will be limited without molecular and ecological metadata to make more sense of patterns revealed by genomic sequences.

Acknowledgements

This work was supported by National Science Foundation grants (DEB 1026095, 1240049) and the Bill and Melinda Gates Foundation/UK Department for International Development. It was enriched by discussions with Mya Breitbart, Karyna Rosario and members of the Duffy Lab: Raffia Ahmed, Allison Hicks, Li Li, Daniel Stern Cardinale and Yee Mey Seah.

References

Altschul, S.F., Gish, W., Miller, W., Myers, E.W., and Lipman, D.J. (1990). Basic local alignment search tool. J Mol Biol *215*, 403-410.

Anisimova, M. (2012). Parametric models of codon evolution. In Codon Evolution Mechanisms and Models, G.M. Cannarozzi, and A. Schneider, eds. (Oxford, Oxford Univeristy Press), pp. 12-33.

Anisimova, M., Cannarozzi, G.M., and Liberles, D.A. (2010). Finding the balance between the mathematical and biological optima in multiple sequence alignment. Trends in Evolutionary Biology *2*, e7.

Anisimova, M., and Liberles, D.A. (2012). Detecting and understanding natural selection. In Codon Evolution Mechanisms and Models, G.M. Cannarozzi, and A. Schneider, eds. (Oxford, Oxford Univeristy Press), pp. 73-96.

Arenas, M., Patricio, M., Posada, D., and Valiente, G. (2010). Characterization of phylogenetic networks with NetTest. BMC Bioinformatics *11*, 268.

Auton, A., and McVean, G. (2007). Recombination rate estimation in the presence of hotspots. Genome research *17*, 1219-1227.

Baccam, P., Thompson, R.J., Fedrigo, O., Carpenter, S., and Cornette, J.L. (2001). PAQ: Partition Analysis of Quasispecies. Bioinformatics *17*, 16-22.

Bahir, I., Fromer, M., Prat, Y., and Linial, M. (2009). Viral adaptation to host: a proteome-based analysis of codon usage and amino acid preferences. Molecular systems biology *5*, 311.

Bahl, J., Nelson, M.I., Chan, K.H., Chen, R., Vijaykrishna, D., Halpin, R.A., Stockwell, T.B., Lin, X., Wentworth, D.E., Ghedin, E., *et al.* (2011). Temporally structured metapopulation dynamics and persistence of influenza A H3N2 virus in humans. Proc Natl Acad Sci U S A *108*, 19359-19364.

Baltimore, D. (1971). Expression of animal virus genomes. Bacteriological reviews *35*, 235-241.

Bao, Y., Bolotov, P., Dernovoy, D., Kiryutin, B., Zaslavsky, L., Tatusova, T., Ostell, J., and Lipman, D. (2008). The influenza virus resource at the National Center for Biotechnology Information. J Virol *82*, 596-601.

Batorsky, R., Kearney, M.F., Palmer, S.E., Maldarelli, F., Rouzine, I.M., and Coffin, J.M. (2011). Estimate of effective recombination rate and average selection coefficient for HIV in chronic infection. Proc Natl Acad Sci U S A *108*, 5661-5666.

Bergsten, J. (2005). A review of long-branch attraction. Cladistics *21*, 163-193.

Bertolotti, L., Kitron, U., and Goldberg, T.L. (2007). Diversity and evolution of West Nile virus in Illinois and the United States, 2002-2005. Virology *360*, 143-149.

Bhatt, S., Katzourakis, A., and Pybus, O.G. (2010). Detecting natural selection in RNA virus populations using sequence summary statistics. Infect Genet Evol *10*, 421-430.

Bloom, J.D. (2014). An experimentally determined evolutionary model dramatically improves phylogenetic fit. Mol Biol Evol *31*, 1956-1978.

Blouin, C., Perry, S., Lavell, A., Susko, E., and Roger, A.J. (2009). Reproducing the manual annotation of multiple sequence alignments using a SVM classifier. Bioinformatics *25*, 3093-3098.

Bouckaert, R., Heled, J., Kuhnert, D., Vaughan, T., Wu, C.H., Xie, D., Suchard, M.A., Rambaut, A., and Drummond, A.J. (2014). BEAST 2: a software platform for Bayesian evolutionary analysis. PLoS Computational Biology 10, e1003537.

Bowen, M.D., Rollin, P.E., Ksiazek, T.G., Hustad, H.L., Bausch, D.G., Demby, A.H., Bajani, M.D., Peters, C.J., and Nichol, S.T. (2000). Genetic diversity among Lassa virus strains. J Virol *74*, 6992-7004.

Bradley, R.K., Roberts, A., Smoot, M., Juvekar, S., Do, J., Dewey, C., Holmes, I., and Pachter, L. (2009). Fast statistical alignment. PLoS computational biology *5*, e1000392.

Bulla, I., Schultz, A.K., Schreiber, F., Zhang, M., Leitner, T., Korber, B., Morgenstern, B., and Stanke, M. (2010). HIV classification using the coalescent theory. Bioinformatics *26*, 1409-1415.

Caporaso, J.G., Knight, R., and Kelley, S.T. (2011). Host-associated and free-living phage communities differ profoundly in phylogenetic composition. PLoS One *6*, e16900.

Cardinale, D.J., DeRosa, K., and Duffy, S. (2013). Base composition and translational selection are insufficient to explain codon usage bias in plant viruses. Viruses *5*, 162-181.

Carrillo-Tripp, M., Shepherd, C.M., Borelli, I.A., Venkataraman, S., Lander, G., Natarajan, P., Johnson, J.E., Brooks, C.L., 3rd, and Reddy, V.S. (2009). VIPERdb2: an enhanced and web API enabled relational database for structural virology. Nucleic Acids Res *37*, D436-442.

Castro-Nallar, E., Pérez-Losada, M., Burton, G.F., and Crandall, K.A. (2012). The evolution of HIV: inferences using phylogenetics. Molecular phylogenetics and evolution *62*, 777-792.

Cattoli, G., Fusaro, A., Monne, I., Coven, F., Joannis, T., El-Hamid, H.S., Hussein, A.A., Cornelius, C., Amarin, N.M., Mancin, M., *et al.* (2011). Evidence for differing evolutionary dynamics of A/H5N1 viruses among countries applying or not applying avian influenza vaccination in poultry. Vaccine *29*, 9368-9375.

Cerny, J., Cerna Bolfikova, B., Valdes, J.J., Grubhoffer, L., and Ruzek, D. (2014). Evolution of tertiary structure of viral RNA dependent polymerases. PLoS One *9*, e96070.

Chare, E.R., and Holmes, E.C. (2004). Selection pressures in the capsid genes of plant RNA viruses reflect mode of transmission. Journal of General Virology *85*, 3149-3157.

Charleston, M.A., and Robertson, D.L. (2002). Preferential host switching by primate lentiviruses can account for phylogenetic similarity with the primate phylogeny. Systematic biology *51*, 528-535.

Chirico, N., Vianelli, A., and Belshaw, R. (2010). Why genes overlap in viruses. Proceedings *277*, 3809-3817.

Chithambaram, S., Prabhakaran, R., and Xia, X. (2014). Differential codon adaptation between dsDNA and ssDNA phages in Escherichia coli. Mol Biol Evol *31*, 1606-1617.

Clement, M., Posada, D., and Crandall, K.A. (2000). TCS: a computer program to estimate gene genealogies. Mol Ecol *9*, 1657-1659.

Comeau, A.M., and Krisch, H.M. (2008). The capsid of the T4 phage superfamily: the evolution, diversity, and structure of some of the most prevalent proteins in the biosphere. Mol Biol Evol *25*, 1321-1332.

Cornman, R.S., Boncristiani, H., Dainat, B., Chen, Y., vanEngelsdorp, D., Weaver, D., and Evans, J.D. (2014). Population-genomic variation within RNA viruses of the Western honey bee, Apis mellifera, inferred from deep sequencing. BMC genomics *14*, 154.

Cummings, M.P., Handley, S.A., Myers, D.S., Reed, D.L., Rokas, A., and Winka, K. (2003). Comparing bootstrap and posterior probability values in the four-taxon case. Systematic biology *52*, 477-487.

Dang, C.C., Le, Q.S., Gascuel, O., and Le, V.S. (2010). FLU, an amino acid substitution model for influenza proteins. BMC evolutionary biology *10*, 99.

Darriba, D., Taboada, G.L., Doallo, R., and Posada, D. (2012). jModelTest 2: more models, new heuristics and parallel computing. Nature methods *9*, 772.

De Bruyn, A., Villemot, J., Lefeuvre, P., Villar, E., Hoareau, M., Harimalala, M., Abdoul-Karime, A.L., Abdou-Chakour, C., Reynaud, B., Harkins, G.W., *et al.* (2012). East African cassava mosaic-like viruses from Africa to Indian ocean islands: molecular diversity, evolutionary history and geographical dissemination of a bipartite begomovirus. BMC evolutionary biology *12*, 228.

de Oliveira, T., Pybus, O.G., Rambaut, A., Salemi, M., Cassol, S., Ciccozzi, M., Rezza, G., Gattinara, G.C., D'Arrigo, R., Amicosante, M., *et al.* (2006). Molecular epidemiology: HIV-1 and HCV sequences from Libyan outbreak. Nature *444*, 836-837.

Di Tommaso, P., Moretti, S., Xenarios, I., Orobitg, M., Montanyola, A., Chang, J.M., Taly, J.F., and Notredame, C. (2011). T-Coffee: a web server for the multiple sequence alignment of protein and RNA sequences using structural information and homology extension. Nucleic Acids Res *39*, W13-17.

Dimmic, M.W., Rest, J.S., Mindell, D.P., and Goldstein, R.A. (2002). rtREV: an amino acid substitution matrix for inference of retrovirus and reverse transcriptase phylogeny. J Mol Evol *55*, 65-73.

Ding, Y.Z., You, Y.N., Sun, D.J., Chen, H.T., Wang, Y.L., Chang, H.Y., Pan, L., Fang, Y.Z., Zhang, Z.W., Zhou, P., *et al.* (2014). The effects of the context-dependent codon usage bias on the structure of the nsp1alpha of porcine reproductive and respiratory syndrome virus. BioMed research international *2014*, 765320.

Domingo, E., Sheldon, J., and Perales, C. (2012). Viral quasispecies evolution. Microbiol Mol Biol Rev *76*, 159-216.

Dorit, R.L., and Riley, M.A. (2002). When comparative information leads us astray: the receptor-binding region of colicin E9. J Mol Evol *55*, 347-355.

Drummond, A.J., and Rambaut, A. (2007). BEAST: Bayesian evolutionary analysis by sampling trees. BMC evolutionary biology *7*, 214.

Duchene, S., Holmes, E.C., and Ho, S.Y. (2014). Analyses of evolutionary dynamics in viruses are hindered by a time-dependent bias in rate estimates. Proc. R. Soc. B *281,* 20140732 .

Dudas, G., and Rambaut, A. (2014). Phylogenetic Analysis of Guinea 2014 EBOV Ebolavirus Outbreak. PLoS Currents Outbreaks. May 2. *Edition 1*.

Duffy, S., and Holmes, E.C. (2007). Multiple introductions of the Old World begomovirus Tomato yellow leaf curl virus into the New World. Appl Environ Microbiol *73*, 7114-7117.

Duffy, S., and Holmes, E.C. (2008). Phylogenetic evidence for rapid rates of molecular evolution in the single-stranded DNA begomovirus Tomato yellow leaf curl virus (TYLCV). Journal of Virology *82*, 957-965.

Duffy, S., and Holmes, E.C. (2009). Validation of high rates of nucleotide substitution in geminiviruses: phylogenetic evidence from East African cassava mosaic viruses. J Gen Virol *90*, 1539-1547.

Duffy, S., Shackelton, L.A., and Holmes, E.C. (2008). Rates of evolutionary change in viruses: patterns and determinants. Nat Rev Genet *9*, 267-276.

Duffy, S., Turner, P.E., and Burch, C.L. (2006). Pleiotropic costs of niche expansion in the RNA bacteriophage phi 6. Genetics *172*, 751-757.

Dunn, R.R. (2008). Every Living Thing: Man's Obsessive Quest to Catalog Life, from Nanobacteria to New Monkeys (Washington, DC, Smithsonian).

Edgar, R.C. (2004). MUSCLE: a multiple sequence alignment method with reduced time and space complexity. Bmc Bioinformatics *5*, 1-19.

Edwards, R.A., and Rohwer, F. (2005). Viral metagenomics. Nat Rev Microbiol *3*, 504-510.
Emerson, J.B., Thomas, B.C., Andrade, K., Heidelberg, K.B., and Banfield, J.F. (2013). New approaches indicate constant viral diversity despite shifts in assemblage structure in an Australian hypersaline lake. Appl Environ Microbiol *79*, 6755-6764.
Excoffier, L., and Lischer, H.E. (2010). Arlequin suite ver 3.5: a new series of programs to perform population genetics analyses under Linux and Windows. Molecular ecology resources *10*, 564-567.
Fancello, L., Raoult, D., and Desnues, C. (2012). Computational tools for viral metagenomics and their application in clinical research. Virology *434*, 162-174.
Felsenstein, J. (1981). Evolutionary trees from DNA sequences: a maximum likelihood approach. J Mol Evol *17*, 368-376.
Felsenstein, J. (1989). PHYLIP -- Phylogeny Inference Package (Version 3.2). Cladistics *5*, 164-166.
Fernandez-Suarez, X.M., Rigden, D.J., and Galperin, M.Y. (2014). The 2014 Nucleic Acids Research Database Issue and an updated NAR online Molecular Biology Database Collection. Nucleic Acids Res *42*, D1-6.
Firth, C., Kitchen, A., Shapiro, B., Suchard, M.A., Holmes, E.C., and Rambaut, A. (2010). Using time-structured data to estimate evolutionary rates of double-stranded DNA viruses. Mol Biol Evol *27*, 2038-2051.
Froissart, R., Roze, D., Uzest, M., Galibert, L., Blanc, S., and Michalakis, Y. (2005). Recombination every day: abundant recombination in a virus during a single multi-cellular host infection. PLoS Biol *3*, e89.
Frost, S.D.W., and Volz, E.M. (2010). Viral phylodynamics and the search for an 'effective number of infections'. Philos T R Soc B *365*, 1879-1890.
Fu, Y.X. (2001). Neutrality and Selection in Molecular Evolution: Statistical Tests. In Encyclopedia of Life Sciences (John Wiley and Sons).
Gao, Y., and Luo, L. (2012). Genome-based phylogeny of dsDNA viruses by a novel alignment-free method. Gene *492*, 309-314.
Garcia-Villada, L., and Drake, J.W. (2012). The three faces of riboviral spontaneous mutation: spectrum, mode of genome replication, and mutation rate. PLoS Genet *8*, e1002832.
Giardine, B., Riemer, C., Hardison, R.C., Burhans, R., Elnitski, L., Shah, P., Zhang, Y., Blankenberg, D., Albert, I., Taylor, J., *et al.* (2005). Galaxy: a platform for interactive large-scale genome analysis. Genome research *15*, 1451-1455.
Gibbs, M.J., Armstrong, J.S., and Gibbs, A.J. (2000). Sister-scanning: a Monte Carlo procedure for assessing signals in recombinant sequences. Bioinformatics *16*, 573-582.
Goldman, N. (1998). Phylogenetic information and experimental design in molecular systematics. Proceedings *265*, 1779-1786.
Gorbachev, V.V. (2012). Effect of random sample size on the accuracy of nucleotide diversity estimation. Genetika *48*, 880-884.
Gorbalenya, A.E. (2014). ICTV Newsletter *11*.
Gould, S.J., and Lewontin, R.C. (1979). The spandrels of San Marco and the Panglossian paradigm: a critique of the adaptationist programme. Proceedings of the Royal Society of London Series B, *205*, 581-598.
Greene, S.E., and Reid, A. (2014). Viruses throughout life and time: friends, foes, change agents (American Academy of Microbiology).
Grenfell, B.T., Pybus, O.G., Gog, J.R., Wood, J.L., Daly, J.M., Mumford, J.A., and Holmes, E.C. (2004). Unifying the epidemiological and evolutionary dynamics of pathogens. Science *303*, 327-332.
Gu, W., Zhou, T., and Wilke, C.O. (2010). A universal trend of reduced mRNA stability near the translation-initiation site in prokaryotes and eukaryotes. PLoS computational biology *6*, e1000664.

Guindon, S., Dufayard, J.F., Lefort, V., Anisimova, M., Hordijk, W., and Gascuel, O. (2010). New algorithms and methods to estimate maximum-likelihood phylogenies: assessing the performance of PhyML 3.0. Systematic biology *59*, 307-321.

Hartl, D.L., and Clark, A.G. (1997). Principles of population genetics, 2nd edn (Sutherland, MA, Sinauer Associates, Inc.).

Hicks, A.L., and Duffy, S. (2012). One misdated sequence of rabbit hemorrhagic disease virus prevents accurate estimation of its nucleotide substitution rate. BMC evolutionary biology *12*, 74.

Hicks, A.L., and Duffy, S. (2014). Cell Tropism Predicts Long-term Nucleotide Substitution Rates of Mammalian RNA Viruses. PLoS Pathog *10*, e1003838.

Ho, S.Y., Shapiro, B., Phillips, M.J., Cooper, A., and Drummond, A.J. (2007). Evidence for time dependency of molecular rate estimates. Systematic biology *56*, 515-522.

Holland, J.J., de la Torre, J.C., Steinhauer, D.A., Clarke, D., Duarte, E., and Domingo, E. (1989). Virus mutation frequencies can be greatly underestimated by monoclonal antibody neutralization of virions. J Virol *63*, 5030-5036.

Holmes, E.C. (2008). Comparative studies of RNA virus evolution. In Origin and Evolution of Viruses, E. Domingo, C.R. Parrish, and J.J. Holland, eds. (London, Academic Press), pp. 119-134.

Holsinger, K.E., and Weir, B.S. (2009). Genetics in geographically structured populations: defining, estimating and interpreting F(ST). Nat Rev Genet *10*, 639-650.

Hopkins, M., Kailasan, S., Cohen, A., Roux, S., Tucker, K.P., Shevenell, A., Agbandje-McKenna, M., and Breitbart, M. (2014). Diversity of environmental single-stranded DNA phages revealed by PCR amplification of the partial major capsid protein. The ISME Journal *8*, 2093-2103.

Huelsenbeck, J.P., and Ronquist, F. (2001). MRBAYES: Bayesian inference of phylogenetic trees. Bioinformatics *17*, 754-755.

Huelsenbeck, J.P., Ronquist, F., Nielsen, R., and Bollback, J.P. (2001). Bayesian inference of phylogeny and its impact on evolutionary biology. Science *294*, 2310-2314.

Huson, D.H., and Bryant, D. (2006). Application of phylogenetic networks in evolutionary studies. Mol Biol Evol *23*, 254-267.

Iantorno, S., Gori, K., Goldman, N., Gil, M., and Dessimoz, C. (2013). Who watches the watchmen? An appraisal of benchmarks for multiple sequence alignment. In Multiple Sequence Alignmnet Methods, D.J. Russell, ed. (Humana Press), pp. 59-73.

Jenkins, G.M., and Holmes, E.C. (2003). The extent of codon usage bias in human RNA viruses and its evolutionary origin. Virus Res *92*, 1-7.

Jukes, T.H., and Cantor, C.R. (1969). Evolution of protein molecules. In Mammalian Protein Metabolism, H.N. Munro, ed. (New York, Academic Press), pp. 21-132.

Just, W. (2001). Computational complexity of multiple sequence alignment with SP-score. J Comput Biol *8*, 615-623.

Karlin, S. (2001). Detecting anomalous gene clusters and pathogenicity islands in diverse bacterial genomes. Trends Microbiol *9*, 335-343.

Katoh, K., and Standley, D.M. (2013). MAFFT multiple sequence alignment software version 7: improvements in performance and usability. Mol Biol Evol *30*, 772-780.

Kern, A.D., and Haussler, D. (2010). A population genetic hidden Markov model for detecting genomic regions under selection. Mol Biol Evol *27*, 1673-1685.

Kieft, J.S. (2008). Viral IRES RNA structures and ribosome interactions. Trends Biochem Sci *33*, 274-283.

Kimura, M. (1980). A simple method for estimating evolutionary rate of base substitutions through comparative studies of nucleotide sequences. Journal of molecular evolution *16*, 111-120.

King, A.M.Q., Adams, M.J., Carstens, E.B., and Lefkowitz, E.J., eds. (2012). Virus taxonomy: classification and nomenclature of viruses: Ninth Report of the International Committee on Taxonomy of Viruses (San Diego, Elsevier).

Koonin, E.V. (2009). Evolution of genome architecture. The international journal of biochemistry and cell biology *41*, 298-306.

Kosakovsky Pond, S.L., Frost, S.D., and Muse, S.V. (2005). HyPhy: hypothesis testing using phylogenies. Bioinformatics *21*, 676-679.
Kosakovsky Pond, S.L., Murrell, B., and Poon, A.F.Y. (2012). Evolution of viral genomes: interplay between selection, recombination and other forces. In Evolutionary Genomics, M. Anisimova, ed. (Humana Press), pp. 239-272.
Kosakovsky Pond, S.L., Posada, D., Gravenor, M.B., Woelk, C.H., and Frost, S.D. (2006). GARD: a genetic algorithm for recombination detection. Bioinformatics *22*, 3096-3098.
Kryazhimskiy, S., Bazykin, G.A., and Dushoff, J. (2008). Natural selection for nucleotide usage at synonymous and nonsynonymous sites in influenza A virus genes. J Virol *82*, 4938-4945.
Kryazhimskiy, S., and Plotkin, J.B. (2008). The population genetics of dN/dS. PLoS Genet *4*, e1000304.
Kuhn, E., Ichimura, A.S., Peng, V., Fritsen, C.H., Trubl, G., Doran, P.T., and Murray, A.E. (2014). Brine assemblages of ultrasmall microbial cells within the ice cover of Lake Vida, Antarctica. Appl Environ Microbiol *80*, 3687-3698.
Kuhner, M.K. (2006). LAMARC 2.0: maximum likelihood and Bayesian estimation of population parameters. Bioinformatics *22*, 768-770.
Kuzmin, I.V., Shi, M., Orciari, L.A., Yager, P.A., Velasco-Villa, A., Kuzmina, N.A., Streicker, D.G., Bergman, D.L., and Rupprecht, C.E. (2012). Molecular inferences suggest multiple host shifts of rabies viruses from bats to mesocarnivores in Arizona during 2001-2009. PLoS Pathog *8*, e1002786.
Lam, T.T., Hon, C.C., and Tang, J.W. (2010). Use of phylogenetics in the molecular epidemiology and evolutionary studies of viral infections. Critical reviews in clinical laboratory sciences *47*, 5-49.
Lanfear, R., Calcott, B., Ho, S.Y., and Guindon, S. (2012). Partitionfinder: combined selection of partitioning schemes and substitution models for phylogenetic analyses. Mol Biol Evol *29*, 1695-1701.
Lapidot, M., Gelbart, D., Gal-On, A., Sela, N., Anfoka, G., Ahmed, F.H., Abou-Jawdah, Y., Sobh, H., Mazyad, H., Aboul-Ata, A.E., *et al.* (2014). Frequent migration of introduced cucurbit-infecting begomoviruses among Middle Eastern countries. Virology Journal *11,* 181.
Larkin, M.A., Blackshields, G., Brown, N.P., Chenna, R., McGettigan, P.A., McWilliam, H., Valentin, F., Wallace, I.M., Wilm, A., Lopez, R., *et al.* (2007). Clustal W and Clustal X version 2.0. Bioinformatics *23*, 2947-2948.
Lauber, C., Goeman, J.J., Parquet Mdel, C., Nga, P.T., Snijder, E.J., Morita, K., and Gorbalenya, A.E. (2013). The footprint of genome architecture in the largest genome expansion in RNA viruses. PLoS Pathog *9*, e1003500.
Lauber, C., and Gorbalenya, A.E. (2012). Toward genetics-based virus taxonomy: comparative analysis of a genetics-based classification and the taxonomy of picornaviruses. J Virol *86*, 3905-3915.
Lefeuvre, P., Lett, J.M., Varsani, A., and Martin, D.P. (2009). Widely conserved recombination patterns among single-stranded DNA viruses. J Virol *83*, 2697-2707.
Lefeuvre, P., Martin, D.P., Harkins, G., Lemey, P., Gray, A.J., Meredith, S., Lakay, F., Monjane, A., Lett, J.M., Varsani, A., *et al.* (2010). The spread of tomato yellow leaf curl virus from the Middle East to the world. PLoS Pathog *6*, e1001164.
Lemey, P., Rambaut, A., Drummond, A.J., and Suchard, M.A. (2009). Bayesian phylogeography finds its roots. PLoS computational biology *5*, e1000520.
Lima, A.T., Sobrinho, R.R., Gonzalez-Aguilera, J., Rocha, C.S., Silva, S.J., Xavier, C.A., Silva, F.N., Duffy, S., and Zerbini, F.M. (2013). Synonymous site variation due to recombination explains higher genetic variability in begomovirus populations infecting non-cultivated hosts. J Gen Virol *94*, 418-431.
Lole, K.S., Bollinger, R.C., Paranjape, R.S., Gadkari, D., Kulkarni, S.S., Novak, N.G., Ingersoll, R., Sheppard, H.W., and Ray, S.C. (1999). Full-length human immunodeficiency virus type 1 genomes from subtype C-infected seroconverters in India, with evidence of intersubtype recombination. J Virol *73*, 152-160.

Luksza, M., and Lassig, M. (2014). A predictive fitness model for influenza. Nature *507*, 57-61.

Lyngsø, R., Song, Y.S., and Hein, J. (2005). Minimum recombination histories by branch and bound. Proceedings of WABI 2005, Lecture Notes in Computer Science *3692*, 239-250.

Maddison, W.P., and Maddison, D.R. (2014). Mesquite: a modular system for evolutionary analysis.

Martin, D., and Rybicki, E. (2000). RDP: detection of recombination amongst aligned sequences. Bioinformatics *16*, 562-563.

Martin, D.P., Lemey, P., Lott, M., Moulton, V., Posada, D., and Lefeuvre, P. (2010). RDP3: a flexible and fast computer program for analyzing recombination. Bioinformatics *26*, 2462-2463.

McVean, G.A., and Cardin, N.J. (2005). Approximating the coalescent with recombination. Philosophical transactions of the Royal Society of London *360*, 1387-1393.

Michely, S., Toulza, E., Subirana, L., John, U., Cognat, V., Marechal-Drouard, L., Grimsley, N., Moreau, H., and Piganeau, G. (2013). Evolution of codon usage in the smallest photosynthetic eukaryotes and their giant viruses. Genome Biol Evol *5*, 848-859.

Minin, V.N., Bloomquist, E.W., and Suchard, M.A. (2008). Smooth skyride through a rough skyline: Bayesian coalescent-based inference of population dynamics. Mol Biol Evol *25*, 1459-1471.

Mir, K., and Schober, S. (2014). Selection pressure in alternative reading frames. PLoS One *9*, e108768.

Mohabatkar, H., Keyhanfar, M., and Behbahani, M. (2012). Protein bioinformatics applied to virology. Current Protein and Peptide Science *13*, 547-559.

Morrison, D.A. (2009). Why would phylogeneticists ignore computerized sequence alignment? Systematic biology *58*, 150-158.

Moszer, I., Rocha, E.P., and Danchin, A. (1999). Codon usage and lateral gene transfer in Bacillus subtilis. Curr Opin Microbiol *2*, 524-528.

Muhire, B., Martin, D.P., Brown, J.K., Navas-Castillo, J., Moriones, E., Zerbini, F.M., Rivera-Bustamante, R., Malathi, V.G., Briddon, R.W., and Varsani, A. (2013). A genome-wide pairwise-identity-based proposal for the classification of viruses in the genus Mastrevirus (family Geminiviridae). Arch Virol *158*, 1411-1424.

Nakleh, L. (2011). Evolutionary phylogenetic networks: models and issues. In Problem Solving Handbooks in Computational Biology and Bioinformatics, L.S. Heath, and N. Ramakrishnan, eds. (New York, Springer), pp. 125-158.

Nei, M. (2013). Mutation-Driven Evolution (Oxford, Oxford University Press).

Nei, M., and Gojobori, T. (1986). Simple methods for estimating the numbers of synonymous and nonsynonymous nucleotide substitutions. Mol Biol Evol *3*, 418-426.

Nei, M., and Li, W.H. (1979). Mathematical model for studying genetic variation in terms of restriction endonucleases. Proc Natl Acad Sci U S A *76*, 5269-5273.

Nielsen, R., and Yang, Z. (1998). Likelihood models for detecting positively selected amino acid sites and applications to the HIV-1 envelope gene. Genetics *148*, 929-936.

Ojosnegros, S., Agudo, R., Sierra, M., Briones, C., Sierra, A., González-López, C., Domingo, E., and Cristina, J. (2008). Topology of evolving, mutagenized viral populations: quasispecies expansion, compression, and operation of negative selection. BMC Evolutionary Biology *8*, 207.

Padidam, M., Sawyer, S., and Fauquet, C.M. (1999). Possible emergence of new geminiviruses by frequent recombination. Virology *265*, 218-225.

Page, R.D.M., and Holmes, E.C. (1998). Molecular Evolution A Phylogenetic Approach (Malden, Blackwell).

Parameswaran, P., Charlebois, P., Tellez, Y., Nunez, A., Ryan, E.M., Malboeuf, C.M., Levin, J.Z., Lennon, N.J., Balmaseda, A., Harris, E., *et al.* (2012). Genome-wide

patterns of intrahuman dengue virus diversity reveal associations with viral phylogenetic clade and interhost diversity. J Virol *86*, 8546-8558.
Parker, J., Rambaut, A., and Pybus, O.G. (2008). Correlating viral phenotypes with phylogeny: accounting for phylogenetic uncertainty. Infect Genet Evol *8*, 239-246.
Phillips, A.J. (2006). Homology assessment and molecular sequence alignment. Journal of biomedical informatics *39*, 18-33.
Piel, W.H., Donoghue, M.J., and Sanderson, M.J. (2002). TreeBASE: a database of phylogentic knowledge. In To the interoperable "Catalog of Life" with partners Species 2000 Asia Oceanea, J. Shimura, K.L. Wilson, and D. Gordon, eds. (Tsukuba, National Institute for Environmental Studies), pp. 41-47.
Pierce, L.R., and Stepien, C.A. (2012). Evolution and biogeography of an emerging quasispecies: diversity patterns of the fish Viral Hemorrhagic Septicemia virus (VHSv). Molecular phylogenetics and evolution *63*, 327-341.
Poon, A.F., Lewis, F.I., Pond, S.L., and Frost, S.D. (2007). An evolutionary-network model reveals stratified interactions in the V3 loop of the HIV-1 envelope. PLoS computational biology *3*, e231.
Price, E.W., and Carbone, I. (2005). SNAP: workbench management tool for evolutionary population genetic analysis. Bioinformatics *21*, 402-404.
Pybus, O.G., and Rambaut, A. (2009). Evolutionary analysis of the dynamics of viral infectious disease. Nat Rev Genet *10*, 540-550.
R Development Core Team (2012). R: a language and environment for statistical computing (Vienna, R Foundation for Statistical Computing).
Rambaut, A., Suchard, M.A., Xie, D., and Drummond, A.J. (2014). Tracer. http://tree.bio.ed.ac.uk/software/tracer/
Ramsden, C., Holmes, E.C., and Charleston, M.A. (2009). Hantavirus evolution in relation to its rodent and insectivore hosts: no evidence for codivergence. Mol Biol Evol *26*, 143-153.
Rasmussen, D.A., Boni, M.F., and Koelle, K. (2014a). Reconciling phylodynamics with epidemiology: the case of dengue virus in southern Vietnam. Mol Biol Evol *31*, 258-271.
Rasmussen, M.D., Hubisz, M.J., Gronau, I., and Siepel, A. (2014b). Genome-wide inference of ancestral recombination graphs. PLoS Genet *10*, e1004342.
Remold, S.K., Rambaut, A., and Turner, P.E. (2008). Evolutionary genomics of host adaptation in vesicular stomatitis virus. Mol Biol Evol *25*, 1138-1147.
Resch, W., Zaslavsky, L., Kiryutin, B., Rozanov, M., Bao, Y., and Tatusova, T.A. (2009). Virus variation resources at the National Center for Biotechnology Information: dengue virus. BMC Microbiol *9*, 65.
Rokyta, D.R., Joyce, P., Caudle, S.B., and Wichman, H.A. (2005). An empirical test of the mutational landscape model of adaptation using a single-stranded DNA virus. Nature genetics *37*, 441-444.
Roth, A., Anisimova, M., and Cannarozzi, G.M. (2012). Measuring codon usage bias. In Codon evolution mechanisms and models, G.M. Cannarozzi, and A. Schneider, eds. (Oxford, Oxford University Press).
Roux, S., Tournayre, J., Mahul, A., Debroas, D., and Enault, F. (2014). Metavir 2: new tools for viral metagenome comparison and assembled virome analysis. BMC Bioinformatics *15*, 76.
Rozas, J., Sanchez-DelBarrio, J.C., Messeguer, X., and Rozas, R. (2003). DnaSP, DNA polymorphism analyses by the coalescent and other methods. Bioinformatics *19*, 2496-2497.
Sabath, N., Landan, G., and Graur, D. (2008). A method for the simultaneous estimation of selection intensities in overlapping genes. PLoS One *3*, e3996.
Sanjuán, R. (2012). From molecular genetics to phylodynamics: evolutionary relevance of mutation rates across viruses. PLoS Pathog *8*, e1002685.
Sanjuán, R., Nebot, M.R., Chirico, N., Mansky, L.M., and Belshaw, R. (2010). Viral mutation rates. J Virol *84*, 9733-9748.

Sau, K., Gupta, S.K., Sau, S., Mandal, S.C., and Ghosh, T.C. (2007). Studies on synonymous codon and amino acid usage biases in the broad-host range bacteriophage KVP40. Journal of microbiology (Seoul, Korea) *45*, 58-63.
Schneemann, A. (2006). The structural and functional role of RNA in icosahedral virus assembly. Annu Rev Microbiol *60*, 51-67.
Schultz, A.K., Zhang, M., Bulla, I., Leitner, T., Korber, B., Morgenstern, B., and Stanke, M. (2009). jpHMM: improving the reliability of recombination prediction in HIV-1. Nucleic Acids Res *37*, W647-651.
Seo, T.K., and Kishino, H. (2008). Synonymous substitutions substantially improve evolutionary inference from highly diverged proteins. Systematic biology *57*, 367-377.
Seoighe, C., Ketwaroo, F., Pillay, V., Scheffler, K., Wood, N., Duffet, R., Zvelebil, M., Martinson, N., McIntyre, J., Morris, L., *et al.* (2007). A model of directional selection applied to the evolution of drug resistance in HIV-1. Mol Biol Evol *24*, 1025-1031.
Seto, D. (2010). Viral genomics and bioinformatics. Viruses *2*, 2587-2593.
Shackelton, L.A., Parrish, C.R., Truyen, U., and Holmes, E.C. (2005). High rate of viral evolution associated with the emergence of carnivore parvovirus. Proc Natl Acad Sci U S A *102*, 379-384.
Shao, W., Kearney, M.F., Boltz, V.F., Spindler, J.E., Mellors, J.W., Maldarelli, F., and Coffin, J.M. (2014). PAPNC, a novel method to calculate nucleotide diversity from large scale next generation sequencing data. Journal of Virological Methods *203*, 73-80.
Shapiro, B., Rambaut, A., and Drummond, A.J. (2006a). Choosing appropriate substitution models for the phylogenetic analysis of protein-coding sequences. Mol Biol Evol *23*, 7-9.
Shapiro, B., Rambaut, A., Pybus, O.G., and Holmes, E.C. (2006b). A phylogenetic method for detecting positive epistasis in gene sequences and its application to RNA virus evolution. Mol Biol Evol *23*, 1724-1730.
Sharma, D., Priyadarshini, P., and Vrati, S. (2015). Unraveling the web of viroinformatics: computational tools and databases in virus research. J Virol *89*, 1489-1501.
Sharp, P.M., and Li, W.H. (1987). The codon Adaptation Index--a measure of directional synonymous codon usage bias, and its potential applications. Nucleic Acids Res *15*, 1281-1295.
Sharp, P.M., Tuohy, T.M., and Mosurski, K.R. (1986). Codon usage in yeast: cluster analysis clearly differentiates highly and lowly expressed genes. Nucleic Acids Res *14*, 5125-5143.
Siebenga, J.J., Lemey, P., Kosakovsky Pond, S.L., Rambaut, A., Vennema, H., and Koopmans, M. (2010). Phylodynamic reconstruction reveals norovirus GII.4 epidemic expansions and their molecular determinants. PLoS Pathog *6*, e1000884.
Sievers, F., Wilm, A., Dineen, D., Gibson, T.J., Karplus, K., Li, W., Lopez, R., McWilliam, H., Remmert, M., Soding, J., *et al.* (2011). Fast, scalable generation of high-quality protein multiple sequence alignments using Clustal Omega. Molecular systems biology *7*, 539.
Simmonds, P., and Smith, D.B. (1999). Structural constraints on RNA virus evolution. J Virol *73*, 5787-5794.
Slatkin, M., and Maddison, W.P. (1989). A cladistic measure of gene flow inferred from the phylogenies of alleles. Genetics *123*, 603-613.
Smith, D.P., and Peay, K.G. (2014). Sequence depth, not PCR replication, improves ecological inference from next generation DNA sequencing. PLoS One *9*, e90234.
Smith, G.J., Vijaykrishna, D., Bahl, J., Lycett, S.J., Worobey, M., Pybus, O.G., Ma, S.K., Cheung, C.L., Raghwani, J., Bhatt, S., *et al.* (2009). Origins and evolutionary genomics of the 2009 swine-origin H1N1 influenza A epidemic. Nature *459*, 1122-1125.
Stainton, D., Kraberger, S., Walters, M., Wiltshire, E.J., Rosario, K., Halafihi, M., Lolohea, S., Katoa, I., Faitua, T.H., Aholelei, W., *et al.* (2012). Evidence of inter-

component recombination, intra-component recombination and reassortment in banana bunchy top virus. J Gen Virol *93*, 1103-1119.
Stamatakis, A. (2006). RAxML-VI-HPC: maximum likelihood-based phylogenetic analyses with thousands of taxa and mixed models. Bioinformatics *22*, 2688-2690.
Stamatakis, A. (2014). RAxML version 8: a tool for phylogenetic analysis and post-analysis of large phylogenies. Bioinformatics *30*, 1312-1313.
Stevens, H. (2013). Life Out of Sequence A Data-Driven History of Bioinformatics (Chicago, University of Chicago Press).
Strimmer, K., Forslund, K., Holland, B., and Moulton, V. (2003). A novel exploratory method for visual recombination detection. Genome biology *4*, R33.
Stucky, B.J. (2012). SeqTrace: a graphical tool for rapidly processing DNA sequencing chromatograms. J Biomol Tech *23*, 90-93.
Suzuki, Y. (2006). Natural selection on the influenza virus genome. Mol Biol Evol *23*, 1902-1911.
Swofford, D.L. (2003). PAUP* Phylogenetic analysis using parsimony (*and other methods) (Sunderland, MA, Sinauer Associates).
Tajima, F. (1989). Statistical method for testing the neutral mutation hypothesis by DNA polymorphism. Genetics *123*, 585-595.
Talavera, G., and Castresana, J. (2007). Improvement of phylogenies after removing divergent and ambiguously aligned blocks from protein sequence alignments. Systematic biology *56*, 564-577.
Tamura, K., Stecher, G., Peterson, D., Filipski, A., and Kumar, S. (2013). MEGA6: Molecular Evolutionary Genetics Analysis version 6.0. Mol Biol Evol *30*, 2725-2729.
Vision, T.J. (2010). Open data and the social contract of scientific publishing. BioScience *60*, 330-331.
Wang, J., Sharma, A.M., Duffy, S., and Almeida, R.P. (2011). Genetic diversity in the 3' terminal 4.7-kb region of grapevine leafroll-associated virus 3. Phytopathology *101*, 445-450.
Wei, X., and Zhang, J. (2015). A simple method for estimating the strength of natural selection on overlapping genes. Genome Biol Evol *7*, 381-390.
Wilker, P.R., Dinis, J.M., Starrett, G., Imai, M., Hatta, M., Nelson, C.W., O'Connor, D.H., Hughes, A.L., Neumann, G., Kawaoka, Y., *et al.* (2013). Selection on haemagglutinin imposes a bottleneck during mammalian transmission of reassortant H5N1 influenza viruses. Nature communications *4*, 2636.
Wommack, K.E., Bhavsar, J., Polson, S.W., Chen, J., Dumas, M., Srinivasiah, S., Furman, M., Jamindar, S., and Nasko, D.J. (2012). VIROME: a standard operating procedure for analysis of viral metagenome sequences. Standards in genomic sciences *6*, 427-439.
Wong, W.S., and Nielsen, R. (2004). Detecting selection in noncoding regions of nucleotide sequences. Genetics *167*, 949-958.
Wright, F. (1990). The 'effective number of codons' used in a gene. Gene *87*, 23-29.
Wu, C.H., Suchard, M.A., and Drummond, A.J. (2013). Bayesian selection of nucleotide substitution models and their site assignments. Mol Biol Evol *30*, 669-688.
Xia, X. (2013). DAMBE5: a comprehensive software package for data analysis in molecular biology and evolution. Mol Biol Evol *30*, 1720-1728.
Yang, Z. (1994). Estimating the pattern of nucleotide substitution. J Mol Evol *39*, 105-111.
Yang, Z. (2007). PAML 4: phylogenetic analysis by maximum likelihood. Mol Biol Evol *24*, 1586-1591.
Yoon, H.S., Price, D.C., Stepanauskas, R., Rajah, V.D., Sieracki, M.E., Wilson, W.H., Yang, E.C., Duffy, S., and Bhattacharya, D. (2011). Single-cell genomics reveals organismal interactions in uncultivated marine protists. Science *332*, 714-717.
Zaheri, M., Dib, L., and Salamin, N. (2014). A generalized mechanistic codon model. Mol Biol Evol *31*, 2528-2541.
Zanini, F., and Neher, R.A. (2013a). Deleterious synonymous mutations hitchhike to high frequency in HIV-1 env evolution (arXiv: 1303.0805 [q-bio.PE]).

Zanini, F., and Neher, R.A. (2013b). Quantifying selection against synonymous mutations in HIV-1 env evolution. J Virol *87*, 11843-11850.

Zwickl, D.J. (2006). Genetic algorithm approaches for the phylogenetic analysis of large biological sequence datasets under the maximum likelihood criterion (The University of Texas at Austin).

All URLs were accessed on 9/1/14.

3

The Evolution and Transmission of Vector-borne Viruses

Naomi L. Forrester[1], Serafín Gutiérrez[2] and Lark L. Coffey[3]*

[1]Department of Pathology, University of Texas Medical Branch, Galveston, TX 77555, USA; [2]CMAEE Unit, CIRAD, 34398 Montpellier Cedex 5, France; [3]Department of Pathology, Microbiology and Immunology, University of California, Davis, CA 95616, USA

**Corresponding author: Lark L. Coffey, lcoffey@ucdavis.edu*

Abstract

Arthropod-borne viruses (arboviruses) are those that are transmitted between vertebrate or plant hosts by arthropod vectors such as mosquitoes, ticks or aphids. Here we discuss the implications of this particular life cycle on virus evolution and examine the different mechanisms by which these viruses have adapted to it. Different transmission strategies exist, some requiring replication within the vector and others not, and each strategy may impose different selection pressures and requires specific adaptations. We discuss the latest literature on the evolution of arboviruses at different scales (i.e. within-host and between-host evolution) and the strategies viruses have in countering the different barriers in their cycles, whether these barriers are anatomical barriers such as plant cell walls, or a host response such as an activated immune system. Finally we present the latest literature about how the viruses may affect the host and impact their behavior. The importance of understanding the evolution of these viruses particularly within their hosts cannot be understated as any host shift or adaptation to a marginal host can result in rapid viral expansion.

Introduction

What is a vector-borne virus

Vector-borne viruses cycle between parasitic arthropods and their vertebrate or plant hosts. To simplify, we will use here the terms vector and host to refer respectively to the arthropods and the vertebrate/plant species parasitized by the arthropods. Vector-borne viruses (also called arthropod borne viruses, or arboviruses) represent a significant public health burden with more than 100 species causing human disease. Transmission typically involves alternating vector-host-vector cycling, although for vertical transmission the virus is

Table 3.1. The major animal viruses transmitted by arthropods, viruses transmitted by mosquitoes are italicized, those transmitted by ticks are emboldened and those transmitted by other arthropods noted in normal type.

Family	Viruses	Reference
Asfarviridae	**African swine fever virus**	Tulman *et al.*, 2009
Bunyarviridae	**Bhanja virus**	Matsuno *et al.*, 2013
	Nairobi sheep disease	Marczinke and Nichol, 2002
	Soldado virus	Hubalek and Rudolf, 2012
	Cache valley virus	Edwards, 1994
	La Crosse virus	Borucki *et al.*, 2002
	Rift valley fever virus	Flick and Bouloy, 2005
	Snowshoe hare virus	Hewlett *et al.*, 1992
	Aino virus	Ali *et al.*, 2012
	Akabane virus	Kirkland, 2002
	Main Drain virus	Emmons *et al.*, 1983
	Schmallenburg virus	Doceul *et al.*, 2013
	Shuni virus	van Eeden *et al.*, 2014
Flaviviridae	*Israel turkey meningoencephalitis virus*	Gould *et al.*, 2003
	Japanese encephalitis virus	Unni *et al.*, 2011
	Murray Valley Encephalitis virus	Selvey *et al.*, 2014
	Tembusu virus	Hubalek *et al.*, 2014
	Usutu virus	Ashraf *et al.*, 2015
	Wesselsbron virus	Ali *et al.*, 2012
	West Nile virus	Pesko and Ebel, 2012
	Yellow fever virus	Beasley *et al.*, 2014
	Kyasanur Forest disease virus	Holbrook, 2012
	Louping Ill virus	Jeffries *et al.*, 2014
	Omsk hemorrahgic fever virus	Růžek *et al.*, 2010
	Tick-borne encephalitis virus	LaSala and Holbrook, 2010
	Tyuleniy virus	Hubalek and Rudolf, 2012
Orthomyxoviridae	**Thogoto virus**	Hubalek and Rudolf, 2012
Reoviridae	African Horse Sickness virus	MacLachlan and Guthrie, 2010
	Kasba virus	Ohashi *et al.*, 2004
	Bluetongue virus	Coetzee *et al.*, 2012
	Epizootic hemorrhagic disease of deer virus	Savini *et al.*, 2011
	Equine encephalosis virus	Oura *et al.*, 2012
	Ibaraki virus	Ohashi *et al.*, 2002
	Peruvian horse sickness virus	Attoul *et al.*, 2009
	Yunnan virus	Attoul *et al.*, 2009
Rhabdoviridae	*Bovine ephemeral fever virus*	Nandi and Negi, 1999
	Vesicular stomatitis virus *Alagoas* *Indiana* *New Jersey*	Novella, 2003
Togaviridae	*Eastern equine encephalitis virus*	Zacks and Paessler, 2010
	Western equine encephalitis virus	Allison *et al.*, 2015
	Venezuelan equine encephalitis virus	Weaver *et al.*, 2004
	Sindbis virus	Laine *et al.*, 2004
	Middelburg virus	Hubalek *et al.*, 2014
	Getah virus	Fukunaga *et al.*, 2000
	Semliki Forest virus	Hubalek *et al.*, 2014

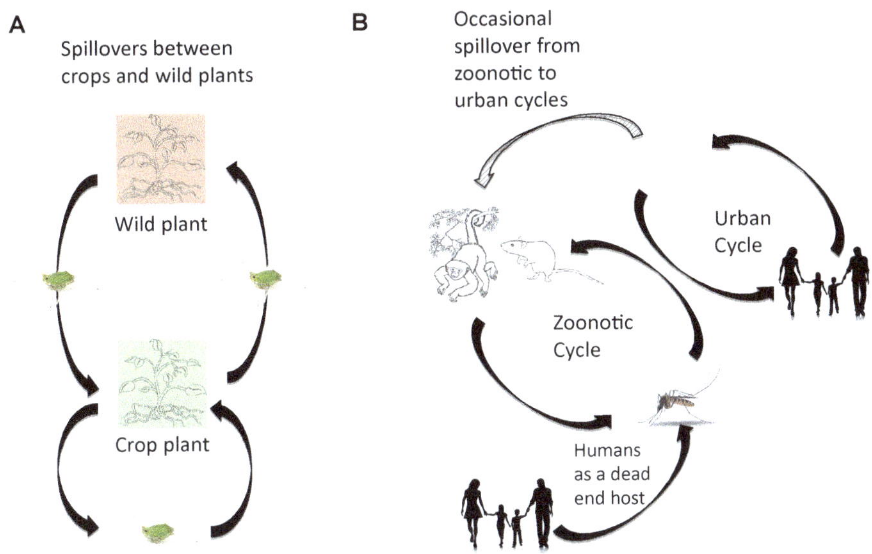

Figure 3.1. Transmission cycles of plant or vertebrate arboviruses. **A**. Plant viruses can be transmitted between wild and domestic or crop plants by vectors. **B**. Zoonotic cycles where the virus cycles between non-human vertebrate and mosquitoes/ticks. Humans are incidentally infected due to encounters with the zoonotic cycle (e.g. West Nile virus). Occasionally, evolutionary change occurs and the virus can establish in a new urban cycle between humans and mosquitoes. This usually involves adaptation to mosquitoes that are primarily anthropophilic such as *Ae. aegyptii* or *Ae. albopictus* (e.g. chikungunya virus and dengue virus).

transferred from adult arthropod to offspring and venereal transmission between insects can occur during copulation (Figure 3.1). Vector-borne viruses of vertebrates belong to numerous RNA viral families, including the *Bunyaviridae*, *Flaviviridae*, *Reoviridae*, *Rhabdoviridae*, and *Togaviridae*, as well as a single genus in the family *Orthomyxoviridae* (Thogotovirus) and the DNA virus African swine fever virus that belongs to the *Asfarviridae* family. These viruses share a near global distribution and they use many different arthropod hosts, including mosquitoes, midges, sandflies, cimicid bugs, ticks and thrips.

Vector-borne viruses and disease causing potential

Animals

Most vector-borne viruses that affect humans also have non-human vertebrate reservoirs. In addition to exacting a significant burden on public health worldwide, at least fifty arboviruses are pathogens in their non-human vertebrate reservoirs [reviewed extensively in (Hubalek *et al.*, 2014)]. These viruses belong to 7 families, some of which are highlighted in Table 3.1. Disease in vertebrate reservoirs caused by these viruses ranges from non-specific febrile illness, sometimes with gastrointestinal involvement, to hemorrhagic or neurologic disease, depending on infecting virus. Death in infected animals and their fetuses is a common outcome and can be a significant cause of economic loss, especially for viruses that affect agriculturally important animals, including African horse sickness, Akabane, bluetongue, Eastern, Western and Venezuelan equine encephalitides, Nairobi sheep disease, Rift Valley fever, Schmallenberg, vesicular stomatitis, and West Nile viruses. Other arboviruses may also cause disease in animals but have not been recognized as pathogens due to non-specific presentations, restricted distributions, or limited.

Human

Infectious diseases of humans are a significant disease burden particularly in developing nations. Of these,17% are caused by vector-borne pathogens (WHO, 2014). The infection that is increasing the most dramatically is dengue virus (DENV), which has increased 30-fold in the past 50 years, resulting in 50-100 million annual cases. For human arboviruses, there are two main ways of transmission to humans. The first is urban human-mosquito-human transmission that is perpetuated without any need for a non-human vertebrate host. There are three significant pathogens for which this is the mode of transmission, DENV, yellow fever virus (YFV) and chikungunya virus (CHIKV) (Bryant *et al.,* 2003; Vasilakis *et al.,* 2010; Volk *et al.,* 2010). Although most human disease caused by CHIKV or DENV occurs as a result of these urban cycles, there are also sylvatic cycles from which the urban cycles are thought to have arisen. These sylvatic cycles involve non-human primates and forest dwelling mosquitoes, with occasional spillover into humans (Vasilakis *et al.,* 2011). The other mode of transmission associated with most other arboviruses is zoonotic where humans fed on by infected arthropods do not develop viremias sufficient to support transmission back to vectors. In these settings, humans therefore serve as dead end hosts resulting in no perpetuation of virus transmission. Human infections by zoonotic arboviruses tend to occur as focal clusters of infections. These foci are often seasonal and dependent on the abundance of vectors and zoonotic hosts. Due to the fact that they are a result of human exposure to the zoonotic transmission cycle, incidence rates tend to be consistent year to year. Occasionally, there is an increase in frequency due to environmental factors that result in an increased number of infected vectors and/or vertebrate hosts, or an increase in frequency because of augmented human activity in endemic areas. This is exemplified by Rift Valley fever virus (RVFV), associated with warm seasons or increased precipitation, which results in increased mosquito abundance and therefore increased viral transmission with continuous zoonotic spillover (Balenghien *et al.,* 2013). Viruses such as tick-borne encephalitis virus (TBEV), RVFV or Crimean-Congo hemorrhagic fever virus (CCHFV) are therefore consistent burdens on the economies of the countries where they are endemic.

Plants

Plants viruses, or phytoviruses, are abundant in nature and rather diverse. They use a large variety of transmission modes, with examples of transmission through seeds, pollen, grafts, tubers or contact between plants. Nevertheless, there is one transmission mode that prevails: transmission by biological vectors. In fact, insect vectors transmit more than 75% of approximately 700 plant viruses officially recognized by the ICTV (Hogenhout *et al.,* 2008). The overrepresentation of vector transmission in plant viruses compared to animal viruses is usually explained by two main factors. These factors are the sessile nature of plants, limiting encounters between infected and healthy hosts, and the cell wall, the strong pectin and cellulose barrier surrounding and protecting plant cells. Thus, entry, transfer, and exit of viruses from one plant to another are usually achieved during feeding and dispersal of plant-feeding flying insects - mainly aphids (*Aphididae*), whiteflies (*Aleyrodidae*), leafhoppers (*Cicadellidae*), planthoppers (*Delphacidae*) and thrips (*Thysanoptera*). However, nematodes, fungi and mites also transmit some plant viruses. In general, vector-borne phytoviruses have restricted host and vector ranges. Nevertheless, some of these viruses can infect plant species from unrelated families and are transmitted by a large number of vector species, usually from the same family. For example, cucumber mosaic virus (CMV; not to be confused with the animal cytomegalovirus also known as CMV), a single stranded (ss) RNA virus transmitted by aphids, is a paradigm of generalism in viruses and close to the description of a Darwinian demon. The different strains of this virus have a collective host range that includes more than 1200 plant species in over 100 families, with examples of both monocots (plants whose embryo has a single cotyledon) and eudicots (plants whose embryos have a double cotyledon). Moreover, CMV can be transmitted by more than 80 aphid species from 33 genera (Scholthof *et al.,* 2011).

Virtually all plant species are infected by viruses and crops are no exception. Viruses are the main cause of emerging diseases of cultivated plants worldwide (Anderson *et al.,* 2004). Given their abundance among known phytoviruses, it is not surprising that vector-borne phytoviruses are responsible for a large proportion of the crop diseases caused by viruses. Virus infection can lead to a reduction, sometimes dramatic, in plant growth, yield, and product quality, eventually leading to economic loss. Famine can be the outcome of viral diseases when infected crops are major staple food in developing countries (e.g. rice, wheat, maize, cassava or bananas) where family farming is the rule (Rybicki and Pietersen, 1999). A clear example of the latter is African cassava mosaic disease, caused by a complex of whitefly-transmitted geminiviruses including *African* cassava mosaic virus. Besides economic losses in the range of US$ 2 billion annually, outbreaks of this disease have led to famine episodes in East and Central Africa and represent a threat to food security in the region (Scholthof *et al.,* 2011). Examples of vector-transmitted viruses leading to important economic losses worldwide are the aphid-transmitted Barley yellow dwarf luteoviruses in cereals, the thrip-transmitted tomato spotted wilt virus (TSWV) affecting a wide range of vegetables, the whitefly-transmitted tomato yellow leaf curl virus (TYLCV) and related geminiviruses in tomato. This short list includes DNA and RNA viruses using unrelated vector species, highlighting the diversity of vector-borne viruses causing crop disease.

As one would expect from the diversity of vector-borne phytoviruses, different vector transmission modes have evolved in plant viruses, including transmission modes not observed in animal viruses. Many important vectors including aphids and whiteflies move from plant to plant, doing short test feedings until they find a suitable host. Then, they proceed to sustained feeding on sap from the vasculature, eventually moving to a new plant or settling and raising offspring on the plant. Some viruses have evolved a transmission strategy targeting the initial foraging phase, whereas others are only transmitted if the sap-feeding phase is achieved. Three main transmission modes have been defined based on the virus-vector interaction: 1) non-circulative, 2) circulative and 3) propagative modes (Blanc, 2007). The non-circulative mode is the simplest, as the virus does not need to infect vector cells to be transmitted. This mode is similar to mechanical transmission in animal viruses. The virus attaches to putative receptors on the cuticle of the mouthparts of vectors during feeding on an infected plant (Uzest *et al.*, 2007). The vector can transmit the virus shortly after acquisition, within a few minutes, upon salivation or egestion on a healthy plant. However, retention times do not exceed minutes or hours. Hence, vector individuals foraging for suitable hosts and doing short test feedings on a plant before rapidly moving to a new one are considered the most efficient vectors. In fact, transmission is generally greatly reduced or abolished if the vector goes into sustained feeding before moving to a healthy plant. The second mode, circulative transmission, is not known to occur in animal viruses. The virus is ingested along with sap and plant material, it crosses the gut epithelium to enter the haemocoel and then it reaches and accumulates in the salivary glands. From there, it will be injected into new host plants together with saliva during feeding. Hence, although the virus circulates and accumulates within the vector, there is no viral replication involved. The third mode of transmission, the propagative mode, is similar, if not identical, to biological transmission of vector-borne viruses of vertebrates. After ingestion, the virus replicates in the gut and the salivary glands of the vector and is inoculated in new host plants with the saliva during feeding. In both circulative and propagative transmission, an incubation period of 1 to several days is often required before the vector can transmit. In addition, once this incubation is over, vectors can usually transmit the virus for extended periods of time (i.e. days to months), eventually until their death. Thus, contrary to non-circulative transmission, these two last modes require settling and sustained feeding to maximize transmission rates.

There are virus families exclusively transmitted using a single transmission mode, like the circulative *Geminiviridae*, and other families with examples of different modes, like the *Potyviridae*. In the latter case, the transmission mode seems to be conserved at the genus level (Nault, 1997). The three transmission modes can be found in largely unrelated families, suggesting convergent evolution among different families. For example, the whitefly-transmitted ssDNA geminiviruses and the aphid-transmitted ssRNA luteoviruses follow a circulative mode. In line with the hypothesis of convergent evolution, different vector/virus associations can use a common mode. For example, fungi, mites, and aphids can vector in a non-circulative mode *Tombusviridae*, *Tritimovirus* and *Caulimovirus*, respectively.

Types of viruses that use vectors

Using vectors as transmission strategy has evolved in many different genera. As shown in Table 3.1, various families of animal and human viruses have evolved to use arthropods as vectors. Most are RNA viruses, with the notable exception of African Horse Sickness virus (AHSV), suggesting that a property inherent to RNA viruses allows the vector-borne life cycle. In plant viruses, vector-borne transmission may have evolved to allow transmission between immobile hosts. This need to transmit by vectors might explain why families of vector-borne phytoviruses are distributed all along the genome-based classification of viruses infecting plants. Although there are examples of families with genomes made of dsRNA (*Reoviridae*), negative ssRNA (*Bunyaviridae* and *Rhabdoviridae*) and ssDNA (*Geminiviridae* and *Nanoviridae*), more than half of the families have positive ssRNA genomes. A single family of pararetroviruses (i.e. dsDNA viruses using a viral reverse transcriptase and an RNA genome intermediate for replication), the *Caulimoviridae*, also infect plants and most members are transmitted by insects. Strikingly, there is not a single example of true vector-borne dsDNA virus. Whether this particular distribution is correlated to vector transmission or just follows the overall distribution of plant viruses is far from clear as vector-borne phytoviruses and RNA viruses are both predominant among known plant viruses. It must be noted that our knowledge on plant virus diversity is probably biased by the extreme paucity of data from non-crop hosts. Metagenomic approaches have been recently launched to further explore this diversity (Roossinck, 2012) and they may help to better understand the evolution of vector transmission in plant viruses and its potential association with specific virus types.

The evolution of vector-borne transmission in vertebrates is much more likely to have developed from either plant viruses or from the arthropod-only viruses. The idea that vector-borne viruses might have evolved from plants was suggested back in the 1970s (Johnson *et al.,* 1969). Recently, the discovery of mosquito-only viruses, both within the genera already described in Table 3.1, and the discovery of several mosquito-borne viruses that are related to plant viruses, suggest a viable mechanism by which this strategy might evolve in vertebrates. However, other potential origins of vector-borne strategies cannot be ruled out given the current paucity of data on virus diversity. Additionally, the ubiquitous nature of the arthropod vectors means that utilizing this vector-borne strategy has the potential to result in the expansion of range and hosts as shown by West Nile virus (WNV) and its spread into North America by utilizing a wide range of native vertebrate and mosquito species (McMullen *et al.,* 2013). While the use of vectors appears to have some benefits, arboviruses do not evolve as fast as single host viruses, suggesting that the need to infect both hosts may result from constraints that limit evolution (Forrester *et al.,* 2012a). Thus, the arthropod-only viruses that are related to the dual-cycling arboviruses may show evidence of increased evolution compared to those that cycle between arthropod and vertebrate or plant.

Evolution of vector-borne viruses

Evolutionary processes associated with:

Arthropod-only viruses

Next generation sequencing has led to the discovery of viruses that are restricted to mosquitoes or ticks with no recognized vertebrate host. These viruses are transovarially transmitted from insect to insect. This allows the viruses to be transmitted from generation to generation without any need for other hosts. These viruses have been found in many different genera including *Alphaviridae*, *Flaviviridae, Nidoviridae* (Junglen *et al.,* 2009; Nasar *et al.,* 2012; Vasilakis *et al.,* 2014), as well as many plant-like genera such as *Tymoviridae* (unpublished data), and the newly discovered Negeviruses (Vasilakis *et al.,* 2013). While these viruses are not true arboviruses in that they do not circulate between arthropod and vertebrate hosts, they might be the potential ancestors of arboviruses, as was discussed for the Rhabdoviruses (Johnson *et al.,* 1969). Recently, some insect-only viruses that appear to have secondarily lost the ability to infect vertebrates have been discovered among the *Flaviviridae* and the *Alphaviridae* (Evangelista *et al.,* 2013; Huhtamo *et al.,* 2014; Junglen *et al.,* 2009; Kolodziejek *et al.,* 2013; Lee *et al.,* 2013; Nasar *et al.,* 2012; Vazquez *et al.,* 2012). These discoveries suggest that the ability to infect vertebrates was probably acquired by these insect-only viruses and then secondarily lost to give rise to viruses such as Lammi virus in the *Flaviviridae* and Eilat virus in the *Alphaviridae*.

Mosquito-borne viruses

Mosquitoes are ubiquitous in all continents except Antarctica and are among the most deadly animals in the world, causing more human deaths than any other animal. The reason for this is that female mosquitoes require a bloodmeal to produce eggs, and humans are bloodmeal sources. As a result, viruses or parasites that are transmitted by mosquitoes often infect humans. The need to obtain blood to facilitate the production of eggs means that mosquitoes are capable of being infected and transmit the viruses within a short time frame (3-5 days depending on the mosquito). In addition, mosquitoes remain infected with arboviruses for the length of their life, increasing the potential for transmission.

Many arbovirus species have adapted to specific mosquito hosts such that infection of other mosquitoes does not necessarily result in productive transmission. Infection of mosquitoes requires the navigation of several anatomical barriers. Once a mosquito ingests an infected bloodmeal the virus must first infect the midgut cells. A certain quantity of virus particles must be ingested to allow the virus to cross the midgut barrier (Hardy, 1988; Weaver, 1997). Infection of midgut cells is dependent on the mosquito and the virus. Infection of *Aedes taeniorhynchus,* a vector of epizootic VEEV with replicon particles capable of only one round of replication showed that few midgut cells are susceptible (Smith *et al.,* 2008). However, a similar study using the endemic mosquito *Culex taeniopus* with its associated endemic virus revealed that most midgut cells were infected (Kenney *et al.,* 2012), suggesting that the amount of virus required to infect the midgut is mosquito and virus dependent. The blood meal is initially taken into the posterior midgut, where the mosquito releases proteolytic enzymes and also alters the pH and the temperature, resulting in a hostile environment for the virus. The foregut and the hindgut of the mosquito are lined with cuticle and are therefore assumed to be impervious to infection. Thus, it is the midgut of the mosquito that is susceptible to infection. The midgut is composed of a single layer of epithelial cells lined with microvilli (Houk *et al.,* 1990). There is some evidence that in *Ae. aegypti,* cells lacking microvilli termed 'bare cells' are present either singly or in clusters throughout the region of the posterior midgut (Zieler *et al.,* 2000). However, the significance of their contribution to viral infection is unknown. Infection of the midgut is aided by blood coagulation, as the serum tends to be concentrated adjacent to the posterior midgut epithelium (Weaver *et al.,* 1993; Weaver *et al.,* 1991). Following infection of the midgut, the virus disseminates throughout the haemocoel, either via the basal lamina or the tracheal system of the mosquito (Romoser *et al.,* 2004). The virus then infects the salivary glands and is excreted into saliva for transmission to a new host. Escape from the midgut and colonisation of the haemolymph does not necessarily guarantee the infection of the salivary glands (Hardy, 1988). The amount of virus transmitted by mosquitoes is highly variable, ranging from a few virus particles to 10^5 in a volume of only a few μL^3 (Hardy, 1988; Smith *et al.,* 2005; Vanlandingham *et al.,* 2008) (Figure 3.2). All of these barriers to productive infection of mosquitoes affect the transmission of arboviruses.

Tick-borne viruses

The lifestyle of ticks heavily influences the transmission of the arboviruses that they carry. Ticks generally develop over a period of 1-3 years, and for ticks that serve as vectors for viruses, these can be persistently infected for the entirety of their lives. Southern ticks in warmer climates such as *Rhipicephalus sanguineas* have up to two life cycles a year, whereas ticks that inhabit northern latitudes such as *Ixodes ricinus* can take up to three years to complete a life cycle (Labuda and Nuttall, 2004). Ticks are blood-feeders and require repeated bloodmeals to complete their life cycles. Once hatched, larvae require a bloodmeal to facilitate transmission to the next stage, the nymphal stage. A second bloodmeal is required to transition to the adult stage, and then a third bloodmeal facilitates egg laying. The necessity to bloodfeed means that ticks must place themselves in a position to encounter vertebrate hosts. Many ticks adopt a questing strategy by positioning themselves on the tips of branches and leaves and waiting for a passing host. Depending on species, once they have found a host, the tick will either immediately attach or search the host for a feeding site. Once attached, a tick will burrow into the host and secrete cement to ensure secure attachment for the duration of bloodfeeding. Unlike mosquitoes that feed in minutes, ticks require hours to weeks to feed on these hosts. This extended period means that the tick has to circumvent blood coagulation at the feeding site, as well as the anti-inflammatory and adaptive responses from the vertebrate host immune response. As a result, ticks have evolved salivary proteins that are deposited at the site of attachment to prevent coagulation and to counteract the host immune response (Wikel, 1999). Additionally, these proteins can also enhance infectivity of the pathogens that are deposited by the tick (Brossard and Wikel, 2004). The temporal dynamics of the tick-virus relationship vary across viruses, where viruses such as tick-borne encephalitis virus (TBEV) and Powassan virus (POWV) *Flaviviridae* are present in the salivary glands before feeding commences and are deposited after only 1 hour of tick feeding (Ebel and Kramer, 2004); whereas other pathogens, such as Thogoto virus and Dugbe virus, do not accumulate in the saliva until after feeding commences and are therefore likely deposited later than TBEV or POWV (Labuda and Nuttall, 2004). This wide range of time to infection means that different tick salivary proteins are associated with different pathogen infections and the effect of these is a major focus of investigation (Hermance and Thangamani, 2014; Nuttall, 1999). Since infection is lifelong, a tick infected with an arbovirus at the larval stage will be able to transmit the virus when it takes bloodmeals at the nymphal and adult stages. This persistent infection by ticks makes them both a transmission vehicle for the virus and a reservoir host for the virus (Labuda and Nuttall, 2004). Transmission of the virus is limited by feeding preferences and the hosts that the ticks preferentially feeding choose. Generally, a small population of hosts carry a disproportionate number of ticks, with about 20% of hosts carrying 80% of the tick burden (Randolph *et al.,* 1999), increasing the potential for viral transmission and decreasing the range of hosts that the virus infects. There is some potential for vertical transmission, particularly from adult to offspring, but this is rare in nature (Labuda and Nuttall, 2004). Non-viremic transmission, first decribed using Thogoto virus when uninfected ticks were infected when co-feeding with infected ticks on an uninfected host (Jones *et al.,* 1987), removes the need for an infected host and therefore increases the opportunity for transmission of the virus to a new tick. Since it was first described, non-viremic transmission has been described in numerous tick-virus relationships and has been well reviewed (Havlíková *et al.,* 2013).

Like mosquitoes, ticks also have barriers to viral infections. There is an anatomical barrier at the midgut that can be artificially bypassed by injecting ticks outside of the midgut to produce infections (Steele and Nuttall, 1989). This barrier is significantly different from the midgut barrier in mosquitoes, since ticks reabsorb their midgut when they moult to the next stage.

Culicoides-borne viruses

The other major group of vectors of vertebrate arboviruses are Culicoides, the biting midges. Midges transmit Bluetongue virus (BTV) and Akabane virus (AKAV). Primarily, these species transmit viruses that cause diseases of livestock and other animals. These viruses have become more prominent in the past few years, because of the recent expansion of BTV in Europe and the more recent discovery of Schmallenberg virus, which has resulted in increased studies of these viruses. The study of the evolution and transmission of these viruses has been limited by the difficulty in maintaining the vector colonies and working with them. As a result there are few viral evolutionary studies. One notable exception is Bluetongue virus (BTV). An experimental transmission cycle was established by infecting *Culicoides sonorensis* midges and then allowing them to infect the sheep host, followed by *C. sonorensis* infection and then infection of a calf vertebrate host, and then infection of *C. sonorensis*. The evolution of BTV throughout this alternating transmission cycle was then studied. BTV may be subject to bottlenecks and founder effects that resulted in titer throughout the transmission cycle. Additionally, each of the 10 segments in the virus genome evolved separately, suggesting that different segments have varied evolutionary pressures during infection of both midges and the vertebrate host (Bonneau *et al.,* 2001). Differential evolution of the 10 BTV segments may be responsible for the wide range of reassortant strains observed in naturally-infected animals. This study underscores the need to understand virus-specific evolutionary processes.

Plant viruses

Did vector transmission originate in the aquatic environment of algae, the plant taxa that probably preceded land plants? Algae are not necessarily immobile and they do not possess the robust cell walls of land plants, the two factors generally proposed to explain the large representation of vector transmission in plant viruses. Hence, a terrestrial origin for vector transmission seems more plausible. In addition, under the "aquatic origin" hypothesis, one would expect to find vector-borne viruses infecting algae or, at least, phylogenetic relationships between current alga viruses and vector-borne plant viruses. However, no vector-borne virus infecting algae has been described. Moreover, known alga viruses are mostly dsDNA viruses which are absent in land plants. From the few known RNA and ssDNA viruses infecting algae, only two are related families including vector-borne viruses: a reovirus (Brussaard *et al.,* 2004) and a putative nanovirus-like virus (Yoon *et al.,* 2011). However, although current data generally do not support an aquatic origin for vector transmission in plant viruses, any conclusion remains largely speculative given the scarcity of information on algal viruses (Steward *et al.,* 2013). In any case, a terrestrial origin seems plausible for vector transmission in the *Bunyaviridae*, *Rhabdoviridae* and perhaps even the *Reoviridae* (that is, all the negative ssRNA and dsRNA viruses in plants). All these families contain examples of viruses infecting vertebrates, insects and plants. Given the larger diversity of animal viruses in those families and the fact that their vectors are animals, the most likely hypothesis is that horizontal transfers of viruses from insects to plants are at the origin of such plant viruses (Dolja and Koonin, 2011). Finally, ssDNA geminiviruses may have evolved from plasmids found in phytoplasmas, parasitic bacteria of plants transmitted by insects, (Krupovic *et al.*, 2009) although a fungal-virus origin has also been proposed (Liu *et al.*, 2011).

Host associations with vector-borne viruses and their ancient origins

The increase in sequencing of vertebrate and invertebrate genomes and the augmented capacity of computational power has led to an enhanced ability to search genomes and find comparable sequences. Such searches have led to the identification of endogenous viral elements (EVEs), regions of genomes that share similarity with known RNA or DNA viruses that have integrated into the germ line DNA of a host. While EVE integration has been known to be common to retroviruses, where between 5-8% of the human genome consists of integrated EVEs (Katzourakis and Tristem, 2005), recently integrated positive and negative stranded RNA viruses have been identified in host genomes. This is surprising as there is no DNA intermediate in RNA replication and given that most RNA virus replication is cytoplasmic (Holmes, 2011). There has been evidence of integration of flaviviruses such as Cell fusing agent virus (CFAV) (Crochu *et al.*, 2004) and other insect flaviviruses (Roiz *et al.*, 2009). In addition, other integrated bunyaviruses (Ballinger *et al.*, 2014; Katzourakis and Gifford, 2010), reoviruses and rhabdoviruses (Katzourakis and Gifford, 2010) have been identified.

Similar studies in plants have identified sequences of viral origin in plant genomes suggesting an ancient association of some vector-borne viruses and their hosts (e.g. geminivirus-like sequences in *Nicotiana* species; (Bejarano *et al.*, 1996; Lefeuvre *et al.*, 2011). However, comparative phylogenetic studies have shown that there is often no strong correlation between the phylogeny at the virus family level and host plant phylogeny (Gibbs *et al.*, 2008; Pagan *et al.*, 2010). These observations suggest short co-evolution and easy host jumps, at least for some plant viruses. For example, a host jump between monocot orchids and eudicot brassicas around a thousand years ago could be at the origin of the aphid-transmitted *Turnip mosaic potyvirus* (TuMV), a major problem of brassica crops worldwide (Nguyen *et al.*, 2013). One explanation for the frequency of these host jumps may be due to the relatively poor homing flight of many insect vectors, a vector behavior favoring encounters between viruses and new plant hosts. If such behaviors take place in ecosystems with a high diversity of plants, the probability of transmission among heterologous hosts could be high, a scenario facilitating the evolution of generalism in viruses (and any other parasite) (Wilke *et al.*, 2006).

Given that most plant viruses do not replicate within their vectors, one could predict less specificity in the association between plant viruses and their vectors than between the same viruses and their plant hosts. However, this does not seem to be the case. A study analyzing the plant and vector ranges of around 500 vector-borne phytoviruses has found narrow vector ranges (most viruses had a single known vector species) and low levels of host specificity (Power and Flecker, 2003). This trend seems to prevail no matter whether the virus uses a non-circulative, circulative or propagative mode, underlining higher constraints during vector transmission than during plant infection, a factor common to all arthropod transmissions.

The presence of EVEs in genomes of both plant and vertebrate hosts suggests a long association between these viruses and their respective hosts. How integration events occur remains a mystery, but at present there are two theories: 1) EVEs are small pieces of host RNA that evolved to become viruses, or, 2) EVEs are incorporated like retroviral elements but then lost their retroviral properties.

Evolution

Alphaviruses

Alphaviruses occur on every continent except Antarctica and they infect a range of vertebrate hosts including birds, fish, mammals, and potentially amphibians. Arthropod vectors include mosquitoes, mites and possibly sea lice, although experimental transmission studies have not validated lice as vectors of aquatic alphaviruses. The medically important alphaviruses use mosquitoes as vectors.

Alphaviruses are classified phenotypically and phylogenetically into the Old World viruses that typically cause arthritic disease and the New World viruses that produce encephalitis. Sindbis virus (SINV) is an exception in that it occurs in the Old World but is genetically more similar to New World viruses. Mayaro and Una viruses (MAYV and UNAV) group genetically with the Old World viruses, but occur in South America. Analysis of full genomes of all alphavirus species places the marine viruses southern elephant seal virus (SESV) and salmon pancreatic disease virus (SPDV) basal in the phylogeny, suggesting that alphaviruses possess a marine origin (Forrester *et al.,* 2012b). Ancestral positioning of both South American and African/Oceanic viruses at the base of separate clades also suggests that alphaviruses were at some point independently introduced in both areas. Limited attempts at worldwide surveillance for alphaviruses, as well as the recent discovery of mosquito-specific Eilat virus (EILV) (Nasar *et al.,* 2012), indicate that more alphaviruses remain undiscovered, and therefore the existing phylogeny could be incomplete. Unlike the flaviviruses described below, there appears to be no evolutionary association with specific types of vectors, and Alphaviruses tend to be more promiscuous in the type of vectors utilized. How this affects their evolution compared to the flaviviruses is still to be determined.

The evolutionary patterns of alphaviruses are dependent on the vertebrate host and their vectors. Phylogenetic analyses sheds light on the vertebrate host associated with the virus. For example, Eastern equine encephalitis virus (EEEV) that circulates in North America shows very little geographic clustering. This pattern is a result of the virus being transmitted by birds (Arrigo *et al.,* 2010). Conversely, Venezuelan equine encephalitis virus (VEEV) and South American EEEV show geographical clustering due to the limited movement of their small rodent hosts (Adams *et al.,* 2012; Arrigo *et al.,* 2010). For the zoonotic alphaviruses for which vertebrate hosts are unknown, using phylogenetics can help narrow down the range of potential vertebrate hosts.

More recently, the expansion of CHIKV in the Indian Ocean has resulted in significant advances in the understanding of CHIKV evolution. The increased circulation of CHIKV was determined to result in part from a host jump from the sylvatic cycle of CHIKV to an urban cycle that includes *Ae. albopictus* and humans resulting from A226V mutation in the E1 gene (Schuffenecker *et al.,* 2006; Tsetsarkin *et al.,* 2007). This mutation has arisen on multiple separate occasions

(Tsetsarkin *et al.,* 2014). Moreover, since this initial mutation, there have been subsequent mutations that secondarily increase the infectivity and transmissibility of CHIKV for *Ae. albopictus* (Tsetsarkin *et al.,* 2014). These secondary mutations do not confer increased infectivity of *Ae. albopictus* in the absence of the A226V mutation; and therefore they have only been observed in the lineages containing the A226V mutation. The recent expansion of CHIKV into the Caribbean and elsewhere in the Americas was expected to be a result of the strains containing the A226V mutation, but instead is part of an older Asian lineage that is primarily circulated by *Ae. aegyptii.* Recent work has also demonstrated that unlike the lineage in which E1A226V occurred, there are epistatic mutations in the Asian lineage that prevent the virus from accumulating the A226V (Tsetsarkin *et al.,* 2011), making it unlikely that there will be a similar adaptation to the *Ae. albopictus* vector.

Flaviviruses

The *Flaviviridae* are a complex group of viruses that are primarily vector-borne and are responsible for some of the most prevalent human arboviral diseases including DENV, YFV, WNV, and TBEV. They comprise a large number of viruses that can be subdivided into several different groups based on their transmission and life cycles; the mosquito-borne viruses, the tick-borne viruses, the no known vector viruses and the mosquito-only viruses. The evolution of the flaviviruses has been widely discussed in several reviews (Gould *et al.,* 2003; Kuno *et al.,* 1998). The major driving force of evolution is likely the adaptation to different vectors, primarily mediated through the envelope protein and the untranslated regions (Yamanaka *et al.,* 2013). When phylogenies of the flaviviruses are performed, the base of the tree is found to be mosquito-only viruses such as CFAV (Cook *et al.,* 2009), as well as numerous viruses identified in *Culex* and *Aedes* mosquitoes (Huhtamo *et al.,* 2014; Junglen *et al.,* 2009). The rest of the phylogeny can be subdivided into thee clades: the no-known vector group, which includes viruses such as Apoi virus and Modoc virus (Gould *et al.,* 2003; Varelas-Wesley and Calisher, 1982) and then the tick-borne or mosquito-borne groups. These groups are distinct phylogenetically and they have evolutionary rates suggesting different evolutionary paths, most likely due to the life-cycles of their vectors (McMullen *et al.,* 2013; Uzcategui *et al.,* 2012). The mosquito-borne viruses can be sub-divided into two groups; those associated with *Aedes* species mosquitoes such as DENV and YFV, and those associated with *Culex* species mosquitoes such as WNV.

Mosquito-only viruses

The mosquito-only viruses are transmitted from mosquito to mosquito without a vertebrate host. They are thought to be mainly transovarially transmitted through the germ line. Some of these viruses have been isolated from mosquito cell lines; given that cell lines are generally isolated from mosquito larvae, this supports the theory of transovarial transmission. Many mosquito-only viruses have been isolated. The first was CFAV that was isolated in 1975 (Stollar and Thomas, 1975), and was only identified as a flavivirus when its genome was sequenced 17 years later (Cammisa-Parks *et al.,* 1992). Since then numerous isolations of CFAV have been detected from cell lines and mosquitoes collected from different locations (Cammisa-Parks *et al.,* 1992; Cook *et al.,* 2006; Hoshino *et al.,* 2009). In addition, there have been numerous other mosquito-only viruses isolated from many different mosquitoes,from both *Culex* and *Aedes,* and which are described in detail elsewhere (Calzolari *et al.*, 2012). The origins of these viruses is still controversial. It is possible that these viruses were originally traditional arthropod-borne viruses that secondarily lost the ability to infect vertebrates, or it is possible that the flaviviruses were originally mosquito-only viruses that acquired the ability to infect vertebrates. Whichever explanation is true, it appears that these viruses are ubiquitous among mosquitoes (Cook *et al.*, 2012), suggesting a long-evolutionary association between mosquitoes and viruses, and that the mosquito-borne viruses are the origin of the flaviviruses.

The initial discovery of Flaviviruses in mosquitoes found that the majority formed a single clade basal to the rest of the *Flaviviridae*. Recent discoveries place several mosquito-only viruses within the clade of the mosquito-borne flaviviruses, and some are closely related to human pathogens (Huhtamo *et al.*, 2014; Kolodziejek *et al.*, 2013).These new topologies indicate that it could be possible for flaviviruses to lose their capacity to infect humans or vertebrates and 'go back' to insect-only viruses, in a type of reversion to an earlier evolutionary state. What is still unclear is what restricts these viruses to mosquito tissues, and makes them unable to infect the vertebrate hosts. Recent work has suggested that there is a conserved frameshift in the non-structural protein 2 of the mosquito-only viruses that may be involved in host restriction (Blitvich *et al.*, 1999; Firth and Atkins, 2009; Firth *et al.*, 2010; Huhtamo *et al.*, 2014).

No-known vector viruses

Flaviviruses that are classified as no-known vector viruses are those that have been isolated from vertebrates, generally either rodents or bats (Gould *et al.*, 2003), with no known mosquito or tick host identified. These viruses have no significant human pathogens associated with them and are rarely isolated. However, they are interesting from an evolutionary point of view. A recent study found that Modoc virus (MODV) could infect mosquito cells, but was not able to replicate in them (Charlier *et al.*, 2010). These viruses may have lost the ability to infect arthropod hosts, and given that the ancestral flaviviruses are probably mosquito-only, the ability to infect vertebrates could have been secondarily acquired and then the ability to infect arthropods lost.

Tick-borne viruses

The tick-borne viruses are found primarily in the northern hemisphere, with the exception of some of the seabird-borne viruses, which occur in remote islands across the globe (Gould *et al.*, 2003). The major human pathogen in the tick-borne virus group is TBEV, which ranges from Western Europe to the far east of Russia (Estrada-Pena and de la Fuente, 2014). Other viruses of note are Omsk hemorrhagic fever virus, which is closely related to TBEV and Kyanasur forest disease virus. POWV and its close relative deer tick virus (DTV) are the only two tick-borne viruses found in the Americas.

The evolution of these viruses has been studied using phylogenetic analyses. The first comprehensive analysis based E gene sequences identified a geographical association between distance and rate of change, suggesting a clinal distribution with the virus evolving geographically east to west (Zanotto *et al.*, 1995). That study also concluded that there was a decline in virulence as the virus spread west since the most virulent forms of TBE are found in the Eastern part of Eurasia

(Zanotto *et al.,* 1995). An updated study of these viruses used the evolutionary dating software BEAST to identify the origins of TBEV. This study used complete genomes of 58 isolates to validate the geographical cline hypothesis. Geographical and temporal analysis of these viruses showed that there was an origin in central Asia with subsequent spread of the TBEV group both east and west. In addition, this study suggested that POWV was carried to North America from Eastern Asia across the Bering Strait during the time of the Beringian land bridge during the last ice age that occurred 15,000-11,000 years ago and that POWV has subsequently evolved separately from the rest of the TBFV. Although POWV is found in Russia, it is likely that its presence there is due to the fur trade between Russia and Alaska and Canada during the late 1800's and early 1900's. This would have re-introduced POWV back into Russia and would explain the long isolation of POWV in the Americas and with co-circulation of the Russian strains. While this work has some merit being a parsimonious explanation, recent work with coalescent methods suggests that these types of analysis may be significantly flawed to its inability to accurately estimate dates that far back (Wertheim *et al.,* 2013; Wertheim and Kosakovsky Pond, 2011).

Mosquito-borne viruses

The mosquito-borne group of flavivirus has been historically divided into the *Culex*-associated viruses and the *Aedes*-associated viruses. For the *Aedes*-associated viruses this also contains those viruses that have secondarily lost their association with vectors, such as Yokose virus (YOKV) and Entebbe bat virus (ENTV). Although the viruses are split into these two groups, the *Aedes* group is not monophyletic as DENV share a more recent common ancestor with the *Culex* group than with the *Aedes* group, while still being transmitted by *Aedes* mosquitoes (Cook and Holmes, 2006). Thus, it appears that adaptation to *Culex* mosquitoes occurred from viruses that were first adapted to *Aedes* species.

The first flavivirus analysis indicated that compared to the tick-borne flaviviruses, the mosquito-borne flaviviruses showed evidence of recent expansion in the branch tips of the tree (de A. Zanotto *et al.,* 1996), suggesting that many of these viruses may have been recent adaptations to mosquitoes that interacted with humans. Recent work has shown that DENV and YFV have emerged from sylvatic cycles (Diallo *et al.,* 2014; Vasilakis *et al.,* 2007) in forests without human involvement and have subsequently adapted to use humans as primary vertebrate hosts. In general, this change in host usage has been associated with the rise of large cities that allow sufficient humans in a localized area. Adaptation to humans explains this radial expansion observed in the trees and the subsequent movement of DENV and YFV around the world. Detailed descriptions of DENV and YFV evolution have been reviewed elsewhere (Carrington and Auguste, 2013; Chen and Vasilakis, 2011).

Aside from DENV and YFV, the other major human arboviral pathogens vectored by mosquitoes are WNV, Murray valley enecephalitis virus (MVEV) and Japanese encephalitis virus (JEV). These viruses are enzootic and cycle primarily between vertebrates and mosquitoes. Humans are infected by spillover from infected mosquitoes and as a consequence are viewed as dead-end hosts as they do not usually contribute to transmission by mosquitoes. The evolution of these viruses is therefore dependent on the distribution and movement of their non-human vertebrate hosts and the mosquitoes. While it is possible for these viruses to move into new environments as exemplified by WNV establishing new transmission cycles in the Americas beginning in the United States in 1999 (Davis *et al.*, 2003), viable hosts must be present in the new environment for the establishment of transmission. Thus, these viruses are limited in their capacities to spread by the presence of competent hosts and mosquito vectors.

Plant arboviruses

Many RNA and DNA plant viruses have measurable evolving populations and long-term evolutionary rates similar to those of animal viruses that are about ~10^4 nucleotide substitutions per site per year (Gibbs *et al.*, 2010). Thus, given the appropriate conditions, these viruses should rapidly evolve. Agriculture has provided such conditions. Agriculture was invented in at least 9 different regions of the world around 8,000-13,000 years ago and it led to the domestication of different plant species depending on the region (Levetin and McMahon, 2011). Since then, high densities of a single plant species are generally grown periodically at the same site. Such conditions probably favored the adaptation of vectors and viruses with specific life cycles. In accordance, different genera of vector-borne viruses causing major crop diseases worldwide radiated within the last few millennia (e.g. *Luteovirus*: 4,000 years ago (Pagan and Holmes, 2010) or *Potyvirus*: ~6,500 years ago (Gibbs *et al.*, 2008). Subsequent to their domestication, crops were spread to other regions. In those new ecosystems, imported plants faced new viruses that, in certain cases, adapted to them. More recently, the intensification of agricultural exchanges at a global scale has also led to the spread of viruses and their vectors. An example of these phenomena is the on-going mergence of tomato yellow leaf curl (TYLC) disease, a major problem for tomato production worldwide. This disease is provoked by a complex of whitefly-trasmitted begomoviruses, the TYLCV-like viruses. These viruses emerged as pathogens of tomato plants, a species that originated in the New World,, in the Middle East between 1930-1950 and reached a global distribution in just a few decades (Lefeuvre *et al.*, 2010). This rapid expansion seems mainly due to many accidental introductions and subsequent territory invasions of specific biotypes of their vector *Bemisia tabaci* [reviewed in (Perefarres *et al.*, 2012)].

Within-host evolution

Vertebrate virus studies

RNA virus heterogeneity: source and nature

RNA viral genetic heterogeneity arises from frequent nucleotide misincorporation during replication due to the inability of the viral RNA-dependent RNA polymerase to correct errors. Coupled with rapid replication and exponential growth, arboviruses exist as heterogenous populations called mutant swarms. Given a ≈10 kb arbovirus genome and a mutation frequency of approximately 10^{-4}, each RNA genome possesses about one mutation. This mutation frequency is several $\log_{10}$ higher than for double-stranded DNA viruses (Holmes, 2009). Most mutations are probably detrimental and thus removed from the swarm via purifying selection. This dynamic mutation-selection balance determines the size and genetic diversity of a mutant swarm. Mutant swarms are difficult to study given challenges to characterizing large and dynamic populations. Despite this, an increasing number of studies, mostly from non-vector borne RNA viruses including HIV, picornaviruses and rhabdoviruses (reviewed comprehensively elsewhere)

(Domingo *et al.*, 2012), reveal that viral genetic diversity is critical to virus evolution. Specifically, the composition and size of a mutant swarm fundamentally affects the ability of an RNA virus to adapt.

Intrahost genetic bottlenecks and their impact on evolution

Arboviruses must circumvent bottlenecks that interrupt transmission in order to survive. Bottlenecks reduce arbovirus population size and can modify the composition of the mutant swarm, potentially influencing viral fitness by restricting phenotypic plasticity that accompanies high genetic diversity. The evolutionary theory of Muller's ratchet postulates that asexual organisms of small population size with high mutation rates irreversibly accumulate deleterious mutations unless compensatory mutations restore mutation-free genomes to the population (Muller, 1964). Model studies support this idea (Lynch and Gabriel, 1990) and experimental data with the alphavirus Eastern equine encephalitis virus (Weaver *et al.*, 1999) and the sandfly-borne rhabdovirus vesicular stomatitis virus (VSV) (Novella *et al.*, 2004) validate this concept; fitness decreases after serial bottleneck passages can be rescued by subsequent large population passages, albeit with much replication needed to overcome the ratchet (Duarte *et al.*, 1993).

Genetic bottlenecks to productive transmission by mosquito vectors are relatively well defined for members of several arbovirus families (Figure 3.2). As a first step, arboviruses in a bloodmeal ingested by a mosquito must infect the midgut epithelium. Some arboviruses encounter secondary impediments to dissemination, where they are unable to escape from the midgut epithelium, infect salivary glands, and escape from salivary gland cells into saliva for transmission to vertebrates. Genetic bottlenecks in vectors may be circumvented by re-establishment of diversity and population size via subsequent replication.

Venezuelan equine encephalitis virus

Venezuelan equine encephalitis virus (VEEV) is in the alphavirus family. The evolution of this virus has been widely reviewed (Coffey *et al.*, 2013; Weaver and Barrett, 2004; Weaver *et al.*, 2012). However, recently, there has been a focus on the within-host evolution of this virus, reviewed below.

Intra-host diversity within VEEV

Like other RNA viruses, VEEV contains intra-host diversity due to error-prone replication, resulting in swarms of mutants. Experiments with enzootic VEEV strains have shown that significant amounts of variation within the genome exists and that certain genomic regions contain higher levels (hotspots) of variance (Forrester *et al.*, 2011). This has also been demonstrated in WNV, and is thought to be a result of the RNA interference (RNAi) response in mosquitoes (Brackney *et al.*, 2009). The RNAi response in mosquitoes that transmit enzootic VEEV strains have not been determined, but it is likely that the presence of these responses in mosquitoes also interact with VEEV, resulting in this differential diversity along the genome.

Evidence for a bottleneck in the midgut of mosquitoes infected with VEEV

Anatomical bottlenecks at both midgut entry and escape have been widely postulated for VEEV. To determine if these physical barriers correspond to genetic bottlenecks at midgut infection, mosquitoes were presented infectious bloodmeals

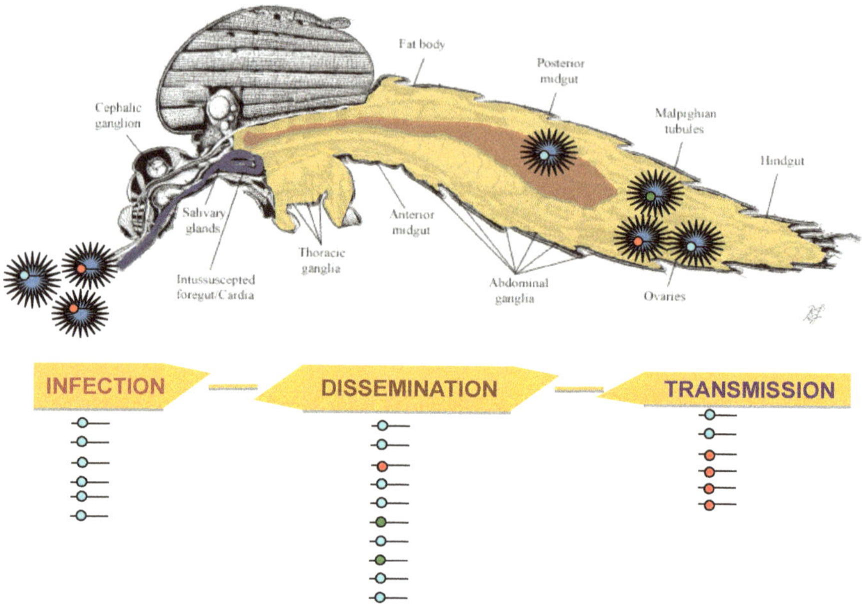

Figure 3.2. Evolution of a RNA viral mutant swarm beginning with a founding genome (line) with mutations (dots) in a mosquito host following ingestion of a bloodmeal containing an arbovirus. Individual RNAs observed after infection of the midgut, dissemination into the hemocoel, and transmission in saliva during re-feeding are noted. Anatomical bottlenecks between infection and dissemination, and dissemination and transmission also serve as genetic bottlenecks that decrease genetic diversity of swarms. Mutations with deleterious effects (green dot) are negatively selected and do not survive to serve as templates for successive replication cycles and transmission. Mutations that confer a phenotypic advantage are positively selected and increase in frequency (red dot). Positively selected mutations that increase transmissibility by vectors can rise from minority frequencies to dominate populations. Only dominant mutants have conventionally been identified and characterized.

containing labelled replicon particles, which are able to establish an infection but not escape from cells. Infected cells therefore express the fluorescent protein incorporated into the replicon particle and the number of infected cells can be determined. These experiments were carried out with an endemic virus and endemic mosquito and found that almost all of the midgut cells were infected (Kenney *et al.,* 2012). However, when the epidemic mosquito vector was infected with epidemic strains of VEEV, only a few cells were infected; these cells were likely 'portals' of virus entry since replicons expressing red and green fluorescent proteins co-localized to the same cells when both replicons were co-presented in bloodmeals. Thus, VEEV experienced a bottleneck when infecting midgut cells (Gutierrez *et al.,* 2015; Smith *et al.,* 2008). Interestingly, although bottlenecks in other arboviruses appear to reduce viruses to one or two genomes (Gutiérrez *et al.,* 2012a), the estimation of both epizootic and enzootic VEEV bottlenecks was 100-1000 genomes. Moreover, the cells showing dual infection occurred either singly or in clusters, suggesting that the midgut is not uniformly susceptible. The small number of susceptible midgut cells may therefore act as an effective bottleneck in arbovirus vectors. Studies with WNV in *Cx. pipiens quinquefasciatus* showed that <15 midgut cells were infected (Scholle *et al.,* 2004). Similar estimates have also been obtained with SINV in *Ae. Aegypti* and other mosquito and lepidopteran species (Foy *et al.*, 2004; Olson *et al.*, 2000).

Further evidence of bottlenecks in VEEV was shown using marked viruses and transmission of endemic strains (Forrester *et al.*, 2012a). Marked clones representing different viral sequences created from endemic VEEV were fed at low and high concentrations to endemic *Cx. taeniopus* mosquitoes, and the number of clones was identified at each potential bottleneck. Mosquitoes that ingested high titered bloodmeals showed no bottleneck at midgut entry, however, at the lower titer, there was a bottleneck observed as a significant reduction in the number of clones infecting the midgut. At both doses, there was always bottleneck at midgut escape, and it was more severe when the titer was higher; this is probably due to the increase in the number of midgut cells infected. There was also an additional bottleneck at the entry into salivary glands. The presence of this bottleneck was much lower than that of the midgut entry or exit, but still observed. Although the presence of diversity within VEEV strains has been identified, the effect of these bottlenecks on the level of diversity and how this affects transmission has yet to be determined.

Chikungunya virus

In vivo studies with CHIKV support the idea that arboviruses can circumvent genetic bottlenecks in mosquito vectors. Although population diversity in the midgut and salivary glands is reduced compared to the bloodmeal input or midgut population, respectively, it recovers downstream of each barrier and the average nucleotide at each genomic position (consensus sequence) remains unchanged (Stapleford *et al.,* 2014). Despite the absence of consensus mutations, deep sequencing the CHIKV E1 gene in saliva expectorants from *Aedes albopictus* reveals the presence of common minority variants. Specifically, the alanine to valine E1 mutation at position 226 that augments infection and transmission by *Ae. albopictus* (Tsetsarkin *et al.*, 2007) and which likely potentiated Indian Ocean island outbreaks in 2005-6 can be detected in a minority of viral RNAs in saliva from experimentally infected mosquitoes (Stapleford *et al.,* 2014). This indicates that experimental infection studies combined with deep sequencing of transmitted viruses can identify arbovirus mutants with epidemiologic relevance before they would otherwise be detected via standard consensus sequencing, and suggests that positively selected minority variants circulate as evolutionary intermediates before rising to dominance and sweeping epidemics. The mechanisms by which minority variants with epidemiologic relevance are maintained through alternating transmission is not well understood; other CHIKV minority variants fluctuate in frequency in alternating mosquito-mouse-mosquito cycling in laboratory experiments and sometimes become the dominant transmitted variants, perhaps due to enhanced stability in mosquito saliva (Stapleford *et al.*, 2014).

West Nile virus

WNV has been widely studied since its introduction into the Americas in 1999 (May *et al.,* 2011). WNV virus was first isolated in Uganda in 1937 and between then and 1999 caused periodic outbreaks, but was not considered an important pathogen of humans (May *et al.,* 2011). However, the emergence of the virus in the Americas raised its profile and since then it has been widely studied with its epidemiology and pathology now well understood (Mann *et al.,* 2013; McMullen *et al.*, 2013). WNV evolved in Africa and has spread around the globe. There are several lineages of WNV, one of which (lineage 1), has been responsible for circumglobal spread. WNV has become endemic in the US, with periodic spillovers into humans as exemplified by the large number of human cases in an epidemic in Texas 2012 (Mann *et al.*, 2013). In Europe, WNV circulates primarily between passerine birds and *Culex* mosquitoes, and the presence of related species in the Americas made the introduction and maintenance of WNV more likely than if there had been no similar species in the US. WNV infects many species of Culicine mosquitoes in the US including *Culex pipiens, C. tarsalis,* and *C. quinquefasciatus*. Studies have shown that the vector/virus relationship is determined by several factors, including environmental variables such as temperature, the amount of virus ingested, and modulation of the mosquito life cycle by the virus, all of which are reviewed elsewhere (Ciota and Kramer, 2013). These studies indicate that the evolution of WNV is partially dependent on the strain of WNV that infects them. As for VEEV, infection of mosquitoes by WNV creates genetic bottlenecks on viral populations. First, at midgut infection, only a few cells are infected as determined using virus-like particles (VLPs) (Scholle *et al.*, 2004). and mixed viral populations in *C. pipiens* (Ciota *et al.*, 2012) and using sequencing of clones in C. *quinquefasciatus* (Brackney *et al.*, 2011). A second bottleneck exists at the salivary glands and in transmission by *C. pipiens*, but not in *C. quinquefasciatus* mosquitoes (Brackney *et al.*, 2011; Ciota *et al.*, 2012). Interestingly, unlike VEEV, there is also a temporal decline in diversity, with less diversity observed at later stages of infection, particularly the haemocoel and the salivary glands compared to the initial infection. The reason for this observation is unclear, but it may be due to the presence of the RNAi response that acts as an antiviral mechanism in the mosquito reducing the burden of the virus and therefore the viral genetic diversity. When *C. quinqefasciatus* were infected with WNV, the RNAi response was activated and a parallel increase in viral diversity was observed. The presence of the RNAi response is a new and interesting aspect of research and will surely help understand within-vector arbovirus evolution.

Plant viruses

The within-host evolution of vector-borne phytoviruses has been studied principally in non-circulative and circulative modes, (that is, in viruses that do not replicate within their vectors), and generally follows similar patterns to those found in many animal RNA viruses.

Mutation

As in their animal counterparts, mutation rates of both RNA and ssDNA phytoviruses range between from 10^{-3} to 10^{-6} mutations per site and replication cycle (Sanjuán *et al.*, 2010). There is also experimental evidence in different vector-borne phytoviruses, although sometimes scarce, supporting (i) the existence of antagonistic pleiotropy, where the same mutation is beneficial or deleterious depending on the host genotype, (ii) a predominance of deleterious mutations during mutation generation, and (iii) negative epistasis where the fitness reduction of two deleterious mutations together is lower than the addition of the effect when each mutation is expressed alone [reviewed in Elena *et al.* (2014)].

Genetic exchange

Genetic exchange through recombination and re-assortment also takes place in phytoviruses. Interspecific recombination is common in DNA plant viruses, with a high proportion (e.g. ≥ 50%; (Urbino *et al.*, 2013)) of recombinants arising during infection of a host. A high recombination rate *in vivo*, in the range of 10^{-5} recombination events per nucleotide site per generation, has been estimated for a dsDNA pararetrovirus, Cauliflower mosaic virus (CaMV) (Froissart *et al.*, 2005). Although RNA phytoviruses have been shown to recombine in the field, the proportion of recombinants generated during a single-host infection does not seem to reach the levels of their DNA counterparts, probably because cell co-infection by different genotypes is often limited (further discussed below) (Tromas *et al.*, 2014a). A recent estimate of the *in vivo* recombination rate in an aphid-transmitted ssRNA potyvirus showed a rate similar to that of CaMV, but only when the number of co-infected cells was taken into account (Tromas *et al.*, 2014b). Besides recombination, many DNA and RNA phytoviruses have segmented or multipartite genomes allowing for re-assortment and this phenomenon occurs in nature, playing a role in virus evolution (*e.g.* White *et al.*, 1995).

Within-host population genetics

In plant viruses, most work has focused on viral genetics within plants as data generated from vector infections of propagative viruses are very rare (*e.g.* Pu *et al.*, 2011). Many vector-borne phytoviruses generate large and diverse populations within plants. These populations are not usually panmictic. Instead, they tend to form distinct subpopulations in different host compartments (e.g. leaves). Spatial segregation of genotypes during plant infection has been repeatedly observed in different virus-plant systems, both between leaves of a plant (Jridi *et al.*, 2006) and between cells within a leaf (Dietrich and Maiss, 2003; Diveki *et al.*, 2002; Takahashi *et al.*, 2007; Takeshita *et al.*, 2004), although exceptions to this rule have been observed (Gutiérrez *et al.*, 2010; Monsion *et al.*, 2008). This unequal distribution of populations within the plant influences which viruses are ingested by arthropod hosts. In general, viruses using a non-circulative mode are acquired from mesophyll and epidermal cells of leaves whereas viruses using the circulative and the propagative modes are acquired from elaborated sap in veins. Moreover, within-host population genetics can also change with time (Gutiérrez *et al.*, 2010; Urbino *et al.*, 2013), further complicating interpretation. Vectors feeding in different locations on a plant or at different time during infection can acquire different viral populations (Figure 3.3). Nevertheless, experimental demonstrations of this prediction are rare for vector-borne phytoviruses (see below).

There are two main factors that can limit the population diversity during the infection of an organ from the vasculature, leading to spatially structured viral populations (Gutiérrez *et al.*, 2012a). The first factor is barriers to virus infection imposed by the plant [e.g. anti-viral defenses (Voinnet, 2001)] or the virus itself [e.g. viral mechanisms limiting superinfection of cells (Folimonova, 2012)]. The second factor is the viral load in the vasculature during organ infection. Low viral loads can lead to small colonizing populations with limited diversity even in the absence of specific barriers. Although largely overlooked, this factor determines within-host population bottlenecks in different phytovirus models (Gutiérrez *et al.*, 2012b; Zwart *et al.*, 2011). For example, studies with the aphid-transmitted CaMV in turnip plants have shown that the dynamics along infection of the viral load in the vasculature define the size of the colonizing population in leaves infected at

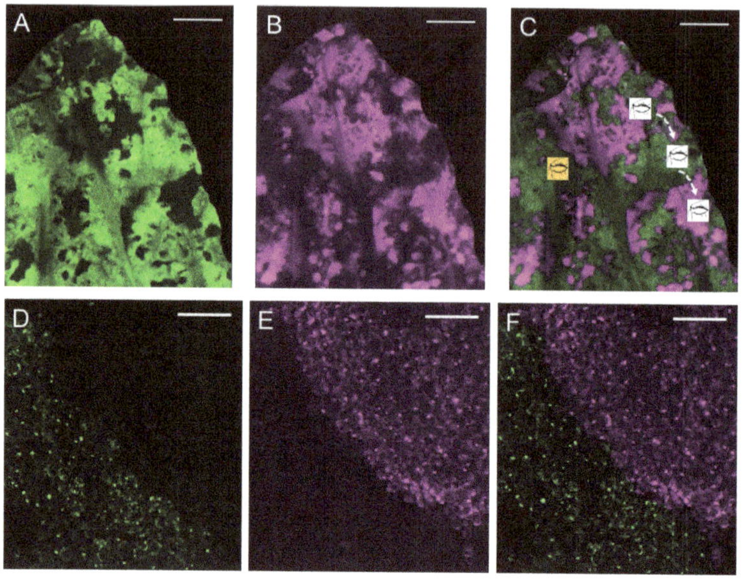

Figure 3.3. Spatial segregation of Turnip mosaic virus (TuMV) and relation with different vector feeding behaviors. A turnip host plant was co-inoculated by two TuMV variants, expressing different fluorescent proteins (GFP in green – left panels- and RFP in red –central panels). Spatial segregation is particularly easy to visualize here through the mutual exclusion of the two colors either at the leaf level (**A-C**) or at the cell level (**D-F**). Each clone can be observed to separately infect leaf cells, yielding a patchwork of infected regions with a single fluorescence. The aphids in **C** represent two different feeding behaviors. The yellow aphid is making a single probe before departing and transporting only one of the two variants, whereas the white aphid is making several probes in different places on the leaf, potentially ingesting both variants.. Scale bars: 0.5 cm in **A-C** and 200 µm in **D-E**.

different time points (Gutiérrez *et al.,* 2012b). The peak levels of viremia correlated with large colonizing populations (hundreds of genomes) and high cell co-infection (up to 100%) in leaves infected at that moment (Gutiérrez *et al.,* 2012b; Gutiérrez *et al.,* 2010). Thus, aphid individuals feeding from leaves infected at different phases of the viral-load dynamics should have access to different CaMV population diversities.

Genetic drift

For plants, within-host bottlenecks can be caused by low viral loads and barriers during organ colonization. These phenomena reduce population sizes and can intensify genetic drift. Although still scarce, we dispose of more formal quantifications of population bottlenecks during host infection for vector-borne phytoviruses than for vertebrate viruses. Severe bottlenecks have been found in non-circulative (French and Stenger, 2003) and circulative viruses (Urbino *et al.,* 2013) that reduce population size during plant infection - between inoculation (or soon after) and systemic infection. Moreover, although not directly estimating population bottlenecks, experimental studies have observed losses in genetic diversity along plant colonization of CMV (non-circulative mode) in different host plants (Ali and Roossinck, 2010). However, other quantitative studies found tenfold reductions in population size due to within-host bottlenecks depending on the viral load, either at inoculation (Zwart *et al.*, 2012) or during systemic infection (Gutiérrez *et al.*, 2012b). The latter studies were done with unrelated non-circulative viruses and their results soften the current view on the intensity of drift effects during host invasion. Beyond host infection, other studies have focused on population bottlenecks at the level of infection of individual cells. Given that viruses multiply in cells, the number of genomes that enter and replicate in a cell is a key parameter influencing within-host evolution. This parameter, also known as the multiplicity of infection (MOI) (i.e. the number of viral particles that infect a single cell), impacts several important phenomena. First, the MOI defines bottlenecks at higher levels within the host because populations colonizing organs derive from a finite number of primary infected cells. Secondly, the MOI determines the intensity of genetic exchange, competition and complementation during cell infection, all major phenomena in virus evolution. Moreover, the MOI also defines the copy number of viral genes at the onset of cell infection that, although largely overlooked, can dramatically change the virus phenotype (Elde *et al.*, 2012; Weitz *et al.*, 2008). Finally, as the population acquired by vectors during non-circulative transmission comes from a few cells, the MOI can greatly influence the diversity of the transmitted population that is sampled by the vector. Three formal estimations of the MOI currently exist for vector-borne phytoviruses. A MOI of between 5-6 genomes per cell was estimated in a circulative-like fungus-borne virus (Miyashita and Kishino, 2010). A MOI of one genome per cell in leaves infected at different time points has been recently found for the non-circulative tobacco etch potyvirus (TEV) (Tromas *et al.*, 2014a). Finally, an MOI of between 2 and 13 genomes per cell was observed for CaMV in different leaves; this variation was probably viral-load dependent (Gutiérrez *et al.*, 2010). All these estimates are relatively small and suggest random effects in population genetics during cell infection. However, high MOI values also promote the maintenance of defective genotypes through complementation (Turner and Chao, 1998) and a reduced MOI has been proposed as a strategy to avoid such genotypes (Miyashita and Kishino, 2010). Although the differences between estimates may not seem important, they lead to different compartmentalization scenarios. Co-inoculation of a plant with two clones, identical except for a neutral marker, leads to rates of cell co-infection by the two genotypes lower or equal to 10% in the TEV model whereas cell co-infection can reach 100% in leaves infected at the highest MOI of CaMV.

As described above, contrasting MOIs should lead to different viral genetic diversities accessible for the vector during non-circulative transmission as the vector acquires the virus from one or a few cells. Under severe compartmentalization of a non-circulative virus, as in the TEV case above, the number of genotypes transmitted (i.e. the genetic bottleneck) should be close to one, no matter how many genomes are acquired and transmitted (i.e. the population bottleneck). In accordance with this prediction, the genetic bottleneck during aphid transmission of the non-circulative CMV, a virus showing strong spatial segregation of genotypes among leaf cells (Takeshita *et al.*, 2004), is around one genotype (Ali *et al.*, 2006; Betancourt *et al.*, 2008). Similar data under less severe compartmentalization, like in CaMV-infected plants, are not available. A second factor is the population bottleneck. Two studies have provided quantitative estimates of population bottlenecks during vector transmission of phytoviruses, one for a non-circulative potyvirus transmitted by aphids (Moury *et al.*, 2007) and the other for a circulative geminivirus transmitted by whiteflies (Perefarres *et al.*, 2014). In both cases, the effective population size is extremely small, with between 1 and 2 genomes effectively transmitted. Under field conditions, plants might be colonized by large numbers of virus-bearing vectors, softening bottleneck severity (Escriu *et al.*, 2003).

The predominance of severe bottlenecks in most studies regardless of the step of the life cycle suggests that these viruses should become extinct following repeated infections. Additionally, both theoretical and experimental approaches predict fitness losses in populations enduring cyclic severe bottlenecks [reviewed in Elena and Sanjuán (2007)] as predicted by Muller's Raqtchet (Muller, 1964). However, populations of vector-borne phytoviruses in nature seem genetically stable under relatively constant environments (e.g. agricultural ecosystems) and can rapidly adapt under new pressures (Garcia-Arenal *et al.*, 2003), two phenomena that seem incompatible with a dominance of genetic drift. This apparent contradiction between experimental and field studies may arise from a bias in our view derived from the paucity of available data. Other explanations for this apparent incompatibility between the field and experimental data are (i) the purge of deleterious mutations or their fitness effects during replication cycles between bottlenecks (through mutation or genetic exchange) (e.g. Novella *et al.*, 2004), (ii) or the non-random nature of population bottlenecks in nature (i.e. strong purifying selection could drive severe population reductions).

A fascinating feature of many vector-borne phytoviruses is their multipartite genome, a particular genomic architecture mainly found in plant viruses and apparently absent in arboviruses of vertebrates. Multipartite genomes are divided in several molecules or segments but, contrary to segmented viruses, each genomic segment is encapsidated alone in a virus particle. That is, genomic segments seem to be physically independent from one another during migration within plants or vectors. Depending on virus species, the number of genomic segments ranges from 2 to 10. This genomic architecture puzzles both virologists and evolutionists as it should imply a cost limiting its evolution resulting from the probability of losing segments and the genetic information they encode during population bottlenecks. Recent data on an aphid-transmitted nanovirus have unveiled a scenario that should further increase the cost: some segments can be more abundant than others (by a differential of 10-fold) in within-host populations, a situation increasing the probability of losing rarer segments (Sicard *et al.*, 2013). Nevertheless, multipartite viruses are abundant and diverse, with examples in both RNA and DNA viruses, in what seems a repeated convergent evolution towards this particular genome organization. The question is then whether these viruses really encounter severe bottlenecks during cycling. Only two formal estimations of bottlenecks of multipartite viruses are available: 1) during aphid transmission of CMV [3 segments; (Betancourt *et al.*, 2008)] and 2) during cell-to-cell movement of the fungus-borne furovirus [2 segments; (Miyashita and Kishino, 2010)]. In both cases, bottlenecks were severe, with only 2 to 6 full genomes transmitted. If such severe bottlenecks are a general feature in multipartite viruses, one cannot but wonder which benefits are associated with multipartite architecture (Iranzo and Manrubia, 2012; Sicard *et al.*, 2013) and whether these viruses have developed adaptative mechanisms to limit potential costs resulting from loss of genetic information.

Modulation of hosts/vectors by viral infection

So far all examples from modifications in vector behaviour driven by viruses come from plant viruses. Phytoviruses can modulate both viruses and their vectors. Some of the ways are discussed below.

Strategies modulating virus biology

A recent study has uncovered a fascinating feature of a vector-borne phytovirus: CaMV can "perceive" the vector feeding on the plant and launch mechanisms directed to promote its transmission (Martiniere *et al.,* 2013). Contrary to many animal viruses, plant viruses generally stop their replication before cell death through lysis. They also often accumulate virions in specific structures within cells (e.g. inclusion bodies). Such structures in CaMV-infected cells are latent forms of the virus that react when an aphid's stylet operates in cells or their surroundings, liberating virions and facilitating transmission. This captivating mechanism is reversible: once the triggering stimuli disappear, the latent structures are formed again. Whether this strategy applies to other viruses is yet unknown but it seems an elegant way to safely store an important viral load in the cell, without the need for continuous replication and the associated virulence. Moreover, the capacity to react to vector presence, already known in other parasite systems (Matthews, 2011), opens new perspectives in the ways viral transmission strategies can operate.

The genetics or relatedness of the transmissible population is another parameter that the virus may control. As described above, different non-circulative phytoviruses have distinct population structures within a leaf. Such contrasting population structures suggest the existence of different strategies during cell colonization among viruses: either by solitary genomes or by groups of genomes (Gutiérrez *et al.,* 2012a). Such strategies would be controlled by altering the MOI, which, in turn, could be controlled by, among other factors, the viral load - as in the CaMV case described above - or by altering the speed of superinfection exclusion mechanisms. Superinfection exclusion is a territorial behavior almost omnipresent in viruses. During this phenomenon, viral mechanisms are triggered after cell infection to block new infections of the cell by related genotypes. The mechanisms behind this are diverse and depend on the virus species (Christen *et al.,* 1990; Huang *et al.,* 2008; Karpf *et al.,* 1997; Lu and Henning, 1994; Nethe *et al.,* 2005; Simon *et al.,* 1990; Tscherne *et al.,* 2007; Zou *et al.,* 2009). Although no such mechanism has been yet been deciphered in plant viruses, proof of their existence is abundant (Dietrich and Maiss, 2003; Diveki *et al.,* 2002; Folimonova, 2012; Takahashi *et al.,* 2007; Takeshita *et al.,* 2004; Ziebell *et al.,* 2007). Those putative mechanisms could lead to different MOI values just by modulating the time to be operational after initial infection: the larger the time window allowing superinfection, the higher the MOI. Different MOIs would lead, in turn, to different co-infection rates and transmissible populations with distinct genetic diversities in the non-circulative mode.

Another strategy proposed to modulate virus diversity during vector acquisition is the helper-component strategy, again, in non-circulative viruses (Pirone and Blanc, 1996). The capsids of some non-circulative viruses can attach directly to their putative receptors on the vector. However, others use a viral protein as a "bridge" between the viral particle and the receptor. This viral protein, named the helper component, is not part of the capsid and it can be acquired by the vector earlier than the virions it bridges to the receptor. Hence, helper components acquired in a cell can be used for receptor attachment by viruses acquired in other cells. This feature could enable trans-complementation of, for example, genotypes defective in the helper component. The existence of a helper component has also been suggested in circulative nanoviruses (Franz *et al.,* 1999). Beyond trans-complementation during transmission, alternative explanations have been proposed to explain the evolution of the helper component strategy in RNA and DNA non-circulative viruses (e.g. competition with other viral species using the same receptor through receptor saturation with the helper component).

Strategies modulating vector biology

Plant viruses can change the physiology and behavior of their vectors, directly or indirectly through their impact on the plant, often favoring transmission. A clear example of a direct effect is the modification of the feeding behavior of vector thrips by the propagative TSWV (Stafford *et al.,* 2012). TSWV infection of male thrips promotes non-cell-destructive feeding behaviors associated with salivation, a modification that is likely to increase transmission rate. However, in non-circulative and circulative viruses, described changes in vector biology are due to virus-induced changes in plant physiology, a situation probably reflecting the absence of replication within the vector of these viruses. Although vectors ingest virions and different molecules with a viral origin, a putative influence of such viral products on vector biology has not been studied in those transmission modes.

Viruses can modify many factors influencing plant-vectors interactions, including plant defenses (Ziebell *et al.,* 2011), amino acids in sap (Ajayi and Dewar, 1982), emission of volatile compounds (de Vos and Jander, 2010) and plant color (Ajayi and Dewar, 1982). Plant infection can lead to increased vector preference for infected plants (Mauck *et al.,* 2010a, b) and higher growth (Belliure *et al.,* 2008), survival (Hodge and Powell, 2008a) or fecundity of vectors (Hodge and Powell, 2008b). Negative effects on vector fitness have also been described (Fereres and Moreno, 2009). The main question that remains here is whether all these changes are due to adaptations to promote transmission.

Evidence for such adaptations comes from an increasing number of studies showing that, for a given transmission mode, virus-induced modifications of vector behavior follow predictable strategies (Mauck *et al.,* 2012). For a non-circulative virus, an optimal strategy would imply to attract and then rapidly repel vector individual vectors since acquisition is fast and retention short. This manipulation would maximize transmission rates while reducing vector-virus competition for plant resources. Such a complex behavior manipulation has been described during the interactions of non-circulative CMV, aphids and squash plants (Mauck *et al.,* 2010a). In circulative and propagative modes, plant infection should favor vector attraction and settlement and this is usually the case (Fereres and Moreno, 2009). Moreover, given that parasites can kill their hosts when beneficial (Matthews, 2011), viruses may trigger higher virulence levels to kill the plant and force vector dispersal when vector density on the plant is high (Gutiérrez *et al.,* 2012a).

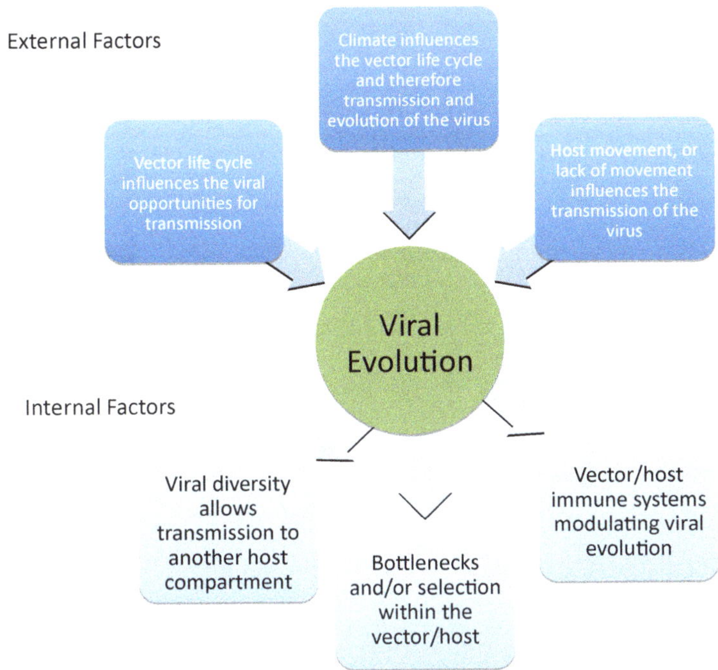

Figure 3.4. Factors influencing viral evolution, external host pressures such as vector life cycles, and internal host pressures.

Beyond the plant-vector-virus triangle of interactions, phytoviruses may have developed transmission strategies at higher ecological levels, through interactions with other organisms. Changes in transmission rates can take place due to interactions with the microbiome of vectors [i.e. *Buchnera* in aphids (Van den Heuvel *et al.*, 1994) and *Hamiltonella* and *Wolbachia* in whiteflies (Gottlieb *et al.*, 2010)] or with other virus species during multiple infections of plants, a common situation in nature (e.g. Srinivasan and Alvarez, 2007).

Conclusion

The evolution and transmission of arboviruses is largely influenced by the vectors that transmit them. The great diversity of arthropod-borne viruses in vertebrate sand plants suggests that this is an effective transmission strategy and results in significant benefits for viral propagation. The investigation of arthropod-borne viruses in plants is more advanced than in vertebrates as witnessed by our understanding of how plant virus modulate their plant hosts and vectors. However, we are still learning the ways vertebrate viruses interact with the host. As has been seen with the expansion of vector-borne plant viruses, the rapid spread of WNV across the United States as well as explosive outbreaks of CHIKV in the Americas, that alterations in arbovirus transmission dynamics can lead to major virus outbreaks. Therefore understanding the relationships between vector-borne viruses and their vectors as well as the within host viral evolution. In both vector and vertebrate or plant hosts is imperative to design novel control strategies for these important diseases that cause significant economic impacts as well as animal and human morbidity and mortality.

References

Adams, A.P., Navarro-Lopez, R., Ramirez-Aguilar, F.J., Lopez-Gonzalez, I., Leal, G., Flores-Mayorga, J.M., Travassos da Rosa, A.P., Saxton-Shaw, K.D., Singh, A.J., Borland, E.M.*, et al.* (2012). Venezuelan equine encephalitis virus activity in the Gulf Coast region of Mexico, 2003-2010. PLoS Neglected Tropical Diseases *6*, e1875.

Ajayi, O., and Dewar, A.M. (1982). The effect of Barley yellow dwarf virus on honeydew production by the cereal aphids Sitobion avenae and Metopolophium dirhodum. Annals of Applied Biology *100*, 203–212.

Ali, A., Li, H., Schneider, W.L., Sherman, D.J., Gray, S., Smith, D., and Roossinck, M.J. (2006). Analysis of genetic bottlenecks during horizontal transmission of Cucumber mosaic virus. J Virol *80*, 8345-8350.

Ali, A., and Roossinck, M.J. (2010). Genetic bottlenecks during systemic movement of Cucumber mosaic virus vary in different host plants. Virology *404*, 279-283.

Ali, H., Ali, A.A., Atta, M.S., and Cepica, A. (2012). Common, emerging, vector-borne and infrequent abortogenic virus infections of cattle. Transbound Emerg Dis *59*, 11-25.

Allison, A.B., Stallknecht, D.E., and Holmes, E.C. (2015). Evolutionary genetics and vector adapation of recombinant viruses of the western equine encephalitis antigenic complex provides new insights into alphavirus diversity and host switching. Virology *474*, 154-162.

Anderson, P.K., Cunningham, A.A., Patel, N.G., Morales, F.J., Epstein, P.R., and Daszak, P. (2004). Emerging infectious diseases of plants: pathogen pollution, climate change and agrotechnology drivers. Trends in ecology and evolution *19*, 535-544.

Arrigo, N.C., Adams, A.P., and Weaver, S.C. (2010). Evolutionary patterns of eastern equine encephalitis virus in North versus South America suggest ecological differences and taxonomic revision. Journal of Virology *84*, 1014-1025.

Ashraf, U., Ye, J., Ruan, X., Wan, S., Zhu, B., and Cao, S. (2015). Usutu virus: an emerging flavivirus in Europe. Viruses *7*, 219-238.

Attoul, H., Mendez-Lopez, M.R., Rao, S., Hurtado-Alendes, A., Lizaraso-Caparo, F., Jaafer, F.M., Samuel, A.R., Belhouchet, M., Pritchard, L.I., Melville, L.*, et al.* (2009). Peruvian horse sickness virus and Yunnan orbivirus isolated from vertebrates and mosquitoes in Peru and Australia. Virology *394*, 298-310.

Balenghien, T., Cardinale, E., Chevalier, V., Elissa, N., Failloux, A.-B., Nipomichene, T.N.J.J., Nicolas, G., Rakotoharinome, V.M., Roger, M., and Zumbo, B. (2013). Towards a better understanding of Rift Valley fever epidemiology in the south-west of the Indian Ocean. Veterinary Research *44*, 78.

Ballinger, M.J., Bruenn, J.A., Hay, J., Czechowski, D., and Taylor, D.J. (2014). Discovery and evolution of Bunyavirid-like sequences in insect genomes. Journal of Virology *16*, 8783-8794.

Beasley, D.W., McAuley, A.J., and Bente, D.A. (2014). Yellow fever virus: Genetic and phenotypic diversity and implications for detection, prevention and therapy. Antiviral Research *115C*, 48-70.

Bejarano, E.R., Khashoggi, A., Witty, M., and Lichtenstein, C. (1996). Integration of multiple repeats of geminiviral DNA into the nuclear genome of tobacco during evolution. Proceedings of the National Academy of Sciences of the United States of America *93*, 759-764.

Belliure, B., Janssen, A., and Sabelis, M.W. (2008). Herbivore benefits from vectoring plant virus through reduction of period of vulnerability to predation. Oecologia *156*, 797-806.

Betancourt, M., Fereres, A., Fraile, A., and Garcia-Arenal, F. (2008). Estimation of the effective number of founders that initiate an infection after aphid transmission of a multipartite plant virus. J Virol *82*, 12416-12421.

Blanc, S. (2007). Virus transmission-getting out and in., Vol 7 (Berlin-Heidelberg: Springer-Verlag).

Blitvich, B.J., Scanlon, D., Shiell, B.J., Mackenzie, J.S., and Hall, R.A. (1999). Identification and analysis of truncated and elongated species of the flavivirus NS1 protein. Virus Research *60*, 67-79.

Bonneau, K.R., Mullens, B.A., and MacLachlan, N.J. (2001). Occurrence of genetic drift and founder effect during quasispecies evolution of the VP2 and NS3/NS3A genes of bluetongue virus upon passage between sheep, cattle, and *Culicoides sonorensis*. Journal of Virology *80*, 3173-3179.

Borucki, M.K., Kempf, B.J., Blitvich, B.J., Blair, C.D., and Beaty, B.J. (2002). La Crosse virus: replication in vertebrate and invertebrate hosts. Microbes and Infection *4*, 341-350.

Brackney, D.E., Beane, J.E., and Ebel, G.D. (2009). RNAi targeting of West Nile virus in mosquito midguts promotes virus diversitification. PLoS Pathogens *5*, e1000502.

Brackney, D.E., Pesko, K.N., Brown, I.K., Deardorff, E.R., Kawatachi, J., and Ebel, G.D. (2011). West Nile virus genetic diversity is maintained during transmission by *Culex pipiens quinquefasciatus* PloS one *6*, e24466.

Brossard, M., and Wikel, S.K. (2004). Tick immunobiology. Parasitology *129*, S161-176.

Brussaard, C.P., Short, S.M., Frederickson, C.M., and Suttle, C.A. (2004). Isolation and phylogenetic analysis of novel viruses infecting the phytoplankton *Phaeocystis globosa* (*Prymnesiophyceae*). Applied and environmental microbiology *70*, 3700-3705.

Bryant, J., Wang, H., Cabezas, C., Ramirez, G., Watts, D., Russell, K., and Barrett, A.D. (2003). Enzootic transmission of Yellow fever virus in Peru. Emerging Infectious Diseases *9*, 926-933.

Calzolari, M., Ze-Ze, L., Ruzek, D., Vazquez, A., Jeffries, C., Defliippo, F., Osorio, H.C., Kilian, P., Ruiz, S., Fooks, A.R., *et al.* (2012). Detection of mosquito-only flaviviruses in Europe. Journal of General Virology *93*, 1215-1225.

Cammisa-Parks, H., Cisar, L.A., Kane, A., and Stollar, V. (1992). The compete nucleotide sequence of cell fusing agent (CFA): homology between the non-structural proteins encoded by CFA nad the nonstructural proteins encoded by arthropod-borne flaviviruses. Virology *189*, 511-524.

Carrington, C.V., and Auguste, A.J. (2013). Evolutionary and ecological factors underlying the tempo and distribution of Yellow fever virus activity. Infection Genetics and Evolution *13*, 198-210.

Charlier, N., Davidson, A., Dallmeier, K., Molenkamp, R., De Clercq, E., and Neyts, J. (2010). Replication of not-known-vector flaviviruses in mosquito cells is restricted by intracellular host factors rather than by the viral envelope proteins. Journal of General Virology *91*, 1693-1697.

Chen, R., and Vasilakis, N. (2011). Dengue-quo tu et quo vadis? Viruses *3*, 1562-1608.

Christen, L., Seto, J., and Niles, E.G. (1990). Superinfection exclusion of vaccinia virus in virus-infected cell cultures. Virology *174*, 35-42.

Ciota, A.T., Ehrbar, D.J., van Slyke, G.A., Payne, A.F., Willsey, G.G., Viscio, R.E., and Kramer, L.D. (2012). Quantification of intrahost bottlenecks of West Nile virus in *Culex pipiens* mosqquitoes using an artificial swarm. Infection Genetics and Evolution *12*, 557-564.

Ciota, A.T., and Kramer, L.D. (2013). Vector-virus interactions and transmission dynamics of West Nile virus. Viruses *5*, 3021-3027.

Coetzee, P., Van Vuuren, M., Stokstad, M., Myrmel, M., and Venter, E.H. (2012). Bluetongue virus genetic and phenotypic diversity: towards identifying the molecular

determinants that influence virulence and transmission potential. Veterinary Microbiology *161*, 1-12.
Coffey, L.L., Forrester, N., Tsetsarkin, K., Vasilakis, N., and Weaver, S.C. (2013). Factors shaping the adaptive landscape for arboviruses: implications for hte emergence of disease. Future Microbiology *8*, 155-176.
Cook, S., Bennett, C.J., Holmes, E.C., De Chesse, R., Moureau, G., and de Lamballerie, X. (2006). Isolation of a new strain of the flavivirus cell fusing agent virus in a natural mosquito population from Puerto Rico. Journal of General Virology *87*, 735-748.
Cook, S., and Holmes, E.C. (2006). A multigene analysis of the phylogenetic relationships among the flaviviruses (Family: *Flaviviridae*) and the evolution of vector transmission. Archives of Virology *151*, 309-325.
Cook, S., Moreau, G., Harbach, R.E., Mukwaya, L., Goodger, K., Ssenfuka, F., Gould, E.A., Holmes, E.C., and de Lamballerie, X. (2009). Isolation of a novel species of flavivirus and a new strain of *Culex flavivirus* (*Flaviviridae*) from a natural mosquito population in Uganda. Journal of General Virology *90*, 2669-2678.
Cook, S., Moureau, G., Kitchen, A., Gould, E.A., de Lamballerie, X., Holmes, E.C., and Harbach, R.E. (2012). Molecular evolution of the insect-specific flaviviruses. Journal of General Virology *93*, 223-224.
Crochu, S., Cook, S., Attoui, H., Charrel, R.N., De Chesse, R., Belhouchet, M., Lemasson, J.-J., de Micco, P., and de Lamballerie, X. (2004). Sequences of flaviviruses-related RNA viruses persist in DNA form integrated in the genome of *Aedes* spp. mosquitoes. Journal of General Virology *85*, 1971-1980.
Davis, C.T., Beasley, D.W., Guzman, H., Raj, R., D'Anton, M., Novak, R.J., Unnasch, T.R., Tesh, R.B., and Barrett, A.D. (2003). Genetic variation among temporally and geographically distinct West Nile virus isolates, United States, 2001, 2002. Emerging Infectious Diseases *9*, 1423-1429.
de A. Zanotto, P.M., Gould, E.A., Gao, G.F., Harvey, P.H., and Holmes, E.C. (1996). Population dynamics of flaviviruses revealed by molecular phylogenies. Proceedings of the national Academy of Sciences USA *93*, 548-553.
de Vos, M., and Jander, G. (2010). Volatile communication in plant-aphid interactions. Current opinion in plant biology *13*, 366-371.
Diallo, D., Sall, A.A., Diagne, C.T., Fayo, O., Hanley, K.A., Buenemann, M., Ba, Y., Faye, O., Weaver, S.C., and Diallo, M. (2014). Patterns of a sylvatic yellow fever virus amplification in southeastern Senegal, 2010. American Journal of Tropical Medicine and Hygiene *90*, 1003-1013.
Dietrich, C., and Maiss, E. (2003). Fluorescent labelling reveals spatial separation of potyvirus populations in mixed infected *Nicotiana benthamiana* plants. The Journal of general virology *84*, 2871-2876.
Diveki, Z., Salanki, K., and Balazs, E. (2002). Limited utility of blue fluorescent protein (BFP) in monitoring plant virus movement. Biochimie *84*, 997-1002.
Doceul, V., Lara, E., Sailleau, C., Belbis, G., Richardson, J., Breard, E., Viarouge, C., Dominguez, M., Hendrikx, P., Calavas, D., *et al.* (2013). Epidemiology, molecular virology and diagnostics of Schmallenberg virus, an emerging orthobunyavirus in Europe. Veterinary Research *44*, 31.
Dolja, V.V., and Koonin, E.V. (2011). Common origins and host-dependent diversity of plant and animal viromes. Current opinion in virology *1*, 322-331.
Domingo, E., Sheldon, J., and Perales, C. (2012). Viral quasispecies evolution. Microbiology and Molecular Biology Reviews *40*, 353-372.
Duarte, E.A., Clarke, D.K., Moya, A., Elena, S.F., Domingo, E., and Holland, J.J. (1993). Many-trillionfold amplication of single RNA virus particles fails to overcome the Muller's ratchet effect. Journal of Virology *67*, 3620-3623.
Ebel, G., and Kramer, L. (2004). Short report: duration of tick attachment required for transmission of Powassan virus by deer ticks. American Journal of Tropical Medicine and Hygiene *71*, 268-271.

Edwards, J.F. (1994). Cache valley virus. Vet Clin North Am Food Anim Pract *10*, 515-524.
Elde, N.C., Child, S.J., Eickbush, M.T., Kitzman, J.O., Rogers, K.S., Shendure, J., Geballe, A.P., and Malik, H.S. (2012). Poxviruses deploy genomic accordions to adapt rapidly against host antiviral defenses. Cell *150*, 831-841.
Elena, S.F., Fraile, A., and Garcia-Arenal, F. (2014). Evolution and emergence of plant viruses. Adv Virus Res *88*, 161-191.
Elena, S.F., and Sanjuán, R. (2007). Virus evolution: Insights from an experimental approach. In Annual Review of Ecology Evolution and Systematics (Palo Alto: Annual Reviews), pp. 27-52.
Emmons, R.W., Woodie, J.D., Laub, R.L., and Oshiro, L.S. (1983). Main Drain virus as a cause of equine encephalomyelitis. Journal of the American Veterinary Medical Association *183*, 555-558.
Escriu, F., Fraile, A., and Garcia-Arenal, F. (2003). The evolution of virulence in a plant virus. Evolution; international journal of organic evolution *57*, 755-765.
Estrada-Pena A. and de la Fuente, J. (2014) The ecology of ticks and epidemiology of tick-borne viral diseases. Antiviral Res. *108*, 104-28.
Evangelista, J., Cruz, C., Guevara, C., Astete, H., Carey, C., Kochel, T.J., Morrison, A.C., Williams, M., Halsey, E.S., and Forshey, B.M. (2013). Characterization of a novel flavivirus isolated from *Culex* (*Melanoconion*) *ocossa* mosquitoes from Iquitoes, Peru. Journal of General Virology *94*, 1266-1272.
Fereres, A., and Moreno, A. (2009). Behavioral aspects influencing plant virus transmission by homopteran insects. Virus Res *141*, 158-168.
Firth, A.E., and Atkins, J.F. (2009). A conserved predicted pseudoknot in the NS2A-encoding sequence of West Nile and Japanese encephalitis flaviviruses suggests NS1' may derive from ribosomal frameshifting. Virology Journal *6*, 14.
Firth, A.E., Blitvich, B.J., Wills, N.M., Miller, C.L., and Atkins, J.F. (2010). Evidence for ribosomal frameshifting and a novel overlapping gene in the genomes of insect-specific flaviviruses. Virology *399*, 153-166.
Flick, R., and Bouloy, M. (2005). Rift Valley Fever virus. Current Molecular Medicine *5*, 827-834.
Folimonova, S.Y. (2012). Superinfection exclusion is an active virus-controlled function that requires a specific viral protein. J Virol *86*, 5554-5561.
Forrester, N.L., Guerbois, M., Adams, A.P., Liang, X., and Weaver, S.C. (2011). Analysis of intrahost variation in Venezuelan equine encephalitis virus reveals repeated deletions in the 6-kilodalton protein gene. Journal of Virology *85*, 8709-8717.
Forrester, N.L., Guerbois, M., Seymour, R., Spatt, H., and Weaver, S.C. (2012a). Vector-borne transmissions imposes a severe bottleneck on an RNA virus population. PLoS Pathogens *8*, e1002897.
Forrester, N.L., Palacios, G., Tesh, R.B., Savji, N., Guzman, H., Sherman, M., Weaver, S.C., and Lipkin, W.I. (2012b). Genome scale phylogeny of the Alphavirus genus suggests a marine origin. Journal of Virology *85*, 8709-8717.
Foy, B.D., Myles, K.M., Pierro, D.J., Sanchez-Vargas, I., Uhlirova, M., Jindra, M., Beaty, B.J., and Olson, K.E. (2004). Develoment of a new Sindbis virus transducing system and its characterization in three Culicine mosquitoes and two Lepidopteran species. Insect Molecular Biology *13*, 89-100.
Franz, A.W., van der Wilk, F., Verbeek, M., Dullemans, A.M., and van den Heuvel, J.F. (1999). Faba bean necrotic yellows virus (genus Nanovirus) requires a helper factor for its aphid transmission. Virology *262*, 210-219.
French, R., and Stenger, D.C. (2003). Evolution of Wheat streak mosaic virus: dynamics of population growth within plants may explain limited variation. Annual review of phytopathology *41*, 199-214.
Froissart, R., Roze, D., Uzest, M., Galibert, L., Blanc, S., and Michalakis, Y. (2005). Recombination every day: abundant recombination in a virus during a single multi-cellular host infection. PLoS Biol *3*, e89.

Fukunaga, Y., Kumanomido, T., and Kamada, M. (2000). Getah virus as an equine pathogen. Vet Clin North Am Equine Pract *16*, 605-617.

Garcia-Arenal, F., Fraile, A., and Malpica, J.M. (2003). Variation and evolution of plant virus population. Int Microbiol *in press*.

Gibbs, A.J., Fargette, D., Garcia-Arenal, F., and Gibbs, M.J. (2010). Time--the emerging dimension of plant virus studies. The Journal of general virology *91*, 13-22.

Gibbs, A.J., Ohshima, K., Phillips, M.J., and Gibbs, M.J. (2008). The prehistory of potyviruses: their initial radiation was during the dawn of agriculture. PloS one *3*, e2523.

Gottlieb, Y., Zchori-Fein, E., Mozes-Daube, N., Kontsedalov, S., Skaljac, M., Brumin, M., Sobol, I., Czosnek, H., Vavre, F., Fleury, F., *et al.* (2010). The transmission efficiency of tomato yellow leaf curl virus by the whitefly *Bemisia tabaci* is correlated with the presence of a specific symbiotic bacterium species. J Virol *84*, 9310-9317.

Gould, E.A., Lamballerie, X., Zanotto, P.M., and Holmes, E.C. (2003). Origins, evolution and vector/host coadaptations within the genus *Flavivirus*. Advances in Virus Research *59*, 277-314.

Gutiérrez, S., Michalakis, Y., and Blanc, S. (2012a). Virus population bottlenecks during within-host progression and host-to-host transmission. Current opinion in virology *2*, 546-555.

Gutierrez, S., Thébaud, G., Smith, D.R., Kenney, J.L., and Weaver, S.C. (2015). Demographics of natural oral infection of mosquitoes by Venezuelan equine encephalitis virus. Journal of Virology *epub ahead of print*.

Gutiérrez, S., Yvon, M., Pirolles, E., Garzo, E., Fereres, A., Michalakis, Y., and Blanc, S. (2012b). Circulating Virus Load Determines the Size of Bottlenecks in Viral Populations Progressing within a Host. PLoS Pathog *8*, e1003009.

Gutiérrez, S., Yvon, M., Thébaud, G., Monsion, B., Michalakis, Y., and Blanc, S. (2010). Dynamics of the Multiplicity of Cellular Infection in a Plant Virus. PLoS Pathog *6*, e1001113.

Haigh, J. (1978). The accumulation of deleterious genes in a population—Muller's Ratchet. Theor Pop Biol *14*.

Hardy, J.L. (1988). Susceptibility and resistance of vector mosquitoes. In Arboviruses: Epidemiology and Ecology, T.P. Monath, ed. (Boca Raton, Florida: CRC Press), pp. 87-126.

Havlíková, S., Ličková, M., and Klempa, B. (2013). Non-viraemic transmission of tick-borne viruses. Acta Virologica *57*, 123-129.

Hecker, H. (1977). Structure and function of midgut epithelial cells in culicidae mosquitoes (Insecta, Diptera). Cell Tissue Research *184*, 321-341.

Hermance, M.E., and Thangamani, S. (2014). Proinflammatory cytokines and chemokines at the skin interface during Powassan virus transmission. Journal of Investigational Dermatology *134*, 2280-2283.

Hewlett, M.J., Clerx, J.P., Clerx- van Haaster, C.M., Chandler, L.J., McLean, D.M., and Beaty, B.J. (1992). Genomic and biologic analyses of snowshoe hare virus field and laboratory strains. American Journal of Tropical Medicine and Hygiene *46*, 542-532.

Hodge, S., and Powell, G. (2008a). Complex interactions between a plant pathogen and insect parasitoid via the shared vector-host: consequences for host plant infection. Oecologia *157*, 387-397.

Hodge, S., and Powell, G. (2008b). Do plant viruses facilitate their aphid vectors by inducing symptoms that alter behavior and performance? Environmental entomology *37*, 1573-1581.

Hogenhout, S.A., Ammar el, D., Whitfield, A.E., and Redinbaugh, M.G. (2008). Insect vector interactions with persistently transmitted viruses. Annual review of phytopathology *46*, 327-359.

Holbrook, M. (2012). Kyasanur forest disease. Antiviral Research *96*, 353-362.

Holmes, E.C. (2009). The evolutionary genetics of emerging viruses. Annual Review of Ecology, Evolution and Systematics *40*, 353-372.

Holmes, E.C. (2011). The evolution of endogenous viral elements. Cell Host and Microbe *10*, 368-377.
Hoshino, K., Isawa, H., Tsuda, Y., Sawabe, K., and Kobayashi, M. (2009). Isolation and characterization of a new insect flavivirus form Aedes albopictus and *Aedes flavopictus* mosquitoes in Japan. Virology *391*, 119-129.
Houk, E.J., Arcus, Y.M., Hardy, J.L., and Kramer, L.D. (1990). Binding of Western equine encephalomyelitis virus to brush border fragments isolated from mesenteronal epithelial cells of mosquitoes. Virus Research *17*, 105-117.
Huang, I.C., Li, W., Sui, J., Marasco, W., Choe, H., and Farzan, M. (2008). Influenza A virus neuraminidase limits viral superinfection. J Virol *82*, 4834-4843.
Hubalek, Z., and Rudolf, I. (2012). Tick-borne viruses in Europe. Parasitology Research *111*, 9-36.
Hubalek, Z., Rudolf, I., and Nowotny, N. (2014). Arboviruses pathogenic for domestic and wild animals. Advances in Virus Research *89*, 201-275.
Huhtamo, E., Cook, S., Moureau, G., Uzcategui, N.Y., Sironen, T., Kuivanen, S., Putkuri, N., Kurkela, S., Harbach, R.E., Firth, A.E., *et al.* (2014). Novel flaviviruses from mosquitoes: Mosquito-specific evolutionary lineages within the phylogenetic group of mosquito-borne flaviviruses. Virology *464-465*, 320-329.
Iranzo, J., and Manrubia, S.C. (2012). Evolutionary dynamics of genome segmentation in multipartite viruses. Proceedings Biological sciences / The Royal Society *279*, 3812-3819.
Jeffries, C.L., Mansfield, K.L., Phipps, L.P., Wakeley, P.R., Mearns, R., Schock, A., Bell, S., Breed, A.C., Fooks, A.R., and Johnson, N. (2014). Louping Ill virus: an endemic tick-borne disease of Great Britain. Journal of General Virology *95*, 1005-1014.
Johnson, K.M., Tesh, R.B., and Peralta, P.H. (1969). The epidemiology of vesicular stomatitis virus: some new data and a hypothesis for transmission of the Indiana serotype. Journal of the American Veterinary Medical Association *115*, 2133-2140.
Jones, L.D., Davies, C.R., Steele, G.M., and Nuttall, P.A. (1987). A novel mode of arbovirus transmission involving a nonviremic host. Science *237*, 775-777.
Jridi, C., Martin, J.F., Marie-Jeanne, V., Labonne, G., and Blanc, S. (2006). Distinct viral populations differentiate and evolve independently in a single perennial host plant. J Virol *80*, 2349-2357.
Junglen, S., Kopp, A., Kurth, A., Pauli, G., Ellerbrok, H., and Leendertz, F.H. (2009). A new flavivirus and a new vector: characterization of a novel flavivirus isolated from uranotaenia mosquitoes from a tropical rain forest. Journal of Virology *83*, 4462-4468.
Karpf, A.R., Lenches, E., Strauss, E.G., Strauss, J.H., and Brown, D.T. (1997). Superinfection exclusion of alphaviruses in three mosquito cell lines persistently infected with Sindbis virus. J Virol *71*, 7119-7123.
Katzourakis, A., and Gifford, R.J. (2010). Endogenous viral elements in animal genomes. PLoS Genetics *6*, e1001191.
Katzourakis, A., and Tristem, M. (2005). Phylogeny of human endogenous and exogenous retroviruses. In Retroviruses and Primate Genome Evolution, E.D. Sverdlov, ed. (Austin, TX: Landes Biosciences).
Kenney, J.L., Adams, A.P., Gorchakov, R., Leal, G., and Weaver, S.C. (2012). Genetic and anatomic determinants of enzootic Venezuelan equine encephalitis virus infection of *Culex (Melanoconion) taeniopus*. Plos Neglected Tropical Diseases *6*, e1606.
Kirkland, P.D. (2002). Akabane and bovine ephemeral fever virus infections. Vet Clin North Am Equine Pract *18*, 501-514.
Kolodziejek, J., Pachler, K., Bin, H., Mendelson, E., Shulman, L., Orshan, L., and Nowotny, N. (2013). Barkedji virus, a novel mosquito-borne flavivirus identified in Culex perexiguus mosquitoes, Israel, 2011. Journal of General Virology *94*, 2449-2457.
Krupovic, M., Ravantti, J.J., and Bamford, D.H. (2009). Geminiviruses: a tale of a plasmid becoming a virus. BMC evolutionary biology *9*, 112.

Kuno, G., Chang, G.-J.J., Tsuchiya, K.R., Karabatsos, N., and Cropp, B.C. (1998). Phylogeny of the genus Flavivirus. Journal of Virology *72*, 73-83.

Labuda, M., and Nuttall, P.A. (2004). Tick-borne viruses. Parasitology *129*, S221-245.

Laine, M., Luukkainen, R., and Toivanen, A. (2004). Sindbis viruses and other alphaviruses as cause of human arthritic disease. Journal of Internal Medicine *256*, 457-471.

LaSala, P.R., and Holbrook, M. (2010). Tick-borne flaviviruses. Clinical Laboratory Medicine *30*, 221-235.

Lee, J.S., Grubaugh, N.D., Kondig, J.P., Turrell, M.J., Kim, H.C., Klein, T.A., and O'Guinn, M.L. (2013). Isolation and genomic characterization of Chaoyang virus strain ROK144 from *Aedes vexans* nipponii from the Republic of Korea. Virology *435*, 220-224.

Lefeuvre, P., Harkins, G.W., Lett, J.M., Briddon, R.W., Chase, M.W., Moury, B., and Martin, D.P. (2011). Evolutionary time-scale of the begomoviruses: evidence from integrated sequences in the *Nicotiana* genome. PloS one *6*, e19193.

Lefeuvre, P., Martin, D.P., Harkins, G., Lemey, P., Gray, A.J., Meredith, S., Lakay, F., Monjane, A., Lett, J.M., Varsani, A., *et al.* (2010). The spread of tomato yellow leaf curl virus from the Middle East to the world. PLoS Pathog *6*, e1001164.

Levetin, E., and McMahon, K. (2011). Plants and Society, 6th edn (McGraw-Hill Science/Engineering/Math).

Liu, H., Fu, Y., Li, B., Yu, X., Xie, J., Cheng, J., Ghabrial, S.A., Li, G., Yi, X., and Jiang, D. (2011). Widespread horizontal gene transfer from circular single-stranded DNA viruses to eukaryotic genomes. BMC evolutionary biology *11*, 276.

Lu, M.J., and Henning, U. (1994). Superinfection exclusion by T-even-type coliphages. Trends in microbiology *2*, 137-139.

Lynch, M., and Gabriel, W. (1990). Mutation load and the survival of small populations. Evolution; international journal of organic evolution *44*, 1725-1737.

MacLachlan, N.J., and Guthrie, A.J. (2010). Re-emergence of bluetongue, African horse sickness, and other orbivirus diseases. Veterinary Research *41*, 35.

Mann, B.R., McMullen, A.R., Swetnam, D.M., Salvato, V., Reyna, M., Guzman, H., Bueno, R.J., Dennett, J.A., Tesh, R.B., and Barrett, A.D. (2013). Continued evolution of West Nile virus, Houston, Texas, USA, 2002-2012. Emerging Infectious Diseases *19*, 1418-1427.

Marczinke, B.I., and Nichol, S.T. (2002). Nairobi sheep disease virus, an important tick-borne pathogen of sheep and goats in Africa, is also present in Asia. Virology *303*, 146-151.

Martiniere, A., Bak, A., Macia, J.L., Lautredou, N., Gargani, D., Doumayrou, J., Garzo, E., Moreno, A., Fereres, A., Blanc, S., *et al.* (2013). A virus responds instantly to the presence of the vector on the host and forms transmission morphs. eLife *2*, e00183.

Matsuno, K., Weisend, C., Travassos da Rosa, A.P.A., Anzick, S.L., Dahlstrom, E., Porcella, S.F., David, D.W., Yu, X., Tesh, R.B., and Ebihara, H. (2013). Characterization of the Bhanja serogroup viruses (Bunyarviridae): A novel species of the genus *Phlebovirus* and its relationship with other emerging tick-borne Phleboviruses. Journal of Virology *87*, 3719-3728.

Matthews, K.R. (2011). Controlling and coordinating development in vector-transmitted parasites. Science *331*, 1149-1153.

Mauck, K.E., Bosque-Pérez, S.D., Eigenbrode, C.M., De Moraes, C.M., and Mescher, M.C. (2012). Transmission mechanisms shape pathogen effects on host-vector interactions: evidence from plant viruses. . Functional Ecology, doi: 10.1111/j. 1365-2435.2012.02026.

Mauck, K.E., De Moraes, C.M., and Mescher, M.C. (2010a). Deceptive chemical signals induced by a plant virus attract insect vectors to inferior hosts. Proceedings of the National Academy of Sciences of the United States of America *107*, 3600-3605.

Mauck, K.E., De Moraes, C.M., and Mescher, M.C. (2010b). Effects of Cucumber mosaic virus infection on vector and non-vector herbivores of squash. Communicative and integrative biology *3*, 579-582.
May, F.J., Davis, C.T., Tesh, R.B., and Barrett, A.D. (2011). Phylogeography of West Nile virus: from the cradle of evolution in Africa to Eurasia, Australia and the Americas. Journal of Virology *85*, 2964-2974.
McMullen, A.R., Albayrak, H., May, F.J., Davis, C.T., Beasley, D.W.C., and Barrett, A.D.T. (2013). Molecular evolution of lineage 2 West Nile virus. Journal of General Virology *94*, 318-325.
Miyashita, S., and Kishino, H. (2010). Estimation of the size of genetic bottlenecks in cell-to-cell movement of soil-borne wheat mosaic virus and the possible role of the bottlenecks in speeding up selection of variations in trans-acting genes or elements. J Virol *84*, 1828-1837.
Monsion, B., Froissart, R., Michalakis, Y., and Blanc, S. (2008). Large bottleneck size in Cauliflower Mosaic Virus populations during host plant colonization. PLoS Pathog *4*, e1000174.
Moury, B., Fabre, F., and Senoussi, R. (2007). Estimation of the number of virus particles transmitted by an insect vector. Proceedings of the National Academy of Sciences of the United States of America *104*, 17891-17896.
Muller, H.J. (1964). The relation of recombination to mutational advance. Mutation Research *1*, 2-9.
Nandi, S., and Negi, B.S. (1999). Bovine ephemeral fever: a review. Comparative Immunology and Microbiology and Infectious Disease *22*, 81-91.
Nasar, F., Palacios, G., Gorchakov, R.V., Guzman, H., Da Rosa, A.P., Savji, N., Popov, V.L., Sherman, M.B., Lipkin, W.I., Tesh, R.B., *et al.* (2012). Eilat virus, a unique alphavirus with host range restricted to insects by RNA replication. Proceedings of the National Academy of Sciences USA *109*, 14622-15627.
Nault, L.R. (1997). Arthropod transmission of plant viruses : a new synthesis. Ann Entomol Soc Am *90*, 521-541.
Nethe, M., Berkhout, B., and van der Kuyl, A.C. (2005). Retroviral superinfection resistance. Retrovirology *2*, 52.
Nguyen, H.D., Tomitaka, Y., Ho, S.Y., Duchene, S., Vetten, H.J., Lesemann, D., Walsh, J.A., Gibbs, A.J., and Ohshima, K. (2013). Turnip mosaic potyvirus probably first spread to Eurasian brassica crops from wild orchids about 1000 years ago. PloS one *8*, e55336.
Novella, I.S. (2003). Contributions of vesicular stomatitis virus to the understanding of RNA virus evolution. Current Opinion in Microbiology *6*, 399-405.
Novella, I.S., Reissig, D.D., and Wilke, C.O. (2004). Density-dependent selection in vesicular stomatitis virus. J Virol *78*, 5799-5804.
Nuttall, P.A. (1999). Pathogen-tick-host interactions: *Borrelia burgorferi* and TBE virus. Zentralbl Bakteriol *289*, 492-505.
Ohashi, S., Matsumori, Y., Yanase, T., Yamakawa, M., Kato, T., and Tsuda, T. (2004). Evidence of an antigenic shift among Palyam serogroup orbiviruses. Journal of Clinical Microbiology *42*, 4610-4614.
Ohashi, S., Yoshida, K., Yanase, T., and Tsuda, T. (2002). Analysis of intratypic variation evident in Ibaraki virus strain and its epizootic hemorrhagic disease virus serogroup. Journal of Clinical Microbiology *40*, 3684-3688.
Olson, K.F., Myles, K.M., Seabaugh, R.C., Higgs, S., Carlson, J.O., and Beaty, B.J. (2000). Development of a Sindbis virus expression system that efficiently expresses green fluorescent protein in midguts of *Aedes aegypti* following per os infection. Insect Molecular Biology *9*, 57-65.
Oura, C.A., Batten, C.A., Ivens, P.A., Balcha, M., Alhassan, A., Gizaw, D., Elharrak, M., Jallow, D.B., Sahle, M., Maan, N., *et al.* (2012). Equine encephalosis virus: evidence for circulation beyond southern Africa. Epidemiology and Infection *140*, 1982-1986.

Pagan, I., Firth, C., and Holmes, E.C. (2010). Phylogenetic analysis reveals rapid evolutionary dynamics in the plant RNA virus genus tobamovirus. Journal of molecular evolution *71*, 298-307.
Pagan, I., and Holmes, E.C. (2010). Long-term evolution of the *Luteoviridae*: time scale and mode of virus speciation. J Virol *84*, 6177-6187.
Perefarres, F., Thébaud, G., Lefeuvre, P., Chiroleu, F., Rimbaud, L., Hoareau, M., Reynaud, B., and Lett, J.M. (2014). Frequency-dependent assistance as a way out of competitive exclusion between two strains of an emerging virus. Proceedings Biological sciences / The Royal Society *281*, 20133374.
Perefarres, F., Thierry, M., Becker, N., Lefeuvre, P., Reynaud, B., Delatte, H., and Lett, J.M. (2012). Biological invasions of geminiviruses: case study of TYLCV and *Bemisia tabaci* in Reunion Island. Viruses *4*, 3665-3688.
Pesko, K.N., and Ebel, G.D. (2012). West Nile virus population genetics and evolution. Infection Genetics and Evolution *12*, 181-190.
Pirone, T., and Blanc, S. (1996). Helper-dependent vector transmission of plant viruses. Annu Rev Phytopathol *34*, 227-247.
Power, A.G., and Flecker, A.S. (2003). Virus specificity in disease systems: are species redundant? In The importance of species, P. Kareiva, and S.A. Levin, eds. (Princeton University Press), pp. 330-347.
Pu, Y., Kikuchi, A., Moriyasu, Y., Tomaru, M., Jin, Y., Suga, H., Hagiwara, K., Akita, F., Shimizu, T., Netsu, O.*, et al.* (2011). Rice dwarf viruses with dysfunctional genomes generated in plants are filtered out in vector insects: implications for the origin of the virus. J Virol *85*, 2975-2979.
Randolph, S.E., Miklisová, D., Lysy, J., Rogers, D.J., and Labuda, M. (1999). Incidence from coincidence: patterns of tick infestations on rodents facilitate transmission of tick-borne encephalitis virus. Parasitology *118*, 177-186.
Roiz, D., Vazquez, A., Seco, M.P., Tenorio, A., and Rizzoli, A. (2009). Detection of novel insect flavivirus sequences integrated in *Aedes albopictus* (Diptera: Culicidae) in Northern Italy. Virology Journal *6*, 93.
Romoser, W.S., Wasieloski, J., L. P., Pushkon, P., Kondig, J.P., Lerdthusnee, K., Neira, M., and Ludwig, G.V. (2004). Evidence for arbovirus dissemination conduits from the mosquito (Diptera: Culicidae) midgut. Journal of Medical Entomology *41*, 467-475.
Roossinck, M.J. (2012). Plant virus metagenomics: biodiversity and ecology. Annual review of genetics *46*, 359-369.
Růžek, D., Yakimenko, V.V., Karan, L.S., and Tkachev, S.E. (2010). Omsk haemorrhagic fever. The Lancet *376*, 2104-2113.
Rybicki, E.P., and Pietersen, G. (1999). Plant virus disease problems in the developing world. Adv Virus Res *53*, 127-175.
Sanjuán, R., Nebot, M.R., Chirico, N., Mansky, L.M., and Belshaw, R. (2010). Viral mutation rates. J Virol *84*, 9733-9748.
Savini, G., Afonso, C., Mellor, P., Aradaib, I., Yadin, H., Sanaa, M., Wilson, W., Monaco, F., and Domingo, M. (2011). Epizootic haemorrhagic disease. Research in Veterinary Science *91*, 1-17.
Scholle, F., Girard, Y.A., Zhao, Q., Higgs, S., and Mason, P.W. (2004). Trans-packaged West Nile virus-like particles: infectious properties *in vitro* and in infected mosquito vectors. Journal of Virology *78*, 11605-11614.
Scholthof, K.B., Adkins, S., Czosnek, H., Palukaitis, P., Jacquot, E., Hohn, T., Hohn, B., Saunders, K., Candresse, T., Ahlquist, P.*, et al.* (2011). Top 10 plant viruses in molecular plant pathology. Molecular plant pathology *12*, 938-954.
Schuffenecker, I., Iteman, I., Michault, A., Murri, S., Frangeul, L., Vaney, M.-C., Lavenir, R., Pardigon, N., Reynes, J.-M., Pettinelli, F.*, et al.* (2006). Genome microevolution of Chikungunya viruses causing the Indian Ocean outbreak. PLOS Med *3*, e263-e270.
Selvey, L.A., Dailey, L., Lindsay, M., Armstrong, P., Tobin, S., Koehler, A.P., Markey, P.G., and Smith, D.W. (2014). The changing epidemiology of Murray Valley encephalitis in Australia: the 2011 outbreak and a review of the literature. Plos Neglected Tropical Diseases *8*, e2656.

Sicard, A., Yvon, M., Timchenko, T., Gronenborn, B., Michalakis, Y., Gutierrez, S., and Blanc, S. (2013). Gene copy number is differentially regulated in a multipartite virus. Nature communications *4*, 2248.
Simon, K.O., Cardamone, J.J., Jr., Whitaker-Dowling, P.A., Youngner, J.S., and Widnell, C.C. (1990). Cellular mechanisms in the superinfection exclusion of vesicular stomatitis virus. Virology *177*, 375-379.
Smith, D., Adams, A.P., Kenney, J.L., Wang, E., and Weaver, S.C. (2008). Venezuelan equine encephalitis virus in the mosquito vector *Aedes taeniorhynchus*: Infection initiated by a small number of susceptible epithelial cells and a population bottleneck. Virology *372*, 176-186.
Smith, D.R., Carrara, A.S., Aguilar, P.V., and Weaver, S.C. (2005). Evaluation of methods to assess transmission potential of Venezuelan equine encephalitis virus by mosquitoes and estimation of mosquito saliva titers. American Journal of Tropical Medicine and Hygiene *73*, 33-39.
Srinivasan, R., and Alvarez, J.M. (2007). Effect of mixed viral infections (potato virus Y-potato leafroll virus) on biology and preference of vectors *Myzus persicae* and *Macrosiphum euphorbiae* (Hemiptera: *Aphididae*). Journal of economic entomology *100*, 646-655.
Stafford, C.A., Walker, G.P., and Ullman, D.E. (2012). Hitching a ride: Vector feeding and virus transmission. Communicative and integrative biology *5*, 43-49.
Stapleford, K.A., Coffey, L.L., Lay, S., Borderia, A.V., Duong, V., Isakov, O., Rozen-Gagnon, K., Arias-Goeta, C., Blanc, H., Beaucourt, S., *et al.* (2014). Emergence and transmission of arbovirus evolutionary intermediates with epidemic potential. Cell Host and Microbe *15*, 706-716.
Steele, G.M., and Nuttall, P.A. (1989). Difference in vector competence of two species of sympatric ticks, *Amblyomma variegatum* and *Rhicephalus appendiculatus* for Dugbe virus (*Nairovirus: Bunyaviridae*). Virus Research *14*, 73-84.
Steward, G.F., Culley, A.I., Mueller, J.A., Wood-Charlson, E.M., Belcaid, M., and Poisson, G. (2013). Are we missing half of the viruses in the ocean? The ISME journal *7*, 672-679.
Stollar, V., and Thomas, V.L. (1975). An agent in the Aedes aegypti cell line (Peleg) which causes fusion of Aedes albopictus cells. Virology *64*, 367-377.
Takahashi, T., Sugawara, t., Yamatsuta, T., Isogai, M., Natsuaki, T., and Yoshikawa, N. (2007). Analysis of the Spatial Distribution of Identical and Two Distinct Virus Populations Differently Labeled with Cyan and Yellow Fluorescent Proteins in Coinfected Plants. Phytopathology *97* 1200-1206.
Takeshita, M., Shigemune, N., Kikuhara, K., Furuya, N., and Takanami, Y. (2004). Spatial analysis for exclusive interactions between subgroups I and II of Cucumber mosaic virus in cowpea. Virology *328*, 45-51.
Tromas, N., Zwart, M.P., Lafforgue, G., and Elena, S.F. (2014a). Within-host spatiotemporal dynamics of plant virus infection at the cellular level. PLoS Genet *10*, e1004186.
Tromas, N., Zwart, M.P., Poulain, M., and Elena, S.F. (2014b). Estimation of the *in vivo* recombination rate for a plant RNA virus. The Journal of general virology *95*, 724-732.
Tscherne, D.M., Evans, M.J., von Hahn, T., Jones, C.T., Stamataki, Z., McKeating, J.A., Lindenbach, B.D., and Rice, C.M. (2007). Superinfection exclusion in cells infected with hepatitis C virus. J Virol *81*, 3693-3703.
Tsetsarkin, K.A., Chen, R., Leal, G., Forrester, N.L., Higgs, S., Huang, J., and Weaver, S.C. (2011). Chikungunya virus emergence is constrained in Asia by lineage-specific adaptative landscapes. Proceedings of the National Academy of Sciences *108*, 7872-7877.
Tsetsarkin, K.A., Chen, R., Yun, R., Rossi, S.L., Plante, K.S., Guerbois, M., Forrester, N., Perng, G.C., Sreekumar, E., Leal, G., *et al.* (2014). Multi-peaked adaptive landscape for chikungunya virus evolution predicts continued fitness optimization in Aedes albopictus mosquitoes. Nature Communications *5*, 4084.

Tsetsarkin, K.A., Vanlandingham, D.L., McGee, C.E., and Higgs, S. (2007). A single mutation in chikungunya virus affects vector specificity and epidemic control. PLoS Pathogens *3*, e201.
Tulman, E.R., Delhon, G.A., Ku, B.K., and Rock, D.L. (2009). African Swine Fever virus. Current topics in microbiology and immunology *328*, 43-87.
Turner, P.E., and Chao, L. (1998). Sex and the evolution of intrahost competition in RNA virus phi6. Genetics *150*, 523-532.
Unni, S.K., Ruzek, D., Chhatbar, C., Mishra, R., Johri, M.K., and Singh, S.K. (2011). Japanese encephalitis virus: from genome to infectome. Microbes and Infection *13*, 312-321.
Urbino, C., Gutierrez, S., Antolik, A., Bouazza, N., Doumayrou, J., Granier, M., Martin, D.P., and Peterschmitt, M. (2013). Within-host dynamics of the emergence of Tomato yellow leaf curl virus recombinants. PloS one *8*, e58375.
Uzcategui, N.Y., Sironen, T., Golovjiova, I., Jaaskelainen, A.E., Valimaa, H., Lundkvist, A., Plyusnin, A., Vaheri, A., and Vapalahti, O. (2012). Rate of evolution and molecular epidemiology of tick-borne encephalitis virus in Europe, including two isolations from the same focus 44 years apart. Journal of General Virology *93*, 786-796.
Uzest, M., Gargani, D., Drucker, M., Hebrard, E., Garzo, E., Candresse, T., Fereres, A., and Blanc, S. (2007). A protein key to plant virus transmission at the tip of the insect vector stylet. Proceedings of the National Academy of Sciences of the United States of America *104*, 17959-17964.
Van den Heuvel, J., Verbeek, M., and van der Wilk, F. (1994). Endosymbiotic bacteria associated with circulative transmission of Potato Leafroll Virus by *Myzus persicae*. Journal of Virology *75*, 2559-2565.
van Eeden, C., Harders, F., Kortekaas, J., Bossers, A., and Venter, M. (2014). Genomic and phylogenetic characterization of Shuni virus. Archives of Virology *159*, 2883-2892.
Vanlandingham, D.L., McGee, C.E., Klingler, K.A., Galbraith, S.E., Barrett, A.D., and Higgs, S. (2008). Short report: comparison of oral infection dose of West Nile virus isolates representing three distinct genotypes in *Culex quinquefasciatus*. American Journal of Tropical Medicine and Hygiene *79*, 951-954.
Varelas-Wesley, I., and Calisher, C.H. (1982). Antigenic relationship of flaiviviruses with undetermined arthropod-borne status. American Journal of Tropical Medicine and Hygiene *31*, 1273-1284.
Vasilakis, N., Cardosa, J., Hanley, K.A., Holmes, E.C., and Weaver, S.C. (2011). Fever fromt he forest: prospects for the continued emergence of sylvatic dengue virus and its impact on public health. Nature Reviews Microbiology *9*, 532-541.
Vasilakis, N., Forrester, N.L., Palacios, G., Nasar, F., Savji, N., Rossi, S.L., Guzman, H., Wood, T.G., Popov, V., Gorchakov, R., *et al.* (2013). Negevirus: a proposed new taxon of insect-specific viruses with wide geographic distribution. Journal of Virology *87*, 2475-2488.
Vasilakis, N., Guzman, H., Firth, C., Forrester, N.L., Widen, S.G., Wood, T.G., Rossi, S.L., Ghedin, E., Popov, V., Blasdell, K.R., *et al.* (2014). Mesoniviruses are mosquito-specific viruses with extensive geographic distribution and host range. Virology Journal *11*, 97.
Vasilakis, N., Hanley, K.A., and Weaver, S.C. (2010). Dengue virus emergence from its sylvatic cycle. In Frontiers in dengue virus research, K.A. Hanley, and S.C. Weaver, eds. (Norfolk, United Kingdom: Caister Academic Press), pp. 183-220.
Vasilakis, N., Shell, E., Fokam, E., Mason, P., Hanley, K., Estes, D., and Weaver, S. (2007). Potential of ancestral sylvatic dengue-2 viruses to re-emerge. Virology *358*, 402-412.
Vazquez, A., Sanchez-Seco, M.P., Palacios, G., Molero, F., Reyes, N., Ruiz, S., Aranda, C., Marques, E., Escosa, R., Moreno, J., *et al.* (2012). Novel flaviviruses detected in different species of mosquitoes in Spain. Vector Borne and Zoonotic Diseases *12*, 223-229.

Voinnet, O. (2001). RNA silencing as a plant immune system against viruses. Trends Genet *17*, 449-459.
Volk, S.M., Chen, R., Tsetsarkin, K.A., Adams, A.P., Garcia, T.I., Sall, A.A., Nasar, F., Schuh, A.J., Holmes, E.C., Higgs, S.*, et al.* (2010). Genome-scale phylogenetic analyses of Chikungunya virus reveal independent emergences of recent epidemic and various evolutionary rates. Journal of Virology *84*, 6497-6504.
Weaver, S., and Barrett, A. (2004). Transmission cycles, host range, evolution and emergence of arboviral disease. Nat Rev *2*, 789-801.
Weaver, S.C. (1997). Vector biology in viral pathogenesis. In Viral Pathogenesis, N. Nathanson, ed. (New York: Lippincott-Raven), pp. 329-352.
Weaver, S.C., Anischenko, M., Bowen, R.A., Brault, A.C., Estrada-Franco, J.G., Fernandez, Z., Greene, I., Ortiz, D., Paessler, S., and Powers, A.M. (2004). Genetic determinants of Venezuelan equine encephalitis emergence. Archive of Virology Supplement *18*, 43-64.
Weaver, S.C., Brault, A.C., Kang, W., and Holland, J.J. (1999). Genetic and fitness changes accompanying adaptation of an arbovirus to vertebrate and invertebrate cells. Journal of Virology *73*, 4316-4326.
Weaver, S.C., Lorenz, L.H., and Scott, T.W. (1993). Distribution of Western equine encephalomyelitis virus in the alimentary tract of *Culex tarsalis* (Diptera:Culicidae) following natural and artificial blood meals. Journal of Medical Entomology *30*, 391-397.
Weaver, S.C., Scott, T.W., Lorenz, L.H., and Repik, P.M. (1991). Detection of eastern equine encephalomyelitis virus deposition in *Culiseta melanura* following ingestion of radiolabeled virus in blood meals. American Journal of Tropical Medicine and Hygiene *44*, 250-259.
Weaver, S.C., Winegar, R., Manger, I.D., and Forrester, N.L. (2012). Alphaviruses: Population genetics and determinants of emergence. Antiviral Research *94*, 242-257.
Weitz, J.S., Mileyko, Y., Joh, R.I., and Voit, E.O. (2008). Collective decision making in bacterial viruses. Biophysical journal *95*, 2673-2680.
Wertheim, J.O., Chu, D.K., Peiris, J.S., Kosakovsky Pond, S.L., and Poon, L.L. (2013). A case for the ancient origin of coronaviruses. Journal of Virology *87*, 7039-7045.
Wertheim, J.O., and Kosakovsky Pond, S.L. (2011). Purifying selection can obscure the ancient age of viral lineages. Molecular Biology and Evolution *28*, 3355-3365.
White, P.S., Morales, F., and Roossinck, M.J. (1995). Interspecific reassortment of genomic segments in the evolution of cucumoviruses. Virology *207*, 334-337.
WHO (2014). About vector-borne diseases.
Wikel, S.K. (1999). Tick modulation of host immunity: an important factor in pathogen transmission. International Journal of Parasitology *29*, 851-859.
Wilke, C.O., Forster, R., and Novella, I.S. (2006). Quasispecies in time-dependent environments. Current topics in microbiology and immunology *299*, 33-50.
Yamanaka, A., Thongrungkiat, S., Ramasoota, P., and Konishi, E. (2013). Genetic and evolutionary analysis of cell-fusing agent virus based on Thai strains isolated in 2008 and 2012. Infection Genetics and Evolution *19*, 188-194.
Yoon, H.S., Price, D.C., Stepanauskas, R., Rajah, V.D., Sieracki, M.E., Wilson, W.H., Yang, E.C., Duffy, S., and Bhattacharya, D. (2011). Single-cell genomics reveals organismal interactions in uncultivated marine protists. Science *332*, 714-717.
Zacks, M.A., and Paessler, S. (2010). Encephalitis alphaviruses. Veterinary Microbiology *140*, 281-286.
Zanotto, P.M., Gao, G.F., Gritsun, T.S., Marin, M.S., Jian, W.R., Venugopal, K., Reid, H.W., and Gould, E.A. (1995). An arbovirus cline across the northern hemisphere. Virology *210*, 152-159.
Ziebell, H., Murphy, A.M., Groen, S.C., Tungadi, T., Westwood, J.H., Lewsey, M.G., Moulin, M., Kleczkowski, A., Smith, A.G., Stevens, M.*, et al.* (2011). Cucumber mosaic virus and its 2b RNA silencing suppressor modify plant-aphid interactions in tobacco. Scientific reports *1*, 187.

Ziebell, H., Payne, T., Berry, J.O., Walsh, J.A., and Carr, J.P. (2007). A cucumber mosaic virus mutant lacking the 2b counter-defence protein gene provides protection against wild-type strains. The Journal of general virology *88*, 2862-2871.

Zieler, H., Garon, C.F., Fischer, E.R., and Shahabuddin, M. (2000). A tubular network associated with the brush-border surface of the *Aedes aegypti* midgut: implications for pathogen transmission by mosquitoes. Journal of Experimental Biology *203*, 1599-1611.

Zou, G., Zhang, B., Lim, P.Y., Yuan, Z., Bernard, K.A., and Shi, P.Y. (2009). Exclusion of West Nile virus superinfection through RNA replication. J Virol *83*, 11765-11776.

Zwart, M.P., Daròs, J.A., and Elena, S.F. (2011). One is enough: *in vivo* effective population size is dose-dependent for a plant RNA virus. PLoS Pathog *7*, e1002122.

Zwart, M.P., Daròs, J.A., and Elena, S.F. (2012). Effects of potyvirus effective population size in inoculated leaves on viral accumulation and the onset of symptoms. J Virol *86*, 9737-9747.

4

Choose Your Weapons: Origins and Evolution of Innate Host Defenses and Viral Counterstrategies

Welkin E. Johnson

Biology Department, Boston College, Chestnut Hill, MA 02467, USA, welkin.johnson@bc.edu

Abstract

Viruses comprise the most abundant and genetically most diverse entities in the biosphere, and can be found in every environment and every niche occupied by cellular organisms. Viruses are common to all three domains of life, parasitizing the cells of eukaryotic, bacterial and archaeal hosts, and several lines of evidence suggest that viruses have been ever-present throughout the evolutionary history of modern, extant organisms. As a consequence of millions of years of virus-host coevolution, host organisms from all three domains have adapted by acquiring a variety of genes encoding dedicated antiviral effector mechanisms, and viruses, in turn, have evolved to encode an equally impressive array of accessory functions to counteract these defenses. Comparative approaches that draw on virus-host interactions from all three domains of life are beginning to reveal common macro-evolutionary processes by which these host and viral genes arise and evolve, and will hopefully lead to a fundamental understanding of how viruses influence host evolution (and vice versa).

Overview

For most of the first 100 years of virology as a scientific discipline, the functions of intrinsic host defenses and viral counter-defenses remained elusive or understudied. As an illustrative example, while it has long been appreciated that interferon treatment of cells induces an antiviral state (Isaacs and Lindenmann, 1957), more than 50 years on the effector functions of a majority of the interferon-stimulated genes (ISGs) still remain to be discovered (Pestka, 2007) – in fact, comprehensive lists of ISGs have only recently been produced, and these primarily for human cells and model organisms (Schoggins *et al.*, 2011). Likewise, despite the use of bacterial model systems by virologists for a century, the CRISPR/CAS systems of bacteria and archaea were only recently recognized as dynamic and rapidly evolving systems of antiviral defense (Horvath and

Barrangou, 2010). As the list of genes involved on both sides of the virus-host interaction grows, and as mechanistic details emerge, it is impossible to escape the conclusion that viruses must have had a major (and largely underappreciated) impact on the evolution of modern species. A comprehensive review of virus-host interactions that encompasses eukaryotic, bacterial, and archaeal viruses is beyond the scope of a single book chapter, but drawing on examples of viruses of all three domains serves to illustrate common themes and provides a starting point to discover fundamental principles that guide virus-host interactions on macro-evolutionary timescales.

Motive and Opportunity: Inferring the Impact of Viruses on Organismal Evolution

Four observations lead to the inescapable conclusion that viruses must have had a major and continuous impact throughout the evolutionary history of extant, living organisms: 1) viruses are plentiful in all environments and niches where cellular life is also found; 2) viral genetic diversity exceeds diversity represented by all living things from all three domains of life; 3) comparative phylogenetics and endogenous viral elements reveal that viruses have been present (and remained remarkably stable) across very deep evolutionary timescales; 4) the coupling of functional assays with quantitative measures of positive selection to study virus-host molecular interactions has uncovered many examples of sequence evolution consistent with long-term cycles of virus-host evolutionary conflict (so-called evolutionary "arms-races") playing out across tens-to-hundreds of millions of years (Emerman and Malik, 2010; Patel *et al.,* 2011).

Viral Abundance

Environmental sampling, metagenomic surveys, and genomics approaches to non-model organisms continue to reveal that viruses are abundant throughout nature, parasitizing cellular organisms from all three domains of life (Angly *et al.,* 2006; Avrani *et al.,* 2012; Edwards and Rohwer, 2005; Youle *et al.,* 2012). Among viruses currently classified by the International Committee for Taxonomy of Viruses (ICTV), the list of known plant viruses exceeds the list of all known vertebrate viruses (King *et al.,* 2012; Miller and Whitham, 2013). Some estimates put the number of bacteriophage particles in the biosphere at astronomical levels (on the order of $>10^{30}$) (Brüssow and Hendrix, 2002; Youle *et al.,* 2012). Metagenomic assays of the gut, of feces from humans and other organisms, and from various environmental samples typically yield up hundreds or thousands of unique viral genomes (Bench *et al.,* 2007; Breitbart *et al.,* 2004a; 2008; 2004b; Cann *et al.,* 2005; Culley *et al.,* 2003; 2006; Desnues *et al.,* 2008). Given that most organisms appear to harbor unique viruses, and that the majority of earth's species remain unidentified or unstudied, it is reasonable to predict that hundreds of thousands to millions of new viral species remain to be discovered (Anthony *et al.,* 2013).

Viral Diversity

While shared ancestral features unite all living cellular organisms into a single tree of life, attempts to lump all viruses together based on phylogenetic criteria may be pointless. Unlike cellular-based organisms, which uniformly share the use of double-stranded DNA as genomic material (among other shared, ancestral features), at least seven different genetic strategies are exhibited by viruses, only three of which involve double-stranded DNA in some form (these are RNA reverse-transcribing viruses, which produce a double-stranded DNA intermediate during replication; DNA reverse-transcribing viruses, which go through an RNA intermediate in producing progeny genomes; and double-stranded DNA viruses, whose replication mechanisms most closely parallel cellular mechanisms of DNA replication). All other viruses have replication cycles based on RNA genomes (positive-sense, negative sense, or ambisense) or single-stranded DNA genomes. It is difficult to conceive of a traditional, bifurcating phylogenetic tree that would relate all viruses to one another, and even attempts to reconstruct ancestral connections between different kinds of RNA viruses are highly speculative, at best (Holmes, 2009). In fact, a popular hypothesis is that the replication strategies of many viruses harken back several billion years to a pre-cellular world (Forterre, 2005; Holmes, 2009; Jalasvuori and Bamford, 2008; Koonin, 2009).

Thus, while all living things share the same basic genetic strategy, viruses represent a minimum of seven genetic systems (King *et al.,* 2012). This diversity of replication strategies has implications for the nature of the virus-host relationship, because each viral replication strategy implies the existence of unique ways to interact with and exploit cellular macromolecular processes to achieve production of progeny virus, and because features that distinguish viral replicative intermediates from cellular macromolecules provide a means for innate detection of non-self.

Viruses of the deep past: the endogenous fossil record

Among animal viruses, the retroviruses have long been known for their ability to leave behind large numbers of endogenous elements – DNA proviruses in this case – inserted in the germline DNA of their hosts (Gifford and Tristem, 2003; Weiss, 2006; Weiss *et al.,* 1999). Consequently, vertebrate genomes have hundreds of thousands of these endogenous retrovirus sequences (ERV) scattered throughout their chromosomes, and the ERV sequences themselves can be used as molecular "fossils" to recapitulate the natural history of the *Retroviridae* (Coffin, 2004; Gifford, 2012; Hayward *et al.,* 2014; Jern and Coffin, 2008). The formation of ERV is a direct result of reverse-transcription and integration of the provirus as part of the normal replication cycle of retroviruses – when infection occurs in a germline cell, a heritable insertion (the ERV) is formed. While other kinds of viruses do not undergo this same process, they have nonetheless occasionally deposited sequences in host germlines via other mechanisms, and examples can be found in a variety of different organisms (Feschotte and Gilbert, 2012; Holmes, 2011; Johnson, 2010).

These endogenous viral elements (EVE) provide a glimpse into the past and an opportunity to compare ancient viral sequences to modern, extant viral sequences. A striking result of such comparisons is how readily ancient viral sequences can be recognized as close relatives of modern viruses. In the case of retroviruses, phylogenetic trees based on the reverse-transcriptase protein (RT) show ERV sequences as interleaved among modern retroviruses, such that extinct retroviruses from millions of years in the past have clear relationships to extant genera in the *Retroviridae*. For example, the existence of ancient lentiviral ERV reveals that the modern HIV/AIDS pandemic in humans is just the most recent manifestation of a lineage that traces back millions of years (Gifford, 2012). The distribution of lentivirus-related ERVs differs from their modern descendants – to date, endogenous lentivirus sequences have been discovered in the genomes of rabbits, hares, weasels, ferrets, colugos, and lemurs (Cui and Holmes, 2012; Gifford *et al.*, 2008; Gilbert *et al.*, 2009; Han and Worobey, 2012; Hron *et al.*, 2014; Katzourakis *et al.*, 2007; van der Loo *et al.*, 2009), whereas modern lentiviruses are found in a variety of primates, small ungulates (such as sheep and goats), cows, horses and cats (Evans *et al.*, 2013). Importantly, these lentiviral ERV, which represent lentiviruses from several million years in the past, do not form a separate, basal clade within the lentivirus tree, but rather are interleaved with the modern lineages of lentiviruses (Gifford, 2012). Thus, despite birth and death of lineages and a prehistory rife with cross-species transmission events, the lentiviral genome and its core proteins cluster in phylogenetic trees as extinct relatives, not predecessors, of modern lentiviruses. Interestingly, the capsid proteins of two of these ancient lentiviruses (from rabbit genome and lemur genome) have been resurrected, expressed and characterized, and found to have strikingly similar structures and biological interactions to the capsid proteins of modern lentiviruses (Goldstone *et al.*, 2010).

It is important to note that the ERV/EVE record comprising hundreds of thousands of loci found in vertebrate genomes is probably a biased sampling, and that it represents a vast underestimate of the actual diversity and abundance of prehistoric viruses and their temporal distribution with respect to organismal evolution. Given the abundance of the “fossil” record (and the underrepresentation of many viruses in the record), it is clear that viruses were a constant presence throughout the evolution of modern organisms, and therefore viruses represent a major and often overlooked source of selective pressure.

The molecular evolutionary consequences of virus-host interactions

The influence of virus-host interactions on evolution of both has been described as “antagonistic”, usually with reference to the Red-Queen Hypothesis or as a special case of predator-prey dynamics (Nowak and May, 2000; Thrall *et al.*, 2012). Similarly, it has become increasingly common to describe cases of interacting virus and host proteins as being engaged in an evolutionary “arms-race” (Meyerson and Sawyer, 2011; Stern and Sorek, 2011; Takeuchi *et al.*, 2012). The arms-race scenario can be viewed as a special instance of the Red-Queen Hypothesis, in which reciprocal bouts of selection imposed by virus on host and by host on virus results in continuing cycles of diversification of both parties. However, in natural situations the arms-race scenario may be too simplistic (and other models have been proposed) (Avrani *et al.*, 201;Tack *et al.*, 2012). Despite ongoing debates over models, quantitative methods for identifying and measuring positive selection have proven useful for examining long-term evolution of protein-coding sequences directly involved in virus-host interactions. Codon-specific methods in particular are a powerful approach for identifying sites in proteins that have undergone episodic bouts of selection over timescales spanning multiple speciation events (i.e., millions of years) (Yang, 2005; 2006). The approach uses an alignment and an evolutionary model (a gene tree) to identify sites wherein fixation of nonsynonymous substitutions (i.e., those that alter the specified amino-acid) exceeds the level expected under neutrality (the latter being inferred from the level of synonymous change at the same site). The result is typically reported as a ratio, such as dN/dS (or as Ka/Ks or ω), where dN/dS>1 indicates positive selection (selection for change), dN/dS=1 reflects neutral evolution (random fixation), and dN/dS<1 indicates purifying selection (elimination of change; Hurst, 2002). The method can also be used to discern lineages (branches within a phylogeny) that have likely encountered positive selection acting on the gene of interest (Yang, 2006). The approach is best suited for studying situations where selection has been frequent across long stretches of evolutionary time. For more recent, species-specific selection, approaches based on examining evolution of the genomic landscape surrounding the gene of interest are becoming more common; and more practical, thanks to high-throughput sequencing methods and deep re-sequencing within species (Andersen *et al.,* 2012; Sabeti *et al.,* 2002).

As obligate intracellular parasites, all viruses engage in multiple interactions with host macromolecules and macromolecular complexes. With respect to viruses, host factors can be split into two categories, those with positive-acting effects (i.e., *cofactors* for viral replication) and those with negative-acting effects (i.e., *restriction factors*). Cofactors are host-encoded macromolecules with cellular functions, and as such are under strong purifying selection to maintain those functions. In contrast, restriction factors are host factors that have undergone natural selection for antiviral activity, and often display strong signals of positive selection (Emerman and Malik, 2010; Johnson, 2013; Patel *et al.,* 2011). Recent, detailed examination of structural interaction networks have provided atomic level comparisons of virus-protein/host-protein and host-protein/host-protein interactions, revealing that viral proteins often converge on the same binding domains engaged in host–protein/host-protein interactions, and that sites in host proteins that interact with viral proteins (including sites involved in both types of interactions) display levels of variation consistent with antagonistic co-evolution(Franzosa and Xia, 2011). Interestingly, viral proteins involved in these virus-host interactions display little or no homology with the host factors they mimic, and tend to involve smaller "footprints" on their targets than the corresponding cellular binding partner (Garamszegi *et al.,* 2013).

The *cofactor* and *restriction factor* categories are not mutually exclusive, and the distinctions can be blurred in some cases. For example, selection acting on the coding sequence of a viral cofactor can lead to fixation of resistance mutations in the cofactor gene, thereby creating a recessive genetic barrier to viral replication (that is, the variant cofactor is now a recessive-acting restriction factor). Perhaps the best documented examples of this phenomenon involve selection acting on cellular proteins that act as receptors for viral entry, and complementary counter-adaptations arising in viral glycoproteins (Coffin, 2013; Kozak, 2010). Recently, Demogines and colleagues reported a direct correlation between codons under positive selection in the mammalian transferrin receptor (TfR) gene and those TfR residues used by several unrelated viruses for which TfR functions as the primary cellular attachment receptor (Demogines *et al.,* 2013). The phenomenon is not exclusive to viral receptors, and in theory any host factor that has been coopted by a viral pathogen could be subject to positive selection. In plants, for example, many resistance mutations have been mapped to host translation initiation factors (e.g., eIF4E) that are hijacked by viral Vpg proteins for cap-independent translation (Miller and Whitham, 2013; Truniger and Aranda, 2009).

Likewise, it may be possible that restriction factors are sometimes used as cofactors by viruses. For example, it has been reported that the interferon-inducible tetherin protein can serve as an entry factor for cytomegalovirus (CMV) (Viswanathan *et al.*, 2011), and the ADAR enzymes (Adenosine Deaminases Acting on RNA), which are known to have antiviral activities, have also in some cases been implicated as cofactors for viral infection (Jayan, 2004; Samuel, 2011; Tomaselli *et al.*, 2014). Whether these represent rare exceptions, or whether examples of restriction factors being exploited as positive cofactors by viruses are common, remains to be seen.

Host Defense: some unifying themes

For each stage of the archetypal viral replication cycle (attachment, entry, replication, expression, assembly, and egress), there are numerous, corresponding examples of host factors that have evolved by natural selection to inhibit that particular stage. Whether all stages are equally exploited (or exploitable) in nature remains an open question, but three aspects are particularly well-covered in the scientific literature –1) resistance to viral entry (either due to selection of resistance mutations or expression of dominant-acting entry inhibitors), 2) recognition of viral nucleic acid (and subsequent inhibition of replication through direct or indirect mechanisms), and 3) self-destruction of the infected cell to prevent further spread of the virus. With respect to the impact of viruses on the evolution of host defense systems, it is notable that strikingly similar mechanisms have evolved independently in organisms representing all three of life's domains - the bacteria, archaea, and eukaryota.

Entry

All viruses must maintain an unbroken chain of transmission from one infected host cell to the next, and the first step in colonizing each cell in the chain of transmission is transporting the viral genome across the cellular membrane. As a defense strategy, inhibition of viral entry is well-documented in many unrelated virus-host systems, and examples can be grouped roughly into "recessive" mechanisms, such as receptor variants driven to fixation by selection for resistance to viral attachment, or phase variation in bacteriophage receptors (Kozak, 2010; Labrie *et al.*, 2010), and "dominant" mechanisms, such as expression of extracellular polymers by bacteria (Labrie *et al.*, 2010) or dedicated entry inhibitors of mammalian cells [e.g., the interferon-inducible IFITM proteins (Zhang *et al.*, 2012)]. Some of the best examples of host organisms co-opting viral genes involve endogenized viral glycoproteins that become host-expressed, dominant-acting inhibitors of viral entry (described in a later section).

Nucleic acid

Host defense systems that employ recognition of viral nucleic acid (genomes or replicative intermediates, DNA or RNA) are among the best-documented in the scientific literature, and examples can be found in host organisms from all three of life's domains (bacteria, archaea, and eukaryota). These include (but are by no means limited to) the very well characterized restriction-modification systems (RM) of bacteria and the CRISPR-CAS (Clustered regularly interspaced short palindromic repeats) systems found among many bacterial and archaeal organisms (Horvath and Barrangou, 2010; Labrie *et al.,* 2010); the interfering RNA (RNAi) systems of plants, invertebrates and vertebrates (Li *et al.,* 2013); epigenetic silencing systems found in all domains of life; the APOBEC3 DNA-editing enzymes and ADAR RNA-editing enzymes of mammals (Cullen, 2006; Samuel, 2012); and the various nucleic-acid sensing receptors of vertebrates (including, but not limited to, the Toll-Like Receptors (TLRs), RIG-I/MDA5, IFI16, cGAS, and others) (Hornung, 2014a; 2014b). That nucleic acid recognition is a dominant theme may derive from the fact that other viral macromolecules, such as proteins, glycoproteins and lipid membranes, are products of the host cell and are biochemically identical to "self", whereas many viruses can be detected by the host cells they colonize due to the unique nature of their nucleic acid products – this may be obvious for RNA viruses with unusual genome types (such as non-capped RNAs or double-stranded RNA intermediates), but can also be true for double stranded DNA viruses when DNA replication occurs in sub-cellular compartments other than the nucleus – for example, large cytoplasmic DNA viruses that undergo DNA replication in the cytoplasm.

Apoptosis and abortive infection

Although the underlying mechanisms are distinct, systems exist in both eukaryotes and bacteria for preventing viral spread via self-destruction of infected cells. In eukaryotic hosts, including vertebrates and invertebrates, this is represented by mechanisms related to programmed cell-death, or apoptosis, which can be triggered in response to viral infection (Benedict *et al.,* 2002). In bacteria, a set of related phenomena analogous to apoptosis go under the collective name Abortive Infection (Abi) (Labrie *et al.,* 2010; Samson *et al.,* 2013), and programmed cell death is also often part of the hypersensitive response (HR) to viral infections in plants (Mandadi and Scholthof, 2013). An important distinction is that apoptosis in multicellular organisms leads to destruction of the infected cell, thereby saving the host, whereas Abi in bacteria necessarily results in death of the single-celled host, and may therefore have evolved as a way to prevent spread to neighboring cells (on the surface, Abi has the appearance of altruism and evokes the possibility of evolution by group/kin selection). Not surprisingly, there are examples of viruses from all three domains that have acquired counterstrategies to delay or block programmed cell death via indirect or direct mechanisms. In fact, the first inhibitors of apoptosis (IAPs) were discovered in Baculoviruses, and homologs of these genes are found in at least seven other families of viruses that infect arthropod hosts/vectors (Clem, 2015).

Evolutionary origins of host restriction factors

Host organisms have acquired, across evolutionary timescales, a vast arsenal of genes that have evolved by natural selection to act as barriers to viral replication at the level of the individual cell. These represent intrinsic, or innate, defense functions (also sometimes called restriction factors, or in the case of plants, *R* [resistance] genes) that are distinct from the adaptive immunity of higher vertebrates, and unlike adaptive immunity, intrinsic defense mechanisms are widespread and abundant throughout all three domains of life and at all levels of complexity (unicellular organisms, organized microbial communities, and multicellular organisms of all shapes and sizes). A cursory comparison of some of these mechanisms reveals some common themes, and suggests directions for further exploration.

Restriction factors: generalists and specialists

At a minimum, obligate parasitism (by viruses) necessitates 1) entry into a host cell, 2) replication of viral nucleic acid, and 3) spread of viral progeny to additional host cells. Thus, it may not be surprising that the three host defense strategies described in the previous section (inhibiting viral entry, detecting and targeting viral nucleic acid, and auto-destruction of infected cells) - dominate the literature on antiviral defenses, as these are generalizable strategies that could, in theory, evolve by natural selection in response to many different kinds of virus-host interactions. However, viral genomes and replication mechanisms come in many different forms, including double-stranded and single-stranded DNA and RNA genomes (with RNA genomes further subdivided into sense, antisense, and ambisense genomes), circular and multipartite genomes, and reverse transcribing genomes, and viruses differ in many other aspects of replication, such as entry pathway (penetration versus membrane-fusion, entry at the plasma membrane versus endocytosis, clathrin-dependent endocytosis versus clathrin-independent mechanisms, etc), replication compartments (e.g., cytosolic versus nuclear), means of trafficking of viral proteins to sites of assembly, and release (e.g., lysis of host cells versus either budding from the plasma membrane or release via the secretory machinery). In theory, any of these biological distinctions between viruses can result in evolution of specialized host defense strategies uniquely adapted to the replication strategies of a particular order, family, genus, or species of virus. In this light, it is possible to group host-encoded restriction factors into 1) those with broad antiviral mechanisms and 2) those with specific mechanisms that target a particular subset of viruses (either related viruses, or viruses which share some aspect of their replication strategy).

There are too many examples of restriction mechanisms with very broad or general effects to mention here, but examples would include mechanisms that result in abortive infection through apoptosis of the infected cell, detection/ signaling pathways that induce shut-down of cellular protein translation, the restriction-modification systems of bacteria (which can target everything from bacteriophage to other forms of parasitic nucleic acid), and RNAi. In some cases, effector proteins target a particular property common to many (but not all) viruses – a premier example being the interferon-stimulated tetherin protein (also known as BST-2); tetherin is a membrane protein that effectively “traps” enveloped virions as they bud from the plasma membrane(Neil *et al.,* 2008; Van Damme *et al.,* 2008), and because tetherin's mechanism of action does not require recognition of a viral protein (Pérez-Caballero *et al.,* 2009), it has the potential to inhibit release of any enveloped virus (and as a consequence, many unrelated viruses have evolved separate mechanisms for overcoming tetherin) (Evans *et al.,* 2010).

Other mammalian restriction factors that appear to affect a broad range of viruses include the Mx proteins, which target post-entry replicative stages of many unrelated viruses (it is unclear if inhibition of different viruses by Mx proteins represents the same or distinct mechanisms), and the IFITM proteins, which collectively inhibit entry of several different kinds of virus by an as yet poorly defined mechanism (Verhelst *et al.,* 2013; Zhang *et al.,* 2012). While the

APOBEC3 DNA-editing enzymes are mostly known as inhibitors of retrovirus replication, and can restrict retroviruses from distinct genera, they also reportedly inhibit other reverse-transcribing viruses (e.g., Hepadnaviruses and retroelements– the common "target" being reverse transcription intermediates) (Chiu and Greene, 2008; Cullen, 2006; Harris and Liddament, 2004).

In contrast, there are also many examples of restriction factors that seem to have evolved specificity for a particular family or group of viruses. There are far too many examples to list here, but these include the cytosolic TRIM5 proteins of mammals which recognize retroviral capsid cores shortly after viral entry, and reportedly restrict retroviruses from multiple genera within the *Retroviridae* family (Johnson and Sawyer, 2009). Similarly, the *Flv* gene of mice encodes an OAS variant that inhibits replication of a diverse array of flaviruses (Perelygin *et al.*, 2002), and an RNAi screen of drosophila cells identified at least two proteins with pan-flavivirus inhibitory activity (Yasunaga *et al.*, 2014).

As the list of known restriction factors grows, an open area of inquiry is related to the evolutionary origins of restriction factor genes – by what mechanisms do they arise, and from what sources? In light of current evolutionary theory, it is possible that many restriction factor genes arise by a process of duplication and exaptation of other genes, which themselves could be genes encoding cellular factors with other, non-immune functions, modified versions of pre-existing genes encoding antiviral factors, or in some cases, viral genes captured by the host and coopted for antiviral function.

Gene duplication

In the 1970's, S. Ohno proposed that new genes originate by a process of gene duplication and functional divergence (Ohno, 2014). A requirement of such a mechanism is that the additional copies must rapidly acquire novel function, in order for selection to favor fixation of the new gene over loss (due to drift) (Bergthorsson *et al.*, 2007). Gene duplication has clearly played a role on both sides of the virus-host conflict, giving rise to novel host defense genes, as well as new accessory functions in viruses. Because genes encoding antiviral defenses are among the most rapidly evolving genes, it is not always possible to identify exact origins. Nonetheless, many can be grouped into protein families or superfamilies, and in some cases, structure-function similarities can be found that span all three domains of life. For example, the HEPN domain superfamily includes examples of bacterial, archaeal and eukaryotic antiviral defense proteins and a significant association with mechanisms involving recognition and destruction of viral nucleic acid, toxin-antitoxin systems, and ABI, among many others (Anantharaman *et al.*, 2013).

Functions for most of the 100 or more vertebrate TRIM (Tri-partite motif) proteins are still unknown, but several have been shown to function as components of innate/intrinsic immunity, including TRIM5, TRIM19, TRIM25, and TRIM28, and many TRIM loci display signs of positive selection consistent with roles in host defense (Boudinot *et al.*, 2014; Malfavon-Borja *et al.*, 2013a; Nisole *et al.*, 2005; Sawyer *et al.*, 2005; Towers, 2005; van der Aa *et al.*, 2012). The canonical TRIM motif, beginning at the N-terminus, comprises a RING domain, a B-Box, and a coiled-coil, and members with this arrangement are found in a wide variety of species (Boudinot *et al.*, 2014; Nisole *et al.*, 2005; van der Aa *et al.*, 2012; Zhang *et al.*, 2014). The C-terminal domains of TRIM proteins are variable, and include B30.2/SPRY, PHD, Fibronection III and CypA domains, among others (Nisole *et al.*, 2005). This modularity suggests that the TRIM proteins may share common mechanistic properties conferred by the tripartite motif, but that the unique functional interactions of each TRIM map to the C-terminal domains. Interestingly, the antiretroviral TRIM5 locus, which normally encodes a C-terminal PRYSPRY domain, has thrice independently acquired a new exon encoding a C-terminal cyclophilin-A (CypA) domain - in each case, the CypA exon arose by LINE-mediated retrotransposition of a CypA coding sequence, once in old world monkeys (Asian macaques), once in new world monkeys (owl monkeys) and once in tree shrews (Brennan *et al.*, 2008; Liao *et al.*, 2007; Mu *et al.*, 2014; Newman *et al.*, 2008; Nisole *et al.*, 2004; Sayah *et al.*, 2004; Virgen *et al.*, 2008; Wilson *et al.*, 2008). At least one study suggests that TRIM-CypA fusions may have occurred multiple times during evolution of old world primates (Malfavon-Borja *et al.*, 2013b). Thus, the combination of gene expansion followed by neofunctionalization (by sequence divergence as well as exon-shuffling and exon-capture) may be a common pathway for formation of novel TRIM genes (Han *et al.*, 2011; Sawyer *et al.*, 2007; Tareen *et al.*, 2009).

An exhaustive list is well beyond the scope of this chapter, but just a few diverse examples (in addition to those described above) make it clear that Ohno's hypothesis applies widely to the origins and evolution of gene families with antiviral activity. For example, the KIR (Killer Inhibitory Receptor) loci in primates are known to undergo rapid evolution, including expansions and contractions in copy number (Bashirova *et al.*, 2006). Comparative genomics reveals that the same is true of many innate immune effectors in Drosophila (Sackton *et al.*, 2007), an ancient series of duplication events produced the present arrays of Peptidoglycan Recognition Protein (PGRP) genes in vertebrates and invertebrates (Montaño *et al.*, 2011), and similar processes gave rise to the interferon-inducible Mx and IFITM proteins (Verhelst *et al.*, 2013; Zhang *et al.*, 2012). Many key enzymes involved in nucleic acid recognition/destruction also come in clusters of related genes, including those that encode the ADARS (Samuel, 2011), the Dicer proteins involved in RNAi-related inhibitory mechanisms of organisms as diverse as arthropods and vertebrates (Mukherjee *et al.*, 2012), and the mammalian APOBEC3 proteins (Harris and Liddament, 2004; Münk *et al.*, 2012). Thus, gene duplication is a major source of novel innate/intrinsic immunity effectors and restriction factors throughout the Tree of Life, providing a rich source of examples for elucidating fundamental principles related to the origins and evolution of new genes.

Exaptation of viral proteins

In 1982, Vrba and Gould proposed use of the term exaptation when referring to characters (including genes) that originally evolved for one function and have been repurposed by evolution, even going so far as to posit that "junk" DNA in organismal genomes might amount to a rich source of exapted function (Gould and Vrba, 1982). As described in an earlier section, retroviruses are especially prone to contributing sequences to host cell genomes, and the germlines of almost all vertebrate species contain tens to hundreds of thousands of endogenous retrovirus (ERV) loci. Given this abundance, it may not be surprising that virologists have uncovered a number of examples wherein ERV sequences – former viral genes – have been exapted in service of the host organism. While the best documented examples are the syncytin proteins, which are ERV-derived fusion proteins involved in placental development (Lavialle *et al.,* 2013), most instances of the phenomenon involve defense of the cell against infection by exogenous retroviruses (Jern and Coffin, 2008). In such cases, the function of the ERV-encoded factor is often related to its former function as a viral protein. For example, the *Fv4* locus found in some species of mice encodes a glycoprotein derived from an endogenous retrovirus, and expression of this endogenous version interferes with infection by exogenous viruses that use the same cell-surface receptor (Gardner *et al.,* 1991). Similar mechanisms have been attributed to other mouse ERV loci(Jung *et al.,* 2002; Wu *et al.,* 2005), as well as endogenized viral envelope proteins in the genomes of chickens and cats (Ito *et al.,* 2013; Robinson *et al.,* 1981). There are also examples of viral structural proteins encoded by ERV in mice (*Fv1* locus) and sheep (*enJSRV* loci) that function as inhibitors of early and late steps of the retroviral replication cycle, respectively (Arnaud *et al.,* 2007; Best *et al.,* 1996). The *sag* gene of mouse mammary tumor viruses (MMTV) encodes a superantigen that normally augments viral replication through the expansion of non-specific T-lymphocytes in the infected host; expression of endogenized MMTV loci encoding *sag* genes have the effect of deleting specific T-cell subsets in the host (thereby interfering with exogenous infection by strains with the same Sag specificity) (Coffin, 1992).

In addition to endogenized retrovirus sequences, the phenomenon may extend to endogenous viral elements derived from other kinds of viruses, although these have not been as well studied. A recent example is the endogenous Bornavirus-like elements (EBLs) (Horie *et al.,* 2010; 2013). EBLs with intact open reading frames and high sequence identity to the nucleoproteins of exogenous bornaviruses (called EBLNs) are found in the genome of the thirteen-lined ground squirrel, and expression of these elements in cell-culture inhibit Borna disease virus (BDV) replication (Fujino *et al.,* 2014; Horie *et al.,* 2010).

The CRISPR/Cas system is often described as a bacterial counterpart to adaptive immunity in vertebrates (Horvath and Barrangou, 2010). In the sense that the CRIPSR-Cas system "records" past encounters and can mount a more effective defense, the analogy holds. However, the CRISPR-Cas system differs from adaptive immunity in being heritable (the archived spacers that provide specificity are replicated as part of the cellular genome and inherited by subsequent generations, which is not true for B- and T-cell receptors involved in adaptive immunity). Moreover, adaptive immunity in vertebrates is based on chance complementarity – the selection and amplification of cells bearing rare B-cell and T-cell receptors that have affinity for viral antigens. In contrast, the CRISPR-Cas system works by archiving viral sequences (and sequences from other forms of parasitic nucleic acid), and using these sequences for homology-guided identification and destruction of the same or similar viruses in future encounters. In this regard, the CRISPR-Cas mechanism can be viewed as a dedicated exaptation system that actively captures viral sequences and turns them into components of host antiviral defense.

Origins and functions of viral accessory proteins

In response to dominant-acting restrictions imposed by host cells, viruses can counter-adapt through 1) selection of resistance mutations in the targeted molecules, or 2) through expression of accessory proteins dedicated to antagonizing or circumventing host restrictions. The two are not mutually exclusive, although the former can be viewed primarily as a microevolutionary process occurring at the level of the individual infected host or host population, while the latter reflects macroevolutionary processes giving rise to new viral lineages with new biological properties. The use of dedicated antagonists may be necessitated by host restrictions factors that function in a general way rather than by targeting a specific viral protein. For example, because the interferon-inducible tetherin protein functions by "trapping" enveloped virions on the cell surface, and not by binding to any known viral protein, many enveloped viruses have evolved specific antagonists that bind tetherin directly and either downmodulate it from the cell-surface or induce its degradation (Evans *et al.,* 2010).

Just as we now understand that hosts of all kinds have genes dedicated specifically to defending cells against hijacking by viruses, we also have come to appreciate that many viruses dedicate a significant proportion of their limited genome space to encoding accessory functions. As sampling of diversity within viral families and genera has increased, one pattern that has emerged is that related viruses (e.g., viruses of the same *Family* or *Genus*) share a core set of canonical genes involved in the basic mode of replication – these include virally-encoded polymerases and essential structural proteins, for example – but differ widely in the distribution and functions of non-canonical proteins embedded within polyproteins or encoded by stand-alone genes. Although these are sometimes called non-structural proteins (often abbreviated as Ns), it is sometimes the case that viral structural proteins will also encode additional subdomains with accessory functions, so I will use the broader term "accessory genes" to encompass all possibilities. In many cases, accessory genes are among the most variable coding sequences found in viral genomes, displaying divergence in both sequence as well as overall distribution of homologs among related viral lineages. Importantly, viral genomes are typically orders of magnitude smaller than those of their hosts, which is likely to impose unique constraints on the *de novo* acquisition and subsequent evolution of accessory functions.

In cases where the functions of accessory sequences have been examined in detail, it is often the case that these genes are dispensable for the inherent replicative functions of the virus, as measured in cell-culture or model systems, yet are nevertheless found to be necessary for optimal replication and spread of the virus *in vivo* [e.g., see (Gibbs *et al.,* 1995; Koff *et al.,* 2006)]. In many cases the functions of accessory genes remain to be discovered, except for some of the better studied viruses (such as human or agricultural pathogens or viruses of model organisms) where a growing number of accessory genes have been found to modify or antagonizing the innate defense systems of their host cells. Due to the intense focus on understanding the origins and spread of the human immunodeficiency viruses (HIV), much is known about the accessory genes of the primate lentiviruses (i.e., HIV-1, HIV-2, and the related simian immunodeficiency viruses, or SIVs), making these viruses useful for illustrating accessory protein diversity. For example, while all the primate lentiviruses share the core genes common to all retroviruses (*gag*, *pro*, *pol*, *env*), they also encode a variety of unique accessory genes that are not found in other kinds of retroviruses (e.g. *vif*, whose product antagonizes the antiviral APOBEC3 enzymes, is unique to lentiviruses), or even in other retroviruses of the lentivirus genus (e.g., *vpu*, *vpr*, *vpx* and *nef* are exclusive to the primate lentivirus clade, and are not found in non-primate lentiviruses) (Evans *et al.*, 2013). Even within the primate lentivirus clade, the accessory genes are not evenly distributed – for example, a subset of viruses (including HIV-1) have a *vpu* gene, but lack a *vpx* gene, while other viruses (such as HIV-2) have *vpx* but not *vpu*; and yet other primate lentiviruses lack both *vpu* and *vpx* (Evans *et al.*, 2013).

Because viruses interact extensively with the macromolecular machinery of the host cell, the genetic distance between reservoir and spillover hosts may influence the likelihood of an exposure leading to interspecies transmission and viral emergence (Holmes and Drummond, 2007). While virologists have traditionally focused on the role of conserved core functions (such as receptor binding and entry) as key determinants of host-range, it is likely that ability to overcome intrinsic/innate effector systems (directly or through acquisition of adaptive changes) is also a critical determinant of spillover transmission and emergence of viruses with new host-ranges. In fact, because intrinsic/innate antiviral systems harbor some of the most divergent host genes (due to long-term positive selection) (Johnson, 2013; Meyerson and Sawyer, 2011), it may be that adaptation to overcome host defenses during spillover transmission and emergence is as important as adaptation to accommodate other kinds of virus-host interactions (e.g., binding to entry receptors). A comparison of homologous protein sequences between different primate lentiviruses also reveals that the accessory protein sequences are noticeably more divergent than the core replicative proteins, suggesting that they may be important players in cross-species transmission and adaptation to new hosts (McCarthy and Johnson, 2014). Consistent with this idea, recent studies focusing on emergence of primate lentiviruses have revealed a role for adaptations in the Vif protein to antagonize APOBEC3 function in the new host (D'arc *et al.*, 2015; Krupp *et al.*, 2013) and that emergence of HIV in humans likely required changes to overcome restriction by the human ortholog of the IFN-inducible restriction factor tetherin/BST-2 (Sauter *et al.*, 2009).

Accessory functions are a common feature of many viruses, regardless of the type of host organism (eukaryotic, bacterial or archaeal; unicellular or multicellular), including both DNA and RNA viruses. Just as host defenses can either be general (i.e., RNAi, Restriction-Modification, Apoptosis) or specific to a particular virus or viral process, viral accessory proteins can work broadly, by inhibiting innate sensing and signaling pathways, or specifically, by antagonizing a particular effector protein or pattern-recognition molecule. Examples of the former include the Viral Suppressors of RNAi (VSRs) found in many plant viruses. VSR is a collective term that includes many different viral proteins that implement a variety of mechanisms to interfere with RNAi – for example, some VSRs function by binding or cleaving the small RNA intermediates that feed into the silencing machinery, while others have evolved to bind and antagonize host proteins that mediate different steps in the RNAi pathway or the systemic propagation of the RNAi signal (Li and Ding, 2006; Li *et al.*, 2013; Miller and Whitham, 2013). Similarly, viral inhibitors of apopotosis (vIAPs), first discovered in insect viruses, include viral proteins that disrupt apoptotic signal transduction, as well as proteins that interfere with downstream cellular caspases that initiate and effect the apoptotic response (Clem, 2015; Friesen, 2013). As with VSRs, the IAPs include a broad range of proteins that differ among viruses and function by different mechanisms. Other examples of generalized antagonism of host defenses include the V proteins of paramyxoviruses, which collectively interfere with innate responses to infection by targeting upstream steps in the IFN and STAT signaling pathways (Horvath, 2004). Despite being closely related to one another, two well-studied Rhabdoviruses - rabies virus (RV) and vesicular stomatitis virus (VSV) - use different proteins and different mechanisms to interfere with interferon signaling by the infected cell (Rieder and Conzelmann, 2009).

In contrast to accessory proteins that subvert signaling pathways or general antiviral processes, there are also accessory proteins that work by directly targeting specific effector proteins. For example, at least two kinds of retroviruses, the lentiviruses and foamyviruses, encode antagonists of the APOBEC3 enzymes (using the Vif and Bet proteins, respectively) (Löchelt *et al.,* 2005; Malim, 2009; Russell *et al.,* 2005), and a subset of lentiviruses encode a dedicated antagonist of SAMHD-1 (the Vpx accessory protein) (Laguette and Benkirane, 2012). The P0 protein of a plant virus (Potato leaf roll virus) uses a mechanism analogous to that used by Vif to degrade Apobec3 to promote degradation of AGO (Argonaut - the effector component of RNAi in plants), and antagonists of AGO have also been reported in insect viruses (Miller and Whitham, 2013; van Mierlo *et al.,* 2012). Primate Tetherin proteins are directly antagonized by at least two accessory proteins encoded by lentiviruses (Vpu and Nef) and by the K5 protein of the KSHV gammaherpesvirus (Evans *et al.,* 2010; Neil and Bieniasz, 2009). As more effector functions of hosts are identified, the expectation is that additional examples of effector-specific viral antagonists will also be discovered.

All of these examples serve to illustrate the plastic nature of accessory functions in viruses – even viruses of the same family or genus will differ in their complement of accessory functions (for example, not all plant viruses encode the same VSRs or use the same mechanisms to overcome RNAi, and not all IAPs are shared equally among insect viruses or work by the same mechanism), and when orthologs of accessory genes are shared by multiple viral species, they are often among the most divergent in sequence. These observations strongly argue that the accessory functions of viruses are major sources of phenotypic diversity of viruses – to put it another way, while the canonical genes that define a viral family are shared by most or all of its member species, the accessory functions are what give each viral species or lineage its unique biological properties.

Origin of viral accessory functions by gene capture and exaptation

Just as viral sequences can sometimes be captured and exapted for host cell functions, many viruses clearly have acquired novel accessory genes from their hosts. Perhaps the first reported example of a captured host gene in a replicating virus was the viral *Src* gene (v-Src) of an avian retrovirus, Rous sarcoma virus (RSV) (Stehelin *et al.,* 1976). However, the phenomenon of gene-capture is most apparent in large DNA viruses (e.g., poxviruses and herpesviruses), which typically encode dozens of genes with detectable homology to cellular genes, particularly genes encoding cytokines, chemokines and the related receptors (Alcami, 2003; Elde *et al.,* 2008). Bacteriophage likewise are known to carry host-derived genes, including genes that influence host quorum-sensing pathways (Hargreaves *et al.,* 2014a; 2014b); whether any of these serve to counteract host defensive strategies remains to be determined. Possibly the richest and most varied source of gene-capture examples are the recently discovered giant viruses of amoebae, whose large genomes (often ≥1,000Kb) include numerous open reading frames with similarity to genes found in bacteria, archaea and eukaryotes; the conglomeration of diverse genes in these viruses may reflect the fact that their amoebic hosts interact with or prey upon bacteria and archaea (Colson and Raoult, 2010).

Due to rapid evolution of viruses, many accessory genes are not readily recognizable as the products of gene capture, although structural/functional homology to families of cellular proteins can sometimes be deduced – for example, cellular genes with functional and structural homology to viral Inhibitors of Apoptosis (vIAPs) encoded by insect baculoviruses have been described, suggesting that some of these viral IAPs may have originated by capture of host IAP genes (which subsequently acquired virus-specific adaptations) (Cerio *et al.,* 2010). Similarly, a variety of unrelated viruses, including both RNA and DNA viruses, have genes predicted or known to encode dUTPases; these likely represent parallel, independent acquisition of dUTPase-coding sequences (Baldo and McClure, 1999). The Vpu protein of HIV and a subset of the related primate lentiviruses serves to antagonize the antiviral function of the IFN-inducible tetherin protein (Neil *et al.,* 2008) - Vpu is a very small Type I membrane protein (~16kd) that is predicted to oligomerize to form an ion channel/porin. Such proteins have been found in a variety of other viruses, including the M2 protein of Influenza viruses, and are sometimes called viroporins (Nieva *et al.,* 2012). Viroporins may represent ancient capture of host porin-encoding genes whose specific origins are no longer detectable by standard homology searches due to extensive evolution as viral genes.

A recent report described a remarkable, clear case of *bona fide* gene capture involving the CRISPR antiviral system of bacteria – in this particular case, a bacteriophage of *Vibrio cholerae* was discovered to have its own CRISPR/Cas system, which it uses to target an antiviral locus of its bacterial host using the same sequence-identity mechanism that host CRISPRs use to target viral genomes (Seed *et al.,* 2013). Moreover, the authors reported that the phage-encoded CRISPR/Cas system was capable of “re-adapting” (under experimental conditions) via de novo acquisition of new host-derived spacers.

Mimicry

The term *molecular mimicry* has been applied to viral genes in a number of different forms, including genes captured (exapted) from the host, such as the myriad cytokine and chemokine modifying proteins encoded by poxviruses and herpesviruses (see previous section), as well as genes whose interactions with host proteins may represent true cases of convergent evolution to produce functional mimicry (Alcami, 2003; Barber and Elde, 2013). The difficulty in identifying the origins of many genes involved in mimicry lies in the rapid evolution of viral genomes – while some mimics are clearly “captured” homologs of host proteins (with sequence identity to cellular homologs and mechanisms of action derived directly from the original cellular functions), others have little or no sequence similarity to the factors they mimic. For the latter, it may be difficult or impossible to determine whether a gene represents mimicry by convergence, or if time has simply obscured shared ancestry with a cellular homolog. One (of many) possible examples of molecular mimicry involves anti-RNAi proteins of some plant viruses, such as the P1 protein of sweet potato mottle virus and the p38 protein of TCV – both proteins have glycine-tryptophan (GW) motifs that interact with AGO, mimicking host proteins that are involved in RISC assembly (via AGO) using similar motifs (Miller and Whitham, 2013). One of the most striking examples of true mimicry is the Ocr protein of bacteriophage T7, which functions to thwart restriction enzymes of the *e.coli* host. Structural analysis revealed that dimers of Ocr bear a striking physical resemblance to double-stranded DNA – about the equivalent of two dozen base-pairs in length – suggesting that Ocr acts as a molecular decoy for restriction endonuclease binding (Walkinshaw *et al.,* 2002) (Cox, 2002).

Gene duplication and multi-functional proteins

There are many clear cases of gene duplication in viruses, particularly in large DNA viruses containing expanded copy numbers of accessory genes (Colson and Raoult, 2010; Moss, 2013; Pellet and Roizman, 2013). Using a cell-culture model of directed evolution, Elde *et al.* demonstrated that poxviruses can undergo rapid adaptation by a process that involves expansion and contraction of viral genes (Elde *et al.,* 2012), and analysis of myxomavirus (MYXV) evolution in Australian rabbit populations reveals multiple examples of adaptation by duplication of viral genes (Kerr *et al.,* 2013a; 2013b). Gene duplication has also been seen in viruses with smaller RNA genomes, such as the *vpx-vpr* gene pair found in HIV-2 and related SIVs (Tristem *et al.,* 1992). It is also common for viral proteins to encode multiple, often independent functions, and in some cases viral proteins with conserved roles in replication (i.e. canonical proteins) also double as accessory proteins – for example, the matrix (M) protein of Vesicular Stomatitis Virus (VSV) is also involved in antagonizing interferon responses (Rieder and Conzelmann, 2009), and the hepacivirus NS3-4a protease complexes, which are essential for processing of the viral polyprotein, also cleave the host MAVS protein and disrupt innate immune signaling (Morikawa *et al.,* 2011; Patel *et al.,* 2012). Given the strong conservation of essential functions of these proteins, the accessory functions are likely to represent secondary adaptations acquired independently. While the use of multi-functional proteins may be a consequence of the smaller genome sizes of viruses (genome economy), the combining of distinct functions in single proteins may also impose additional constraints on evolution of the protein sequences.

Conclusions and questions

It is evident that viruses have been a constant and abundant presence throughout the evolutionary history leading to modern organisms, and that viruses influence the evolution of their hosts as agents of natural selection and as sources of genetic novelty. It is also increasingly clear that host organisms from all three of life's domains have adapted by acquiring a wide spectrum of genes dedicated to suppressing or preventing viral replication at the level of the cell, and that viruses, in turn, encode an equally impressive array of genes to counteract these innate/ intrinsic host defenses. Comparative approaches that incorporate knowledge from many systems, including model and non-model systems from the eukaryotic, bacterial, and archaeal realms, can reveal the macroevolutionary processes that give rise to novel genes (host genes and virus genes) at the heart of this evolutionary conflict, and may lead to elucidation of fundamental principles related to virus-host interactions as they unfold across evolutionary timescales. Some open questions for further investigation include (but are by no means limited to):

- How significant has the impact of viruses (as a selective force) been on the evolution of modern organisms?
- Can a better understanding of virus-host coevolution and the genes involved lead to better models for understanding interspecies transmission and emergence of viral pathogens?
- To what extent have endogenous retroviruses and other endogenous viral elements contributed novel genes to the innate/intrinsic defense repertoire of their hosts?
- In defense of the host cell, have some viral processes been exploited more often than others? Can nature thereby serve as a guide to the rational development of more effective, broadly acting antiviral drugs?
- Which host processes are most likely to be targeted by viral accessory functions (and do virus accessory proteins tend to converge on similar or identical targets)?
- When new genes with novel antiviral functions are born, how do they become networked into existing regulatory systems (i.e., how do coordinated, multigenic regulatory systems, such as the IFN response, evolve)?
- To what extent does genomic context constrain evolution of viral accessory genes?
- How do viruses acquire new genes, and how do these become embedded in existing viral genomes and polyproteins (and is it a one-step or multi-step process)?
- Does the birth of new accessory genes correlate with host-switching events?

Acknowledgements

I am grateful to all members of my laboratory for their patience during the writing of this chapter, and to A. Kirmaier, J. Henzy and the students in BIOL807 (Fall 2014) for helpful conversations on the topic of virus-host co-evolution. I sincerely apologize to all those hard working virologists whose contributions could not be mentioned specifically (due to limitations on space). Work in my laboratory is supported by Boston College and by grants from the NIH (AI095092 and AI083118).

References

Alcami, A. (2003). Viral mimicry of cytokines, chemokines and their receptors. Nat Rev Immunol *3*, 36–50.

Anantharaman, V., Makarova, K.S., Burroughs, A.M., Koonin, E.V., and Aravind, L. (2013). Comprehensive analysis of the HEPN superfamily: identification of novel roles in intra-genomic conflicts, defense, pathogenesis and RNA processing. Biol. Direct *8*, 15.

Andersen, K.G., Shylakhter, I., Tabrizi, S., Grossman, S.R., Happi, C.T., and Sabeti, P.C. (2012). Genome-wide scans provide evidence for positive selection of genes implicated in Lassa fever. Philos Trans R Soc Lond B Biol Sci *367*, 868–877.

Angly, F.E., Felts, B., Breitbart, M., Salamon, P., Edwards, R.A., Carlson, C., Chan, A.M., Haynes, M., Kelley, S., Liu, H., et al. (2006). The marine viromes of four oceanic regions. PLoS Biol. *4*, e368.

Anthony, S.J., Epstein, J.H., Murray, K.A., Navarrete-Macias, I., Zambrana-Torrelio, C.M., Solovyov, A., Ojeda-Flores, R., Arrigo, N.C., Islam, A., Ali Khan, S., et al. (2013). A strategy to estimate unknown viral diversity in mammals. MBio *4*, e00598–13.

Arnaud, F., Murcia, P.R., and Palmarini, M. (2007). Mechanisms of Late Restriction Induced by an Endogenous Retrovirus. J. Virol. *81*, 11441–11451.

Avrani, S., Schwartz, D.A., and Lindell, D. (2012). Virus-host swinging party in the oceans: Incorporating biological complexity into paradigms of antagonistic coexistence. Mob Genet Elements *2*, 88–95.

Baldo, A.M., and McClure, M.A. (1999). Evolution and horizontal transfer of dUTPase-encoding genes in viruses and their hosts. J. Virol. *73*, 7710–7721.

Barber, M.F., and Elde, N.C. (2013). Evolutionary biology: Mimicry all the way down. Nature *501*, 38–39.

Bashirova, A., Martin, M., McVicar, D., and Carrington, M. (2006). The Killer Immunoglobulin-Like Receptor Gene Cluster: Tuning the Genome for Defense (*). Annu Rev Genomics Hum Genet *7*, 277–300.

Bench, S.R., Hanson, T.E., Williamson, K.E., Ghosh, D., Radosovich, M., Wang, K., and Wommack, K.E. (2007). Metagenomic characterization of Chesapeake Bay virioplankton. Appl. Environ. Microbiol. *73*, 7629–7641.

Benedict, C.A., Norris, P.S., and Ware, C.F. (2002). To kill or be killed: viral evasion of apoptosis. Nat Immunol *3*, 1013–1018.

Bergthorsson, U., Andersson, D.I., and Roth, J.R. (2007). Ohno's dilemma: evolution of new genes under continuous selection. Proc. Natl. Acad. Sci. U.S.a. *104*, 17004–17009.

Best, S., Tissier, P.L., Towers, G., and Stoye, J.P. (1996). Positional cloning of the mouse retrovirus restriction gene Fvl. Nature *382*, 826–829.

Boudinot, P., Zou, J., Ota, T., Buonocore, F., Scapigliati, G., Canapa, A., Cannon, J., Litman, G., and Hansen, J.D. (2014). A tetrapod-like repertoire of innate immune receptors and effectors for coelacanths. J. Exp. Zool. B Mol. Dev. Evol. *322*, 415–437.

Breitbart, M., Felts, B., Kelley, S., Mahaffy, J.M., Nulton, J., Salamon, P., and Rohwer, F. (2004a). Diversity and population structure of a near-shore marine-sediment viral community. Proc. Biol. Sci. *271*, 565–574.

Breitbart, M., Haynes, M., Kelley, S., Angly, F., Edwards, R.A., Felts, B., Mahaffy, J.M., Mueller, J., Nulton, J., Rayhawk, S., et al. (2008). Viral diversity and dynamics in an infant gut. Res. Microbiol. *159*, 367–373.

Breitbart, M., Wegley, L., Leeds, S., Schoenfeld, T., and Rohwer, F. (2004b). Phage community dynamics in hot springs. Appl. Environ. Microbiol. *70*, 1633–1640.

Brennan, G., Kozyrev, Y., and Hu, S.L. (2008). From the Cover: TRIMCyp expression in Old World primates Macaca nemestrina and Macaca fascicularis. Proceedings of the National Academy of Sciences *105*, 3569–3574.

Brüssow, H., and Hendrix, R.W. (2002). Phage genomics: small is beautiful. Cell *108*, 13–16.

Cann, A.J., Fandrich, S.E., and Heaphy, S. (2005). Analysis of the virus population present in equine faeces indicates the presence of hundreds of uncharacterized virus genomes. Virus Genes *30*, 151–156.

Cerio, R.J., Vandergaast, R., and Friesen, P.D. (2010). Host insect inhibitor-of-apoptosis SfIAP functionally replaces baculovirus IAP but is differentially regulated by Its N-terminal leader. J. Virol. *84*, 11448–11460.

Chiu, Y.-L., and Greene, W.C. (2008). The APOBEC3 cytidine deaminases: an innate defensive network opposing exogenous retroviruses and endogenous retroelements. Annu Rev Immunol *26*, 317–353.
Clem, R.J. (2015). Viral IAPs, then and now. Semin. Cell Dev. Biol.
Coffin, J. (2004). Evolution of retroviruses: fossils in our DNA. Proc Am Philos Soc *148*, 264–280.
Coffin, J.M. (1992). Superantigens and endogenous retroviruses: a confluence of puzzles. Science *255*, 411–413.
Coffin, J.M. (2013). Virions at the gates: receptors and the host-virus arms race. PLoS Biol. *11*, e1001574.
Colson, P., and Raoult, D. (2010). Gene repertoire of amoeba-associated giant viruses. Intervirology *53*, 330–343.
Cox, E.H. (2002). Picture story. A DNA look-alike. Nat. Struct. Biol. *9*, 163.
Cui, J., and Holmes, E.C. (2012). Endogenous lentiviruses in the ferret genome. J. Virol. *86*, 3383–3385.
Cullen, B.R. (2006). Role and mechanism of action of the APOBEC3 family of antiretroviral resistance factors. J. Virol. *80*, 1067–1076.
Culley, A.I., Lang, A.S., and Suttle, C.A. (2003). High diversity of unknown picorna-like viruses in the sea. Nature *424*, 1054–1057.
Culley, A.I., Lang, A.S., and Suttle, C.A. (2006). Metagenomic analysis of coastal RNA virus communities. Science *312*, 1795–1798.
D'arc, M., Ayouba, A., Esteban, A., Learn, G.H., Boué, V., Liegeois, F., Etienne, L., Tagg, N., Leendertz, F.H., Boesch, C., et al. (2015). Origin of the HIV-1 group O epidemic in western lowland gorillas. Proceedings of the National Academy of Sciences.
Demogines, A., Abraham, J., Choe, H., Farzan, M., and Sawyer, S.L. (2013). Dual host-virus arms races shape an essential housekeeping protein. PLoS Biol. *11*, e1001571.
Desnues, C., Rodriguez-Brito, B., Rayhawk, S., Kelley, S., Tran, T., Haynes, M., Liu, H., Furlan, M., Wegley, L., Chau, B., et al. (2008). Biodiversity and biogeography of phages in modern stromatolites and thrombolites. Nature *452*, 340–343.
Edwards, R.A., and Rohwer, F. (2005). Viral metagenomics. Nat Rev Micro *3*, 504–510.
Elde, N.C., Child, S.J., Eickbush, M.T., Kitzman, J.O., Rogers, K.S., Shendure, J., Geballe, A.P., and Malik, H.S. (2012). Poxviruses deploy genomic accordions to adapt rapidly against host antiviral defenses. Cell *150*, 831–841.
Elde, N.C., Child, S.J., Geballe, A.P., and Malik, H.S. (2008). Protein kinase R reveals an evolutionary model for defeating viral mimicry. Nature *457*, 485–489.
Emerman, M., and Malik, H.S. (2010). Paleovirology--modern consequences of ancient viruses. PLoS Biol. *8*, e1000301.
Evans, D.T., Elder, J.H., and Desrosiers, R.C. (2013). Nonhuman lentiviruses. In Field's Virology, D. Knipe, and P. howley, eds. pp. 1584–1612.
Evans, D.T., Serra-Moreno, R., Singh, R.K., and Guatelli, J.C. (2010). BST-2/tetherin: a new component of the innate immune response to enveloped viruses. Trends Microbiol *18*, 388–396.
Feschotte, C., and Gilbert, C. (2012). Endogenous viruses: insights into viral evolution and impact on host biology. Nat Rev Genet *13*, 283–296.
Forterre, P. (2005). The two ages of the RNA world, and the transition to the DNA world: a story of viruses and cells. Biochimie *87*, 793–803.
Franzosa, E.A., and Xia, Y. (2011). Structural principles within the human-virus protein-protein interaction network. Proceedings of the National Academy of Sciences *108*, 10538–10543.
Friesen, P.D. (2013). Insect Viruses. In Fields Virology, D. Knipe, and P. howley, eds.
Fujino, K., Horie, M., Honda, T., Merriman, D.K., and Tomonaga, K. (2014). Inhibition of Borna disease virus replication by an endogenous bornavirus-like element in the ground squirrel genome. Proceedings of the National Academy of Sciences *111*, 13175–13180.

Garamszegi, S., Franzosa, E.A., and Xia, Y. (2013). Signatures of pleiotropy, economy and convergent evolution in a domain-resolved map of human-virus protein-protein interaction networks. PLoS Pathog. *9*, e1003778.

Gardner, M.B., Kozak, C.A., and O'Brien, S.J. (1991). The Lake Casitas wild mouse: evolving genetic resistance to retroviral disease. Trends Genet *7*, 22–27.

Gibbs, J.S., Lackner, A.A., Lang, S.M., Simon, M.A., Sehgal, P.K., Daniel, M.D., and Desrosiers, R.C. (1995). Progression to AIDS in the absence of a gene for vpr or vpx. J. Virol. *69*, 2378–2383.

Gifford, R.J., Katzourakis, A., Tristem, M., Pybus, O.G., Winters, M., and Shafer, R.W. (2008). From the Cover: A transitional endogenous lentivirus from the genome of a basal primate and implications for lentivirus evolution. Proceedings of the National Academy of Sciences *105*, 20362–20367.

Gifford, R., and Tristem, M. (2003). The evolution, distribution and diversity of endogenous retroviruses. Virus Genes *26*, 291–315.

Gifford, R.J. (2012). Viral evolution in deep time: lentiviruses and mammals. Trends Genet *28*, 89–100.

Gilbert, C., Maxfield, D.G., Goodman, S.M., and Feschotte, C. (2009). Parallel germline infiltration of a lentivirus in two Malagasy lemurs. PLoS Genet. *5*, e1000425.

Goldstone, D.C., Yap, M.W., Robertson, L.E., Haire, L.F., Taylor, W.R., Katzourakis, A., Stoye, J.P., and Taylor, I.A. (2010). Structural and functional analysis of prehistoric lentiviruses uncovers an ancient molecular interface. Cell Host and Microbe *8*, 248–259.

Gould, S.J., and Vrba, E. (1982). Exaptation-A Missing Term in the Science of Form. Paleobiology *8*, 4–15.

Han, G.-Z., and Worobey, M. (2012). Endogenous lentiviral elements in the weasel family (Mustelidae). Mol. Biol. Evol. *29*, 2905–2908.

Han, K., Lou, D.I., and Sawyer, S.L. (2011). Identification of a genomic reservoir for new TRIM genes in primate genomes. PLoS Genet. *7*, e1002388.

Hargreaves, K.R., Kropinski, A.M., and Clokie, M.R.J. (2014a). What does the talking?: quorum sensing signalling genes discovered in a bacteriophage genome. PLoS ONE *9*, e85131.

Hargreaves, K.R., Kropinski, A.M., and Clokie, M.R. (2014b). Bacteriophage behavioral ecology: How phages alter their bacterial host's habits. Bacteriophage *4*, e29866.

Harris, R.S., and Liddament, M.T. (2004). Retroviral restriction by APOBEC proteins. Nat Rev Immunol *4*, 868–877.

Hayward, A., Cornwallis, C.K., and Jern, P. (2014). Pan-vertebrate comparative genomics unmasks retrovirus macroevolution. Proceedings of the National Academy of Sciences.

Holmes, E.C., and Drummond, A.J. (2007). The evolutionary genetics of viral emergence. Curr Top Microbiol Immunol *315*, 51–66.

Holmes, E.C. (2009). The Evolution and Emergence of RNA Viruses (Oxford University Press on Demand).

Holmes, E.C. (2011). The evolution of endogenous viral elements. Cell Host and Microbe *10*, 368–377.

Horie, M., Honda, T., Suzuki, Y., Kobayashi, Y., Daito, T., Oshida, T., Ikuta, K., Jern, P., Gojobori, T., Coffin, J.M., et al. (2010). Endogenous non-retroviral RNA virus elements in mammalian genomes. Nature *463*, 84–87.

Horie, M., Kobayashi, Y., Suzuki, Y., and Tomonaga, K. (2013). Comprehensive analysis of endogenous bornavirus-like elements in eukaryote genomes. Philos Trans R Soc Lond B Biol Sci *368*, 20120499.

Hornung, V. (2014a). SnapShot: nucleic acid immune sensors, part 1. Immunity *41*, 868–868.e1.

Hornung, V. (2014b). SnapShot: Nucleic acid immune sensors, part 2. Immunity *41*, 1066–1066.e1.

Horvath, C.M. (2004). Weapons of STAT destruction. Interferon evasion by paramyxovirus V protein. Eur. J. Biochem. *271*, 4621–4628.

Horvath, P., and Barrangou, R. (2010). CRISPR/Cas, the immune system of bacteria and archaea. Science *327*, 167–170.
Hron, T., Fábryová, H., Pa Es, J., and Elleder, D. (2014). Endogenous lentivirus in Malayan colugo (Galeopterus variegatus), a close relative of primates. Retrovirology *11*, 84.
Hurst, L.D. (2002). The Ka/Ks ratio: diagnosing the form of sequence evolution. Trends Genet *18*, 486–486.
Isaacs, A., and Lindenmann, J. (1957). Virus interference. I. The interferon. Proc. R. Soc. Lond., B, Biol. Sci. *147*, 258–267.
Ito, J., Watanabe, S., Hiratsuka, T., Kuse, K., Odahara, Y., Ochi, H., Kawamura, M., and Nishigaki, K. (2013). Refrex-1, a soluble restriction factor against feline endogenous and exogenous retroviruses. J. Virol. *87*, 12029–12040.
Jalasvuori, M., and Bamford, J.K.H. (2008). Structural co-evolution of viruses and cells in the primordial world. Orig Life Evol Biosph *38*, 165–181.
Jayan, G.C. (2004). RNA editing in hepatitis delta virus: unsolved puzzles. ScientificWorldJournal *4*, 628–637.
Jern, P., and Coffin, J.M. (2008). Effects of Retroviruses on Host Genome Function. Annu Rev Genet *42*, 709–732.
Johnson, W.E. (2010). Endless forms most viral. PLoS Genet. *6*, e1001210.
Johnson, W.E. (2013). Rapid adversarial co-evolution of viruses and cellular restriction factors. Curr Top Microbiol Immunol *371*, 123–151.
Johnson, W.E., and Sawyer, S.L. (2009). Molecular evolution of the antiretroviral TRIM5 gene. Immunogenetics *61*, 163–176.
Jung, Y.T., Lyu, M.S., Buckler-White, A., and Kozak, C.A. (2002). Characterization of a polytropic murine leukemia virus proviral sequence associated with the virus resistance gene Rmcf of DBA/2 mice. J. Virol. *76*, 8218–8224.
Katzourakis, A., Tristem, M., Pybus, O.G., and Gifford, R.J. (2007). Discovery and analysis of the first endogenous lentivirus. Proc. Natl. Acad. Sci. U.S.a. *104*, 6261–6265.
Kerr, P.J., Rogers, M.B., Fitch, A., Depasse, J.V., Cattadori, I.M., Hudson, P.J., Tscharke, D.C., Holmes, E.C., and Ghedin, E. (2013a). Comparative analysis of the complete genome sequence of the California MSW strain of myxoma virus reveals potential host adaptations. J. Virol. *87*, 12080–12089.
Kerr, P.J., Rogers, M.B., Fitch, A., Depasse, J.V., Cattadori, I.M., Twaddle, A.C., Hudson, P.J., Tscharke, D.C., Read, A.F., Holmes, E.C., et al. (2013b). Genome scale evolution of myxoma virus reveals host-pathogen adaptation and rapid geographic spread. J. Virol. *87*, 12900–12915.
King, A., Adams, M.J., and Lefkowitz, E.J. (2012). Virus Taxonomy: Classification and Nomenclature of Viruses : Ninth Report of the ICTV.
Koff, W.C., Johnson, P.R., Watkins, D.I., Burton, D.R., Lifson, J.D., Hasenkrug, K.J., McDermott, A.B., Schultz, A., Zamb, T.J., Boyle, R., et al. (2006). HIV vaccine design: insights from live attenuated SIV vaccines. Nat Immunol *7*, 19–23.
Koonin, E.V. (2009). On the origin of cells and viruses: primordial virus world scenario. Ann. N. Y. Acad. Sci. *1178*, 47–64.
Kozak, C.A. (2010). The mouse “xenotropic” gammaretroviruses and their XPR1 receptor. Retrovirology *7*, 101.
Krupp, A., McCarthy, K.R., Ooms, M., Letko, M., Morgan, J.S., Simon, V., and Johnson, W.E. (2013). APOBEC3G polymorphism as a selective barrier to cross-species transmission and emergence of pathogenic SIV and AIDS in a primate host. PLoS Pathog. *9*, e1003641.
Labrie, S.J., Samson, J.E., and Moineau, S. (2010). Bacteriophage resistance mechanisms. Nat Rev Micro *8*, 317–327.
Laguette, N., and Benkirane, M. (2012). How SAMHD1 changes our view of viral restriction. Trends Immunol *33*, 26–33.

Lavialle, C., Cornelis, G., Dupressoir, A., Esnault, C., Heidmann, O., Vernochet, C., and Heidmann, T. (2013). Paleovirology of "syncytins," retroviral env genes exapted for a role in placentation. Philos Trans R Soc Lond B Biol Sci *368*, 20120507.

Li, F., and Ding, S.-W. (2006). Virus counterdefense: diverse strategies for evading the RNA-silencing immunity. Annu. Rev. Microbiol. *60*, 503–531.

Li, Y., Lu, J., Han, Y., Fan, X., and Ding, S.-W. (2013). RNA interference functions as an antiviral immunity mechanism in mammals. Science *342*, 231–234.

Liao, C.-H., Kuang, Y.-Q., Liu, H.-L., Zheng, Y.-T., and Su, B. (2007). A novel fusion gene, TRIM5-Cyclophilin A in the pig-tailed macaque determines its susceptibility to HIV-1 infection. Aids *21 Suppl 8*, S19–S26.

Löchelt, M., Romen, F., Bastone, P., Muckenfuss, H., Kirchner, N., Kim, Y.-B., Truyen, U., Rösler, U., Battenberg, M., Saïb, A., et al. (2005). The antiretroviral activity of APOBEC3 is inhibited by the foamy virus accessory Bet protein. Proc. Natl. Acad. Sci. U.S.a. *102*, 7982–7987.

Malfavon-Borja, R., Sawyer, S.L., Wu, L.I., Emerman, M., and Malik, H.S. (2013a). An evolutionary screen highlights canonical and noncanonical candidate antiviral genes within the primate TRIM gene family. Genome Biol Evol *5*, 2141–2154.

Malfavon-Borja, R., Wu, L.I., Emerman, M., and Malik, H.S. (2013b). Birth, decay, and reconstruction of an ancient TRIMCyp gene fusion in primate genomes. Proceedings of the National Academy of Sciences.

Malim, M.H. (2009). APOBEC proteins and intrinsic resistance to HIV-1 infection. Philos Trans R Soc Lond B Biol Sci *364*, 675–687.

Mandadi, K.K., and Scholthof, K.-B.G. (2013). Plant immune responses against viruses: how does a virus cause disease? Plant Cell *25*, 1489–1505.

McCarthy, K.R., and Johnson, W.E. (2014). Plastic proteins and monkey blocks: how lentiviruses evolved to replicate in the presence of primate restriction factors. PLoS Pathog. *10*, e1004017.

Meyerson, N.R., and Sawyer, S.L. (2011). Two-stepping through time: mammals and viruses. Trends Microbiol *19*, 286–294.

Miller, W.A., and Whitham, S.A. (2013). Plant Viruses. In Fields Virology, D. Knipe, and P. howley, eds.

Montaño, A.M., Tsujino, F., Takahata, N., and Satta, Y. (2011). Evolutionary origin of peptidoglycan recognition proteins in vertebrate innate immune system. BMC Evol. Biol. *11*, 79.

Morikawa, K., Lange, C.M., Gouttenoire, J., Meylan, E., Brass, V., Penin, F., and Moradpour, D. (2011). Nonstructural protein 3-4A: the Swiss army knife of hepatitis C virus. J. Viral Hepat. *18*, 305–315.

Moss, B. (2013). Poxviridae. In Fields Virology, D. Knipe, and P. howley, eds.

Mu D., Yang H., Zhu J.W., Liu F.L., Tian R.R., Zheng H.Y., Han J.B., Shi P., and Zheng Y.T. (2014). Independent Birth of a Novel TRIMCyp in *Tupaia belangeri* with a Divergent Function from Its Paralog TRIM5. Mol Biol Evol. 31(11): 2985-97

Mukherjee, K., Campos, H., and Kolaczkowski, B. (2012). Evolution of Animal and Plant Dicers: Early Parallel Duplications and Recurrent Adaptation of Antiviral RNA Binding in Plants. Mol. Biol. Evol.

Münk, C., Willemsen, A., and Bravo, I.G. (2012). An ancient history of gene duplications, fusions and losses in the evolution of APOBEC3 mutators in mammals. BMC Evol. Biol. *12*, 71.

Neil, S., and Bieniasz, P. (2009). Human Immunodeficiency Virus, Restriction Factors, and Interferon. Journal of Interferon and Cytokine Research *29*, 569–580.

Neil, S.J.D., Zang, T., and Bieniasz, P.D. (2008). Tetherin inhibits retrovirus release and is antagonized by HIV-1 Vpu. Nature *451*, 425–430.

Newman, R.M., Hall, L., Kirmaier, A., Pozzi, L.-A., Pery, E., Farzan, M., O'Neil, S.P., and Johnson, W. (2008). Evolution of a TRIM5-CypA splice isoform in old world monkeys. PLoS Pathog. *4*, e1000003.

Nieva, J.L., Madan, V., and Carrasco, L. (2012). Viroporins: structure and biological functions. Nat Rev Micro *10*, 563–574.

Nisole, S., Lynch, C., Stoye, J.P., and Yap, M.W. (2004). A Trim5-cyclophilin A fusion protein found in owl monkey kidney cells can restrict HIV-1. Proc. Natl. Acad. Sci. U.S.a. *101*, 13324–13328.
Nisole, S., Stoye, J.P., and Saïb, A. (2005). TRIM family proteins: retroviral restriction and antiviral defence. Nat Rev Micro *3*, 799–808.
Nowak, M., and May, R.M. (2000). Virus Dynamics (Oxford University Press on Demand).
Ohno, S. (2014). Evolution by Gene Duplication (Springer).
Patel, M.R., Emerman, M., and Malik, H.S. (2011). Paleovirology - ghosts and gifts of viruses past. Curr Opin Virol *1*, 304–309.
Patel, M.R., Loo, Y.-M., Horner, S.M., Gale, M., and Malik, H.S. (2012). Convergent evolution of escape from hepaciviral antagonism in primates. PLoS Biol. *10*, e1001282.
Pellet, P.E., and Roizman, B. (2013). Herpesviridae. In Fields Virology, D. Knipe, and P. howley, eds.
Perelygin, A.A., Scherbik, S.V., Zhulin, I.B., Stockman, B.M., Li, Y., and Brinton, M.A. (2002). Positional cloning of the murine flavivirus resistance gene. Proc. Natl. Acad. Sci. U.S.a. *99*, 9322–9327.
Pérez-Caballero, D., Zang, T., Ebrahimi, A., McNatt, M.W., Gregory, D.A., Johnson, M.C., and Bieniasz, P.D. (2009). Tetherin inhibits HIV-1 release by directly tethering virions to cells. Cell *139*, 499–511.
Pestka, S. (2007). The interferons: 50 years after their discovery, there is much more to learn. J Biol Chem *282*, 20047–20051.
Rieder, M., and Conzelmann, K.-K. (2009). Rhabdovirus evasion of the interferon system. Journal of Interferon and Cytokine Research *29*, 499–509.
Robinson, H.L., Astrin, S.M., Senior, A.M., and Salazar, F.H. (1981). Host Susceptibility to endogenous viruses: defective, glycoprotein-expressing proviruses interfere with infections. J. Virol. *40*, 745–751.
Russell, R.A., Wiegand, H.L., Moore, M.D., Schäfer, A., McClure, M.O., and Cullen, B.R. (2005). Foamy virus Bet proteins function as novel inhibitors of the APOBEC3 family of innate antiretroviral defense factors. J. Virol. *79*, 8724–8731.
Sabeti, P.C., Reich, D.E., Higgins, J.M., Levine, H.Z.P., Richter, D.J., Schaffner, S.F., Gabriel, S.B., Platko, J.V., Patterson, N.J., McDonald, G.J., et al. (2002). Detecting recent positive selection in the human genome from haplotype structure. Nature *419*, 832–837.
Sackton, T.B., Lazzaro, B.P., Schlenke, T.A., Evans, J.D., Hultmark, D., and Clark, A.G. (2007). Dynamic evolution of the innate immune system in Drosophila. Nat Genet *39*, 1461–1468.
Samson, J.E., Magadán, A.H., Sabri, M., and Moineau, S. (2013). Revenge of the phages: defeating bacterial defences. Nat Rev Micro *11*, 675–687.
Samuel, C.E. (2011). Adenosine deaminases acting on RNA (ADARs) are both antiviral and proviral. Virology *411*, 180–193.
Samuel, C.E. (2012). ADARs: viruses and innate immunity. Curr Top Microbiol Immunol *353*, 163–195.
Sauter, D., Schindler, M., Specht, A., Landford, W.N., Münch, J., Kim, K.-A., Votteler, J., Schubert, U., Bibollet-Ruche, F., Keele, B.F., et al. (2009). Tetherin-driven adaptation of Vpu and Nef function and the evolution of pandemic and nonpandemic HIV-1 strains. Cell Host and Microbe *6*, 409–421.
Sawyer, S.L., Emerman, M., and Malik, H.S. (2007). Discordant evolution of the adjacent antiretroviral genes TRIM22 and TRIM5 in mammals. PLoS Pathog. *3*, e197.
Sawyer, S.L., Wu, L.I., Emerman, M., and Malik, H.S. (2005). Positive selection of primate TRIM5alpha identifies a critical species-specific retroviral restriction domain. Proc. Natl. Acad. Sci. U.S.a. *102*, 2832–2837.

Sayah, D.M., Sokolskaja, E., Berthoux, L., and Luban, J. (2004). Cyclophilin A retrotransposition into TRIM5 explains owl monkey resistance to HIV-1. Nature *430*, 569–573.

Schoggins, J.W., Wilson, S.J., Panis, M., Murphy, M.Y., Jones, C.T., Bieniasz, P., and Rice, C.M. (2011). A diverse range of gene products are effectors of the type I interferon antiviral response. Nature *472*, 481–485.

Seed, K.D., Lazinski, D.W., Calderwood, S.B., and Camilli, A. (2013). A bacteriophage encodes its own CRISPR/Cas adaptive response to evade host innate immunity. Nature *494*, 489–491.

Stehelin, D., Varmus, H.E., Bishop, J.M., and Vogt, P.K. (1976). DNA related to the transforming gene(s) of avian sarcoma viruses is present in normal avian DNA. Nature *260*, 170–173.

Stern, A., and Sorek, R. (2011). The phage-host arms race: shaping the evolution of microbes. Bioessays *33*, 43–51.

Tack, A.J.M., Thrall, P.H., Barrett, L.G., Burdon, J.J., and Laine, A.-L. (2012). Variation in infectivity and aggressiveness in space and time in wild host-pathogen systems: causes and consequences. J. Evol. Biol. *25*, 1918–1936.

Takeuchi, N., Wolf, Y.I., Makarova, K.S., and Koonin, E.V. (2012). Nature and intensity of selection pressure on CRISPR-associated genes. J. Bacteriol. *194*, 1216–1225.

Tareen, S.U., Sawyer, S.L., Malik, H.S., and Emerman, M. (2009). An expanded clade of rodent Trim5 genes. Virology *385*, 473–483.

Thrall, P.H., Laine, A.-L., Ravensdale, M., Nemri, A., Dodds, P.N., Barrett, L.G., and Burdon, J.J. (2012). Rapid genetic change underpins antagonistic coevolution in a natural host-pathogen metapopulation. Ecol. Lett. *15*, 425–435.

Tomaselli, S., Galeano, F., Locatelli, F., and Gallo, A. (2014). ADARs and the Balance Game between Virus Infection and Innate Immune Cell Response. Curr Issues Mol Biol *17*, 37–52.

Towers, G. (2005). Control of viral infectivity by tripartite motif proteins. Hum Gene Ther *16*, 1125–1132.

Tristem, M., Marshall, C., Karpas, A., and Hill, F. (1992). Evolution of the primate lentiviruses: evidence from vpx and vpr. Embo J *11*, 3405–3412.

Truniger, V., and Aranda, M.A. (2009). Recessive resistance to plant viruses. Adv. Virus Res. *75*, 119–159.

Van Damme, N., Goff, D., Katsura, C., Jorgenson, R.L., Mitchell, R., Johnson, M.C., Stephens, E.B., and Guatelli, J. (2008). The interferon-induced protein BST-2 restricts HIV-1 release and is downregulated from the cell surface by the viral Vpu protein. Cell Host and Microbe *3*, 245–252.

van der Aa, L.M., Jouneau, L., Laplantine, E., Bouchez, O., Van Kemenade, L., and Boudinot, P. (2012). FinTRIMs, fish virus-inducible proteins with E3 ubiquitin ligase activity. Dev. Comp. Immunol. *36*, 433–441.

van der Loo, W., Abrantes, J., and Esteves, P.J. (2009). Sharing of endogenous lentiviral gene fragments among leporid lineages separated for more than 12 million years. J. Virol. *83*, 2386–2388.

van Mierlo, J.T., Bronkhorst, A.W., Overheul, G.J., Sadanandan, S.A., Ekström, J.-O., Heestermans, M., Hultmark, D., Antoniewski, C., and van Rij, R.P. (2012). Convergent evolution of argonaute-2 slicer antagonism in two distinct insect RNA viruses. PLoS Pathog. *8*, e1002872.

Verhelst, J., Hulpiau, P., and Saelens, X. (2013). Mx proteins: antiviral gatekeepers that restrain the uninvited. Microbiol. Mol. Biol. Rev. *77*, 551–566.

Virgen, C.A., Kratovac, Z., Bieniasz, P.D., and Hatziioannou, T. (2008). From the Cover: Independent genesis of chimeric TRIM5-cyclophilin proteins in two primate species. Proceedings of the National Academy of Sciences *105*, 3563–3568.

Viswanathan, K., Smith, M.S., Malouli, D., Mansouri, M., Nelson, J.A., and Früh, K. (2011). BST2/Tetherin enhances entry of human cytomegalovirus. PLoS Pathog. *7*, e1002332.

Walkinshaw, M.D., Taylor, P., Sturrock, S.S., Atanasiu, C., Berge, T., Henderson, R.M., Edwardson, J.M., and Dryden, D.T.F. (2002). Structure of Ocr from bacteriophage T7, a protein that mimics B-form DNA. Mol Cell *9*, 187–194.

Weiss, R.A. (2006). The discovery of endogenous retroviruses. Retrovirology *3*, 67.

Weiss, R.A., Griffiths, D., Takeuchi, Y., Patience, C., and Venables, P. Retroviruses: ancient and modern. Arch Virol Suppl *15*, 171–177.

Wilson, S.J., Webb, B.L.J., Ylinen, L.M.J., Verschoor, E., Heeney, J.L., and Towers, G.J. (2008). From the Cover: Independent evolution of an antiviral TRIMCyp in rhesus macaques. Proceedings of the National Academy of Sciences *105*, 3557–3562.

Wu, T., Yan, Y., and Kozak, C.A. (2005). Rmcf2, a xenotropic provirus in the Asian mouse species Mus castaneus, blocks infection by polytropic mouse gammaretroviruses. J. Virol. *79*, 9677–9684.

Yang, Z. (2005). The power of phylogenetic comparison in revealing protein function. Proc. Natl. Acad. Sci. U.S.a. *102*, 3179–3180.

Yang, Z. (2006). Computational Molecular Evolution (Oxford University Press).

Yasunaga, A., Hanna, S.L., Li, J., Cho, H., Rose, P.P., Spiridigliozzi, A., Gold, B., Diamond, M.S., and Cherry, S. (2014). Genome-wide RNAi screen identifies broadly-acting host factors that inhibit arbovirus infection. PLoS Pathog. *10*, e1003914.

Youle, M., Haynes, M., and Rohwer, F. (2012). Scratching the Surface of Biology's Dark Matter. In VIruses: Essential Agents of Life, G. Witzany, ed. (dordrecht: Springer Science and Business Media).

Zhang, X., Zhao, H., Chen, Y., Luo, H., and Yao, B. (2014). Molecular characterization of a CpTRIM35-like protein and its splice variants from whitespotted bamboo shark (Chiloscyllium plagiosum). Biochem Biophys Res Commun *453*, 425–431.

Zhang, Z., Liu, J., Li, M., Yang, H., and Zhang, C. (2012). Evolutionary dynamics of the interferon-induced transmembrane gene family in vertebrates. PLoS ONE *7*, e49265.

5

Evolution of the Interactions of Viruses With Their Plant Hosts

Israel Pagán, Aurora Fraile and Fernando García-Arenal*

Centro de Biotecnología y Genómica de Plantas (UPM-INIA) and Dpto. de Biotecnología, Campus Montegancedo, Universidad Politécnica de Madrid, Pozuelo de Alarcón (Madrid), Spain

**Corresponding author: fernando.garciaarenal@upm.es*

Abstract

More than a century of studies have shown that viruses are the causal agents of plant diseases. However, the vast majority of our understanding on plant-virus interactions derives from the study of viruses that cause diseases in crops. It has been only recently that scientists have started to explore plant-virus interactions in wild ecosystems. The results of these new studies have shown that in wild ecosystems plant viruses often appear to be commensals or even have beneficial effects for their hosts. Thus, the absence of negative effects of virus infection in wild ecosystems might be the rule, rather than the exception. This observation could change our perception of plant-virus interactions, and may lead to re-visiting our current understanding on how they evolve. Here, we review existing knowledge on three major aspects of the evolution of plant-virus interactions, plant-virus co-evolution, virus emergence and evolution of virus host range, with special attention to differences between wild and agricultural ecosystems.

Introduction

Since Martinus Bijerinck at the end of the 19th century first proposed that a virus was the causal agent of a plant disease, the study of any aspect of plant-virus interactions, including their evolution, has been motivated by the agro-economic impact of the detrimental effects of virus infection in crops. As a consequence, plant viruses are generally perceived as pathogens that negatively affect the fitness of their hosts. The advent of next generation sequencing techniques, which have been used to obtain the virome of entire ecosystems, and the development of bioinformatic tools suitable for the analysis of plant-virus interactions at the community level, has allowed the beginning of the exploration of plant-virus interactions in wild ecosystems. The results of the studies in this emerging area of research have suggested that, more often than not, viruses may be commensals, or even have beneficial effects, for their host plants. Such observations challenge the traditional view of plant viruses, and have opened the question of under what circumstances viruses are pathogens of plants. Answering this question may deeply impact our understanding of plant-virus interactions, and influence scientific approaches to the study of the evolution of plant viruses.

Plant-virus interactions: are viruses pathogens of plants?

Under the most accepted view, plant viruses are pathogens that have a negative impact in the fitness of infected host plants (Read, 1994; D'Arcy *et al.*, 2001; Anderson *et al.*, 2004). However, most of our knowledge on plant-virus interactions comes from the study of viruses that cause diseases in crops. Therefore, our understanding of plant viruses might be biased towards those interactions that have the most obvious agro-economic consequences. Indeed, there are very few surveys for viruses that are asymptomatic in crops (Sherwood, 1997; Anderson *et al.*, 1991; Rodelo-Urrego *et al.*, 2013), and until recently weedy and wild plants were considered of interest only as source/reservoir of virus inoculum for crops (Cooper and Jones, 2006; Jones, 2009; Alexander *et al.*, 2014). The advent of next generation sequencing (NGS) has allowed exploration of the virome of a number of different ecosystems (e.g., Delwart, 2007; Hurwitz and Sullivan, 2013), including wild plant communities (Roossinck, 2012; Stobbe and Roossinck, 2014), and in the past few years the evolutionary and ecological consequences of virus infection in wild plants has also started to be analyzed (Alexander *et al.*, 2014; Roossinck and García-Arenal, 2015). These pioneering studies have shown that plant-virus interactions in wild ecosystems are far more diverse than in agricultural ecosystems. The first interesting observation is that in wild plants virus infections are common, and virus diversity and identity appear to be different from that in crops (Roossinck, 2012; Stobbe and Roossinck, 2014). Indeed, NGS analyses in wild plants have uncovered large (previously) unknown virus diversity (Roossinck, 2011a). However, this observation might be biased because metagenomics of viruses in crops is still to be done. Another interesting observation is that multiple virus infection is frequent in wild plants (Malpica *et al.*, 2006; Roossinck *et al.*, 2010), which underscores the relevance of considering both virus-virus and host-virus interactions in determining the outcome of infection (Syller, 2012). But perhaps the most interesting observation is that virus infection generally does not result in the expression of disease symptoms (Prendeville *et al.*, 2012; Roossinck, 2012). It is true that some examples exist of plant viruses that cause obvious diseases, affect plant population sizes and plant ecosystem composition (Power and Mitchell, 2004; Malmstrom *et al.*, 2005a; Power *et al.*, 2011; Rua *et al.*, 2011; Rodelo-Urrego *et al.*, 2013; Prendeville *et al.*, 2014). However, it has also been documented that in some instances virus infection may even be beneficial for their host plants (Gibbs, 1980; Xu *et al.*, 2008), and that parasites mediate processes that maintain ecosystem equilibrium (Hudson *et al.*, 2006). All this information has led to the proposal that in wild ecosystems plant viruses would be most often mutualistic symbionts of plants (Roossinck, 2005; Wren *et al.*, 2006). This idea could deeply influence our view of plant-virus interactions, and change our perception of what is a "healthy ecosystem", i.e., one that persists, maintains vigor (productivity), organization (biodiversity and predictability) and resilience (time to recovery) (Costanza and Mageau, 2000).

Wild and agricultural ecosystems are not isolated environments. As a consequence of the intensification of agriculture, an increasing proportion of wild ecosystems lie adjacent to agricultural ecosystems. The border between these two ecosystems has been termed "agro-ecological interface" (Burdon and Thrall, 2008; Burdon and Thrall, 2014). The agro-ecological interface can be viewed as a doorway rather than a barrier between ecosystems, with plants, vectors and viruses moving through it (Burdon and Thrall, 2008; Blitzer *et al.*, 2012; Burdon and Thrall, 2014). These movements may be of relevance to understand processes such as transmission dynamics and virus emergence, and therefore are key to plant-virus interactions (Alexander *et al.*, 2014). Because the agro-ecological interface represents the border between heterogeneous environments, a complete understanding of virus diversity and population dynamics at the agro-ecological interface should consider analyses at the landscape scale (Archie *et al.*, 2009; Biek and Real, 2010). The few published analyses at such scale have already provided evidence of changes in interaction patterns across gradients of human ecosystem management affecting infection dynamics, virus evolution and plant-virus co-evolution (e.g. Fabre *et al.*, 2012; Pagán *et al.*, 2012; Rodelo-Urrego *et al.*, 2013). Plant-virus interactions will follow different evolutionary dynamics according to ecosystem complexity, and processes such as co-evolution that may be major driving forces in specialized pairwise plant-pathogen interactions may be less significant at the community level, where interaction matrices are more complex (Nuismer *et al.*, 2013; Karasov *et al.*, 2014; Lively *et al.*, 2014). Also, evolutionary dynamics will differ depending on whether viruses are mutualistic symbionts, endemic pathogens co-evolving with the host, or epidemic emergent pathogens in new host populations (Roossinck, 2011b; Karasov *et al.*, 2014; Lively *et al.*, 2014).

Here, we review existing knowledge on three major aspects of the evolution of plant-virus interactions, including plant-virus co-evolution, virus emergence and virus host range, emphasizing differences between wild and agricultural ecosystems.

Plant-virus co-evolution: Evidence from agricultural to wild ecosystems

Conditions for plant-virus co-evolution

As stated above, it is considered that plant viruses are pathogens that have a negative impact in the fitness of the infected hosts (Read, 1994; D'Arcy *et al.*, 2001; Anderson *et al.*, 2004). If this were so, hosts would have developed different mechanisms to avoid/limit virus infection, or to reduce its detrimental effects (Agnew *et al.*, 2000; Little *et al.*, 2010). The two main mechanisms of plant defense against viruses are resistance, i.e., the host ability to limit virus multiplication (Clarke, 1986); and tolerance, i.e., the host ability to reduce the effect of infection on its fitness (Clarke, 1986). Because pathogen virulence and host resistance/tolerance have reciprocal effects on each other's fitness, it is generally assumed that host-virus interactions result in co-evolutionary processes (Woolhouse *et al.*, 2002). Host-pathogen co-evolution requires four conditions: (i) genetic variation in the relevant host and pathogen traits (e.g. resistance, tolerance, infectivity, virulence); (ii) reciprocal effects of the relevant traits of the interaction on the fitness of host and pathogen; (iii) dependence of the outcome of the host-pathogen interaction on the specific host and pathogen genotypes involved; (iv) changes in genotype frequencies in both the host and the pathogen

Table 5.1. Evidence of plant-virus co-evolution in agricultural and wild ecosystems[1]

Condition for co-evolution	Evidence from agricultural ecosystems	Evidence from wild ecosystems
Genetic variation in the relevant plant and virus traits	+++ *Capsicum* spp. – Tobamovirus *Capsicum* spp. – PVY Pepper – PMMoV Pepper – TWSV *Solanum* spp. – ToMV Tobacco – TMV Rice – RYMV	+ *A.thaliana* – CMV
Reciprocal effects of the relevant traits of the interaction on the fitness of plant and virus	+++ *Solanum* spp. – Tobamovirus Various spp. – Luteovirus Various spp. – Mastrevirus Maize – MSV Melon – CMV Tobacco – CMV Rice – RYMV Squash – PRSV *Musa accuminata* – BSV	++ Wild cabbage – TuMV *Poaceae* – BYDV/CYDV Squash – CMV/ZYMV
Dependence of the outcome of the plant-virus interaction on the specific plant and virus genotypes	+++ *Capsicum* spp. – Tobamovirus *Capsicum* spp. – PVY Pepper – PMMoV Pepper – TWSV *Solanum* spp. – ToMV Tobacco – TMV Rice – RYMV Melon – ZYMV	+++ *A.thaliana* – CMV *Squash* – CMV/ZYMV
Changes in genotype frequencies in both the plant and the virus populations	+++ *Capsicum* spp. – Tobamovirus	-

[1] Strength of the evidence supporting the different conditions for co-evolution in the analyzed systems is indicated as +++ (strong) to + (weak). - indicates no evidence.

populations (Woolhouse *et al.,* 2002). Despite these general assumptions, there is surprisingly little evidence of plant-virus interactions that meet these conditions. Current evidence mostly derives from highly virulent viruses in crops (Table 5.1), which have detrimental effects in crop production not necessarily linked to a host fitness decrease, and from the changes in the genetic structure of virus populations in agricultural ecosystems as a response to human manipulation of the genetics of the host by breeding in resistance factors (Fraile and García-Arenal, 2010). Demonstration of plant-virus co-evolution in wild plant populations, where the genetic composition of the host may change in response to virus infection, is lacking, and evidence in support of plant-virus co-evolution is very scant (but see Pagán *et al.,* 2010a) (Table 5.1). In this section we discuss the available evidence in support of co-evolution in plant-virus systems, both in wild and agricultural ecosystems.

The first condition for plant-virus co-evolution to occur is the genetic variation in host resistance/tolerance and in the pathogen's ability to infect and multiply in the host (i.e., infectivity) and to cause disease (i.e., pathogenicity). As for other plant pathogens (Karasov *et al.,* 2014) there is abundant evidence that this is indeed the case for viruses. The available evidence derives mostly from analysis of the patterns of variability of plant genes determining resistance, and of the viral genes determining infectivity and pathogenicity, in interactions resulting in qualitative resistance.

On the plant side, most characterized dominant qualitative resistance to viruses fits a gene-for-gene (GFG) model of host-pathogen interaction (Chisholm *et al.,* 2000; Whitham *et al.,* 2000; Ishibashi *et al.,* 2007), in which plant resistance genes (*R*) encode R proteins that mediate recognition of virus encoded avirulence proteins (AVR proteins). Recognition triggers a defense reaction, which limits virus multiplication to the infection site (Jones and Dangl, 2006; Brown and Tellier, 2011). In most systems, resistance is expressed as either a hypersensitive reaction, or as extreme resistance (Fraile and García-Arenal, 2010). Changes in *AVR* that impair R-AVR recognition result in resistance-breaking, leading eventually to universal pathogenicity, i.e., to the capacity to infect all host genotypes, which is a trait of GFG systems (Agrawal and Lively, 2002). Thus, overcoming dominant resistance is a case of host range expansion. Genetic variation in *R* genes leading to the recognition of new viral strains has been demonstrated for several *R* genes to viruses. This is the case for the genes *Tm1*, *Tm2* and *Tm2*^{2} in wild *Solanum* species, and of the different alleles of the *L* gene in *Capsicum*, which confer resistance to different genotypes of *Tomato mosaic virus* (ToMV) and to *Tobamovirus* species, respectively, through the accumulation of a few non-synonymous mutations (Lanfermeijer *et al.,* 2005; Tomita *et al.,* 2008; Ishibashi *et al.,* 2012). Importantly, the three-dimensional structure of the *Tm-1*-encoded protein, and its interaction with the AVR protein, which is the ToMV helicase domain of the replication protein, has been elucidated in an unprecedented analysis of R-AVR interactions (Ishibashi *et al.,* 2014).

Another major mechanism in the evolution of *R* genes is duplications and recombination through unequal crossover, which also generate new specificities (Friedman and Baker, 2007), and it has been shown that virus infection promotes recombination in the host plant genome with transgeneration effects (Kovalchuk *et al.,* 2003; Molinier *et al.,* 2006). Hence, the appearance of new recognition specificities through the accumulation of point mutations and of recombination in the *R* genes would be compatible with plant-virus co-evolution (Friedman and Baker, 2007). On the virus side, genetic variation in *AVR* genes of several plant viruses also has been reported. Different mutations in the capsid protein gene (CP) of TMV led to overcoming *N´*-mediated resistance in tobacco (Culver, 2002). Similarly, field isolates of *Pepper mild mottle virus* (PMMoV) that overcome *L*-mediated resistance in pepper differ from the AVR genotype by a few amino acid substitutions in the CP (Berzal-Herranz *et al.,* 1995; Tsuda *et al.,* 1998; Hamada *et al.,* 2002, Hamada *et al.,* 2007). In *Tomato spotted wilt virus* (TSWV) different genotypes overcoming *Tsw*-mediated resistance in pepper have been reported, resulting from amino acid substitutions in the NSs protein (Margaria *et al.,* 2007). Interestingly, for many *R* genes only one or a few resistance-breaking genotypes have been reported. Moreover, mutations resulting in the overcoming of resistance occur only in certain combinations, and/or different resistance-breaking genotypes have limited geographical ranges. Such limited polymorphism in resistance-breaking genotypes may be the consequence of the multifunctional nature of viral proteins, which will result in epistatic and pleitropic effects of the resistance-breaking mutations (see section "*Mechanisms involved in across-host fitness trade-offs*"), and it has not been observed in other plant pathogens (Sacristán and García-Arenal, 2007).

Qualitative resistance of plants to viruses may also be recessive, which is really an impaired susceptibility, and is often expressed as immunity at the cell level (Kang *et al.*, 2005; Maule *et al.*, 2007). Recessive resistance, and its breakdown, may be explained by the Matching Alleles (MA) model, in which the success of infection requires a specific match between host and pathogen genes (Agrawal and Lively, 2002). Mutations leading to resistance in the host can be countered by mutations in the virus, thus restoring compatibility with the mutated host gene. In a pure MA interaction, virus mutations that result in overcoming a specific recessive resistance allele will lead to a host switch, as the mutant virus no longer will be infectious to the "susceptible" original plant genotype. A good example of MA interaction is provided by rice-*Rice yellow mottle virus* (RYMV) interaction. In this system, different mutations have been observed in the RYMV VPg determining infection on rice genotypes carrying different resistance alleles at the *rymv* locus, encoding eIF(iso)4G (Pinel-Galzi *et al.*, 2007; Poulicard *et al.*, 2014). Another well-characterized system is the pepper-*Potato virus Y* (PVY) interaction determined by the *pvr2* locus of *Capsicum*, encoding eIF4E, and the virus VPg (Quenoiulle *et al.*, 2013; Moury *et al.*, 2014). The direct interaction eIF4E-VPg is required for virus multiplication (Robaglia and Caranta, 2006). At odds with GFG systems, high variability and positive selection has been observed at both the VPg and eIF4E, and almost all amino acid changes are linked to gains of function (resistance of the plant or infectivity of the virus). In spite of these mechanistic features, typical of MA interactions, the structure of infectivity matrices between PVY and plant genotypes carrying different *eIF4E* alleles was not modular (Moury *et al.*, 2014), as expected in a MA system, but rather it was nested, as expected in a GFG system (Weitz *et al.*, 2013). This was so because PVY mutations involved in adaptation to a particular plant genotype were often pleoitropic, also conferring adaptation to other *eIF4E* alleles (Moury *et al.*, 2014). This study underscores the complexity of plant-virus interactions, and the limitations of current models. It also illustrates how relaxed, rather than pure, MA models (Agrawal and Lively, 2002), could better capture this complexity. Hence, evidence of genetic variation in plant resistance and in virus infectivity/pathogenicity in interactions conforming to GFG and MA models is abundant for viruses in crops, but no equivalent analysis have been done in wild plant-virus interactions.

The second condition for plant-virus co-evolution is that there must be reciprocal effects of the relevant traits of the interaction on the fitness of host and pathogen. For this to happen, pathogenicity and resistance must negatively affect the fitness of plants and viruses, respectively. Evidence that virus infection can have detrimental effects in the host mainly comes from crop-virus systems (Fraile and García-Arenal, 2010; Froissart *et al.*, 2010). A few experiments have shown such effects in wild plants under controlled conditions (e.g., Kelly, 1994; Friess and Maillet, 1996; Pagán *et al.*, 2007), but evidence that plant viruses have a negative effect on plant fitness in natural ecosystems is still scant (but see Maskell *et al.*, 1999; Power and Mitchell, 2004; Malmstrom *et al.*, 2005a; Prendeville *et al.*, 2014). Also, evidence of negative effects of plant resistance in virus fitness derives from crop-virus interactions (Fraile and García-Arenal, 2010). Most of these studies quantify virus fitness as within-host multiplication, and assume that virus multiplication and transmission are positively correlated. Although this seems to be the case for horizontally transmitted parasites (Ebert and Bull, 2003; Alizon *et al.*, 2009), it has been shown that this positive correlation does not hold for some vertically transmitted parasites (Ebert, 2013; Pagán *et al.*, 2014). A potential consequence of the reciprocal effects of plants and viruses on each other's fitness and evolution is the congruence of their phylogenies (Nieberding and Olivieri, 2007). In various plant-virus systems, such congruence has been interpreted as evidence of co-evolution, and even of co-divergence, co-divergence meaning that the host plant and the virus have shared their whole evolutionary history (Lartey *et al.*, 1996; Wu *et al.*, 2008; Gibbs *et al.*, 2010; Stobbe *et al.*, 2012). Analyses of evolutionary scales of several virus species/families traced back the radiation of these viruses to the origin of agriculture, which suggests that viruses could have co-evolved with their crop hosts (Fargette *et al.*, 2008; Gibbs *et al.*, 2008; Pagán and Holmes, 2010). However, other phenomena such as population geographic isolation may result in host-virus phylogenetic congruence (see the "*Landscape heterogeneity and infection dynamics*" section), and different evolutionary scales have been estimated for some viruses and their hosts, which do not support plant-virus co-divergence over extended periods of time (Pagán *et al.*, 2010b; Rodelo-Urrego *et al.*, 2013). Thus, although there is evidence supporting reciprocal effects of plants and virus traits on each other fitness, how general these effects are across plant-virus interactions is still unclear.

The third condition for plant-virus co-evolution is dependence of the outcome of the interaction on the combination of host and pathogen genotypes involved. The two major models of host-pathogen interaction determining the success of infection (the GFG and MA models) assume this condition to be true, and genetic and molecular genetics evidence support both of these models to explain plant-virus interactions (Kang *et al.*, 2005; Maule *et al.*, 2007; Sacristán and García-Arenal, 2007). As discussed above, in a variety of GFG and MA interactions genetic diversity in genes determining resistance/susceptibility in plants and infectivity/pathogenicity in viruses have been described (Fraile and García-Arenal, 2010). In such interactions, the specific combination of plant and virus alleles determines whether the plant-virus interaction would be a compatible one, i.e., whether the virus is able to establish a successful infection (Tomita *et al.*, 2008; Ishibashi *et al.*, 2012; Ishibashi *et al.*, 2014; Quenouille *et al.*, 2013; Moury *et al.*, 2014). Thus, this is evidence that the outcome of the plant-virus interaction depends on the specific genotypes of the interacting symbionts. Resistance to viruses in plants may also be polygenic, usually expressed as quantitative or partial resistance, in which within-host multiplication of the virus is reduced. Examples of quantitative resistance depending on the plant-virus genotype x genotype interaction are also present in the literature. Such is the case of the interaction between *Cucumber mosaic virus* (CMV) and its natural host *Arabidopsis thaliana* (Pagán *et al.*, 2007; Pagán *et al.*, 2010a; Pagán *et al.*, 2014). Polygenic resistance has been used in crop breeding for virus disease control, and virus genotypes overcoming these resistances have been reported (García-Arenal and McDonald, 2003). The other major defense strategy of plants against pathogens, tolerance, is determined by genotype x genotype interactions. Tolerance of *Arabidopsis thaliana* to CMV, which depends on modification of life-history traits, varies according to the plant and virus genotype (Pagán *et al.*, 2008; Pagán *et al.*, 2009). Also, *Zucchini yellow mosaic virus* (ZYMV) genotypes that overcome melon tolerance become prevalent in the virus population after the extensive use of tolerant varieties (Desbiez *et al.*, 2002; Desbiez *et al.*, 2003). In summary, there is evidence that the outcome of plant-virus interactions depends on genotype x genotype interactions regardless of the type of plant defense or interaction model.

Finally, the fourth condition for plant-virus co-evolution is the existence of changes in genotype frequencies in both the host and the pathogen populations. Again, this evidence derives mostly form the analysis of agricultural systems, in which virus genotype frequency changes according to crop genotype management by humans. Even in agricultural systems, long-term analyses of these dynamics are rather rare, but they have demonstrated that the frequency of virus genotypes changes in response to the deployment of genetic resistance of crop varieties (see Fraile *et al.*, 2011). Note that changes in genotype frequency require the existence of polymorphisms for defense in the host and for infectivity and virulence in the virus, as discussed above, and that these traits may have fitness costs for both the virus and the host.

Fitness costs in plant-virus co-evolution

In GFG systems it is considered that fitness penalties associated with defense and infectivity are a necessary, but not sufficient, condition for the maintenance of such polymorphisms in the host and virus populations (Agrawal and Lively, 2002; Brown and Tellier, 2011; Dybdahl *et al.*, 2014). In spite of abundant analyses, not only with plant host systems, evidence for costs of resistance remains controversial, and there is none for any resistance to viruses (Bergelson and Purrington, 1996; Mauricio, 1998; Bergelson *et al.*, 2001; Brown, 2003; Martins *et al.*, 2014). However, artificial evolution of the resistance gene *Rx* by introducing random point mutations in the LRR domain resulted in the appearance of variants with enlarged recognition of *Potato virus X* (PVX) genotypes or even of distantly related viruses (Farnham and Baulcombe, 2006). Alleles with this broad recognition specificity do not occur in nature, which suggests that fitness penalties constrain the evolution of *R* genes.

Experimental evidence of infectivity costs is more abundant, derives from several systems, and has been discussed in detail by García-Arenal and Fraile (2013). Mutations responsible for overcoming *Ry*-mediated resistance of potato to PVY are lethal (Mestre et al. 2003), and no resistance-breaking genotype has been reported from the field. Resistance-breaking mutations often result in a diminished ability for within-host multiplication, as shown for different plant-virus systems (Goulden *et al.*, 1993; Jenner *et al.*, 2002; Janzac *et al.*, 2010; Fraile *et al.*, 2011; Ishibashi *et al.*, 2012; Khatabi *et al.*, 2013). In most cases data show that high fitness penalties are associated with resistance-breaking, resulting in a diminished competitive ability of the resistance-breaking genotype in the non-resistant host (García-Arenal and Fraile, 2013). Because of the multifunctionality of viral proteins, mutations resulting in resistance breaking may affect also other components of the virus fitness, as transmission by vectors or through the seed (Murant *et al.*, 1968; Hanada and Harrison, 1977). A recently published report documents how CP mutations in PMMoV resulting in overcoming *L*-gene allele resistance also results in changes in the stability of virus particles, which affects survival in the environment (Fraile *et al.*, 2014). It is important to note that these fitness components are not directly related to the plant-virus interactions and, thus, that changes in the genetic structure of the host plant population may drive changes in the virus population through several different mechanisms.

Costs of resistance and infectivity are not predicted in pure MA models (Agrawal and Lively, 2002) but, as discussed in the previous section, it could be that most MA-like interactions would follow a relaxed model, in which mutations resulting in resistance-breaking would allow infection in various host genotypes with different efficiencies and with different levels of associated fitness penalties. Again, evidence for these penalties derives from the analysis of the rice-RYMV and pepper-PVY systems. In RYMV, the different mutations that result in overcoming *rymv1-2* resistance in rice have either no fitness penalties or different fitness penalties in the susceptible host, and these costs may be, according to the mutation, positively correlated with the level of virus multiplication in the resistance host (Poulicard *et al.,* 2010; Poulicard *et al.,* 2012). The relative fitness of various PVY mutants overcoming *pvr2*^{3} was tested in pepper genotypes homozygous for either the resistance allele *pvr2*^{3} or for the susceptibility allele *pvr2*$^{+}$. There were fitness differences among five resistance-breaking PVY genotypes in the resistant host, but not in the susceptible one (Ayme *et al.,* 2006). These experimental data agree with results of field studies in which PVY isolates that were able to overcome recessive resistance in tobacco or pepper, were frequent in the absence of the corresponding host resistance (Lacroix *et al.,* 2010; Ben-Khalifa *et al.,* 2012), suggesting there was no selection against them. Thus, at least some infectivity mutations in PVY can be without cost. Results from these two pathosystems thus agree with expectations under a relaxed MA model (Agrawal and Lively, 2002; Dybdahl *et al.,* 2014).

From plant-virus co-evolution to emergence in new host populations

Landscape heterogeneity and infection dynamics

It has been long postulated by plant pathologists that ecological changes associated with agriculture favor epidemic infection dynamics of highly virulent pathogens, that is, a change from a stable plant-pathogen co-evolutionary dynamic to disease emergence. Three main factors are considered to have a major role in this process: the reduced species diversity of agricultural ecosystems as compared with wild ecosystems, the reduced genetic diversity of crops as compared with their wild relatives, and the higher host density (Burdon and Chilvers, 1982; Thresh, 1982), all three factors being the consequence of ecosystem simplification (Figure 5.1). Support for these hypotheses is scant, and derives mostly from circumstantial or historical evidence (Stukenbrock and McDonald, 2008). Current declines in biodiversity and increasing awareness of pathogen emergence have resulted in a new interest on the relationship between ecosystem simplification and disease risk (Keesing *et al.,* 2010). Two hypotheses representing extreme situations have been proposed that relate ecosystem diversity to disease risk: diversity may be either positively ("Amplification Effect" hypothesis) or negatively ("Dilution Effect" hypothesis) correlated with disease risk, as higher biodiversity may result in either increased abundance of reservoirs for a focal host, or on a decreased abundance of the focal host hindering pathogen transmission (Keesing *et al.,* 2006; Ostfeld and Keesing, 2012). Hence, the effects of diversity on disease risk would be related to the host range of the pathogen.

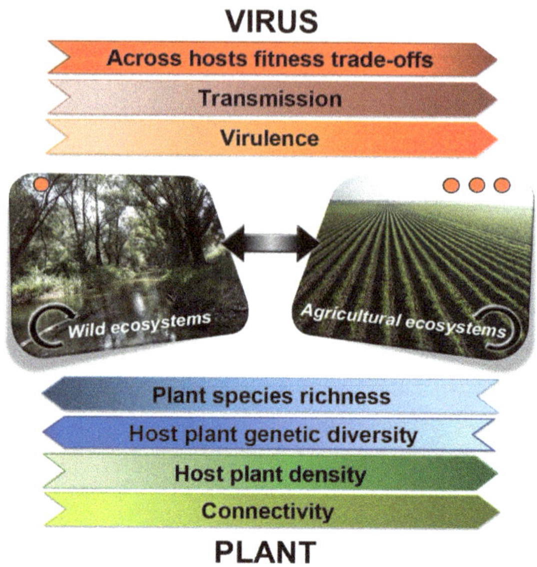

Figure 5.1. Plant and virus traits affecting disease emergence in wild vs. agricultural ecosystems. Each colored arrow represents a different trait, and the direction of the arrow indicates increasing value of the trait. Red dots indicate risk of disease emergence (one dot: low risk; three dots: high risk). Black arrows indicate direction of possible host jumps.

The relationship between biodiversity and disease risk has been analyzed in *Cereal-* and *Barley yellow dwarf viruses* (C/BYDV). These viruses have a broad host range within the *Poaceae*, and their effect in wild grassland ecosystems in western North America has been extensively studied. Most published results are compatible with the amplification effect hypothesis, but conclusions depend on the composition of the analyzed species assemblage, as grasses differ in their ability to support virus multiplication and in their competence as sources for virus transmission (Power and Mitchell, 2004; Malmstrom *et al.,* 2005a; 2005b; Borer *et al.,* 2009; 2010; Hall *et al.,* 2010; Power *et al.,* 2011). A recent study has analyzed virus incidence in Mexican populations of the wild pepper or chiltepin, *Capsicum annuum* var. *glabriusculum*, growing in habitats along a gradient of human management, from wild to cultivated populations (Pagán *et al.,* 2012). Increased human management was associated with an increase of virus infection risk, and the main predictor of disease risk was the biodiversity of the habitat, agreeing with the dilution effect hypothesis. Landscape heterogeneity was also a determinant of the evolution of the two most frequent virus species, belonging to the genus *Begomovirus*, which resulted in phylogenetic congruence between the host and virus populations from wild habitats. Interestingly, cultivation resulted in the loss of the spatial genetic structure of both the host and the viruses, disrupting phylogenetic congruence (Rodelo-Urrego *et al.,* 2013). These results indicate that at the landscape scale, cultivation disrupts the local adaptation of hosts and viruses, and suggest the passage along the agro-ecological interface from an interaction potentially resulting in co-evolution to an interaction resulting in emergence. This effect of cultivation may be explained by translocation of plants and, most probably, viruses and vectors (González-Jara *et al.,* 2011; Rodelo-

Urrego *et al.,* 2013), i.e., by an increased connectivity between plant and virus populations as a result of human activity.

Increased connectivity between host plant populations, and between virus and host populations, has been known to favor emergence for a long time (Figure 5.1), as documented by the effects of introducing host plant species, either cultivated or wild, and viruses or their vectors, into new geographical areas (reviewed in Elena *et al.,* 2014). It should be noted, however, that the sharing of an area by a virus and a potential host does not necessarily result in the connectivity between both populations. Thus, analyses of infection by CMV, a typical generalist virus with a very large host range, along an agro-ecological gradient in Central Spain showed that the infection dynamics varied largely among the hosts plant species occupying different habitats, strongly suggesting that inoculum fluxes across hosts and habitats were restricted (Sacristán *et al.,* 2004).

Landscape heterogeneity and virulence evolution

Ecosystem simplification along the agro-ecological gradient may also determine the evolution of virus virulence. Theory assumes that trade-offs exist between different parasite life-history traits, mostly between transmission and virulence, so that in horizontally transmitted parasites, the level of virulence will be constrained by the optimization of transmission (Frank, 1996; Day and Proulx, 2004; Alizon *et al.,* 2009; Ebert, 2013; Bull and Lauring, 2014). Thus, for specialist viruses, the increase of host density associated with ecosystem simplification will result in higher transmission rates and in the evolution to higher levels of virulence, i.e., more virulent viruses will be present in agricultural than in wild ecosystems (Figure 5.1). Accordingly, the virulence of begomovirus infections in chiltepin was observed to increase with increased human management of the habitat (Pagán *et al.,* 2012; Rodelo-Urrego *et al.,* 2013). Also, both field observations and simulations with models using experimentally estimated parameters of transmission and virulence, have shown that the emergence of highly virulent CMV strains supporting CMV satellite RNAs necrogenic for tomato was dependent on high transmission rates, which required high density of aphid vector populations (Escriu *et al.,* 2000; 2003). Most plant viruses are generalists or multihost parasites (see the "*Evidence of across-host fitness trade offs*" section), i.e., are able to infect different hosts belonging to different taxa (Woolhouse *et al.,* 2001). Theoretical analyses that consider infection in multiple-host populations identify among-host heterogeneity in resistance and virulence, costs of infecting different hosts, and differences in within-host and between-host transmission rates, as the main factors that determine the evolution of generalist parasites. These analyses mostly predict that virulence will evolve to levels below the optima for each host (Ebert and Hamilton, 1996; Regoes *et al.,* 2000; Gandon *et al.,* 2002; Dobson, 2004; Gandon, 2004; Williams, 2012). The analysis of the evolution of CMV virulence in a mixed host population in which two host plant species differed in resistance, tolerance and competence for virus transmission, showed evidence for some but not all of these predictions: evolution towards high virulence levels required high transmission rates and high between-host transmission, and in the less competent host, virulence was highly dependent on transmission from the more competent one. However, at high transmission rates, virulence reached a maximum, and not an intermediate level, in both hosts (Betancourt *et al.,* 2013). This work, thus, emphasizes the relevance of host competence for virus multiplication and transmission in determining the infection dynamics and the evolution of virulence.

The relationship between host competence, virus emergence and virulence has been better analyzed in the C/BYDV-wild grass ecosystems of Western USA mentioned before. It was shown that virus infection had a prominent role in determining the composition and dynamics of the host population. This was so because the grass species that supported higher virus multiplication levels and were more competent for transmission acted as reservoirs for infection of less competent species, in which infection was more virulent (Power and Mitchell, 2004; Mitchell and Power, 2006). The determinants of host reservoir potential, i.e., host susceptibility and competence, were negatively correlated with host lifespan, so that shorter-lived grass species had a higher reservoir potential than longer-lived ones (Cronin *et al.,* 2010; 2014). The relationship between host susceptibility, competence, reservoir potential and lifespan has been shown to hold also at the host intra-species level in analyses of the interaction of *A. thaliana* genotypes and CMV (Hily *et al.,* 2014). Interestingly, since the reservoir potential of a short-lived genotype will be balanced against the longer infectious period of a long-lived genotype, reservoir potential and infection in the heterogeneous host population will depend on the host population demography and genetic composition, and on the density of the vector populations (Hily *et al.,* 2014). Since, for a given vector density, transmission rates depend on the host population density and genetic structure, the interactions among these factors will determine the evolution of virulence, but experimental evidence of this is lacking.

In vertically transmitted parasites, the trade-offs between transmission and virulence will result in evolution towards reduced virulence or even to mutualism, a situation in which transmission-virulence trade-offs will no longer hold (Anderson and May, 1982; Ewald, 1987; Yamamura, 1996; Ebert, 2013). Many plant viruses are transmitted both vertically and horizontally, in different stages of their life cycle (Hull, 2014), and the relative weight of the different transmission mode could affect the evolution of virulence. It has been shown through experimental evolution analyses that vertical transmission of different plant viruses results in lower virulence (Stewart *et al.,* 2005; Pagán *et al.,* 2014). Because different host plant species or different host genotypes differ in the rates of vertical transmission of the different virus species for which they are hosts (Hull, 2014; Pagán *et al.,* 2014) the relative importance of vertical *vs.* horizontal transmission will depend on the genetic structure and density of the host population. The effects of the triple interaction between mode of transmission, host demography and genetic structure of the host population on virulence evolution have not yet been explored, to our knowledge.

Evolution of host range

The survival and reproduction of plant viruses depends on their ability to encounter susceptible hosts and exploit the resources that these hosts provide. Consequently, host range, i.e., the number of species that a parasite can infect, is considered a major determinant of the evolution of parasites, including plant viruses (Frank, 1996; Woolhouse *et al.,* 2001; García-Arenal and Fraile, 2013). As other parasites, plant viruses can be classified in two major categories according to their host range: (i) Specialists, which are plant viruses able to infect one or a few closely related host species, where these viruses are highly transmissible and efficiently exploit host resources (Futuyama and Moreno, 1988); and (ii) Generalists, which are plant viruses able to infect many host species, often from taxonomically distant plant families (Whitlock, 1996), an ability that provides such viruses with more chances of survival and transmission. Because biological fitness varies across environments, virus fitness shall vary across host species (Elena *et al.,* 2014). As a consequence, it is often assumed that generalism comes at the cost of suboptimal virus transmission and host resource exploitation, as generalist viruses cannot maximize their fitness in all hosts. Therefore, the key idea underlying this theory is that across-host fitness trade-offs are common in host-virus interactions, or in other words, that increasing of virus fitness in a new host generally occurs at the expense of reducing its fitness in the original host (García-Arenal and Fraile, 2013). Consequently, for any given host plant species, a specialist virus would have higher fitness than a generalist virus, which has led to the proposal that evolution would favor specialist viruses (Whitlock, 1996; Woolhouse *et al.,* 2001; Kirchner and Roy, 2002; Elena *et al.,* 2009; 2014). However, in a study considering 474 vector-transmitted plant viruses, only 9.9% had one single host (Brunt *et al.,* 1996), and data on natural host range of 29 viruses indicated that it was restricted to a single genus in 17% of them, whereas 35% of the viruses infected hosts species of different taxonomical families (García-Arenal and McDonald, 2003). Hence, many plant viruses seem to be generalists. Is that because across-host fitness trade-offs are uncommon in plant-virus interactions? What are the mechanisms that determine across-host fitness trade-offs? In this section we review work done to tackle these questions.

Evidence of across-host fitness trade-offs

Analyses of across-host fitness trade-offs come mainly from three different approaches. First, adaptation to different hosts has been studied in field isolates of several generalist plant viruses. For instance, as mentioned earlier, incidence, within-host multiplication and transmission of *Barley yellow dwarf virus* PAV strain (BYDV-PAV) varied in seven species of the plant family *Poaceae*, indicating differences in host adaptation (Power and Mitchell, 2004; Cronin *et al.,* 2010; Power *et al.,* 2011). Variation in within-host multiplication across wild and cultivated *Solanum* species also has been reported for *Pepino mosaic virus* (PepMV) as indicative of fitness trade-offs (Moreno-Pérez *et al.,* 2014). Data on the incidence of five plant viruses in 21 plant species showed an association between viruses and hosts, indicating that virus fitness varies across hosts (Malpica *et al.,* 2006). In other cases, such as the analysis of three CMV isolates collected from species of different plant families, evidence of trade-offs has been found only for some virus fitness traits (infectivity), but not for others (within-host multiplication) (Sacristán *et al.,* 2005). All these reports are in agreement with the assumption that virus fitness differs across hosts. On the other hand, some examples of the absence of fitness trade-offs in plant-virus interactions can be also found in the literature. For instance, three CMV genotyes differed in within-host multiplication level, but the genotypes that accumulated to highest or lowest levels in two different hosts, tomato and melon, were the same, indicating no across-host trade-off (Escriu *et al.,* 2003; Betancourt *et al.,* 2011).

The second group of evidence of across-hosts fitness trade-offs derives from phylogenetic and population genetics analyses, as genetic structure of the virus population according to the host may be indicative of host adaptation. Such host-associated structure of plant virus populations has been described for some species of plant viruses (García-Arenal *et al.,* 2001; Jeger *et al.,* 2006; Moury *et al.,* 2006; Elena *et al.,* 2011), and also at higher taxonomic levels like genus or family (Lartey *et al.,* 1996; Wu *et al.,* 2008; Pagán and Holmes, 2010; Thompson *et al.,* 2014). This is not exclusive for plant viruses, and genetic structure of animal and human virus populations at different taxonomic levels according to host has been also described (Kitchen *et al.,* 2011). However, all these studies only provide indirect evidence because, as we discussed in previous sections, other phenomena might also result in the congruence of host and virus phylogenies. Exceptions are the analysis of the adaptation of *Turnip mosaic virus* (TuMV) to species of the genera *Brassica* and *Raphanus*, and of PepMV to species of the genus *Solanum*, for which the genetic structure of the virus population according to host has been shown to be linked to the process of host adaptation (Ohshima *et al.,* 2002; Tomimura *et al.,* 2003; Nguyen *et al.,* 2013; Moreno-Pérez *et al.,* 2014).

The third group of evidence of across-hosts fitness trade-offs, and perhaps the most abundant, is based on serial host passage experiments. Adaptation of PVX to tobacco, of *Hibiscus chlorotic spot virus* (HCSV) to *Chenopodium quinoa*, of *Tobacco etch virus* (TEV) to pepper, and of *Plum pox virus* (PPV) to pea, resulted in a reduction of virus fitness in the original hosts: potato, hibiscus, tobacco and peach, respectively (Yarwood, 1979; Liang *et al.,* 2002; Wallis *et al.,* 2007; Agudelo-Romero *et al.,* 2008). Hence, an important conclusion from this third group of analyses is that, when challenged against only one host, plant viruses tend to act as specialists. There are very few examples of serial passage experiments addressing the behavior of plant viruses when adapting to more than one host. In the most detailed of these multihost experiments, Bedhomme et al. passaged TEV alternating two different natural hosts, and they did not find evidence of cost of adaptation to either of these hosts (Bedhomme *et al.,* 2012). In the same sense, passages of six CMV isolates in their original hosts did not result in fitness improvements, which suggest that this generalist virus had reached its fitness maximum in all hosts (Sacristán *et al.,* 2005). These results are compatible with observations for most animal viruses in similar experimental designs (Elena *et al.,* 2009). Altogether, these studies suggest that the environment, meaning the environment as the host availability, determines host range strategy of plant viruses: when only one host species is available, plant viruses would favor a specialist strategy, generally associated with fitness trade-offs, whereas when more than one host is available, plant viruses would tend to behave as a generalist (Elena *et al.,* 2009; 2014). Note that the bias of studies towards viruses in crops, usually focusing in only one host, may condition our perception of host range evolution and trade-offs in plant viruses.

Mechanisms involved in across-host fitness trade-offs

Two main mechanisms have been proposed to determine across-host fitness trade-offs. The first is the existence of antagonistic pleiotropy associated with adaptation to different hosts (Whitlock, 1996). That is, a mutation that is beneficial in one host may be detrimental in another, which is likely to occur because plant virus genomes are highly compacted, frequently harboring overlapping genes and multifunctional proteins. Hence, in both the original and the new hosts, which mutations get fixed in the virus population is driven by selection (Fry, 1996;

Whitlock, 1996). Evidence of antagonistic pleiotropy associated with host adaptation is abundant for plant viruses. For instance, it has been shown that both gain of HCSV and *Pelargonium flower break virus* (PFBV) fitness in a new host, and loss in the original host, is determined by mutations in a few specific amino acid sites of the virus CP (Liang *et al.*, 2002; Rico *et al.*, 2006). Similarly, a single amino acid substitution in the CP of PVY is responsible of increased fitness in tobacco and reduced in potato (Moury and Simon, 2011). Nucleotide substitutions in PVY, RYMV, ToMV, TuMV and tobamoviruses that are responsible for overcoming genetic resistance of different host species also reduce the competitive ability of the mutant virus in comparison with its wild type in the original susceptible host (Jenner *et al.*, 2002; Poulicard *et al.*, 2010; Ishibashi *et al.*, 2012; Fraile *et al.*, 2011; Quenouille *et al.*, 2013), another indication of antagonistic pleiotropy.

The second mechanism that may be involved in across-host fitness trade-offs is the accumulation of mutations that are neutral in the original host but have positive/negative fitness effects in the new host (Kawecki, 1994). This means that, at odds with antagonistic pleiotropy, in the original host mutations get fixed by genetic drift, whereas in the new host mutation fixation occurs through selection. In a given host, the fitness of a virus depends on the distribution of mutational fitness effect (DMFE); that is, the frequency of mutations that have positive, negative or neutral effects on virus fitness. Due to the compactness of viral genomes it is expected that, in hosts for which a virus is well adapted, most mutations will have negative fitness effects or will be lethal (Elena *et al.*, 2006). Indeed, analyses of the mutation spectrum of TEV in its primary host *N. benthamiana* (Carrasco *et al.*, 2007), and in several animal RNA viruses (Sanjuán *et al.*, 2004), indicated that most mutations were lethal or strongly deleterious. The DMFE may be determined by the environmental context, as mutations in the viral genome that are neutral in one host may have either deleterious or beneficial effects in another, which leads to across-host fitness trade-offs (Martin and Lenormand, 2006). To our knowledge, the only existing analysis of changes in DMFE of a plant virus across hosts refers to TEV. Fitness measures of a set of single-nucleotide mutants, derived from Carrasco *et al.*, 2007 experiments, in eight different hosts indicated that DMFE depended on the host. Moreover, the frequency of mutation with positive fitness effects increased as the host taxonomic proximity with the primary one decreased (Lalić *et al.*, 2011), in agreement with observations in animal viruses (Elena *et al.*, 2009). The DMFE will obviously vary also according to the genetic context in which each mutation occurs. For instance, mutations in regions of gene overlap are more likely to be deleterious (Belshaw *et al.*, 2008). Accordingly, for a wide range of animal and plant viruses the rate of evolution has been shown to be smaller in overlapping than in non-overlapping regions of the same gene (Simon-Loriere *et al.*, 2013). In addition, the effect of a given mutation may depend on its interaction with pre-existing mutations, such that some combinations are more likely to get fixed in the virus genome than others. In other words, mutational effects may be epistatic (Kimura and Maruyama, 1966; Elena *et al.*, 2006). The role of epistasis in shaping the genome of plant viruses has been studied for TEV, showing that epistatic effects were mostly negative, i.e., the joint effect of two deleterious mutations is lesser than expected from the effect of each individual mutation (Lalić and Elena, 2012). Subsequent work with the same experimental system demonstrated that epistatic effects among pairs of mutations depended on the host plant species, and was less the more closely related the host species (Lalić and Elena, 2013). Finally, it is worth noting that the appearance of mutations that increase viral fitness in a new host during the process of adaptation does not exclusively occur in the original and/or in the new hosts. For some beneficial mutations to arise in the virus populations, specific mutational pathways are required that can only occur in a third host, referred to as intermediate host, because intermediate mutants of such pathways have extremely low fitness in both the original and in the new hosts. Analyses of across host fitness trade-offs that include the original, the intermediate and the new hosts are restricted to animal viruses (Allison *et al.*, 2012; Stucker *et al.*, 2012). Equivalent studies in plant viruses would yield invaluable information on the process of host range determination in plant viruses.

Antagonistic pleiotropy and the accumulation of mutations are not mutually exclusive. Both can act at the same time and their relative contribution in determining the virus fitness may vary depending on the host. Hence, a full understanding of the mechanisms that determine across host fitness trade-offs requires analyzing the relative importance of both mechanisms in determining virus fitness across hosts. Unfortunately, very few examples of such type of analyses are available in the literature for either animal or plant viruses (Elena *et al.*, 2009; 2014). For plant viruses, it has been shown that the effect of some mutations in TEV fitness varied according to the host in which they occur, with a fraction of these mutations showing opposite fitness effects (i.e., antagonistic epistasis) (Lalić *et al.*, 2011).

Altogether, the work discussed in this section shows that across host fitness trade-offs are modulated by both the genetic and the environmental (i.e., host) context in which mutations occur. However, the information presented here derives from the study of only a handful of host-virus interactions, in which wild hosts/viruses are scantly represented. Further analyses of other host-virus interactions, specially those occurring in wild ecosystems, may help to determine the generality of trade-offs associated with changes in host range, and the relative importance of antagonistic pleiotropy and the accumulation of mutations.

Concluding remarks

In the past decades, important advances have been achieved in the knowledge of the molecular basis of plant-virus interactions. The large amount of information on this subject is in sharp contrast with our currently fragmentary knowledge on the patterns and evolution of plant-virus interactions, even though molecular genetics and evolution of plant-virus interactions are intrinsically linked. Although there is an abundance of theoretical studies modeling the evolution of host-parasite interactions, only a few are specific to plant-virus interactions. In addition, the scant empirical/experimental analyses accompanying such models is restricted to a handful of specific pairwise interactions that mostly involve crops in agricultural ecosystems (Table 5.1). Because by definition agricultural landscapes are simplified ecosystems, the analysis of crop-virus interactions can yield only a partial view of all the richness and complexity of plant-virus interactions that can be found in nature. Analyses in wild ecosystems have been undertaken only recently and, although they are still in their childhood, these studies have already revealed key information on plant-virus interactions that could not be extracted from crop-virus systems. An increasing effort in the analysis of wild ecosystems may yield highly relevant information to understand key aspects of plant-virus interactions. We think that future studies would pay special attention to aspects that include, but are not restricted to:

- Whether viruses are generally pathogens of plants, and/or in which conditions they behave as such, as commensals, or as beneficial symbionts of plants. To answer this question, the combination of NGS data, which will allow estimating the biodiversity of viruses in wild and agricultural ecosystems, and information on the incidence and virulence of virus infection in these ecosystems will be invaluable (Paterson and Piertney, 2011; Roossinck *et al.,* 2011). Information of how viruses that are pathogens of crops affect the fitness of their wild hosts, and conversely the effect in crops of viruses whose primary host is a wild plant species, will also help us to understand under what conditions virus virulence is increased or reduced, and which factors favor disease emergence (Elena *et al.,* 2014; Roossinck and García-Arenal, 2015).

- Whether observations in support of plant-virus co-evolution, which currently derive mostly from crop-virus interactions, hold in wild ecosystems. In this regard, analyses in wild ecosystems should consider exploring: i) if reciprocal fitness effects of infection and defense exist in wild plant-virus interactions, and if so, how common they are; and ii) if analyses of co-evolution based on pairwise interactions are relevant for understanding plant-virus interactions at the community level, where viruses will be able to infect a variety of hosts, and hosts will be exposed/infected by a variety of viruses. The increasing number of theoretical models and bioinformatic tools designed to analyze mutualistic/infection networks, rather than pairwise interactions, can be very helpful to address these challenges. Network models open the possibility of understanding how plant-virus interactions evolve at scales other than the single individual (Nuismer *et al.,* 2013), and may allow exploration of how viruses shape plant communities, ecosystems, and even how viruses interact with their host plants across the agro-ecological interface (Alexander *et al.,* 2014).

- Whether and how changes in ecosystem complexity and organization will affect infection dynamics and plant-virus interactions, including the evolution of both host range and virulence. Most of our knowledge on plant-virus interactions derives from analysis at local scale. However, plant virus dispersion, invasion and infection, and therefore virus evolution, occurs at broader scales and through landscapes whose architecture is defined by plant communities with different levels of complexity, connectivity and ecological dynamics (Meentemeyer *et al.,* 2012; Alexander *et al.,* 2014). Consequently, studies on the evolution of plant-virus interactions at the landscape scale will help us to gain a better understanding of how such interactions change depending on the context in which they occur. Because human intervention is one of the major factors shaping the organization and complexity of plant communities, analyses at the landscape scale should also include the comparison of the epidemiology and evolution of plant viruses in host populations under different levels of human management (Meentemeyer *et al.,* 2012; Pagán *et al.,* 2012; Rodelo-Urrego *et al.,* 2013). The information that this type of studies may yield would be relevant not only from an academic point of view, it can also contribute to improve existing approaches for disease management and ecosystem conservation (Meentemeyer *et al.,* 2012; Vander Wal *et al.,* 2014).

- How disease risk, and virulence evolution and the associated trade-offs, are influenced by the demography and genetic structure of host populations. Although these are not new questions (see Read *et al.,* 1994), they still remain open. Opposing hypotheses propose that the level of genetic diversity in the host population may either increase or decrease disease risk and pathogen virulence (Burdon and Thrall, 2008; Ostfeld and Keesing, 2012; Lively *et al.,* 2014). Most likely neither of these hypotheses are right or wrong, and risk of infection and pathogen virulence rather depends on the identity and relative abundance of the species/genotypes present in a plant community (Randolph and Dobson, 2012; Karasov *et al.,* 2014). Hence, population diversity in host plant features such as reservoir potential, frequency of resistant genotypes, and the main mode of pathogen transmission in the host population may determine disease risk and virulence evolution (Randolph and Dobson, 2012; Ebert, 2013; Hily *et al.,* 2014; Karasov *et al.,* 2014), and associated virulence-transmission and virulence-multiplication trade-offs (Bull and Lauring, 2014). Most of what is known on the role of these traits in the evolution of plant-virus interactions again derives from laboratory-based experiments. Equivalent analyses in wild plant communities would be an avenue for future research.

Addressing these questions will be a most stimulating challenge in the next few years, and will much contribute to understand plant-virus interactions, and host-parasite interactions in general.

References

Agnew, P., Koella, J.C., and Michalakis, Y. (2000). Host life-history responses to parasitism. Microbes Infect. *2*, 891-896.

Agrawal, A., and Livel,y C.M. (2002). Infection genetics: gene-for-gene versus matching-alleles models and all points in between. Evol. Ecol. Res. *4*, 79-90.

Agudelo-Romero, P., Carbonell, P., Pérez-Amador, M.A., and Elena, S.F. (2008). Virus adaptation by manipulation of host's gene expression. PLoS ONE. *3*, e2397.

Alexander, H.M., Mauck K.E., Whitfield A.E., Garrett K.A., and Malmstrom C.M. (2014). Plant-virus interactions and the agro-ecological interface. Eur. J. Plant Pathol. *138*, 529-547.

Alizon, S., Hurford, A., Mideo, N., and Van, B.M. (2009). Virulence evolution and the trade-off hypothesis: history, current state of affairs and the future. J. Evol. Biol. *22*, 245-259.

Allison, A.B., Harbison, C.E., Pagan, I., Stucker, K.M., Kaelber, J.T., Brown, J.D., Ruder, M.G., Keel, M.K., Dubovi, E.J., Holmes, E.C., and Parrish, C.R. (2012). Role of multiple hosts in the cross-species transmission and emergence of a pandemic parvovirus. J. Virol. *86*, 865-872.

Anderson, J.A., Ghabrial, S.A., and Taylor, N.L. (1991). Natural incidence of peanut stunt virus infection in hybrid populations of *Trifolium ambiguum X T. repens*. Plant Dis. *75*, 156-159.

Anderson, P.K., Cunningham, A.A., Patel, N.G., Morales, F.J., Epstein, P.R., and Daszak, P. (2004). Emerging infectious diseases of plants: pathogen pollution, climate change and agrotechnology drivers. Trends Ecol. Evol. *19*, 535-544.

Anderson, R.M., and May, R.M. (1982). Coevolution of hosts and parasites. Parasitology. *85*, 411-426.

Archie, E.A., Luikart, G., and Ezenwa, V.O. (2009). Infecting epidemiology with genetics: a new frontier in disease ecology. Trends Ecol. Evol. *24*, 21-30.

Ayme, V., Souche, S., Caranta, C., Jacquemond, M., Chadœuf, J., Palloix, A., Moury, B. (2006). Different mutations in the genome-linked protein VPg of *Potato virus Y*

confer virulence on the *pvr2*^{3} resistance in pepper. Mol. Plant-Microbe Interact. *19*, 557-563.
Bedhomme, S., Lafforgue, G., and Elena, S.F. (2012). Multihost experimental evolution of a plant RNA virus reveals local adaptation and host-specific mutations. Mol. Biol. Evol. *29*, 1481-1492.
Belshaw, R., Gardner, A., Rambaut, A., and Pybus, O.G. (2008). Pacing a small cage: mutation and RNA viruses. Trends Ecol. Evol. *23*, 188-193.
Ben-Khalifa, M., Simon, V., Fakhfakh, H., and Moury, B. (2012). Tunisian *Potato virus Y* isolates with unnecessary pathogenicity towards pepper: support for the matching allele model in eIF4E resistance–potyvirus interactions. Plant Pathol. *61*, 441-447.
Bergelson, J., Dwyer, G., and Emerson, J.J. (2001). Models and data on plant-enemy coevolution. Annu. Rev. Genet. *35*, 469-499.
Bergelson, J., and Purrington, C.B. (1996). Surveying patterns in the cost of resistance in plants. Am. Nat. *148*, 536-558.
Berzal-Herranz, A., de la Cruz, A., Tenllado, F., Díaz-Ruíz, J.R., López, L., Sanz, A. I., Vaquero, C., Serra, M.T., and García-Luque, I. (1995). The *Capsicum* L^3 gene-mediated resistance against the tobamoviruses is elicited by the coat protein. Virology. *209*, 498-505.
Betancourt, M., Escriu, F., Fraile, A., and García-Arenal, F. (2013). Virulence evolution of a generalist plant virus in a heterogeneous host system. Evol. Appl. *6*, 875-890.
Betancourt, M., Fraile, A., and García-Arenal, F. (2011). *Cucumber mosaic virus* satellite RNAs that induce similar symptoms in melon plants show large differences in fitness. J. Gen. Virol. *92*, 1930-1938.
Biek, R., and Real, L.A. (2010). The landscape genetics of infectious disease emergence and spread. Mol. Ecol. *19*, 3515-3531.
Blitzer, E.J., Dormann, C.F., Holzschuh, A., Klein, A.M., Rand, T.A., and Tscharntke, T. (2012). Spillover of functionally important organisms between managed and natural habitats. Agric. Ecosyst. Environ. *146*, 34-43.
Borer, E.T., Adams, V.T., Engler, G.A., Adams, A.L., Schumann, C.B., and Seabloom, E.W. (2009). Aphid fecundity and grassland invasion: invader life history is the key. Ecol. Applic. *19*, 1187-1196.
Borer, E.T., Seabloom, E.W., Mitchell, C.E., and Power, A.G. (2010). Local context drives infection of grasses by vector-borne generalist viruses. Ecol. Lett. *13*, 810-818.
Brown, J.K.M. (2003). A cost of disease resistance: Paradigm or peculiarity? Trends Genet. *19*, 667-671.
Brown, J.K.M., and Tellier, A. (2011). Plant-parasite coevolution: bridging the gap between genetics and ecology. Annu. Rev. Phytopathol. *49*, 345-367.
Brunt, A., Crabtree, K., Dallwitz, M., Gibbs, A., Watson, L., and Zurcher, E. (1996). Plant viruses online. Descriptions and list from the VIDE database. [http://pvo.bio-mirror.cn/refs.htm].
Bull, J.J., and Lauring, A.S. (2014). Theory and empiricism in virulence evolution. PLoS Pathog. *10*, e1004387.
Burdon, J.J., and Chilvers, G.A. (1982). Host density as a factor in plant-disease ecology. Annu. Rev. Phytopathol. *20*, 143-166.
Burdon, J.J., and Thrall, P.H. (2008). Pathogen evolution across the agro-ecological interface: implications for disease management. Evol. Appl. *1*, 57-65.
Burdon, J.J., and Thrall, P.H. (2014). What have we learned from studies of wild plant-pathogen associations?—the dynamic interplay of time, space and life-history. Eur. J. Plant Pathol. *138*, 417-429.
Carrasco, P., de la Iglesia, F., and Elena, S.F. (2007). Distribution of fitness and virulence effects caused by single-nucleotide substitutions in *Tobacco etch virus*. J. Virol. *81*, 12979-12984.
Chisholm, S.T., Mahajan, S.K., Whitham, S.A., Yamamoto, M.L., and Carrington, J.C. (2000). Cloning of the Arabidopsis *RTM1* gene, which controls restriction of long-distance movement of tobacco etch virus. Proc. Natl. Acad. Sci. USA. *97*, 489-494.

Clarke, D.D. (1986). Tolerance of parasites and disease in plants and its significance in host-parasite interactions. Adv. Plant Pathol. *5*, 161-198.
Cooper, I., and Jones, R.A.C. (2006). Wild plants and viruses: underinvestigated ecosystems. Adv. Virus Res. *67*, 1-47.
Costanza, R., and Mageau, M. (2000). What is a healthy ecosystem? *Aquat. Ecol.* 33, 105-115.
Cronin, J.P., Rua, M.A., and Mitchell, C.E. (2014). Why is living fast dangerous? Disentangling the roles of resistance and tolerance of disease. Am. Nat. *184*, 172-187.
Cronin, J.P., Welsh, M.E., Dekkers, M.G., Abercrombie, S.T., and Mitchell, C.E. (2010). Host physiological phenotype explains pathogen reservoir potential. Ecol. Lett. *13*, 1221-1232.
Culver, J.N. (2002). *Tobacco mosaic virus* assembly and disassembly: Determinants in pathogenicity and resistance. Annu. Rev. Phytopathol. *40*, 287-308.
Day, T., and Proulx, S.R. (2004). A general theory for the evolutionary dynamics of virulence. Am. Nat. *163*, E40-E63
D'Arcy, C.J., Eastburn, D.M., and Schumann, G.L. (2001). Illustrated glossary of plant pathology. Plant Health Instr. doi: 10.1094/PHI-I-2001-0219-01.
Delwart, E.L. (2007). Viral metagenomics. Rev. Med. Virol. *17*, 115-131.
Desbiez, C., Wipf-Scheibel, C., and Lecoq, H. (2002). Biological and serological variability, evolution and molecular epidemiology of *Zucchini yellow mosaic virus* (ZYMV, Potyvirus) with special reference to Caribbean islands. Virus Res. *85*, 5-16.
Desbiez, C., Gal-On, A., Girard, M., Wipf-Scheibel, C., and Lecoq, H. (2003). Increase in *Zucchini yellow mosaic virus* symptom severity in tolerant zucchini cultivars is related to a point mutation in P3 protein and is associated with a loss of relative fitness on susceptible plants. Phytopathology. *93*, 1478-1484.
Dobson, A. (2004). Population dynamics of pathogens with multiple host species. Am. Nat. *164*, S64-S78.
Dybdahl, M.F., Jenkins, C.E., and Nuismer, S.L. (2014). Identifying the molecular basis of host-parasite coevolution: Merging models and mechanisms. Am. Nat. 184, 1-13.
Ebert, D. (2013). The epidemiology and evolution of symbionts with mixed-mode transmission. Annu. Rev. Ecol. Syst. *44*, 623-643.
Ebert, D., and Bull, J.J. (2003). Challenging the trade-off model for the evolution of virulence: is virulence management feasible? Trends Microbiol. *11*, 15-20.
Ebert, D., and Hamilton, W.D. (1996). Sex against virulence: the coevolution of parasitic diseases. Trends Ecol. Evol. *11*, 79-82.
Elena, S.F., Agudelo-Romero, P., and Lalić, J. (2009). The evolution of viruses in multi-host fitness landscapes. O. Virol. J. *3*, 1-6.
Elena, S.F., Bedhomme, S., Carrasco, P. Cuevas, J.M., de la Iglesia, F., Lafforgue, G., Lalić, J., Pròsper, A., Tromas, N., and Zwart, M.P. (2011). The evolutionary genetics of emerging plant RNA viruses. Mol. Plant–Microbe Interact. *24*, 287-293.
Elena, S.F., Carrasco, P., Darós, J.A., and Sanjuán, R. (2006). Mechanisms of genetic robust- ness in RNA viruses. EMBO Rep. *7*, 168-173.
Elena, S.F., Fraile, A., and García-Arenal, F. (2014). Evolution and emergence of plant viruses. Adv. Virus Res. *88*, 161-191.
Escriu, F., Fraile, A., and García-Arenal, F. (2003). The evolution of virulence in a plant virus. Evolution. *57*, 755-765.
Escriu, F., Perry, K., and García-Arenal, F. (2000). Transmissibility of *Cucumber mosaic virus* by *Aphis gossypii* correlates with viral accumulation and is affected by the presence of its satellite RNA. Phytopathology. *90*, 1068-1072.
Ewald, P.W. (1987). Transmission modes and evolution of the parasitism-mutualism continuum. Ann. NY Acad. Sci. *503*, 295-306.
Fabre, F., Rousseau, E., Mailleret, L., and Moury, B. (2012). Durable strategies to deploy plant resistance in agricultural landscapes. New Phytol. *193*, 1064-1075.
Fargette, D., Pinel-Galzi, A., Sereme, D., Lacombe, S., Hebrard, E., Traore, O., and Konate, G. (2008). Diversification of *Rice yellow mottle virus* and related viruses

spans the history of agriculture from the Neolithic to the present. PLoS Pathog. *4*, e1000125.
Farnham, G., and Baulcombe, D.C. (2006). Artificial evolution extends the spectrum of viruses that are targeted by a disease-resistance gene from potato. Proc. Natl. Acad. Sci. USA. *103*, 18828-18833.
Fraile, A., and García-Arenal, F. (2010). The coevolution of plants and viruses: Resistance and pathogenicity. Adv. Virus Res. *76*, 1-32.
Fraile, A., Hily, J.M., Pagán, I., Pacios, L.F. and García-Arenal, F. (2014). Host resistance selects for traits unrelated to resistance-breaking that affect fitness in a plant virus. Mol. Biol. Evol. *31*, 928-939.
Fraile, A., Pagán, I., Anastasio, G., Saez, E., and García-Arenal, F. (2011). Rapid genetic diversification and high fitness penalties associated with pathogenicity evolution in a plant virus. Mol. Biol. Evol. *28*, 1425-1437.
Frank, S.A. (1996). Models of parasite virulence. Q. Rev. Biol. *71*, 37-78.
Friedman, A.R., and Baker, B.J. (2007). The evolution of resistance genes in multi-protein plant resistance systems. Curr. Opin. Genet. Dev. *17*, 493-499.
Friess, N., and Maillet, J. (1996). Influence of cucumber mosaic virus infection on the intraspecific competitive ability and fitness of purslane (*Portulaca oleracea*). New Phytol. *132*, 103-111.
Froissart, R., Doumayrou, J., Vuillaume, F., Alizon, S., and Michalakis, Y. (2010). The virulence-transmission trade-off in vector-borne plant viruses: a review of (non-)existing studies. Phil. Trans. R. Soc. B. *365*, 1907-1918.
Fry, J.D. (1996). The evolution of host specialization: are tradeoffs over- rated? Am. Nat. *148*, S84-S107.
Futuyma, D.J., and Moreno, G. (1988). The evolution of ecological specialization. Annu. Rev. Ecol. Syst. *19*, 207-233.
Gandon, S. (2004). Evolution of multihost parasites. Evolution. *58*, 455-469.
Gandon, S., Van Baalen, M., and Jansen, V.A. (2002). The evolution of parasite virulence, superinfection, and host resistance. Am. Nat. *159*, 658-669.
García-Arenal, F., and Fraile, A. (2013). Trade-offs in host range evolution of plant viruses. Plant Pathol. *62*, S1-S8.
García-Arenal, F., Fraile, A., and Malpica, J.M. (2001). Variability and genetic structure of plant virus populations. Annu. Rev. Phytopathol. *39*, 157-186.
García-Arenal, F., and McDonald, B.A. (2003). An analysis of the durability of resistance to plant viruses. Phytopathology *93*, 941-952.
Gibbs, A. (1980). Plant virus that partially protects its wild legume host against herbivores. Intervirology. *13*, 42-47.
Gibbs, A.J., Fargette, D., Garcia-Arenal, F., and Gibbs, M.J. (2010). Time–the emerging dimension of plant virus studies. J. Gen. Virol. *91*, 13-22.
Gibbs, A.J., Ohshima, K., Phillips, M.J., and Gibbs, M.J. (2008). The prehistory of potyviruses: their initial radiation was during the dawn of agriculture. PLoS ONE. *3*, e2523.
González-Jara, P., Moreno-Letelier, A., Fraile, A., Piñero, D., García-Arenal, F. (2011). Impact of human management on the genetic variation of wild pepper, *Capsicum annuum* var. *glabriusculum*. PLoS ONE. *6*, e28715.
Goulden, M.G., Köhm, B.A., Santa Cruz, S., Kavanagh, T.A., and Baulcombe, D.C. (1993). A feature of the coat protein of *Potato virus X* affects both induced virus resistance in potato and viral fitness. Virology. *197*, 293-302.
Hall, G.S., Peters, J.S., Little, D.P., and Power, A.G. (2010). Plant community diversity influences vector behavior and Barley yellow dwarf virus population structure. Plant Pathol. *59*, 152-1158.
Hamada, H., Takeuchi, S., Kiba, A., Tsuda, S., Hikichi, Y., and Okuno, T. (2002). Amino acid changes in *Pepper mild mottle virus* coat protein that affect *L*^{3} gene-mediated resistance in pepper. J. Gen. Plant Pathol. *68*, 155-162.
Hamada, H., Tomita, R., Iwadate, Y., Kobayashi, K., Minemura, I., Takeuchi, S., Hikichi, Y., and Suzuki, K. (2007). Cooperative effect of two amino acid mutations in the coat

protein of *Pepper mild mottle virus* overcomes L^3-mediated resistance in *Capsicum* plants. Virus Genes. *34*, 205-214.
Hanada, K., and Harrison, B.D. (1977). Effects of virus genotype and temperature on seed transmission of nepoviruses. Ann. Appl. Biol. *85*, 79-92.
Hily, J.M., García, A., Moreno, A., Plaza, M., Wilkinson, M.D., Fereres, A., Fraile, A., and García-Arenal, F. (2014). The relationship between host lifespan and pathogen reservoir potential: An analysis in the system *Arabidopsis thaliana-Cucumber mosaic virus*. PLoS Pathog. *10*, e1004492.
Hudson, P.J., Dobson, A.P., and Lafferty, K.D. (2006). Is a healthy ecosystem one that is rich in parasites? Trends Ecol. Evol. *21*, 381-385.
Hull, R. (2014). Plant Virology 5th Ed. (San Diego, USA: Academic Press)
Hurwitz, B.L., and Sullivan, M.B. (2013). The Pacific Ocean Virome (POV): A marine viral metagenomic dataset and associated protein clusters for quantitative viral ecology. PLoS ONE. *8*, e57355.
Ishibashi, K., Kezuka, Y., Kobayashi, C., Kato, M., Inoue, T., Nonaka, T., Ishikawa, M., Matsumura, H., and Katoh, E. (2014). Structural basis for the recognition-evasion arms race between *Tomato mosaic virus* and the resistance gene *Tm-1*. Proc. Natl. Acad. Sci. USA. *111*, 3486-3495.
Ishibashi, K., Masuda, K., Naito, S., Meshi, T., and Ishikawa, M. (2007). An inhibitor of viral RNA replication is encoded by a plant resistance gene. Proc. Natl. Acad. Sci. USA. *104*, 13833-13838.
Ishibashi, K., Mawatari, N., Miyashita, S., Kishino, H., Meshi, T., and Ishikawa, M. (2012). Coevolution and hierarchical interactions of *Tomato mosaic virus* and the resistance gene *Tm-1*. PLoS Pathog. *8*, e1002975.
Janzac, B., Montarry, J., Palloix, A., Navaud, O., and Moury, B. (2010). A point mutation in the polymerase of *Potato virus Y* confers virulence toward the *Pvr4* resistance of pepper and a high competitiveness cost in susceptible cultivar. Mol. Plant–Microbe Interact. *23*, 823-830.
Jeger, M.J., Seal, S.E., and Van den Bosch, F. (2006). Evolutionary epidemiology of plant virus disease. Adv. Virus Res. *67*, 163-203.
Jenner, C.E., Wang, X., Ponz, F., and Walsh, J.A. (2002). A fitness cost for *Turnip mosaic virus* to overcome host resistance. Virus Res. *86*, 1-6.
Jones, J.D.G., and Dangl, J.L. (2006). The plant immune system. Nature. *444*, 323-329.
Jones, R.A.C. (2009). Plant virus emergence and evolution: origins, new encounter scenarios, factors driving emergence, effects of changing world conditions, and prospects for control. Virus Res. *141*, 113-130.
Kang, B.C., Yeam, I., Frantz, J.D., Murphy, J.F., and Jahn, M.M. (2005). The *pvr1* locus in pepper encodes a translation initiation factor eIF4E that interacts with *Tobacco etch virus* VPg. Plant J. *41*, 392-405.
Karasov, T.L., Horton, M.W., and Bergelson, J. (2014). Genomic variability as a driver of plant–pathogen coevolution? Curr. Opin. Plant Biol. *18*, 24-30.
Kawecki, T.J., (1994). Accumulation of deleterious mutations and the evolutionary cost of being a generalist. Am. Nat. *144*, 833-838.
Keesing, F., Belden, L.K., Daszak, P., Dobson, A., Harvell, C.D., Holt, R.D., Hudson, P., Jolles, A., Jones, K.E., Mitchell, C.E., Myers, S.S., Bogich, T., and Ostfeld, R.S. (2010). Impacts of biodiversity on the emergence and transmission of infectious diseases. Nature. *468*, 647-652.
Keesing, F., Holt, R.D., and Ostfeld, R.S. (2006). Effects of species diversity on disease risk. Ecol. Lett. *9*, 485-498.
Kelly, S.E. (1994). Viral pathogens and the advantage of sex in the perennial grass *Anthoxantum odoratum*: A review. Philos. Trans. R. Soc. Lond. B. *346*, 295-302.
Khatabi, B., Wen, R.H., Hajimorad, M.R. (2013). Fitness penalty in susceptible host is associated with virulence of Soybean mosaic virus on *Rsv1*-genotype soybean: a consequence of perturbation of HC-Pro and not P3. Mol. Plant Pathol. *14*, 885-897.

Kimura, M., and Maruyama, T. (1966). The mutational load with epistatic gene interactions in fitness. Genetics. *54*, 1337-1351.
Kirchner, J.W., and Roy B.A. (2002). Evolutionary implications of host–pathogen specificity: fitness consequences of pathogen virulence traits. Evol. Ecol. Res. *4*, 27-48.
Kitchen, A., Shackelton, L.A. and Holmes, E.C. (2011). Family level phylogenies reveal modes of macroevolution in RNA viruses. Proc. Natl. Acad. Sci. USA. *108*, 238-243.
Kovalchuk, I., Kovalchuk, O., Kalck, V., Boyko, V., Filkowski, J., Heinlein, M., and Hohn, B. (2003). Pathogen-induced systemic plant signal triggers DNA rearrangements. Nature. *423*, 760-762.
Lacroix, C., Glais, L., Kerlan, C., Verrier, J.L., and Jacquot, E. (2010). Biological characterization of French *Potato virus Y* (PVY) isolates collected from PVY-susceptible or -resistant tobacco plants possessing the recessive resistance gene *va*. Plant Pathol. *59*, 1133-1143.
Lalić, J., Cuevas, J.M., and Elena, S.F. (2011). Effect of host species on the distribution of mutational fitness effects for an RNA virus. PLoS Genet. *7*, e1002378.
Lalić, J., and Elena, S.F. (2012). Magnitude and sign epistasis among deleterious mutations in a positive-sense plant RNA virus. Heredity. *109*, 71-77.
Lalić, J., and Elena, S.F. (2013). Epistasis between mutations is host-dependent for an RNA virus. Biol. Lett. *9*, 20120396.
Lanfermeijer, F.C., Warmink, J., and Hille, J. (2005). The products of the broken *Tm-2* and the durable *Tm-2*^{2} resistance genes from tomato differ in four amino acids. J. Exp. Bot. *56*, 2925-2933.
Lartey, R.T., Voss, T.C., Melcher, U. (1996). Tobamovirus evolution: gene overlaps, recombination, and taxonomic implications. Mol. Biol. Evol. *13*, 1327-1338.
Liang, X.-Z., Lee, B.T.K., and Wong, S.-M. (2002). Covariation in the capsid protein of *Hibiscus chlorotic ringspot virus* induced by serial passaging in a host that restricts movement leads to avirulence in its systemic host. J. Virol. 76, 12320-12324.
Little, T.J., Shuker, D.M., Colegrave, N., Day, N., and Graham, A.L. (2010). The coevolution of virulence: Tolerance in perspective. PLoS Pathog. *6*, e1001006.
Lively, C.M., de Roode, J.C., Duffy, M.A., Graham, A.L., and Koskella, B. (2014). Interesting open questions in disease ecology and evolution. Am. Nat. *184*, S1-S8.
Malmstrom, C.M., Hughes, C.C., Newton, L.A., and Stoner, C.J. (2005a). Virus infection in remnant native bunchgrasses from invaded California grasslands. New Phytol. *168*, 217-230.
Malmstrom, C.M., McCullough, A.J., Johnson, H.A., Newton, L.A., and Borer, E.T. (2005b). Invasive annual grasses indirectly increase virus incidence in California native perennial bunchgrasses. Oecologia *145*, 153-164.
Malpica, J.M., Sacristán, S., Fraile, A., García-Arenal, F. (2006). Association and host selectivity in multi-host pathogens. PLoS ONE. *1*, e41.
Margaria, P., Ciuffo, M., Pacifico, D., and Turina, M. (2007). Evidence that the nonstructural protein of *Tomato spotted wilt virus* is the avirulence determinant in the interaction with resistant pepper carrying the *Tsw* gene. Mol. Plant Microbe Interact. *20*, 547-558.
Martin, G., and Lenormand, T. (2006). The fitness effect of mutations across environments: A survey in light of fitness landscapes models. Evolution. *60*, 2413-2427.
Martins, N.E., Faria, V.G., Nolte, V., Schlötterer, C., Teixeira, L., Sucena, É., Magalhães, S. (2014). Host adaptation to viruses relies on few genes with different cross-resistance properties. Proc. Natl. Acad. Sci. USA. *111*, 5938-5943.
Maskell, L.C., Raybould, A.F., Cooper, J.I., Edwards, M.L., and Gray, A.J. (1999). Effects of turnip mosaic virus and turnip yellow mosaic virus on the survival, growth and reproduction of wild cabbage (*Brassica oleracea*). Ann. Appl. Biol. *135*, 401-407.
Maule, A., Caranta, C., and Boulton, M. (2007). Sources of natural resistance to plant viruses: Status and prospects. Mol. Plant Pathol. *8*, 223-231.

Mauricio, R. (1998). Costs of resistance to natural enemies in field populations of the annual plant *Arabidopsis thaliana*. Am. Nat. *151*, 20-28.

Meentemeyer, R.K., Haas, S.E., and Tomáš, V. (2012). Landscape epidemiology of emerging infectious diseases in natural and human-altered ecosystems. Annu. Rev. Phytopathol. *50*, 379-402.

Mestre, P., Brigneti, G., Durrant, M.C., and Baulcombe, D.C. (2003). *Potato virus Y* NIa protease activity is not sufficient for elicitation of Ry-mediated disease resistance in potato. Plant J. *36*, 755-761.

Mitchell, C.E., and Power, A.G. (2006). Disease dynamics in plant communities. In Disease ecology: community structure and pathogen dynamics, S.K. Collinge and C. Ray, eds. (Oxford, UK: Oxford University Press), pp. 58-72.

Molinier, J., Ries, G., Zipfel, C., and Hohn, B. (2006). Transgeneration memory of stress in plants. Nature. *442*, 1046-1049.

Moreno-Pérez, M.G., Pagán, I., Aragón-Caballero, L., Cáceres, F., Fraile, A., García-Arenal, F. (2014). Ecological and genetic determinants of *Pepino mosaic virus* emergence. J. Virol. *88*, 3359-3368.

Moury, B., Desbiez, C., Jacquemond, M., and Lecoq, H. (2006). Genetic diversity of plant virus populations: towards hypothesis testing in molecular epidemiology. Adv. Virus Res. *67*, 49-87.

Moury, B., Janzac, B., Ruellan, Y., Simon, V., Ben Khalifa, M., Fakhfakh, H., Fabre, F., and Palloix, A. (2014). Interaction patterns between *Potato virus Y* and eIF4E-mediated recessive resistance in the *Solanaceae*. J. Vir. 88, 9799-9807.

Moury, B., and Simon, V. (2011). dN/dS-based methods detect positive selection linked to trade-offs between different fitness traits in the coat protein of *Potato virus Y*. Mol. Biol. Evol. *28*, 2707-2717.

Murant, A., Taylor, C.E., and Chambers, J. (1968). Properties relationships and transmission of a strain of raspberry ringspot virus infecting raspberry cultivars immune to common Scottish strain. Ann. Appl. Biol. *61*, 175-186.

Nguyen, H.D., Tomitaka, Y., Ho, S.Y.W., Duchêne, S., Vetten, H.J., Lesemann, D., Walsh, J.A., Gibbs, A.J., and Ohshima, K. (2013). Turnip mosaic potyvirus probably first spread to Eurasian brassica crops from wild orchids about 1000 years ago. PLoS ONE. *8*, e55336.

Nieberding, C.M., and Olivieri, I. (2007). Parasites: proxies for host genealogy and ecology? Trends Ecol. Evol. *22*, 156-165.

Nuismer, S.L., Jordano, P., and Bascompte, J. (2013). Coevolution and the architecture of mutualistic networks. Evolution. *67*, 338-354.

Ohshima, K., Yamaguchi, Y., Hirota, R., Hamamoto, T., Tomimura, K., Tan, Z., Sano, T., Azuhata, F., Walsh, J.A., Fletcher, J., Chen, J., Gera, A., and Gibbs, A. (2002). Molecular evolution of *Turnip mosaic virus*: evidence of host adaptation, genetic recombination and geographical spread. J. Gen. Virol. *83*, 1511-1521.

Ostfeld, R.S., and Keesing F. (2012). Effects of host diversity on infectious disease. Annu. Rev. Ecol. Evol. Syst. *43*, 157-182

Pagán, I., Alonso-Blanco, C., and García-Arenal, F. (2007). The relationship of within-host multiplication and virulence in a plant-virus system. PLoS ONE. *2*, e786.

Pagán, I., Alonso-Blanco, C., and García-Arenal, F. (2008). Host responses in life-history traits and tolerance to virus infection in *Arabidopsis thaliana*. PLoS Pathog. *4*, e1000124.

Pagán, I., Alonso-Blanco, C., and García-Arenal, F. (2009). Differential tolerance to direct and indirect density-dependent costs of viral infection in *Arabidopsis thaliana*. PLoS Pathog. *5*, e1000531.

Pagán, I., Fraile, A., Fernández-Fuello, E., Montes, N., Alonso-Blanco, C., and García-Arenal, F. (2010a). *Arabidopsis thaliana* as a model for plant-virus co-evolution. Phil. Trans. R. Soc. B. *365*, 1983-1995.

Pagán, I., Firth, C. and Holmes, E.C. (2010b). Phylogenetic analysis reveals rapid evolutionary dynamics in the plant RNA virus genus *Tobamovirus*. J. Mol. Evol. *71*, 298-307.

Pagán, I., González-Jara, P., Moreno-Letelier, A., Rodelo-Urrego, M., Fraile, A., Piñero, D., García-Arenal, F. (2012). Effect of biodiversity changes in disease risk: exploring disease emergence in a plant-virus system. PLoS Pathog. *8*, e1002796.

Pagán, I., and Holmes, E.C. (2010). Long-term evolution of the *Luteoviridae*: time scale and mode of virus speciation. J. Virol. *84*, 6177-6187.

Pagán, I., Montes, N., Milgroom, M.G., and García-Arenal, F. (2014). Vertical transmission selects for reduced virulence in a plant virus and for increased resistance in the host. PLoS Pathog. *10*, e1004293.

Paterson, S., and Piertney, S.B. (2011). Frontiers in host-parasite ecology and evolution. Mol. Ecol. *20*, 869-871.

Pinel-Galzi, A.S., Rakotomalala, M., Sangu, E., Sorho, F., Kanyeka, Z., Traore, O., Sereme, D., Poulicard, N., Rabenantoandro, Y., Sere, Y., Konate, G., Ghesquiere, A., Hébrard, E., and Fargette, D. (2007). Theme and variations in the evolutionary pathways to virulence of an RNA plant virus species. PLoS Pathog. 3, e180.

Poulicard, N., Pinel-Galzi, A., Fargette, D., and Hebrard, E., (2014). Alternative mutational pathways, outside the VPg, of rice yellow mottle virus to overcome eIF(iso)4G-mediated rice resistance under strong genetic constraints. J. Gen. Virol. *95*, 219-224.

Poulicard, N., Pinel-Galzi, A., Hebrard, E., and Fargette, D. (2010). Why *Rice yellow mottle virus*, a rapidly evolving RNA plant virus, is not efficient at breaking *rymv1-2* resistance. Mol. Plant Pathol. *11*, 145-154.

Poulicard, N., Pinel-Galzi, A., Traoré, O., Vignols, F., Ghesquière, A, Konaté, G., Hébrard, E., and Fargette, D. (2012). Historical contingencies modulate the adaptability of *Rice yellow mottle virus*. PLoS Pathog. *8*, e1002482.

Power, A.G., Borer, E.T., Hosseini, P., Mitchell, C.E., and Seabloom, E.W. (2011). The community ecology of barley/ cereal yellow dwarf viruses in Western US grasslands. Virus Res. *159*, 95-100.

Power, A.G., and Mitchell, C.E. (2004). Pathogen spillover in disease epidemics. Am. Nat. *164*, S79-S89.

Prendeville, H.R., Tenhumberg, B., and Pilson, D. (2014). Effects of virus on plant fecundity and population dynamics. New Phytol. *202*, 1346-1356.

Prendeville, H.R., Ye, X.H., Morris, T.J., and Pilson, D. (2012). Virus infections in wild plant populations are both frequent and often unapparent. Am. J. Bot. *99*, 1033-1042.

Quenoiulle, J., Vassilakos, N., and Moury, B. (2013). *Potato virus Y*: a major crop pathogen that has provided major insights into the evolution of viral pathogenicity. Mol. Plant Pathol. *14*, 439-452.

Randolph, S.E., and Dobson, A.D.M. (2012). Pangloss revisited: a critique of the dilution effect and the biodiversity-buffers-disease paradigm. Parasitology. *139*, 847-863.

Read, A.F. (1994). The evolution of virulence. Trends Microbiol. *2*, 73-76.

Regoes, R. R., Nowak, M.A., and Bonhoeffer, S. (2000). Evolution of virulence in a heterogeneous host population. Evolution. *54*, 64-71.

Rico, P., Ivars, P., Elena, S.F., and Hernandez, C. (2006). Insights into the selective pressures restricting *Pelargonium flower break virus* genome variability: Evidence for host adaptation. J. Virol. *80*, 8124-8132.

Robaglia, C. and Caranta, C. (2006). Translation initiation factors: a weak link in plant RNA virus infection. Trends Plant Sci. *11*, 40-45.

Rodelo-Urrego, M., Pagán, I., González-Jara, P., Betancourt, M., Moreno-Letelier, A., Ayllón, M.A., Fraile, A., Piñero, D., and García-Arenal, F. (2013). Landscape heterogeneity shapes host-parasite interactions and results in apparent plantvirus codivergence. Mol. Ecol. *22*, 2325-2340.

Roossinck, M.J. (2005). Symbiosis versus competition in plant virus evolution. Nat. Rev. Microbiol. *3*, 917-924.

Roossinck, M.J. (2011a). The big unknown: Plant virus biodiversity. Curr. Op. Virol. *1*, 63-67.

Roossinck, M.J. (2011b). Changes in population dynamics in mutualistic versus pathogenic viruses. Viruses. *3*, 12-19.
Roossinck, M.J. (2012). Plant virus metagenomics: Biodiversity and ecology. Annu. Rev. Genet. *46*, 357-367.
Roossinck, M.J., and García-Arenal, F. (2015). Ecosystem simplification, biodiversity loss and plant virus emergence. Curr. Op. Virol. *In press*.
Roossinck, M.J., Saha, P., Wiley, G.B., Quan, J., White, J.D., Lai, H., Chavarría, F., Shen, G., and Roe, B.A. (2010). Ecogenomics: Using massively parallel pyrosequencing to understand virus ecology. Mol. Ecol. *19*, S81-S88.
Rua, M.A., Pollina, E.C., Power, A.G., and Mitchell, C.E. (2011). The role of viruses in biological invasions: friend or foe? Curr. Opin. Virol. *1*, 68-72.
Sacristán, S., Fraile, A., and García-Arenal, F. (2004). Population dynamics of *Cucumber mosaic virus* in melon crops and in weeds in Central Spain. Phytopathology. *94*, 992-998.
Sacristán, S., Fraile, A., Malpica, J.M., and García-Arenal, F. (2005). An analysis of host adaptation and its relationship with virulence in *Cucumber mosaic virus*. Phytopathology. *95*, 827-833.
Sacristán, S., and García-Arenal, F. (2007). The evolution of virulence and pathogenicity in plant pathogen populations. Mol. Plant Pathol. *9*, 369-384.
Sanjuán, R., Moya, A., and Elena, S.F. (2004). The distribution of fitness effects caused by single-nucleotide substitutions in an RNA virus. Proc. Nat. Acad. Sci. USA. *101*, 8396-8401.
Sherwood, R.T. (1997). Viruses of white clover in pastures of Pennsylvania, New York, and Vermont. Plant Dis. *81*, 817-820.
Simon-Loriere, E., Holmes, E.C., and Pagán, I. (2013). The effect of gene overlapping on the rate of RNA virus evolution. Mol. Biol. Evol. *30*, 1916-1928.
Stewart, A.D., Logsdon, J.M., and Kelley, S.E. (2005). An empirical study of the evolution of virulence under both horizontal and vertical transmission. Evolution. *59*, 730-739.
Stobbe, A.H., Melcherl, U., Palmer, M.W., Roossinck, M.J., and Shen, G. (2012). Co-divergence and host-switching in the evolution of tobamoviruses. J. Gen. Virol. *93*, 408-418.
Stobbe, A.H., and Roossinck, M.J. (2014). Plant virus metagenomics: what we know and why we need to know more. Front. Plant Sci. *5*, 150.
Stucker, K.M., Pagan, I., Cifuente, J.O., Kaelber, J.T., Lillie, T.D., Hafenstein, S., Holmes, E.C., and Parrish, C.R. (2012). The role of evolutionary intermediates in the host adaptation of canine parvovirus. J. Virol. *86*, 1514-1521.
Stukenbrock, E.H., and McDonald, B.A. (2008). The origin of plant pathogens in agro-ecosystems. Annu. Rev. Phytopathol. *46*, 75-100.
Syller, J. (2012). Facilitative and antagonistic interactions between plant viruses in mixed infections. Mol. Plant Pathol. *13*, 204-216.
Thompson, J.R., Kamath, N., and Perry, K.L. (2014). An evolutionary analysis of the *Secoviridae* family of viruses. PLoS ONE. *9*, e106305.
Thresh, J.M. (1982). Cropping practices and virus spread. Annu. Rev. Phytopathol. *20*, 193-218.
Tomimura, K., Gibbs, A.J., Jenner, C.E., Walsh, J.A., and Ohshima, K., (2003). The phylogeny of *Turnip mosaic virus*; comparisons of 38 genomic sequences reveal a Eurasian origin and a recent 'emergence' in east Asia. Mol. Ecol. *12*, 2099-2111.
Tomita, R., Murai, J., Miura, Y., Ishikara, H., Liu, S., Kubotera, Y., Honda, A., Hatta, R., Kuroda, T., Hamada, H., Sakamoto, M., Munemura, I., Nunomura, O., Ishikawa, K., Genda, Y., Kawasaki, S., Suzuki, K., Meksem, K., and Kobayashi, K. (2008). Fine mapping and DNA fiber FISH analysis locates the tobamovirus resistance gene L^3 of *Capsicum chinense* in a 400-kb region of R-like genes cluster embedded in highly repetitive sequences. Theor. Appl. Genet. *117*, 1107-1118.

Tsuda, S., Kirita, M., and Watanabe, Y. (1998). Characterization of a pepper mild mottle tobamovirus strain capable of overcoming the *L*³ gene-mediated resistance, distinct from the resistance-breaking Italian isolate. Mol. Plant Microbe Interact. *11*, 327-331.

Vander Wal, E., Garant, D., Calmé, S., Chapman, C.A., Festa-Bianchet, M., Millien, V., Rioux-Paquette, S., and Pelletier, F. (2014). Applying evolutionary concepts to wildlife disease ecology and management. Evol. Appl. *7*, 856-868.

Wallis, C.M., Stone, A.L., Sherman, D.J., Damsteegt, V.D., Gildow, F.E., and Schneider, W.L. (2007). Adaptation of *Plum pox virus* to a herbaceous host (*Pisum sativum*) following serial passages. J. Gen. Virol. *88*, 2839-2845.

Weitz, J.S., Poisot, T., Meyer, J.R., Flores, C.O., Valverde, S., Sullivan, M.B., and Hochberg , M.E. (2013). Phage-bacteria interaction networks. Trends Microbiol. *21*, 82-91.

Whitham, S.A., Anderberg, R.J., Chisholm, S.T., and Carrington, J.C. (2000). Arabidopsis *RTM2* gene is necessary for specific restriction of tobacco etch virus and encodes an unusual small heat shock-like protein. Plant Cell. *12*, 569-582.

Whitlock, M.C. (1996). The Red Queen beats the jack-of-all-trades: the limitations on the evolution of phenotypic plasticity and niche breadth. Am. Nat. *148*, S65-S77.

Williams, P. D. (2012). New insights into virulence evolution in multigroup hosts. Am. Nat. *179*, 228-239.

Woolhouse, M.E.J., Taylor, L.H., and Haydon, D.T. (2001). Population biology of multi-host pathogens. Science. *292*, 1109-1112.

Woolhouse, M.E.J., Webster, J.P., Domingo, E., Charlesworth, B., and Levin, B.R. (2002). Biological and biomedical implications of the co-evolution of pathogens and their hosts. Nat. Genet. *32*, 569-577.

Wren, J.D., Roossinck, M.J., Nelson, R.S., Scheets, K., Palmer, M.W., and Melcher, U. (2006). Plant virus biodiversity and ecology. PLoS Biol. *4*, 314-315.

Wu, B., Melcher, U., Guo, X., Wang, X., Fan, L., and Zhou, G. (2008). Assessment of codivergence of *Mastrevirus* with their plant hosts. BMC Evol. Biol. *8*, 335.

Xu, P., Chen, F., Mannas, J.P., Feldman, T., Sumner, L.W., and Roossinck, M.J. (2008). Virus infection improves drought tolerance. New Phytol. *180*, 911-921.

Yamamura, N. (1996). Evolution of mutualistic symbiosis: A differential equation model. Res. Popul. Ecol. *38*, 211-218.

Yarwood, C.E. (1979). Host passage effects with plant viruses. Adv. Virus Res. *25*, 169-190.

6

Evolution of Viral Virulence: Empirical Studies

Gael Kurath[1]* and Andrew R. Wargo[2]

[1]*USGS Western Fisheries Research Center, Seattle, Washington 98115, USA;* [2]*Virginia Institute of Marine Science, Gloucester Point, VA 23062-1346, USA*

**Corresponding author: gkurath@usgs.gov*

Abstract

The concept of virulence as a pathogen trait that can evolve in response to selection has led to a large body of virulence evolution theory developed in the 1980-1990s. Various aspects of this theory predict increased or decreased virulence in response to a complex array of selection pressures including mode of transmission, changes in host, mixed infection, vector-borne transmission, environmental changes, host vaccination, host resistance, and co-evolution of virus and host. A fundamental concept is prediction of trade-offs between the costs and benefits associated with higher virulence, leading to selection of optimal virulence levels. Through a combination of observational and experimental studies, including experimental evolution of viruses during serial passage, many of these predictions have now been explored in systems ranging from bacteriophage to viruses of plants, invertebrates, and vertebrate hosts. This chapter summarizes empirical studies of viral virulence evolution in numerous diverse systems, including the classic models myxomavirus in rabbits, Marek's disease virus in chickens, and HIV in humans. Collectively these studies support some aspects of virulence evolution theory, suggest modifications for other aspects, and show that predictions may apply in some virus:host interactions but not in others. Finally, we consider how virulence evolution theory applies to disease management in the field.

Virulence as a target of viral evolution

As a viral trait, virulence is fundamentally important because of our natural desire to avoid, treat, and control infectious diseases. In general terms, virulence refers to harm or reduced fitness caused to a host due to infection with a pathogen. Within viral species, different strains often vary in virulence: a virus strain that causes only mild disease is referred to as having low virulence, while a strain that causes severe disease has high virulence. The concept of pathogen virulence as a target of evolution was presented in published literature as early as the 1880s (Alizon *et al.,* 2009; Methot, 2012), and many theories have arisen to try to explain the basic mechanism(s) of virulence evolution. Although our focus here is on viruses, these theories address virulence as a trait of all pathogens, including viruses, bacteria, and cellular and multicellular parasites.

The first major theory to gain widespread acceptance among both scientists and the general public was the **avirulance theory**, which hypothesized that pathogens should always evolve toward being benign in order to preserve their hosts as resources for pathogen replication and transmission (Burnet and White, 1972; Ewald, 1983; Schmid-Hempel, 2011; Smith, 1904). Proposed at least as early as 1904 (Smith, 1904), this hypothesis became so dominant in the mid-1900s that it was referred to by Anderson and May in 1982 as the "conventional wisdom" (Anderson and May, 1982a). Although the idea is based on inherently appealing logic, observations of increasing pathogen virulence in some situations, and pathogens that stably maintain high virulence over long time periods, suggested that this theory is not as universal as once believed.

A major shift in the field took place in the 1980s with the development of a body of theory referred to as **Virulence Evolution Theory** (Schmid-Hempel, 2011), which questioned the generality of the avirulence theory, and used mathematical models to explore possible trade-off mechanisms that could explain evolutionary selection for different levels of virulence. In these models virulence is hypothesized to have both fitness costs and benefits for the pathogen. Originally based on a relatively simple equation for the basic reproductive number of a pathogen, R_0 (described below), the success of the theory led to development of many variations on the basic model that address how various parameters and situations might drive evolution of increased or decreased virulence. Although this has been an active field of theory for over three decades, empirical confirmation of published predictions has been relatively slow to follow. In recent years there has been a response to the call for increased empirical work in this field. Several excellent reviews have been published describing the mathematical basis and models that underpin current theory of virulence evolution (Alizon *et al.,* 2009; Bull, 1994; Lipsitch and Moxon, 1997; Sacristan and Garcia-Arenal, 2008; Schmid-Hempel, 2011). Therefore, here we provide only a brief introduction to virulence evolution theory, and then focus on the status of empirical support for various aspects of the theory, specifically as it applies to evolution of viral pathogens.

Definitions of virulence

At the surface, virulence seems to be a simple concept, but in actuality it is surprisingly difficult to define. This has led to the use of many different definitions amongst the scientific community, which have, in some cases, become system-specific (Table 6.1). In the broadest sense virulence can be defined as the magnitude of the **fitness costs that a virus causes to its host** (Bull, 1994; Read, 1994). As such, virulence is at the root of the parasitic relationship, in which the parasitic organism (in this case a virus) obtains fitness benefits at the fitness cost of the host. These fitness costs to the host therefore represent the virulence of the virus. This is the classical positive-negative relationship seen in text books when describing the parasitic type of symbiosis (Campbell, 1996). If we continue with this symbiosis concept, a completely avirulent virus that does not have a negative fitness effect on its host would be a commensal (either host or symbiont gain in fitness, usually symbiont, and the other experiences no fitness impacts) or even a mutualist (both host and symbiont gain in fitness).

A challenge with this broad definition of virulence is that measuring virus impacts on overall host fitness is exceedingly difficult in most systems. To do so would require absolute quantification of how viral infection impacts host fecundity. This has been done mostly in plant and invertebrate host-virus systems (Read, 1994; Sacristan and Garcia-Arenal, 2008; Schmid-Hempel, 2011), where host generation times are relatively short. Due to the difficulty of these types of studies, another commonly used definition of virulence is **harm done to the host as a result of infection with a pathogen** (Alizon *et al.,* 2009; Schmid-Hempel, 2011). Based on this definition, **host mortality** emerged as a common proxy for estimating viral impacts on host fitness (Read, 1994). The idea here is that hosts that die more quickly have a shorter reproductive lifespan, and as such lower fecundity and fitness. Of course there are several inherent assumptions in that hypothesis, largely stemming from the timing of host mortality – i.e. if virus-induced mortality occurs after the end of the host's reproductive age, it will have no impact on fitness (Day, 2003). Regardless, because of its clear and measurable phenotype, host mortality is probably the oldest (Pasteur *et al.,* 1881) and has become the most widely adopted measure of virulence (Alizon *et al.,* 2009; Bull, 1994; Levin, 1996; Lipsitch and Moxon, 1997; Read, 1994; Schmid-Hempel, 2011). This is particularly true in the theoretical literature, which often mathematically parameterizes virulence as mortality caused to the host due to infection (Alizon *et al.,* 2009; Frank, 1996).

Unfortunately, host mortality runs into similar challenges as host fitness, in that it is difficult to measure in many systems. This is largely because experimentally quantifying mortality is impractical in long-lived hosts, and sometimes ethically unjustified, particularly in larger mammalian hosts including humans. The other limitation to this definition of virulence is that many viruses have significant fitness impacts on their host without inducing mortality (Read, 1994; Sacristan and Garcia-Arenal, 2008). For these reasons, various proxies for mortality have emerged for quantifying virulence. These are largely centered around measures of **host morbidity**, which is basically non-lethal effects that also comprise harm done to a host due to infection with a pathogen (Casadevall and Pirofski, 1999; Schmid-Hempel, 2011; Thomas and Elkinton, 2004). In this case "harm" includes both direct damage caused by virus replication, such as necrosis of host tissues, and also in some cases indirect damage resulting from the host's attempt to fight the infection with immune responses. In extreme situations the indirect immune mechanisms can actually contribute to increased damage, referred to as immunopathology (Schmid-Hempel, 2011). Measures of host morbidity, or harm, are very system-specific and can include parameters such as weight loss, anemia, tissue necrosis, and immune function in animals; or leaf degradation, reduced growth, and disease signs such as chlorosis (yellowing) in plants (Table 6.1). Basically, any type of measurable host trait that is associated with viral infection or host fitness might be used as a proxy for virulence. The critical component to using these morbidity metrics is a prior establishment of their relationship with host mortality or host fitness. In some systems such relationships have been established; for example host CD4 cell counts are correlated with survivorship in human immunodeficiency virus (HIV)-infected patients (Levy, 1993), but in many

Table 6.1. Experimental measures of viral virulence used in published literature

Host Type	Experimental measure of virulence	Examples of viruses[a] using this measure	Refs[c]
Bacteria	cell killing	Qβ, λ	1
	time to lysis	Φ2	2
	plaque size	Qβ	3
	reduction of host growth rate	f1	4, 5
	phage replication	Qβ ,T7	6, 3
Plants, fungi	reduction in total number of viable seeds/spores produced relative to non-infected control plants/fungi	CMV, TEV, BSMV, CHV1	7, 8, 9, 10, 11, 12, 13
	reduction in plant/fungal size (weight, biomass, height, leaf area) relative to non-infected controls	TEV, PVY, CMV, CaMV CMV+satRNA, CHV1	7, 10, 11, 12, 14, 15, 16, 17, 18
	severity of visible disease symptoms, e.g. chlorosis, necrosis	CMV, CMV+satRNA	19, 20
	time to onset of symptoms	CMV+satRNA, PPV	16, 21
	ability to infect disease-resistant plant lines, i.e. resistance-breaking	tobamoviruses, SMV, RYMV, TuMV	22, 23, 24
Animals	mortality (percent in experiments, case fatality rate in field observation)	LdNPV, ATV, polio, VSV, WNV, VHSV, IHNV, DENV, AIV, RHDV, Influenza A	25, 26, 27, 28, 29, 30, 31, 32, 33, 34, 35
	time to death	ATV, MYXV, RHDV, SINV	14, 25, 36, 37, 38
	mortality in a mouse model	SFV, CHKV, RRV, VSV, WNV, DENV, Influenza A, polio	39, 40, 41, 42, 43, 44, 45, 46
	severity of visible symptoms	VEEV, FMDV, polio, Influenza A, DENV, AIV, ISAV	32, 34, 39, 47, 48, 49
	time to symptom onset	HIV, influenza A	45, 50, 51
	internal organ pathology	FV, coxsackie B3, MDV	52, 53, 54
	increased host development time	DMelSV	55
	reduced egg numbers or hatch rates	WNV (in mosquitos)	26
	ability to infect vaccinated hosts	MDV	54
	virus accumulation in host	RHDV	36
	virus competitive fitness	HIV	55

Table 6.1 continued

Animal cells	cytotoxicity, cell viability	FMDV, MHV, VSV, Ad5,	28, 56, 57, 58, 59, 60, 61
	plaque size	VSV	61, 62
	virus competitive fitness[b]	VSV	63, 64

[a]Ad5, adenovirus-5; AIV, avian influenza virus; ATV, *Ambrystoma tigrinum* virus; BSMV, barley stripe mosaic virus; CaMV, cauliflower mosaic virus; CHKV, Chikungunya virus; CHV1, Cryphonectria hypovirus-1; CMV, cucumber mosaic virus; DENV, Dengue virus; DMelDS, Drosophila melanogaster sigma virus; FMDV, foot-and-mouth disease virus; FV, Friend virus complex; HIV, human immunodeficiency virus; IHNV, infectious hematopoietic necrosis virus; ISAV, infectious salmon anemia virus; MDV, Marek's disease virus; LdNPV, *Lymantria dispar* nuclear polyhedrosis virus; PiGV, *Plodia interpunctella* granulosis virus; MYXV, myxomavirus; PPV, plum pox virus; PVY, potato virus Y; RHDV, rabbit hemorrhagic disease virus; RRV, Ross River virus; RYMV, rice yellow mottle virus; satRNA, satellie RNA; SFV, Semliki Forest virus; SINV, Sindbis virus; SMV, soybean mosaic virus; TEV, tobacco etch virus; TuMV, turnip mosaic virus; VEEV, Venezuelan equine encephalitis virus ; VHSV, viral hemorrhagic septicemia virus; VSV, vesicular stomatitis virus; WNV, West Nile virus

[b] Competitive fitness is frequently measured for several viruses, but in most cases it is not used as a measure of virulence.

[c] References for this table are: 1, Berngruber *et al.*, 2013; 2, Leggett *et al.*, 2013; 3, Kashiwagi and Yomo, 2011; 4, Bull *et al.*, 1991; 5, Messenger *et al.*, 1999; 6, Heineman and Brown, 2012; 7, Bryner and Rigling 2012; 8, Carrasco *et al.*, 2007; 9, Lafforgue *et al.*, 2011; 10, Pagan *et al.*, 2007; 11, Pagan *et al.*, 2014; 12, Peever *et al.*, 2000; 13, Stewart *et al.*, 2005; 14, Bedhomme *et al.*, 2012; 15, Doumayrou *et al.*, 2013; 16, Escriu *et al.*, 2003; 17, Hillung *et al.*, 2014; 18, Montarry *et al.*, 2012; 19, Betancourt *et al.*, 2011; 20, Escriu 2000; 21, Wallis *et al.*, 2007; 22, Fraile *et al.*, 2011; 23, Hajimorad *et al.*, 2011; 24, Pinel-Galzi *et al.*, 2007; 25, Brunner and Collins, 2009; 26, Ciota *et al.*, 2013; 27, Cooper *et al.*, 2002; 28, Furio *et al.*, 2012; 29, Hodgson *et al.*, 2002; 30, Lancaster and Pfeiffer, 2011; 31, Pascua *et al.*, 2012; 32, Ping *et al.*, 2011; 33, Snow and Cunningham, 2000; 34, Suzuki *et al.*, 2009; 35, Wargo *et al.*, 2010; 36, Elsworth *et al.*, 2014; 37, Fenner, 1983; 38, Olmsted *et al.*, 1984; 39, Jerzak *et al.*, 2007; 40, Novella *et al.*, 1995; 41, Peleg, 1971; 42, Singh, 1971; 43, Taylor and Marshall, 1975b; 44, Twiddy *et al.*, 2002; 45, Vignuzzi *et al.*, 2006; 46, Xu *et al.*, 2011; 47, Aamelfot *et al.*, 2014; 48, Anishchenko *et al.*, 2006; 49, Carrillo *et al.*, 2007, 50, Ewald, 2011; 51, Pfeiffer and Kirkegaard, 2005; 52, Beck, 1997; 53, Kubinak *et al.*, 2012; 54, Witter *et al.*, 1997; 55, Brusini *et al.*, 2013; 56, Arien *et al.*, 2005a; 57, Chen and Baric, 1996; 58, Garijo *et al.*, 2014; 59, Herrera *et al.*, 2007; 60, Ojosnegros *et al.*, 2010; 61, Yan *et al.*, 2003; 62, Zarate and Novella, 2004; 63, Ogbunugafor *et al.*, 2013; 64, Elena *et al.*, 2001.

systems this relationship is simply assumed without concrete evidence. It should be noted that the definitions of the terms virulence and pathogenicity have been a subject of discourse: they are sometimes used interchangeably and other times defined differently in the literature. There are several excellent reviews that deal with this topic (Alizon *et al.*, 2009; Casadevall and Pirofski, 1999; Sacristan and Garcia-Arenal, 2008; Schmid-Hempel, 2011; Thomas and Elkinton, 2004), which will not be further elaborated upon here.

Several other proxy definitions of virulence have also emerged based on what can be easily measured in various host:pathogen study systems. A positive relationship between virus replication and virulence has been demonstrated in many systems (Brunner and Collins, 2009; Furio *et al.*, 2012; Pascua *et al.*, 2012; Schmid-Hempel, 2011; Suzuki *et al.*, 2009; Wargo and Kurath, 2011) (discussed more below). This has led some authors to simply define viral virulence as **virus replication capacity**, even in systems where the correlation between replicative fitness and virulence has not been established. This has led to a plethora of other virulence measures related to the **success of a parasite**, such as transmission, infectivity, and persistence (Alizon *et al.,* 2009; Schmid-Hempel, 2011). However, as with replication, in many systems the correlation between these traits and virulence has not been established. This is concerning given that where examined, these associations have sometimes been found to be non-existent, negative, non-linear, or unstable. (Brunner and Collins, 2009; Kell *et al.,* 2013; Lipsitch and Moxon, 1997; Ten Haaft et al., 1998). As such, it is important in interpreting published work to clearly identify the operational measure of virulence used, and to note that many definitions of virulence involve assumptions that may not always be met.

Regardless of which definition is chosen, the key here is that virulence is considered a phenotypic trait of the virus (Read, 1994). Furthermore, virulence is believed to be under some level of genetic control. However, the environment can also play an important role in the expression of viral virulence, just as say the swimming performance of a fish might depend on genetics as well as environmental factors such as water temperature or dissolved oxygen. As such, comparing the virulence of different strains or species of a virus can only be accomplished by taking into consideration environmental effects. For viruses, the environment is largely determined by the host. Hosts can differ in properties such as their resistance to a virus, which has a profound impact on the expression of viral virulence. Teasing apart the contributions of host resistance and pathogen virulence is a major scientific challenge. The key here is that resistance is considered a trait of the host, and virulence is considered a trait of the virus, particularly in mathematical models, but in reality virulence is often very host-specific, as will be explained in more detail in the section entitled "*Host specificity and generalist-specialist virulence trade-offs*".

Conceptual basis of virulence evolution theory

Mathematical basis: the basic reproductive number, R_0

The basis of virulence evolution theory is rooted in mathematical modeling of epidemiological infectious disease dynamics. The most heavily cited early presentations of these models were by Anderson and May in the late 1970s and 80s (Anderson and May, 1979, 1982a; May and Anderson, 1979, 1983), although their work built upon that of Kermack and McKendrick in 1927 as well as others, and has since seen an explosive expansion (Alizon *et al.,* 2009; Bremermann and Pickering, 1983; Frank, 1996; Kermack and McKendrick, 1927; Schmid-Hempel, 2011). In these models, virulence was parameterized as additional host mortality caused by infection, which was above baseline mortality in uninfected hosts. In addition to modeling the spread of a disease epidemic in a host population these models can be used to determine how infectious disease impacts the growth of a host population (Anderson and May, 1979). Mortality is a logical way to define virulence in these models because it removes the host from the population and prevents it from contributing to transmission and future host reproduction. These models are typically built around a system of differential equations that track the change in the number (or proportion) of host individuals that fall into susceptible (S), infectious (I), or recovered (R) classes (Equation 1). In these models, the increase in the number of susceptible individuals is governed by a birth rate (b). Movement from the susceptible to the infected class is governed by a transmission rate (β), which is the probability that a host becomes infected when contacting an infectious individual. Movement from the infectious class into the recovered class

Equation 1: Basic epidemic SIR model

$$\frac{dS}{dt} = b(S, I, R) - \beta SI - \mu S$$

$$\frac{dI}{dt} = \beta SI - \mu I - \alpha I - \rho I$$

$$\frac{dR}{dt} = \rho I - \mu R$$

Equation 2: Epidemic can occur when

$$\frac{dI}{dt} = \beta SI - \mu I - \alpha I - \rho I > 0$$

Simplified

$$\frac{\beta S}{\mu + \alpha + \rho} > 1$$

Equation 3: Basic reproductive number

$$R_0 = \frac{\beta N}{\mu + \alpha + \rho}$$

See text for parameter definitions

is governed by a recovery rate (ρ), which is the probability of an infected individual becoming no longer infectious and immune to future infection. Other parameters in this simple model include a baseline mortality rate (μ), which is the probability of a susceptible or recovered individual host dying. Lastly we have our virulence term (α), which is the additional mortality rate of infectious individuals due to disease, in addition to the background mortality rate (μ).

These simple disease dynamic models have been surprisingly useful in determining when an epidemic of an emergent disease will occur (Kermack and McKendrick, 1927; Ransome, 1880, 1881). Mathematically, an epidemic is defined as an increase in the number of infected individuals, i.e. when dI/dt is greater than zero in equation 2. Of particular interest is the threshold at which an emergent disease will become epidemic in an entirely susceptible population, which is given by **R_0, or the basic reproductive rate**. The parameter R_0 is defined as the number of newly infected hosts resulting from one primary infected individual that enters an entirely susceptible host population (Diekmann *et al.,* 1990). In this situation, the number of susceptible individuals represents the entire host population (N), as shown in equation 3. As such, an epidemic can only occur when $R_0 > 1$. In other words, each infectious individual must infect greater than one individual for disease incidence to increase in a host population. In addition to providing information on the epidemiological spread of disease through a host population, the parameter R_0 also provides an estimate of pathogen fitness at the population level, which we will refer to hereafter as **overall fitness**. Those pathogen strains with a higher R_0 will spread more quickly and infect a larger proportion of the host population, and therefore have greater transmission success and overall fitness (Wargo and Kurath, 2012).

Many variations of the transmission model have also been formulated. The goal of this chapter is not to discuss the myriad of theoretical developments in disease transmission modeling, but rather to examine how the basic assumptions within this theory as it pertains to virulence evolution have been empirically investigated. Those readers who are interested in a theoretical synthesis are directed towards the following excellent books and reviews (Anderson, 1982; Anderson and May, 1982b; Collinge and Ray, 2006; Frank, 1996; Grenfell and Dobson, 1995; Hudson *et al.,* 2001; Keeling and Rohani, 2008; Schmid-Hempel, 2011; Thomas *et al.,* 2009).

Virulence as a costly viral fitness trait

Given the above model framework, where the basic reproductive number R_0 is an indicator of overall viral fitness within a host population, it is relatively straight forward to see how virulence, taken on its own, would be a costly fitness trait for the pathogen. When examining the R_0 equation (equation 3) it is evident that as virulence (α) decreases, R_0 will increase, all else being held constant. In a biological sense this is intuitive; less virulent viruses that keep their hosts alive for longer will allow their hosts more opportunities to contact naïve hosts, and therefore have greater transmission opportunities. In other words, maximizing the duration of transmission maximizes fitness, which is achieved by minimizing virulence. As such, viruses were hypothesized to evolve a benign mutualist or commensal state with their host, described earlier as the avirulence theory. Certainly observation and experimental evidence for the evolution of decreased virulence exists in many systems, indicating a fitness cost of virulence to the pathogen in some cases (Bao and Roossinck, 2013; Carrillo *et al.,* 2007; Escriu *et al.,* 2000; Kerr *et al.,* 2012; Pagan *et al.,* 2014). However, it is important to note that basic evolutionary theory tells us there are a variety of other mechanisms that can explain the presence or evolution of avirulence in addition to the trait having an inherent fitness cost. These include founder effects, genetic drift, or changes in the environment [e.g. the evolution of host resistance, (Freeman and Herron, 1998]. Few systems have empirically demonstrated a virus fitness cost of virulence, but there is a growing body of convincing and well-cited examples for this cost (Fraile *et al.,* 2011; Heineman and Brown, 2012; Kerr *et al.,* 2012; Laanto et al., 2012).

Virulence as a beneficial viral fitness trait

Although there are many examples of avirulent viruses there are also many examples of virulent viruses (HIV, Ebola, rabies, smallpox, yellow fever). The avirulence theory would argue these viruses are virulent because they recently emerged in their host and as such have not had enough evolutionary time to be selected to become avirulent. However, there are many viruses that have remained virulent through long durations of evolutionary time after emergence in a host species (Ewald, 1983; Toft and Karter, 1990). One example is smallpox virus , which was highly virulent for thousands of years after it emerged in humans, before it was eradicated through vaccination (Li *et al.,* 2007; Lofquist *et al.,* 2003). The maintenance of high virulence for even just a few decades after emergence can represent a huge expanse of evolutionary time when considering that viral generation time can be exceedingly rapid and in the case of RNA viruses, mutation rates are extremely high relative to those of their hosts (Duffy *et al.,* 2008). Furthermore there is growing evidence of viruses evolving increased virulence over time (Atkins *et al.,* 2013a; Bull *et al.,* 1991; Cooper *et al.,* 2002; Ebert, 1998; Elsworth *et al.,* 2014; Gimeno, 2008; Kubinak and Potts, 2013; Kubinak *et al.,* 2012; Nair, 2005; Rodriguez-Roche *et al.,* 2011; Snow and Cunningham, 2000; Stewart *et al.,* 2005; Witter, 1997), although this comes with the same caveats of different mechanisms potentially driving evolution besides viral fitness, such as founder effects or genetic drift, as discussed in the previous section. These observations led to the hypothesis that increased virulence may actually have fitness benefits for viruses and other pathogens (Alizon *et al.,* 2009; Bull, 1994; Read, 1994; Schmid-Hempel, 2011). The classic example is a positive association between viral replication and virulence (Brunner and Collins, 2009;

Furio *et al.*, 2012; Pascua *et al.*, 2012; Schmid-Hempel, 2011; Suzuki *et al.*, 2009; Wargo and Kurath, 2011). The argument goes as follows: virulence is a function of the density of a virus within the host in that higher viral loads cause greater amounts of cellular damage and therefore greater morbidity and mortality (virulence). In other words, the virus is the consumer and the host cells are the resource, so that the higher the viral load, the more cells or cell resources are consumed. The fitness benefit to the virus is in replication such that a more highly replicating virus will be producing more infectious progeny (Lipsitch and Moxon, 1997). If, as is often assumed, transmission probability (β in equation 1) increases with infectious virus production, virus replication should be positively correlated with transmission and ultimately, overall fitness. Clearly, there are two inherent assumptions within this hypothesis: 1) virulence is positively associated with virus replication and 2) transmission increases with infectious particle production. The number of virus systems in which all these assumptions have empirically been tested has grown in recent years (Atkins *et al.*, 2013a; Brunner and Collins, 2009; Bryner and Rigling, 2012; Doumayrou *et al.*, 2013; Kerr *et al.*, 2012; Mackinnon *et al.*, 2008; Wargo and Kurath, 2011), but it is still vastly exceeded by the number of systems in which they are assumed to be true but have not been rigorously tested (Froissart *et al.*, 2010).

Other fitness benefits to increased virulence are also possible. For example studies have indicated that virulence may be associated with benefits in viral fitness traits such as host entry, infectivity, persistence, and host immune evasion (Ogbunugafor *et al.*, 2013; Saiz and Domingo, 1996; Walther and Ewald, 2004; Wargo and Kurath, 2011). A negative association is also hypothesized between virulence and recovery rates (ρ in equation 1), such that hosts infected with high virulence pathogens recover more slowly (Anderson and May, 1982a). Examination of equation 3 reveals that if higher virulence comes with lower recovery rates, this could counteract the negative impact of virulence on R_0. In biological terms, more virulent viruses are predicted to have a longer duration of transmission because their hosts recover more slowly. Ultimately, these fitness benefit-virulence associations are predicted to provide a mechanism for the evolution of increased virus virulence (Alizon *et al.*, 2009; Bull, 1994; Read, 1994; Schmid-Hempel, 2011).

The virulence trade-off hypothesis

The studious reader will by now have come to the conclusion that increased virulence may simultaneously provide fitness costs and benefits to viruses. This is the basis of what is referred to as the trade-off hypothesis (Alizon *et al.*, 2009). This theory ascertains that virulence provides costs to the pathogen, such as truncation of the duration of transmission through mortality, but also benefits through positive association with pathogen fitness traits such as replication and transmission rates, as discussed in the previous two sections. Mathematically, the trade-off theory maintains that the terms in the numerator and denominator of equation 3 do not change independently, but are rather linked, such that increases in one can lead to decreases in the other. The basis of much of current virulence evolution theory is that the parameters in Equation 1 evolve in various trade-off relationships to maintain maximal pathogen fitness in terms of R_0.

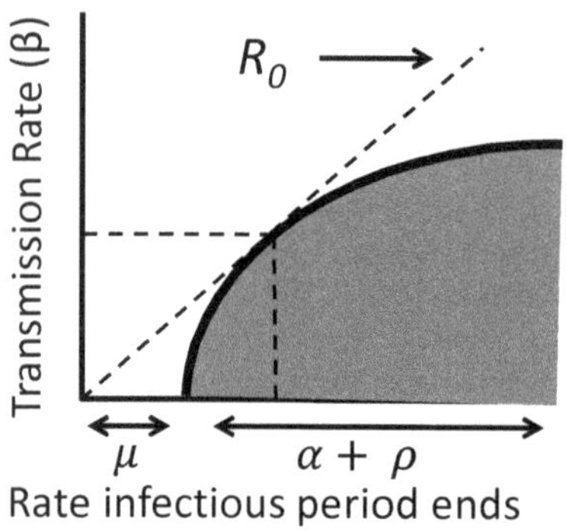

Figure 6.1. Graphical depiction of the relationship between the concave transmission rate (β) and virulence (α) as well as recovery rate (ρ), showing a virulence tradeoff. The tangent line is the values for β, α, and ρ for which R_0 is maximized (see equation 3). Background host mortality (μ) is assumed to be constant. Viruses in the grey area are predicted to evolve upwards and inwards towards the curved black line by either increasing transmission rate or increasing the duration of the infectious period through decreases in infection-induced mortality (virulence) or recovery rates. Once viruses reach the curved black line, they cannot go above it but rather travel along it until they reach the evolutionary stable strategy where R_0 is maximized (tangent line).

One of the first trade-offs published was an association between virulence (α) and recovery (ρ) (Anderson and May, 1982a) such that increases in virulence lead to increases in transmission duration due to lower recovery rates, as well as decreases in transmission duration due to higher host mortality rates. However, the most commonly cited and tested virulence trade-off is between transmission and virulence, such that increases in virulence (α) lead to increases in transmission rate (β) as well as decreases in transmission duration (Bremermann and Pickering, 1983; Massad, 1987), as illustrated in Figure 6.1, through the mechanisms discussed in the two previous sections. Ultimately, this trade-off theory predicts that pathogens should evolve under these counterbalancing influences to some intermediate level of virulence, as can be derived from equation 3 and Figure 6.1. However, it is important to note that the shape of the relationship between virulence and the fitness trait is not known, except in a few systems, and can heavily impact the predicted evolutionary outcome. For example if the relationship between transmission and virulence is linear or convex (i.e. continuously increasing) rather than concave, this would lead to the boundless evolution of increased virulence. Furthermore, host mortality will not necessarily decrease the transmission duration of the pathogen, especially if transmission has ended before mortality occurs. Trade-offs between virulence and other pathogen fitness traits have also been proposed, as will be detailed below.

As always the environmental context can be important in determining how the costs and benefits of virulence are balanced and what level of virulence evolves. For example if transmission opportunities are frequent, say under high host density situations, the cost of killing the host quickly becomes diminished and viral fitness is predicted to be optimized at higher levels of virulence. The inverse is true when transmission opportunities are rare and far between. Another interesting prediction from Figure 6.1 is that, as host background mortality increases (μ), the level of virulence that maximizes fitness will also increase because the tangent point for the R_0 line will move up the curve. However, we know of no empirical tests of this phenomenon in a viral system. Predictions and empirical studies of virulence trade-off hypotheses within various environmental contexts will be discussed in the sections below.

Virulence unlinked from fitness

An often overlooked aspect in virulence evolution theory is that in some cases virulence might not be associated with pathogen fitness and therefore not under selection. A classic example of this would be dead-end infection, such as rabies in humans, which clearly is very lethal. Because rabies virus cannot generally transmit from the human host, it therefore does not undergo more than transient selection in this host. Likewise dead-end infection of host tissues can occur, such as polio virus infection of the central nervous system, which is much more virulent than the typical enteric infection, but represents a dead end for transmission (Levin and Bull, 1994). It is also possible that virulence is not always under the genetic control of the pathogen. For example a virus may cause a high degree of immunopathology, which is under the control of the host but not under the control of virus genetics (Graham *et al.,* 2005). Another possibility often overlooked is that virus transmission may not end at the time of host death, as assumed by the modeling framework. Many viruses, such as Ebola, continue to transmit for some period after host death, and therefore the virulence cost of truncating transmission duration because of host mortality may not apply. If, however, transmission can only occur with host death, or replication is tied to virulence, this could move virulence from an unselected trait to one undergoing positive selection. Of course, in all these cases, it is possible that what may initially be non-adaptive could become adaptive for the virus under the right conditions or host availability, and this is often how the process of disease emergence begins. Finally, it is important to remember that the expression of virulence is heavily dependent on the host environment, which can have substantial levels of variation that could overwhelm other selective forces that might act on virulence. For example when faced with antiviral drug treatment, viral fitness may be maximized by selection for drug resistance, and former viral virulence determinants would no longer be major drivers relevant to viral fitness. Although it is important to remember that there are certain situations where viral virulence is not linked to viral fitness, these tend to be rather infrequent exceptions to the general rule of adaptive virulence.

Overview of predictions of current virulence evolution theory

Current virulence evolution theory is comprised of a basic mathematical model and trade-off concepts as defined above, and variations of the model that describe how various factors may impact selection for virulence in different interactions of host, pathogen, and environment. The specific concepts most commonly explored are summarized in Table 6.2, along with brief comments on the predictions of each concept with regard to selection of increased or decreased virulence, and the extent of empirical testing that has been reported. Some concepts have had a great deal of theoretical analysis that is not yet matched by the extent of empirical testing. Detailed descriptions of the experimental studies reported for individual concepts of virulence evolution are provided below, after comments on the various methods used for empirical analyses.

Empirical approaches for testing virulence evolution theory concepts

The predictions of virulence evolution theory have been tested empirically in a wide range of biological systems involving viral, bacterial, and cellular pathogens. The majority of viral studies involve RNA viruses due to their inherent capacity for

Table 6.2. Predictions of virulence evolution theory

Virulence evolution theory concept	Prediction(s)	Virus[a]-host systems studied empirically	Empirical results
Transmission trade-off	Increased transmission rate associated with faster in-host replication comes at a "cost" of decreased transmission duration due to decreased host lifespan	MYXV-Rabbits HIV-Humans VSV-Cells ATV-Salamander CaMV-Brassica CHV1-Fungus	Evolution to intermediate virulence observed in field Experimental evidence of tradeoff in some systems but not others The benefits of virulence well established, costs not as well characterized
Recovery trade-off	Slower or lower frequency of recovery from viral infection comes at a "cost" of decreased transmission duration due to decreased host lifespan	MYXV-Rabbits SIV-Macaques AIV-Chickens	Less virulent strains typically cleared more quickly Cost of virulence not well documented
Mode of transmission: vertical vs. horizontal	Vertical transmission leads to lower virulence, Horizontal transmission leads to higher virulence	Phage f1-E.coli ZYMV-Cucurbit BSMV-Barley CMV-Arabidopsis	Generally well supported Vertical transmission typically less virulent Horizontal transmission varies in virulence
Mode of transmission: vector-borne	a)vector-borne disease more virulent than non-vector borne to primary host b)virulence in vector lower than in primary host c) cycling between alternate hosts restricts ability to adapt to higher fitness and virulence relative to replication on either host alone	CMV-Brassica SFV-Insect cells, mice CHKV-Insect cells, mice RRV-Insect cells, mice VSV-Cells EEEV-Cells SINV-Cells DENV-Cells VEEV-Mosquito, mice	a) generally true in plants, inconsistent in other hosts b) generally true c) inconsistent
Mixed infections	In mixed infections competition will favor a faster replicating genotype, which has higher virulence	MDV-Chickens IHNV-Salmonids FMDV-Cells Phage-E. coli Phage-Pseudomonas PfNPV-Moth	More virulent strains are typically favored in competition, but not for all systems tested Evolution of lower virulence seen in competition with defective interfering particles
Life history traits	Early transmission favors increased virulence, later transmission favors reduced virulence	LdNPV-moth larva VSV-BHK cells SINV-Cells	Supported in NPV system, partial support or not supported in others
Sit-and-wait hypothesis	Higher virion survival in the external environment should correlate with higher virulence	respiratory pathogens-humans Phages-E. coli VSV-Cells AIV-Ducks	Supported in meta-analysis of several human respiratory pathogens, but no correlation or contradicted in other systems
Non-equilibrium virulence	Transient increase in level of virulence selected early in an epidemic or emergence, with subsequent reduced virulence as number of susceptible hosts becomes limiting	Phage λ-E. coli	Supported, but little empirical testing
Specialist: generalist strategies	a)specialization on one host leads to increased fitness and virulence on that host, with tradeoff costs of reduced fitness and virulence in other hosts b) generalists that infect more than one host are constrained to fitness and virulence levels below the maximum possible on each host. Thus generalists have overall lower virulence than specialists.	IHNV-Salmonids CPV-Dogs and cats VSV-Cells TEV-Tobacco sp., Datura, pepper, Arabidopsis CMV-Cucumber, bean, tomato PPV-Peach, pea FV-Mice	a) Specialization often leads to increased fitness and virulence in one host, but trade-off costs in other hosts less often tested or variable b) Variable results for cost of generalists, not often tested, and no-cost generalism is observed in some systems.

Table 6.2 continued

Co-evolution	Leads to cycles of increased resistance in host and increased virulence in virus	Phage Qβ-E. coli Myxoma-Rabbits Tobamovirus-Pepper PVY-Pepper SMV-Soybean RHDV-Rabbits	Increased resistance seen in host. In several cases led to increased virulence in virus, but decreased virulence also observed.
Vaccine induced virulence	Disease reducing vaccines may select for increased virulence	MDV-Chickens Feline Calicivirus-Cats	Supported
Host spatial structure	Increased spatial structure (patchiness) in host population leads to reduced virulence	PiGV-Moth larvae	Increased spatial structure in a moth larva population led to reduced viral infectivity

[a]AIV, avian influenza virus; ATV, *Ambrystoma tigrinum* virus; BSMV, barley stripe mosaic virus; CaMV, cauliflower mosaic virus; CHKV, Chikungunya virus; CHV1, Cryphonectria hypovirus-1; CMV, cucumber mosaic virus; DENV, Dengue virus; EEEV, Eastern equine encephalitis virus; FMDV, foot-and-mouth disease virus; HIV, human immunodeficiency virus; IHNV, infectious hematopoietic necrosis virus; MDV, Marek's disease virus; LdNPV, *Lymantria dispar* nuclear polyhedrosis virus; PfNPV, *Panolis flammea* nuclear polyhedrosis virus; PiGV, *Plodia interpunctella* granulosis virus; MYXV, myxomavirus; RRV, Ross River virus; SFV, Semliki Forest virus; SINV, Sindbis virus; SIV, simian immunodeficiency; VEEV, Venezuelan equine encephalitis virus ; VSV, vesicular stomatitis virus; ZYMV, zucchini yellow mosaic virus

rapid evolution, based on high mutation rates, large population sizes, and rapid generation times (Duffy *et al.,* 2008). However, both DNA and RNA viruses from taxonomically diverse viral families have been tested in a variety of *in vivo* hosts and *in vitro* cell culture systems. Each virus-host system has specific features that are advantages or limitations for certain types of studies, so they have been tested using different experimental approaches. The three main approaches used in empirical virulence evolution studies are observational studies based on field data, laboratory studies of experimental infections, and serial passage experiments.

A major advantage of observational studies that analyze data from naturally occurring infections is the direct, high relevance to disease as it actually occurs in the field. Although this advantage outweighs many concerns, this is also the least controlled and most complex source of data to obtain, and many unmeasured factors may impact field infections. Ultimately these studies are correlative rather than causative, making it difficult to determine the drivers of observed patterns. Because of these limitations such studies optimally require large amounts of data, which is available in host-viral systems such as myxomavirus in rabbits, pepper tobamoviruses and cucumoviruses in plant hosts, and several human viruses including HIV.

As an alternative to observational studies, much of the empirical literature on virulence evolution involves experimental infections designed specifically to test various predictions detailed in Table 6.2. These have been conducted *in vivo,* using a wide array of viruses in bacterial, fungal, plant, invertebrate, and vertebrate hosts, and *in vitro* in cultured cell lines. Due to ethical and logistic constraints, viral pathogens of humans and large mammalian hosts are most often tested in cell lines or small animal models such as mice. A common strategy is to compare two or more isolates of the same viral species in infections of individual hosts or host populations. Comparing the virulence of viruses in a common host controls for host effects and is referred to as a common garden experiment. Variables such as different host types or treatments can be added within this framework to examine impacts of other factors on virulence evolution. Infections are monitored for outputs such as virulence, viral infection rates, within-host replication levels (replicative fitness), dissemination, or rates of transmission to

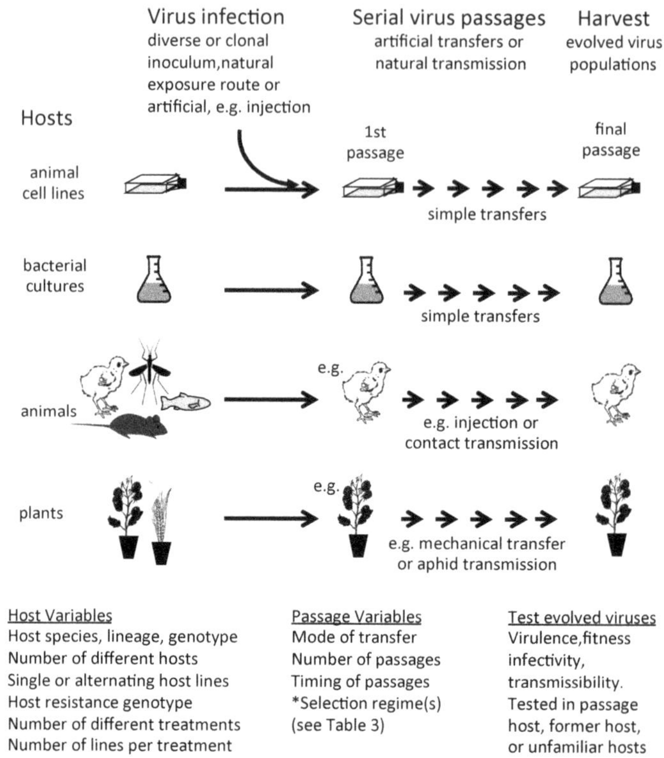

Figure 6.2. Experimental evolution by serial passage of viruses under varying selection pressures.

new susceptible hosts. In addition to testing predictions of virulence evolution theory, experimental infections are used in support of field observations of virulence evolution by assessing differences in the virulence of field isolates, or resistance levels of hosts collected from the field.

Serial passage experiments are an extension of the single infection cycle experiments just described, but they are so fundamental to empirical studies of evolution that they deserve a separate description (Ebert, 1998; Elena *et al.*, 2008). Referred to as "**experimental evolution,**" (Ebert, 1998) this experimental strategy has made major contributions to the field of virulence evolution. The basic concept is to conduct multiple passages of experimental infections under controlled conditions, to allow time for evolution of the virus in response to a defined selection regime. The selection regime is established by varying the virus strains, hosts, or conditions depending on the question addressed, and there are a surprising number of technical variables, such as timing and method of passage transfers, which can influence the outcome (Figure 6.2). In many cases transfers between passages are done artificially by the researchers, which by-passes natural transmission and tests the selective regime of the within-host replication environment (Ebert, 1998). However, natural modes of transmission have also been used between serial passages *in vivo* to assess both in-host and between-host selection. Impacts of selection are revealed by assessing phenotypes of evolved virus populations such as virulence, within-host replicative fitness, or transmissibility, in comparison with the original inoculum virus.

The first published serial passage studies involved animal viruses passaged *in vitro* in cultured cell lines to generate attenuated viruses as vaccine candidates. Although, based on the inherent capacity of RNA viruses for rapid evolution (Ebert, 1998) it is typically assumed that phenotype changes observed during experimental evolution studies are due to evolution of the virus, evolution of host resistance has also been observed. The possible role of chance also needs to be considered, so parallel, replicate passage lines are conducted to look for consistent changes that distinguish true adaptive selection from random effects. In many cases passaged viruses are sequenced to characterize genetic changes, such as convergent mutations associated with observed phenotypic changes. In serial passage studies using either cultured cell lines or live hosts the most common outcome is increased fitness and virulence in the passage host (Ebert, 1998). However, it is important to note that exceptions have been reported (Carrillo *et al.,* 2007). An overview of many serial passage studies relevant to the evolution of viral virulence is presented in Table 6.3, and findings will be described in the appropriate concept-based sections below.

Experimental studies of virulence evolution hypotheses and predictions

Correlation of virulence with viral fitness

The hypothesized correlation of viral replicative fitness and virulence is a basic premise of virulence evolution theory. In practical terms, replicative fitness is assessed as viral accumulation in infected cells or samples from infected hosts, and virulence is assessed using a wide variety of operational definitions as described above. The correlation between within-host fitness and virulence has been experimentally tested in many virus-host systems, with varying results that are generally system-dependent. For several animal viruses, in both cell lines and vertebrate hosts, a strong positive correlation between replicative fitness and virulence is reported in the great majority of cases (Furio *et al.,* 2012; Garijo *et al.,* 2014; Kubinak *et al.,* 2012; McGee and Remold, 2013; Peñaranda *et al.,* 2011; Suzuki *et al.,* 2009; Ten Haaft *et al.,* 1998; Wargo *et al.,* 2010; Xu *et al.,* 2011). However, a small number of exceptions have been described, such as individual clones of vesicular stomatitis virus (VSV) and foot-and-mouth disease virus (FMDV) derived by mutagenesis or plaque-to-plaque transfers that show de-coupled replication and virulence in cell culture (Furio *et al.,* 2012; Herrera *et al.,* 2007). In addition, replicative fitness *in vitro* does not always correlate with fitness *in vivo*, as observed for some live-attenuated vaccine strains (Vignuzzi et al., 2008). For plant viruses there are many studies showing a positive correlation of viral replication with virulence, but there are also many reports of a more general lack of correlation using natural virus strains or large numbers of mutants generated by mutagenesis (Betancourt *et al.,* 2011; Carrasco *et al.,* 2007; Pagan *et al.,* 2007; Sacristán *et al.,* 2005; Sacristan and Garcia-Arenal, 2008). This lack of a general correlation for plant viruses may be due to insufficient time for natural selection for the artificial mutants, the possibility of a higher incidence of tolerance in plant hosts that would obscure the relationship between virus accumulation and virulence, or the use of different virulence measures in plant and animal viruses (Sacristan and Garcia-Arenal, 2008). Regardless, although a general correlation between viral replicative fitness and virulence exists for animal viruses, exceptions

Table 6.3. Experimental evolution studies using serial passage of different viruses under various selection regimes

a) RNA and DNA bacteriophages		
f1	strict horizontal or vertical transmission in *E. coli;* alternating horizontal and vertical transmission with varied length of vertical passages	4, 5
Qβ	co-evolution with E. coli host for 54 passages, mechanisms of evolved host resistance and evolved phage fitness and virulence	3
Φ2	single versus mixed infections, evolution of plastic lysis time	2
λ	competition of latent and virulent phage, transient high non-equilibrium virulence	1
b) Plant RNA viruses		
CMV	adaptation to three host plant species; horizontal or vertical transmission	11, 73
BSMV	horizontal or vertical transmission in barley	13
PPV	adaptation to new host, pea, using mechanical and aphid transmission	21
SMV	soybean cultivars of increasing resistance, evolution of resistance breaking virus	23
TuMV	RNAi-mediated resistance in *Arabidopsis*, evolution of resistance breaking virus	71
PVY	susceptible or polygenic resistant pepper as single or alternating host lines, evolution of resistance breaking virus	18
TEV	multiple plant host species with single or alternating host lines; multiple ecotypes within one host species, generalist-specialist predictions	14, 17
c) Insect DNA and RNA viruses		
LdNPV	early versus late transmission timing during in vivo passages in moth larvae	27
PfNPV	single and mixed infections in moth larvae, impact of host food plant	29
DMelSV	selection for both increased and decreased viral titers in whole Drosophila flies	55
d) Vertebrate RNA arboviruses		
SFV	insect cell line passage, testing virulence in mice for attenuated vaccine candidates	41
CHKV	insect cell line passage, testing virulence in mice for attenuated vaccine candidates; insect and mammalian cells in single and alternating host passages, arbovirus cycling	42, 66
RRV	insect and mammalian cells testing virulence in mice for attenuated vaccine candidate; passage in mice and mosquitos in vivo as single and alternating host lines	43, 74
SINV	mammalian cells, selecting for rapid growth by early passage timing; insect and mammalian cells in single and alternating host passages, arbovirus cycling	38, 68
EEEV	insect and mammalian cells in single and alternating passages, arbovirus cycling	76
VEEV	mice or hamsters and mosquitoes, single and alternating host lines, arbovirus cycling	65
WNV	mosquito and chicken in vivo passage, genetic diversity of virus populations; adaptation to mosquito in vivo, virulence in mosquito vector	26, 39
VSV	persistent infection in insect cells; insect and mammalian cell lines as single and alternating hosts; insect and mammalian cell lines as persistent and acute infections in single and alternating hosts; generalist-specialist evolution in mammalian cell lines with analysis for host specificity and virion stability; early, middle, and late transmission timing in mammalian cells; adaptation to p53-deficient cancer cell lines to generate oncolytic virotherapy candidates	40, 58, 62, 63, 64, 72, 75

e) Vertebrate RNA viruses (non-arboviruses)		
VHSV	adaptation to rainbow trout as a novel host, mimic hypothesized host jump	33
FMDV	in vivo passage in pigs by natural contact transmission	49
FV	congenic mice with different MHC alleles; mouse lines that differ across entire genome; specialization and contribution of MHC to selection	53, 69, 70
f) Vertebrate DNA viruses		
Ad-5	adaptation to cancer cell lines to generate oncolytic virotherapy candidates	61
vaccinia	adaptation to HeLa cancer cells under strong selection to evade host PKR defense	67

[a]Ad5, adenovirus-5; BSMV, barley stripe mosaic virus; CHKV, Chikungunya virus; CMV, cucumber mosaic virus; DMelDS, Drosophila melanogaster sigma virus; FMDV, foot-and-mouth disease virus; FV, Friend virus complex; LdNPV, *Lymantria dispar* nuclear polyhedrosis virus; PPV, plum pox virus; PVY, potato virus Y; RRV, Ross River virus; RYMV, SFV, Semliki Forest virus; SINV, Sindbis virus; SMV, soybean mosaic virus; TEV, tobacco etch virus; TuMV, turnip mosaic virus; VEEV, Venezuelan equine encephalitis virus ; VHSV, viral hemorrhagic septicemia virus; VSV, vesicular stomatitis virus; WNV, West Nile virus
[b] References numbered between 1-64 are listed in footnote for Table 6.1; 65, Coffey *et al.*, 2008; 66, Coffey and Vignuzzi, 2011; 67, Elde *et al.*, 2012; 68, Greene *et al.*, 2005; 69, Kubinak and Potts, 2013; 70, Kubinak *et al.*, 2013; 71, Lafforgue *et al.*, 2011; 72, Novella *et al.*, 1999; 73, Sacristan *et al.*, 2005; 74, Taylor and Marshall, 1975a; 75, Turner and Elena, 2000; 76, Weaver *et al.*, 2006.

have been shown for both animal and plant viruses, suggesting caution should be used where the relationship has not been demonstrated in a specific virus-host system.

In addition, viral fitness is complex and not solely driven by replication (Wargo and Kurath, 2012). Thus other fitness traits may play an important role in virulence evolution (Alizon *et al.,* 2009; Sacristan and Garcia-Arenal, 2008; Wargo and Kurath, 2012). However, few studies have investigated correlations between virulence and viral fitness traits beyond replication. A positive correlation was demonstrated between virulence and contact transmission fitness for influenza virus in a ferret infection and transmission model (Pascua *et al.,* 2012). For the fish rhabdovirus infectious hematopoietic necrosis virus (IHNV), *in vivo* analyses of several components of overall fitness have demonstrated positive correlations between viral virulence and host entry, within-host replication, and transmission potential, the latter measured as shed virus (Figure 6.3) (Wargo *et al.,* 2010; Wargo and Kurath, 2011). Surprisingly, higher virulence did not correlate with higher superinfection fitness tested for IHNV in rainbow trout (Kell et al., 2013).

Transmission virulence trade-off

The transmission virulence trade-off hypothesis states that viruses with higher virulence will have greater transmission rates (due to higher virus replication) but shorter transmission durations (due to killing their host more quickly). Despite being one of the major hypotheses of virulence evolution theory, surprisingly few studies have empirically tested this hypothesis in any host-parasite system. A handful of studies have been conducted in virus systems, the most well-known being myxoma virus in rabbits. In this system it was demonstrated that transmission to mosquito and flea vectors was maximized at intermediate levels of

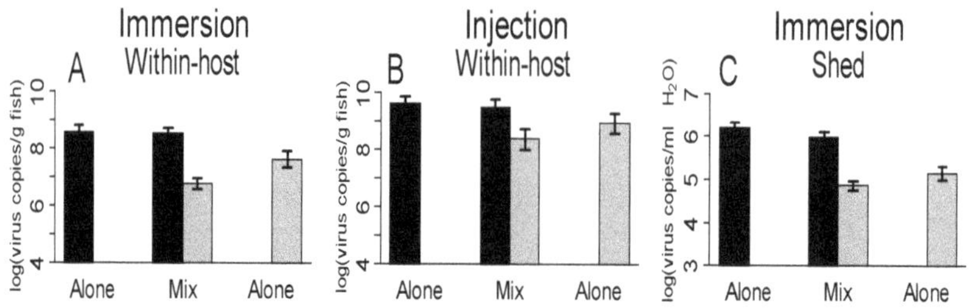

Figure 6.3. Correlation of virulence with various components of viral fitness, demonstrated with infectious hematopoietic necrosis virus (IHNV) in juvenile rainbow trout. A high virulence genotype (HV, black) and a low virulence genotype (LV, gray) were compared in both single and mixed infections, using either the natural immersion route of infection that assesses both host entry and in-host replication (panel A), or injection challenge that by-passes entry and assesses only in-host replication (panel B). HV reaches higher mean peak viral loads in-host than LV after either immersion or injection challenge, but the difference is smaller after injection challenge, indicating that HV has advantages over LV in both host entry and in-host replicative fitness. Virus shed into water during the immersion challenges (panel C) indicates that HV is shed more than LV at equal in-host viral loads (see Wargo *et al.*, 2011), indicating also a transmission advantage for HV. The advantages of HV are maintained in single and mixed infections, suggesting it has higher competitive fitness. Modified with permission from Wargo *et al.*, 2011, copyright © American Society for Microbiology.

virulence (Kerr, 2012; Mead-Briggs and Vaughan, 1975). This provided a mechanism for the virus to evolve intermediate levels of virulence as observed in the field (Fenner, 1983) and supported by theory (Massad, 1987).

The challenge with this and other empirical studies is that the trend of a derived virulence transmission trade-off is typically driven by one or two data points. Furthermore there seems to be some bias in the literature as to whether authors choose to fit a linear or curvilinear association between virulence and fitness [these caveats are discussed in Froissart *et al.,* (2010)]. This is important because a linear relationship predicts that fitness will continuously increase or decrease with virulence (depending on positive or negative slope), whereas the curvilinear relationship would support the trade-off model only if it is concave and saturating (see Figure 6.1). In the few studies where both a linear and curvilinear relationship were fit to the data, there is often inadequate statistical support to favor one relationship over the other, and the linear association is often assumed to be correct because it is the most parsimonious. For example in a study of VSV the authors compared a linear increasing model, a curvilinear saturating model, and an accelerating increasing function, and found little evidence to refute the linear model (Furio *et al.,* 2012). Similarly, a study of *Ambystoma tigrinum* virus in the tiger salamander found that a model of transmissibility being maximized at the lowest levels of virulence had equal fit to the data compared to a curvilinear association where virulence was maximized at an intermediate level of virulence, despite the virus being highly virulent to its host (Brunner and Collins, 2009). In both of these studies the curvilinear association was driven by 2 or 3 data-points. A study of Cucumber mosaic virus on tomatoes found that isolates of intermediate virulence had highest transmissibility in laboratory tests, which supported the observation that the wild-type virus strain of intermediate virulence had increased to become predominant in the field. This conclusion was driven by one isolate of intermediate virulence. However, the relationship was lost, such that the association between virulence and fitness became linear, with the lowest virulence isolates having highest fitness (calculated as R_0) when vector abundance was high (Escriu *et al.*, 2003). A study of two West Nile virus strains showed that more virulent strains have higher infectivity to their mosquito vectors, but this came with a potential transmission cost of a reduced blood feeding rate by infected mosquito vectors (Ciota *et al.*, 2013). Similar results were found in the fungal pathogen Cryphonectria hypovirus 1 (Bryner and Rigling, 2012) and the plant pathogen cauliflower mosaic virus (Doumayrou *et al.*, 2013), where a positive association between virulence and fitness was found, but without sufficient data to discriminate between a curvilinear saturating or a linear ever-increasing relationship.

It is interesting that many of the above systems demonstrate a positive association between virulence and transmission. However, the transmission truncation cost of virulence has not been well documented, although it is often assumed to be a result of host mortality. This seems to contradict the conventional avirulence evolution hypothesis. It is plausible that the virus fitness costs are not observed because they are difficult to quantify, as this requires measuring the overall transmission of many individual virus strains with varying levels of virulence. Some field observational studies seem to support this virulence trade-off hypothesis, as supported by the most common virus strains in the field being of intermediate virulence. The myxoma system mentioned above is the canonical case (Fenner, 1983). In this system a highly virulent strain of the virus was introduced as a biocontrol agent for rabbits in Australia and Europe, and on both continents it evolved reduced virulence to an intermediate value. Similar patterns have been seen in two populations of HIV patients (over 100 patients in each population) where the most common viral load seen in chronic infections (known as the set-point) was an intermediate value (Fraser *et al.*, 2007). This outcome seems to be driven by a virulence trade-off between host lifespan and transmission rate. These examples will be discussed further in sections entitled "*Myxoma virus pest control in rabbits*" and "*HIV virulence evolution in the field*". However, many field studies found evidence of continual increases in virulence in the field, with no apparent costs (Cardenas *et al.*, 2014; Elsworth *et al.*, 2014; Rico-Hesse, 2007). Furthermore, there is almost certainly some degree of sampling bias here in that the viruses of highest concern and study are likely to be those that cause significant disease, so avirulent viruses are likely to be overlooked (Read, 1994).

Recovery virulence trade-off

This virulence trade-off is less intuitive than the transmission trade-off but certainly as plausible. The hypothesis goes as follows: Hosts are better able to clear less virulent pathogens [perhaps because high virulence strains evade host defenses better (Schmid-Hempel, 2011)] and this will truncate the duration of transmission compared to high virulence pathogens. However, with increased virulence host mortality also truncates transmission. As such, pathogens must balance their level of virulence to avoid truncation of transmission through more rapid recovery if they are low virulence, or transmission truncation due to more rapid host mortality if they are high virulence. The recovery trade-off hypothesis has had very few empirical studies, perhaps because it is more difficult to test experimentally. However, this hypothesis was also possibly the first virulence trade-off to be empirically demonstrated, again in the iconic myxoma virus system. It was demonstrated that myxoma virus strains of high virulence indeed truncate their transmission through host mortality, and low virulence strains also have truncated transmission because they are cleared more quickly by the host (Fenner and Ratcliffe, 1965). This led to the classic trade-off model presented by Anderson and May (Anderson and May, 1982a). In a study of simian immunodeficiency virus in macaques it was shown that low virulence virus strains reach lower peak viral loads and are cleared more quickly compared to high virulence strains. However a transmission duration cost of high virulence through host mortality was not demonstrated, because even the most pathogenic virus strains did not kill the host (Ten Haaft *et al.,* 1998). A study of avian influenza virus in chickens comparing high and low virulence strains (referred to as high and low pathogenicity) also found the low virulence strain was cleared more quickly by the host (Van der Goot *et al.,* 2003), and in this case the high virulence strain caused faster and higher mortality. Nevertheless, the transmission duration of the high virulence strain was on average longer than that of the low virulence strain, and R_0 was maximized at the higher level of virulence. However, this system does seem to have the components of a recovery-transmission trade-off. Particularly when we consider that the 2 chickens with the shortest infectious period were infected with the highest virulence strain, because they died on days 2 and 3, whereas the shortest infectious period for an individual chicken infected with the low virulence strain was 4 days. Future tests of more virus strains across a range of virulence levels may more robustly illuminate the possibility of a recovery trade-off in avian influenza.

Horizontal versus vertical transmission

The impact of horizontal versus vertical modes of transmission on pathogen virulence evolution is one of the best-studied predictions of virulence evolution theory, likely because it is amenable to experimental testing in many systems. The simple, intuitive prediction is that vertical transmission will select for lower virulence than horizontal transmission because vertical transmission is tied to host reproductive success, so high virulence would negatively impact the virus as well as the host. However, most viruses that are vertically transmitted are also horizontally transmitted at some frequency. Theoretical models allowing both transmission modes predict that use of predominantly vertical transmission will still favor low virulence, but predominantly horizontal transmission can favor low, medium, or high virulence depending on other factors (Lipsitch and Moxon, 1997). Despite this complexity, results of several empirical studies comparing the two transmission modes are consistent in supporting the evolution of lower virulence under vertical transmission regimes. This was first demonstrated using filamentous phage f1 in E. coli under culture conditions that allowed only strict horizontal or vertical transmission (Bull *et al.,* 1991). A later f1 phage study confirmed that vertical transmission with infrequent horizontal transmission also selected for reduced virulence, compared with passage lines that used alternating transmission modes (Messenger *et al.,* 1999). A meta-analysis of the literature on arbovirus virulence in mosquito vectors found that horizontal infection caused reduced mosquito survival but vertically transmitted infection had no significant effect, correlating with lower virulence (Lambrechts and Scott, 2009). Similarly, a study of the plant potyvirus zucchini yellow mosaic virus reported that seed-borne vertical transmission led to asymptomatic infections, while horizontal aphid-transmission caused distinct symptoms of disease (Simmons *et al.,* 2013). The most convincing demonstrations come from elaborate plant virus experimental evolution studies where virus populations evolved in replicated horizontal or vertical transmission lines were tested for transmission efficiency, viral fitness, and virulence changes. Consistent with theoretical predictions, both barley stripe mosaic virus and cucumber mosaic virus rapidly evolved reduced virulence upon vertical transmission passage, but horizontal transmission led to increased virulence for one virus and no change for the other (Pagan *et al.,* 2014; Stewart *et al.,* 2005). Thus, there is consistent support for evolution of reduced virulence during vertical transmission, but not necessarily increased virulence through horizontal transmission.

Mixed infections

It is theorized that mixed infections, where multiple pathogen strains infect the same host, can drive the evolution of increased virulence. The argument is that the host lifespan reduction costs of virulence can be overcome by what has been termed the tragedy of the commons hypothesis (Hardin, 1968). The theory is that if two strains of a pathogen are infecting a host, one strategy is for both strains is to have equally low virulence, and thus maintain the longevity of the host resource (the commons), maximizing the duration of transmission. However if one virus strain 'cheats' and evolves higher virulence, the 'commons' will be used up more quickly, and the host will die sooner, truncating the transmission of both the high and low virulence strains. If we assume the high virulence strain has higher transmission rates (as discussed in the "*Transmission virulence trade-off*" section), it will also have greater overall fitness than the low virulence strain in this scenario. In other words, both virus strains feel the consequence of the shortened host lifespan, but the low virulence strain suffers more because it produces fewer infectious progeny while the host is alive (Read and Taylor, 2001). This theory has been extended to address kinship and altruism, predicting that if the pathogen strains co-infecting the host are closely related, the likelihood that one will cheat and evolve increased virulence is diminished, since both pathogens want to maximize the reproduction of their shared genome. As such, the more distinct the viral genotypes are from each other, the more likely that high virulence will evolve (Brown *et al.,* 2002; Frank, 1996; Lion, 2013; Nowak and May, 1994; West and Buckling, 2003).

This theory is pertinent because the majority of virus infections are composed of multiple virus genotypes, either due to *de novo* mutation in the host or multiple viral exposure events. An array of empirical work has been done addressing these concepts. For example, in Marek's disease virus infection of chickens, high virulence strains have a competitive advantage over low virulence strains in co-infection *in vivo* (Dunn *et al.,* 2012). Similar results were found for IHNV in trout where a high virulence genotype had higher competitive fitness and transmission potential than a low virulence type when the strains infected the host at the same time (Wargo *et al.,* 2010; Wargo and Kurath, 2011). Interestingly, when the strains infected the host sequentially (superinfection), the advantage of the high virulence strain disappeared, and the strain that infected the host first had higher fitness, regardless of virulence (Kell *et al.,* 2014). If this is true in other viral systems, the advantage of infecting the host first could overwhelm the advantage of high virulence in sequential, mixed infections. However, a different pattern was observed for FMDV in cell culture, where high virulence virus strains were better at displacing a persistent virus during superinfection compared to low virulence strains (Saiz and Domingo, 1996).

An alternate theory proposed that some mixed infections might actually lead to the evolution of decreased virulence. The idea here is that viruses could evolve ways to sequester the replication resources produced by co-infecting virus strains (Chao *et al.,* 2000). This has been the theory behind the wide-spread occurrence of defective interfering (DI) particles, which are viruses that have lost protein production machinery but often have enhanced ability to sequester the proteins produced by co-infecting viruses (Bangham and Kirkwood, 1993; Marriott and Dimmock, 2010). These DI particles cannot replicate on their own, so they have zero fitness in single infections. However, since they do not produce a full complement of proteins and have truncated genomes, they can replicate more quickly in co-infection compared to standard viruses, and thus have greater fitness during co-infection. Ultimately, DI particles lead to reduced total virus replication (interference) in mixed infections because they interfere with the replication of co-infecting viruses by sequestering their resources. Therefore, if replication is positively correlated with virulence, DI particles are also predicted to lower virulence in mixed infections, and this has been demonstrated in many systems. Such a scenario was observed in a bacteriophage system, where co-infection did indeed lead to the evolution of a virus that had higher fitness in co-infection compared to the ancestral virus, but reduced replication and fitness in a single infection (Turner and Chao, 1999). However, the authors did not measure virulence in this study. Similar results were seen in a study of FMDV in cell culture, where the authors passaged the virus for hundreds of generations at a high multiplicity of infection to allow for evolution in a setting with a high probability of mixed infection. It was observed that two types of viruses evolved: colonizers that had higher virulence and fitness in single infection, and competitors, which had higher fitness in co-infection because they interfered with the replication of the colonizers (Ojosnegros *et al.,* 2010). An even more astounding possibility is that viruses might evolve plasticity in their response to single compared to mixed infections. Such results were found in a study of bacteriophage φ2 in *Pseudomonas flourescens*, which when passaged in a mixed infection environment evolved to express a long lysis time in single infections and a short lysis time in mixed infections. This allowed the phage that evolved in mixed infection to have an initial replication advantage in mixed infections, and equal performance in single infections, compared to the phage evolved in single infections (Leggett *et al.,* 2013).

Competition driving the evolution of increased or decreased virulence does not seem to be the only outcome of co-infection. A study of an insect nuclear polyhedrosis virus in a moth host suggested a third possibility, competition leading to co-operation. In this study, virulence was higher in co-infection compared to single infection for both high and low virulence virus genotypes, and viral yields were also equal or higher in co-infection for each genotype (Hodgson *et al.,* 2004). The mechanisms for the observed facilitation were not determined, but the study highlighted that single infections are poor predictors of virus performance in mixed infections. These studies reiterate the amazing capacity viruses have to evolve diverse strategies for dealing with environmental conditions such as mixed infections.

Life-history theory perspective on timing of transmission during infection

Life history theory of infection (Kochin *et al.*, 2010; Schmid-Hempel, 2011) predicts that pathogen transmission earlier in the course of infection selects for higher virulence than later transmission. This is because, in a virulence-transmission trade-off, the benefit of higher virulence in terms of increased transmission occurs temporally before the cost of virulence in terms of mortality. In this hypothesis a longer time lag between transmission and mortality will reduce the cost of higher virulence because death of the host has less impact on the virus after more transmission has been accomplished. The most compelling support for this prediction comes from an experimental study in which a nuclear polyhedrosis virus was serially passaged in gypsy moth larvae using either early (5 days) or late (9 days) transfer timing between *in vivo* passages (Cooper *et al.*, 2002). Virulence assays revealed that early timing lineages had higher virulence than late timing lineages, in that they killed a higher percent of larva, but at a cost of lower overall productivity. In another test of this prediction, VSV was serially passaged in mammalian cells using different passage timing and intermediate transfer time lineages evolved significantly higher *in vitro* competitive replication fitness than early or later transfer lineages (Elena, 2001). Although virulence was not tested in this study, a previous study reported a positive correlation between *in vitro* competitive fitness and *in vivo* virulence for VSV (Novella *et al.*, 1995). Finally, Sindbis virus passaged in mammalian cells with early transfer timing became attenuated in mouse neurovirulence assays (Olmsted *et al.*, 1984), likely due to selection for increased heparin sulfate binding in cell culture. Thus early transmission may lead to increased virulence, but not in all cases.

Pathogen survival in external environments

Survival in the external environment is another important fitness trait for some viruses. It seems intuitive that survival outside the host, also referred to as virion stability or durability, could influence virulence evolution, but empirical studies have generated inconsistent results. An early hypothesis known as "curse of the pharaoh" predicted that pathogens with long-lived propagules should evolve high levels of virulence because they employ a “sit-and-wait” transmission mode that maintains a long transmission duration even after immobilization and mortality of infected hosts (Ewald, 1994). Mathematical models predict that this should be true in some cases, such as systems not at equilibrium (i.e. recently after disease emergence) (Bonhoeffer *et al.*, 1996; Roche *et al.*, 2011). An epidemiological meta-analyses of 16 diverse human respiratory pathogens, including nine viruses, supported the positive correlation of virulence with duration of survival outside of the host (Ewald, 2011; Walther and Ewald, 2004). However, experimental studies involving several species of *E. coli* phage revealed that phages with higher durability had lower replication rates, suggesting a trade-off between reproduction and extracellular durability in phages (De Paepe and Taddei, 2006; Ogbunugafor et al., 2013). Also, for influenza virus no correlation was found between in-host replication and virion durability (Handel et al., 2014). These three studies did not assess virulence, but if one assumes a correlation between replication and virulence, they contradict the original prediction and suggest equal or lower virulence in more durable virions. Similarly, for VSV a selection regime resulting in increased extracellular survival also resulted in decreased replicative fitness and virulence (Heineman and Brown, 2012). Finally, for plant tobamoviruses, selection for increased virulence, in the form of ability to break through host genetic resistance, coincided with decreased virion durability (Fraile *et al.,* 2014). Thus, experimental studies are not consistent with the previous observation of a correlation of extracellular survival with virulence, and it may be that selection acts independently at the within- and between-host scales (Goldhill and Turner, 2014; Handel *et al.,* 2014).

Non-equilibrium virulence in viral epidemics and emergence events

Many mathematical models of virulence evolution involve assumptions that host and pathogen populations are in dynamic equilibrium. Theorists recognize that this may not always be the case, especially for emerging pathogens that have only very recently invaded a new susceptible host population. In this situation theory predicts a transient peak of higher virulence in the beginning epidemic phase, followed by a drop in virulence as the system approaches equilibrium (Bull and Ebert, 2008; Day and Proulx, 2004; Frank, 1996). This transient virulence increase has been observed in model-driven simulations using parameter estimates from published empirical studies (Bolker *et al.,* 2010), and it has also been demonstrated in a controlled experimental study of bacteriophage λ in *E. coli* (Berngruber *et al.,* 2013). In the phage study, competition between latent (avirulent) and virulent variants favored the virulent phage early in an experimental epidemic, but as the epidemic developed the latent phage became dominant. Although more testing is needed, the data available so far support this transient change in virulence over time, which is particularly relevant to emerging infectious diseases.

Host specificity and generalist-specialist virulence trade-offs

Although virulence is considered to be a pathogen trait, it is clear that its expression depends heavily on host competency and resistance; thus viral virulence is inherently host-specific. The divergence of viruses into different strains and species is often accompanied by variation in their ability to infect and cause disease in different hosts. Evidence of host-specific virulence is plentiful for viruses of all host classes. It is illustrated in cross-infection experiments in which one or more virus strains are tested for fitness and virulence in multiple host types. Host-specific variation in virulence has been demonstrated empirically at the level of different host species, lineages within one host species, specific host genotypes, or cultured cell lines. An interesting study testing five *Ambrystoma tigrinum* virus isolates in five different host lineages within one salamander species demonstrated significant virulence variations that were more dependent on the host lineage than the virus isolate (Brunner and Collins, 2009).

Theory suggests that as a virus adapts during continuing replication in one host type, it evolves increased fitness for infection of that host, often with increased virulence when fitness and virulence are positively correlated. It is also hypothesized that a virus that specializes on one host pays a cost in reduced fitness on other hosts, with likely reduced virulence. Thus, a **specialist** virus is a genotype that has asymmetric absolute fitness across different host environments (Remold, 2012). Genetic mechanisms proposed for the cost of specialization include sign pleiotropy (also referred to as antagonistic pleiotropy), in which mutations that confer an advantage in one host create a disadvantage in another host, epistatic effects between mutations, and epistatic pleiotropy (Bedhomme *et al.,* 2015; Ebert, 1998; Remold, 2012). Virus host jumps observed in nature provide an opportunity to test the theoretical predictions of host specialization as the virus adapts to a new host. IHNV isolates obtained after a host jump into rainbow trout in the 1970s had increased replicative fitness and virulence in rainbow trout, but they lost both fitness and virulence in the former host, sockeye salmon (Garver *et al.,* 2006; Peñaranda *et al.,* 2011). This demonstrates a cost associated with specialization on the new host. In another case, canine parvoviruses isolated after a host jump from cats to dogs in the 1970s showed increased infectivity and virulence for dogs, and early strains from dogs had lost the ability to infect the former feline host. Although this also implies the predicted specialist trade-off in the former host, it is interesting that later variants arose that infected both cats and dogs, as well as several non-domestic carnivores (Hoelzer and Parrish, 2010; Truyen *et al.,* 1996).

In contrast to the specialist environment, a virus that exists in a diverse host environment involving frequent transmission between different host types evolves to maintain fitness, and often virulence, in multiple hosts. **Generalist** viruses, defined as having equal absolute fitness across different host environments (Remold, 2012), have an advantage in some situations because they are able to utilize many hosts. However, a **generalist-specialist trade-off** is hypothesized in which use of multiple hosts constrains generalists to compromised fitness levels below the level they could reach if they specialized on any one host (Remold, 2012; Woolhouse *et al.,* 2001). If virulence is also constrained, due to a correlation with fitness, this suggests that specialists are likely to be more virulent in the few hosts they can infect, whereas generalists will have lower virulence in the many hosts they can infect. For example, observational studies have shown that smallpox, which is highly specialized to humans, is highly virulent, whereas cowpox, which infects many host species and is thus a generalist, is relatively avirulent. However it should be noted that whether the virulence differences in these two examples are driven by the generalist-specialist virulence trade-off or other factors, has not been elucidated (Haller *et al.,* 2014).

Serial passage experiments provide a strategy for critically examining predications associated with host specialization and generalist-specialist trade-offs. An excellent review by Ebert summarizing results of experimental evolution studies through 1995 concluded that serial passage most often led to virulence increases in the passage host and decreases in the former host, documenting a cost of specialization for many viruses (Ebert, 1998). The majority of the studies cited involved animal viruses passaged in cultured cell lines leading to *in vivo* attenuation in mammalian hosts, often in the interest of developing vaccines. More recent serial passage studies have provided additional tests of specialization and generalist-specialist fitness trade-offs. The hypothesized cost of generalism, in terms of constrained fitness, is tested by comparing viruses evolved during serial passage in a single host type (specialist lines), with viruses evolved in simultaneous serial passage using alternating host types (generalist lines). A comprehensive study of VSV in three mammalian cell lines found that passaging the virus in a single host cell line resulted in increased competitive fitness in all cases, indicating specialization. However, viruses evolved simultaneously by passage through two alternating cell lines also gained fitness in both host cell lines, to levels comparable with the single host lines, suggesting there was not a generalist-specialist fitness trade-off. For the specialist virus passaged in a single cell line, costs were demonstrated as reduced fitness on the former host line in most cases, but with variable results for one of the treatments (Turner and Elena, 2000). Although virulence was not directly measured in this system, a correlation of competitive fitness with virulence in mice has been documented for VSV, as noted previously (Novella *et al.,* 1995).

Additional case studies can be found in plant viruses. An elaborate study of tobacco etch virus (TEV) using four host species in both single and alternating host passages yielded mixed results for both adaptation to higher fitness and virulence in the passage host(s) and virulence trade-offs in the former host. There was little evidence of a generalist-specialist virulence trade-off in that viruses passaged through alternating hosts were just as virulent to those hosts as viruses passaged through only one of the host species (Bedhomme *et al.,* 2012). Another study of TEV passaged in five different host genotypes within one host species (Hillung *et al.,* 2014) also yielded mixed results for both adaptation to the passage host and trade-offs in the former host, confirming complexity in these phenotypes (Bedhomme *et al.,* 2012; Hillung *et al.,* 2014). For the generalist plant virus cucumber mosaic virus (CMV), serial passage in three different hosts also led to variable results in which virulence did not increase during passage and virulence trade-offs in non-passage hosts were not observed in five out of six cases (Sacristán *et al.,* 2005). However, different results were obtained in a study of two plum pox virus isolates from peach trees that were serially passaged in pea using either mechanical or aphid transmission (Wallis *et al.,* 2007). In this case all passage lines showed increased infectivity, fitness, and virulence in the pea passage host, and the evolved viruses had reduced infectivity in peach, demonstrating specialization with increased virulence in the new host and trade-offs in decreased capability in the former host.

As a vertebrate virus example, fascinating results have been reported for the retrovirus Friend virus complex. Serial passage in three congenic inbred mouse strains that differed only at the major histocompatibility complex (MHC) locus resulted in increased fitness and virulence in the passage host in all cases, with clear virulence reductions in the non-passage host strains in four out of six cases and fitness reductions in all cases (Kubinak *et al.,* 2012). Subsequent studies including mouse strains that differed across the whole genome showed that MHC polymorphisms were responsible for the great majority of the trade-offs in non-passage hosts, and that variability in inherent genetic resistance correlated with the magnitude of the fitness changes observed during passage (Kubinak and Potts, 2013; Kubinak *et al.,* 2013). Altogether these studies find, not surprisingly, that host differences exert strong selection on viral virulence evolution. Overall, although there is some support for virulence trade-offs associated with evolution of specialists and generalists, this support is not strong and there are clear exceptions.

Co-evolution of viral virulence and host resistance

In long-term virus-host associations, the virus creates selection pressure on the host population to develop increased resistance, which in turn selects for increased viral fitness, with potential for impacts on viral virulence evolution. This has been described as the Red Queen hypothesis, or arms race dynamics, in which continuing cycles of evolution of increased host resistance and increased viral fitness result in co-evolution of virus and host populations, with realized levels of virulence that may fluctuate up or down (Schmid-Hempel, 2011). This phenomenon has been observed in many viruses of bacterial, plant, and animal hosts. Bacteriophages have been used successfully as tractable experimental co-evolution models in simplified environments. As an example, continuous co-propagation of the lytic RNA bacteriophage Qβ in *E. coli* for 54 days revealed complex mechanisms of co-evolution in which the bacterial host first evolved partial resistance, and later an accelerated growth rate, and the phage counter-adapted by developing increased release efficiency, with a change in host specificity and decreased virulence (Kashiwagi and Yomo, 2011). Molecular genetic analyses identified mutations associated with these changes, and revealed that both the phage and host had higher rates of evolutionary change during co-evolution than they did in separate control cultures. Among the many examples of virus-host co-evolution the best-studied cases are described below, involving co-evolution of plant viruses in genetically resistant crops, and the co-evolution of myxoma virus in rabbits after its release as a biocontrol agent in Australia and Europe.

Evolution of viral virulence in resistant plant cultivars

The most effective strategy for control of plant viruses has been breeding or engineering of specific disease-resistant lines of crop plants. Although the production and large-scale deployment of these resistant cultivars in agriculture has been successful against some plant viruses, it represents a strongly altered selection pressure on viruses in the field, and resistance-breaking viruses arise in many cases. The genetic basis for specific disease resistance in most crop plants involves mutations in one or a small number of resistance genes that act by a gene-for-gene mechanism to trigger plant defenses in response to virus infection (Sacristan and Garcia-Arenal, 2008). The evolution of viruses that break resistance of plant cultivars is traditionally referred to as increased virulence in plant virus literature, and it is also based on simple genetic changes of one or a few mutations that allow the virus to evade the plant defenses and replicate in previously resistant plants, resulting in disease (Fraile *et al.*, 2011; Sacristan and Garcia-Arenal, 2008). In some major crop plants, such as pepper and soybean, evolution of resistance-breaking viruses in the field has led to repeated cycles of breeding for additional resistance, producing cultivars with 3 or 4 resistance alleles in polygenic resistance.

Pepper cultivars with three different levels of resistance deployed in agriculture in Spain have provided for an excellent observational study of evolution of viral virulence in response to host resistance (Fraile *et al.*, 2011). In this study tobamoviruses isolated from pepper over a 20-year period were tested to determine their pathotypes, defined as their differential abilities to infect the three pepper cultivars. Changes in the frequencies of the different viral pathotypes over time could be explained in the context of shifts in the relative frequencies of pepper cultivars used in agriculture (Figure 6.4). Most interestingly, a shift in pepper plantings during the 1990s from an intermediate resistance cultivar to the older, more susceptible cultivar was followed by a shift in the tobamovirus

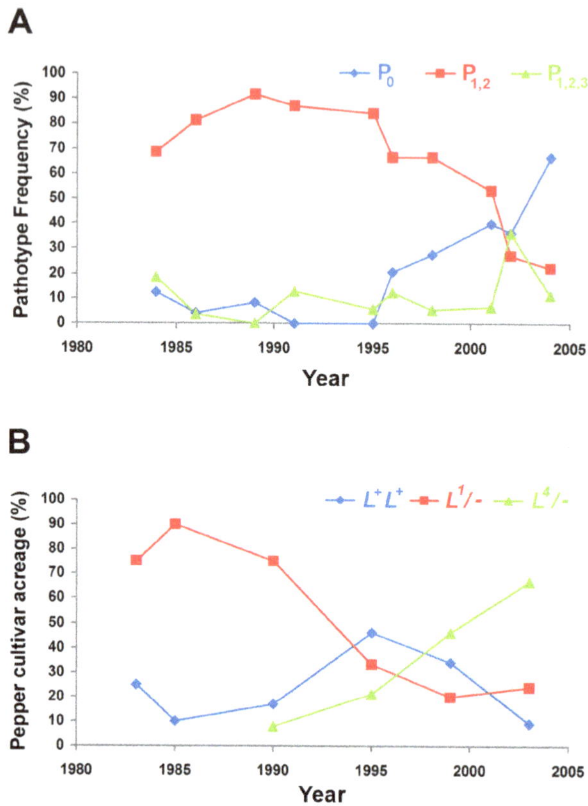

Figure 6.4. Field observation of the evolution of the pathotype composition of the tobamovirus population (panel A) in response to changes in the resistance genotype of pepper crops (panel B) grown in southeast Spain between 1984-2004. In the 1990s there was a reduction in use of the intermediate resistance $L^1/-$ pepper cultivar because it did not protect against the resistance-breaking $P_{1,2}$ pathotype tobamovirus. Growers instead returned to using the older, highly productive, susceptible L^+L^+ pepper during the mid-1990s, until a highly effective L_4 cultivar became available. In response to the shift in from$L^1/-$ to L^+L^+ pepper, the tobamovirus population in the field evolved with a reduction in the $P_{1,2}$ pathotype and an increase in the older, non-resistance-breaking P_0 pathotype after 1995. The decline in $P_{1,2}$ virus suggests it was less fit than P_0 when its specific virulence was not necessary, indicating a fitness cost associated with evolution of resistance-breaking in pepper. Redrawn from Fraile *et al.*, 2011, by permission of Oxford University Press on behalf of the Society for Molecular Biology and Evolution.

population. There was a decline in the virus pathotype that had evolved ability to infect the resistant pepper, and an increase in the older pathotype that was unable to infect the resistant pepper. These findings supported the theoretical prediction that evolution of higher virulence in the resistance-breaking virus had an associated fitness penalty on non-resistant hosts. This fitness cost was subsequently confirmed in experimental infections comparing non-resistance-breaking and resistance breaking viruses (Fraile *et al.,* 2011). A fitness cost associated with resistance-breaking viruses was also observed during experimental evolution of potato virus Y on both susceptible and resistant pepper cultivars. Serial passage on the resistant pepper selected for a resistance-breaking virus that had increased replicative fitness in the resistant pepper, but significantly reduced fitness in the susceptible pepper (Montarry *et al.,* 2012). In this study the evolved resistance-breaking virus showed no change in a more general measure of virulence determined as reduced plant dry weight due to infection, indicating that virulence defined as the ability to break through host resistance alleles is not necessarily correlated with other virulence measures. In other experimental evolution studies passage of soybean mosaic virus in soybean cultivars of increasing resistance levels re-capitulated genetic changes associated with resistance-breaking in the field (Hajimorad *et al.,* 2011), and a high frequency of virus evolution breaking through RNAi-mediated resistance was demonstrated for turnip mosaic virus (Lafforgue *et al.,* 2011a). These studies are consistent in showing that host resistance is a strong selection pressure for increasing viral virulence in terms of resistance-breaking, with relevance to durability of resistant plant cultivars in agriculture.

Myxoma virus pest control in rabbits

Myxoma virus is a one of the most famous examples of co-evolution of host resistance and pathogen virulence. As noted above (sections "*Transmission virulence trade-off*" and "*Recovery virulence trade-off*") when myxoma virus, a DNA poxvirus, was introduced into Europe and Australia in 1950-1952, it originally had very high virulence in rabbits, and subsequently evolved lower virulence, to become predominantly of intermediate, although still very high, virulence (Fenner, 1983). However, the story does not end there. The virus initially caused extremely high levels of mortality to its host, such that on average greater than 98% of exposed individuals succumbed to death. Even the evolved intermediate virulence strain of the virus still caused 70-95% mortality in 17-28 days. As such the rabbit populations were drastically reduced by the virus. This resulted in strong selection pressure on the rabbits, which subsequently evolved resistance to the virus. This was demonstrated through controlled laboratory challenges of naïve rabbits from the field using a single virus strain of intermediate virulence. In these studies it was found that wild rabbit mortality levels caused by the intermediate virulence virus strain had dropped from 80% in 1952 to 30% in 1958 in Australia (Marshall and Fenner, 1958), and from 90% in 1966 to 20% in 1976 in Great Britain (Ross and Sanders, 1977). A logical question is did the virus respond to this increased resistance by evolving increased virulence? Evidence suggests indeed it did although the data are limited. In the mid-1980s there appeared to be a shift in the field isolates of the virus in Australia, such that a larger percentage of circulating virus fell into the higher virulence grades compared to the previous decade (Figure 6.5) (Dwyer *et al.,* 1990; Fenner, 1983). A similar pattern was observed in the early 1980s in Great Britain and France, although the data are less extensive [summarized in (Kerr, 2012)]. The virulence of more modern virus isolates has not been published, but would be interesting.

What is particularly compelling about these data, in addition to the wealth of field and controlled laboratory experiments showing the same pattern after independent introductions on two continents, is that molecular mechanisms of host resistance and viral virulence evolution are being explored (Kerr, 2012). Phylogenetic and genomic analyses show a strong genetic signal of co-evolution between myxoma and its rabbit host, with several potential viral virulence genes identified (Kerr *et al.,* 2013). Likewise, the trajectory of host resistance and viral virulence co-

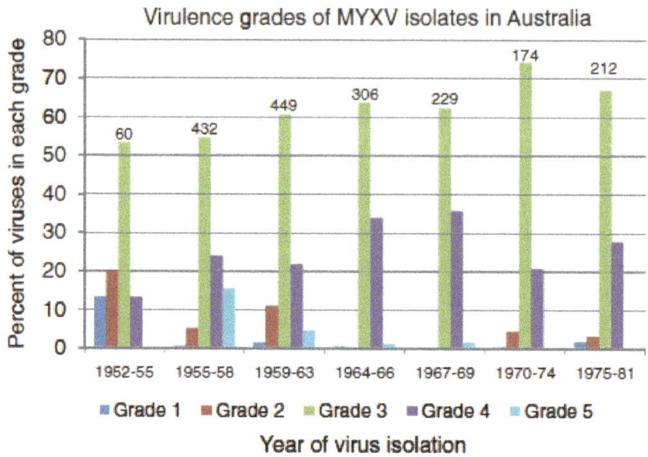

Figure 6.5. Myxoma virus virulence distribution over time after its intentional introduction to control Australian rabbits in 1952. The virulence of the virus is classified in grades from high (1) to low (5). The prevalence of the highest and lowest virulence grades decreased, while the intermediate grade prevalence increased, until the 1970s, suggesting the selection of intermediate virulence. After that time period, the high virulence types begin to re-emerge. This is hypothesized to be due to the increased prevalence of host resistance, which increased in the mid-to-late 1950s (see Kerr *et al.*, 2012). Numbers above the bars are total number of viruses tested for each time period. Data are from Fenner *et al.*, 1983. Reprinted from Kerr *et al.*, 2012 with permission from Elsevier.

evolution inferred with mathematical models, parametrized from historical field and laboratory data, closely matches the patterns observed in nature (Anderson and May, 1982a; Dwyer *et al.*, 1990; Massad, 1987).

A surprisingly less known but compelling story involves rabbit haemorrhagic disease virus (RHDV). Much like myxoma virus, RHDV was introduced into Australia in the 1990s as a form of bio-control with high initial virulence in rabbits. Unlike myxoma virus, RHDV is an RNA calicivirus, and it did not evolve decreased virulence after its release. Instead, the host rapidly evolved resistance. The virus seems to have subsequently evolved increased virulence, as confirmed by *in vivo* infection studies and phylogenetic assessments (Elsworth *et al.*, 2014; Kovaliski *et al.*, 2014). It is unclear why this system resulted in the rapid evolution of increased virus virulence in response to host resistance in contrast to the myxoma virus response, which was much more gradual. One possibility is that myxoma virus is transmitted through vectors feeding on live rabbits (Fenner and Ratcliffe, 1965), whereas RHDV is transmitted either directly between live rabbits or by insects from dead rabbits (Morisse *et al.*, 1991). Thus the cost of virulence might be less for RHDV because death does not terminate transmission. As in previous sections, this highlights how evolutionary patterns can be very system specific, suggesting general conclusions should be made with caution.

Virulence evolution in response to vaccination

In 2001 a theory emerged stating that vaccination could drive the evolution of increased virulence (Gandon *et al.*, 2001). This theory is largely an extension of

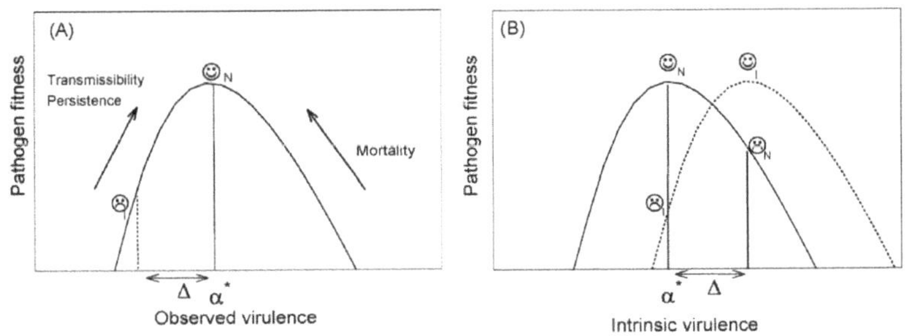

Figure 6.6. Vaccine impact on virulence evolution, where immunity selects for higher levels of intrinsic virulence. Vaccines reduce the disease caused by viruses (delta in panel A). This forces the virus from its optimal fitness peak (smiley face panel A) to a suboptimal level (frowning face panel A). The virulence trade-off theory predicts that the virus will evolve increased virulence so that it causes the same level of disease in the vaccinated host, as it did in a previously unvaccinated host (smile face above dotted line panel B). If this newly evolved virus then infects an unvaccinated host, it will cause increased disease in that host, despite having suboptimal fitness (frowning face, subscript N, panel B). Subscript I stands for immune (vaccinated), subscript N stands for non-vaccinated. Reprinted from Mackinnon *et al.*, 2008, with permission from Elsevier.

the virulence-trade-off hypothesis. The theory is based on the fact that efficacious vaccines reduce the levels of disease in vaccinated compared to non-vaccinated hosts after subsequent exposures to the target pathogen, but they may not prevent subsequent infection, or even onward transmission of the pathogen. As such, vaccines reduce the cost of virulence associated with truncation of transmission duration due to host mortality. Thus, as shown in Figure 6.6, under a vaccination scenario, pathogens infecting a host after vaccination maximize their fitness (R_0) at a higher level of replication, which brings their virulence back toward its pre-vaccination level. This has two important consequences. Firstly, the vaccine is less effective at providing disease protection in vaccinated hosts. Secondly, unvaccinated hosts infected with this evolved pathogen strain will experience elevated levels of disease. It is important to note that only certain types of vaccines are believed to enable virulence evolution, vaccines which protect against disease but do not block infection and transmission upon subsequent exposures to the virus (Gandon *et al.,* 2001). Clearly a vaccine that blocks transmission (either by blocking pathogen entry into, or exit from, the host) will prevent spread of a pathogen that may evolve in vaccinated hosts, and therefore reduces pathogen fitness to zero. Interestingly, few vaccines have been tested for their transmission blocking capacity, and where examined transmission is often reduced but still observed from vaccinated hosts (Atkins *et al.,* 2011; Coutts *et al.,* 1994; Dawson *et al.,* 1991; Gandon and Day, 2008; Group, 2010; Pileri *et al.,* 2015). It is important to note that this theory is not focused on the evolution of the vaccine itself, say in the case of a live-attenuated virus vaccine, but rather the evolution of virus that may infect the host after vaccination.

Empirical evidence for vaccine-driven virulence evolution in viral systems is increasing (Gandon and Day, 2008). The most well studied example is Marek's Disease Virus (MDV) in chickens. The first vaccine for MDV was introduced in the poultry industry in the 1970s against what, at that time, was called virulent MDV

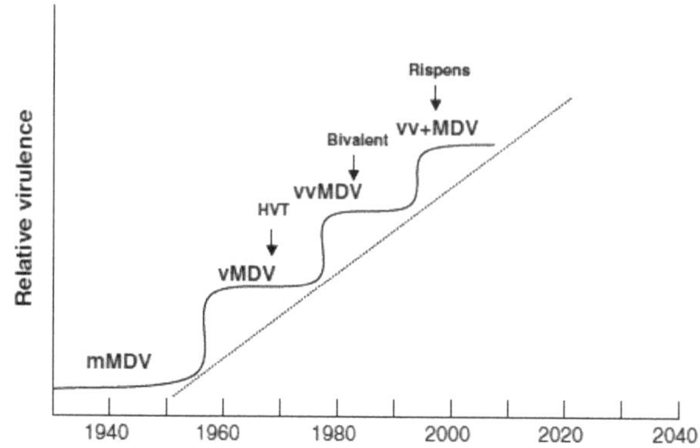

Figure 6.7. Step-wise evolution of virulence of Mareks Disease virus in chickens. Relationship between the virulence increase and the introduction of different vaccines is shown. Virulence phenotypes of the virus are indicated as mild, mMDV; virulent, vMDV; very virulent, vvMDV; and very virulent plus, vv+MDV. The three vaccines used sequentially in the poultry industry are HVT, herpes virus of turkeys; bivalent, HVT and serotype 2 (SB-1) vaccines; Rispens, CVI988 strain. After each vaccine was introduced, the virus evolved increased virulence a decade later. Redrawn with permission from Nair 2005, copyright © 2004 Elsevier Ltd.

(vMDV). A decade later a more virulent strain (vvMDV) emerged against which the vaccine was no longer protective. A new vaccine was introduced, and a decade later an even more virulent virus strain again emerged (vv+MDV) (Figure 6.7) (Gimeno, 2008; Nair, 2005). This system is compelling because the virulence increases of the virus observed in the field were confirmed with laboratory challenges of vaccinated and unvaccinated chickens, showing that vaccinated birds can transmit virus and vaccination results in the evolution of new virus strains (Atkins *et al.,* 2011; Witter, 1997). Data from this system were used to parameterize a mathematical model, demonstrating vaccination could drive virulence evolution in this system by acting on the virulence trade-off (Atkins *et al.,* 2011; Atkins et al., 2013b).

Evidence for risk of vaccine-induced virulence evolution has also emerged from other virus systems. Hepatitis B virus vaccine escape mutants perform better in vaccinated compared to unvaccinated hosts (Gandon and Day, 2008), but virulence changes have not been investigated in this system. In feline calcivirus some vaccines enhance virus shedding from cats (Dawson *et al.,* 1991), and the vaccines do not prevent subsequent infection (Coutts *et al.,* 1994). These were cited as possible mechanisms driving the evolution of virulent strains that emerged in the mid-2000s (Radford *et al.,* 2006). A variety of other non-viral pathogens have also shown evidence of vaccine-induced virulence evolution (Gandon and Day, 2008; Kennedy *et al.,* In press; Mackinnon *et al.,* 2008). It is likely the ingredients are in place for vaccines to drive virulence evolution in other viral systems (Christiansen *et al.,* 2011; Gandon and Day, 2008; Murall *et al.,* 2015; Spackman and Swayne, 2013), but more information on the transmission-blocking efficacy of current and future vaccines is sorely needed to make this determination.

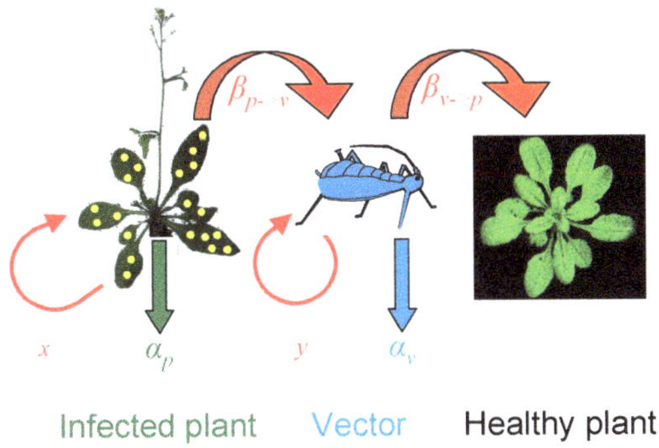

Figure 6.8. Impact of the vector-borne transmission on virulence evolution theory. The transmission rate can be decomposed into two steps: from the infected plant to the vector ($\beta_{p\rightarrow v}$) and from viruliferous vector to healthy plant ($\beta_{p\rightarrow v}$). Within-plant (x) and within vector (y) viral accumulation are thought to have an effect on mortality of the plant (α_p) and vector (α_v), respectively. This added complexity modifies the theoretical expression of R_0 (see Froissart *et al.*, 2010). Figure from Froissart *et al.*, 2010, by permission of the Royal Society.

Impact of vector transmission on virulence evolution

Vector transmitted viruses are transferred between "primary" hosts by an alternate host species, referred to as the vector, which is usually an insect or other small invertebrate. Evolution of these viruses is thus shaped by interactions with two host species that are typically very different. Vector-borne virus transmission provides significant advantages in ease of dispersal for viruses of both plants and animals. In terms of virulence evolution theory, vector transmission adds complexity to the equation for the basic reproductive number R_0 (equation 3 in the "*Mathematical basis: the basic reproductive number, R_0*" section) in that it must incorporate separate transmission rate factors for transfer of virus from an infected host to the vector, and then from the vector to a susceptible host. There are also separate baseline mortality rates, virus-induced mortality rates (virulence), and in-host virus accumulation rates for the primary host and the vector, as illustrated for a typical vector-borne virus in Figure 6.8 (Froissart *et al.*, 2010). In cases where viruses replicate in the vector as well as in the primary host, virulence levels in each host will be determined by a range of selection factors unique to each virus-host interaction. General predictions for virulence evolution in vector-borne diseases have suggested that vector borne diseases will be more virulent overall than directly transmitted diseases, because vectors provide long-distance spread of viruses from immobilized hosts, and host immobilization may enhance vector feeding success, and thus transmission. Virulence is also hypothesized to be lower in the vector than in the primary host because the virus requires the vector to be healthy enough to be mobile for transmission to occur, so it maximizes replication only in the primary host (Elliot *et al.*, 2003; Ewald, 1994). Here we discuss vector-borne transmission for plant viruses and animal viruses (arboviruses), where research has provided insights into virulence evolution.

Vector borne plant viruses

In plants the great majority of known viruses are transmitted horizontally by insect vectors, which are mostly homopteran species such as aphids, leafhoppers, and planthoppers. This provides an important advantage in dispersal of viruses from immobile hosts. Insect-borne plant viruses are transmitted by non-circulative, circulative, or propagative modes, with only the last involving virus replication in the vector (Froissart *et al.,* 2010). The only other natural transmission mode for plant viruses is vertical transmission through pollen or seed, although mechanical inoculation of abraded leaf tissue is commonly used for experimental infections. Field observations support the prediction that vector-borne viruses (horizontal transmission) are in general more virulent in plants than seed-transmitted viruses (vertical transmission), which fits with theory presented in section entitled "*Horizontal versus vertical transmission*" (Power, 1992; Roossinck, 2010). Indeed, plant viruses that cause asymptomatic (low virulence) persistent infections are all seed-transmitted, while acute viruses of well-known agricultural diseases are typically vector-borne (Roossinck, 2010). In terms of mechanisms, benefits of virulence in vector-borne plant viruses are proposed to arise from increased vector transmission based on increased in-plant virus accumulation, and increased disease signs, such as yellowing of leaf tissue, which is known to attract aphid vectors to infected plants (Froissart *et al.,* 2010; Power, 1992). Experimental work with several plant viruses provides support for a generally positive correlation between virus accumulation levels in plants (replicative fitness) and insect-vectored transmission for natural virus isolates, although some exceptions have been noted among individual point mutant viruses (Froissart *et al.,* 2010; Power, 1992; Sacristan and Garcia-Arenal, 2008). However, the second important correlation, between virus accumulation levels and virulence, has less support. This may simply reflect less study, but it has also been suggested that some definitions of virulence used in plant virus research, particularly those based on reduced seed production, may not be appropriate for assessing impact on horizontal transmission (Froissart et al., 2010). At present we know of only one plant virus study designed specifically to test the transmission trade-off theory, using nine isolates of cauliflower mosaic virus in *Brassica rapa*, and measuring virus accumulation, virulence as reduction of leaf surface area, and aphid transmission (Doumayrou et al., 2013). The results demonstrated a concave shaped virulence:transmission trade-off relationship in this plant virus-host system, as predicted by theory (Figure 6.9).

Arboviruses

Vector-borne transmission also occurs in animal viruses with vertebrate or invertebrate primary hosts. Here, in addition to ease of dispersal, the use of vector-borne transmission provides advantages in reduced dependence on mobility of infected hosts for onward transmission, hypothetically allowing for evolution of higher virulence. Transmission can be "mechanical," with no replication in the vector, such as myxoma virus, or "biological," with virus replication required, often resulting in a life-long persistent infection in the vector. The latter mechanism applies for the great majority of arthropod-borne viruses of vertebrates, known collectively as arboviruses. Nearly all arboviruses are RNA viruses and they cause numerous significant diseases in avian or mammalian primary hosts. Vectors are mainly mosquitoes, but also other flies such as sandflies and Culicoides, and non-insect arthropods such as ticks. Well-known

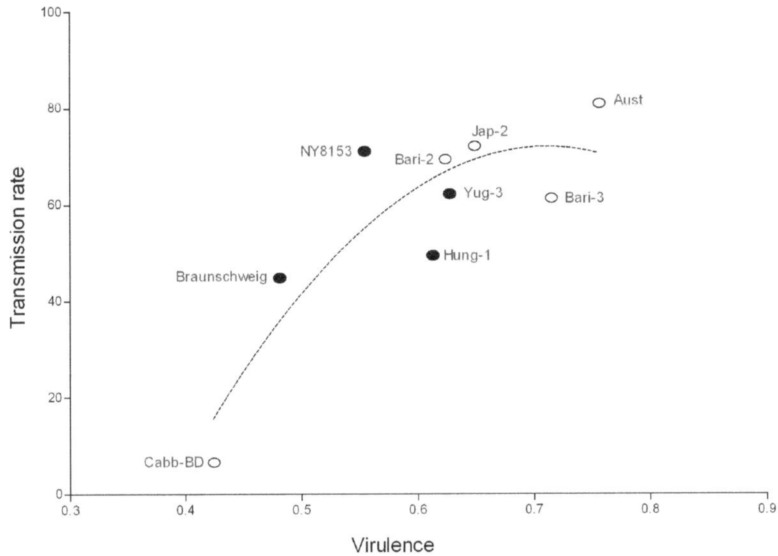

Figure 6.9. Relationship between virulence and transmission rate of *Cauliflower mosaic* virus infecting *Brassica rapa*. White symbols correspond to virus isolates with low levels of within-host viral accumulation, whereas black symbols correspond to high within-host viral accumulation isolates. Each symbol corresponds to the average of virulence and transmission rates at 28 days postinoculation. The dashed line corresponds to the quadratic relationship between the transmission rate and the virulence proxy. Redrawn with permission from Doumayrou *et al.*, 2012, copyright © 2012 The Society for the Study of Evolution.

arboviruses include Sindbis (SINV), Semliki Forest (SFV), Ross River (RRV), eastern equine encephalitis (EEEV), Venezuelan equine encephalitis (VEEV), chikungunya (CHKV), yellow fever (YFV), West Nile (WNV), dengue (DENV) and Rift Valley fever viruses (RVFV), and VSV. Ecologically, arboviruses typically cycle between vertebrate and vector hosts in enzootic sylvatic cycles. Humans or domestic animals, which sometimes serve as intermediate amplification hosts, are most often tangential or dead-end hosts infected from spillover events.

Due to the long history of human diseases caused by arboviruses, vaccine research in the 1970s led to the first publications relevant to their virulence evolution. Serial passage of SFV, CHKV, and RRV on insect cell lines all resulted in viruses that were attenuated, measured as reduced neurovirulence in mice (Peleg, 1971; Singh, 1971; Taylor and Marshall, 1975a). During that time the first reported *in vivo* passaging studies showed that RRV passage in mice increased mouse virulence, while passage in mosquitos showed attenuation of mouse virulence in one of four lineages, and passage in alternating hosts caused no virulence change (Taylor and Marshall, 1975a, b).

A major hypothesis in arbovirus research is that cycling between two disparate hosts selects for generalist viruses that are constrained to fitness levels below the maximum possible in each host. (Weaver, 2006). Recent reviews of arbovirus evolution describe this “trade-off” hypotheses in terms of the cost of host cycling

(Ciota and Kramer, 2010; Weaver, 2006). Under the assumption that replicative fitness correlates with virulence in the primary hosts, this also suggests arbovirus virulence is constrained to a level lower than could be attained if the virus were allowed to specialize in a single host. Strategies to investigate these hypotheses involve comparison of serial passage lines using alternating hosts with passages in single hosts, where the virus is released from the hypothesized constraints of host cycling. The majority of studies have tested *in vitro* passage using mammalian and insect cell lines, presenting results in terms of changes in viral fitness and sequence rather than virulence. Results of such studies for VSV, EEEV, SINV, DENV, VEEV, and CHIKV have shown many cases of fitness increases in single host lines, with decreases in by-passed hosts, but results for alternating lines are inconsistent, and there are many cases where no change, or opposite results, were found (Coffey *et al.,* 2008; Coffey and Vignuzzi, 2011; Greene *et al.,* 2005; Novella *et al.,* 1995; Novella *et al.,* 1999; Vasilakis *et al.,* 2009; Weaver *et al.,* 1999; Zarate and Novella, 2004). Thus at present *in vitro* studies do not provide consistent support for the hypothesis regarding host cycling and it has been suggested that the mechanisms driving arbovirus evolution vary between virus-host systems (Ciota and Kramer, 2010; Weaver, 2006). However, *in vivo* serial passages of VEEV result in increased fitness or infectivity for single host lines in either mosquitos or mice, while alternating host lines show no fitness gain (Coffey *et al.,* 2008). These results support the hypothesis of fitness constraint due to host cycling, and it may be that *in vivo* passage systems are more relevant to evolutionary mechanisms in nature, but the effects on virulence remain to be confirmed.

Another interesting aspect of arbovirus evolution involves the virulence of arboviruses for their vectors. Ewald predicted that arboviruses in their vectors are one of the few situations that should lead to evolution of low virulence or avirulence (Ewald, 1994). However, a meta-analysis of relevant data from 12 publications concluded that arboviruses do reduce survival of their vectors in many cases, but the magnitude of the effect varies with virus, vector, and mode of transmission. For instance, as separately controlled factors, alphaviruses and horizontal transmission were associated with higher virulence impacts in vectors, while vertical transmission and the *Aedes* mosquito vector species showed lower impacts (Lambrechts and Scott, 2009). In addition, a detailed experimental study with WNV found that adaptation to the mosquito by 20 *in vivo* serial passages generated a strain that decreased mosquito survival, while the progenitor wild-type virus had no detectable virulence effects on the vector (Ciota *et al.,* 2013). From a field virulence perspective, thorough mechanistic studies have shown that the spread of WNV across the United States involved evolution of a virus with enhanced vector competence for transmission to larger numbers of avian hosts (Kilpatrick *et al.,* 2008; Moudy *et al.,* 2007).

Virulence evolution in chronic and persistent viral infection

Evolutionary aspects of the distinct viral lifestyles associated with persistent, chronic, and acute infection have been increasingly considered in recent years (Roossinck, 2010; Villarreal, 2009). Persistent or chronic viral infection, as opposed to acute infection, lasts for long periods of time, sometimes for the life of the host. In animal viruses, persistence occurs when early acute infections are not cleared by the host immune responses, and involve the ability of the virus to interfere with or avoid host immune mechanisms. Persistent viral infection is responsible for globally important diseases caused by bovine viral diarrhea virus, HIV, human cytomegalovirus, and varicella-zoster virus. However, many persistent viral infections do not result in overt disease. The latter case generally applies to persistent viruses in plants, most fungal viruses, and most arboviruses in their insect vectors (Bryner and Rigling, 2012; Ciota and Kramer, 2010; Marquez and Roossinck, 2012; Roossinck, 2010). From a virulence evolution perspective, persistent or chronic infections inherently comprise a different selective environment than acute infection with active viral replication. For arboviruses, which alternate between persistent infection in vectors and productive acute infection in vertebrate hosts, the difference in replication strategies (persistent versus productive infection) may be a more important factor shaping the adaptive landscape of the virus than the difference between insect and mammalian cells (Novella *et al.,* 1995; Novella *et al.,* 1999; Zarate and Novella, 2004). Persistently infected cell cultures developed as *in vitro* models of persistence have revealed that viruses such as FMDV and mouse hepatitis virus can evolve increased virulence during persistent infection (Chen and Baric, 1996; Saiz and Domingo, 1996).

HIV virulence evolution in the field

Human immunodeficiency virus (HIV) is clearly a highly virulent virus, with case mortality close to 100% in the absence of medical interventions. The question is does the virus have high virulence because it recently emerged in humans or because it has maximized its fitness at a high level of virulence? In other words, will it evolve towards an avirulent state as it adapts to humans? The virus is estimated to have infected 2.1 million people in 2013, with a global prevalence of 35 million and as such, its general fitness is currently very high (Nations, 2014). Furthermore, although the virus has only been in humans since roughly the 1930s (Korber *et al.,* 2000), like other RNA viruses, during active replication it has short generation times and high mutation rates (Drake, 1993), and thus in the time since its emergence it has had enormous opportunities for virulence evolution (Domingo *et al.,* 1996). As such, one would expect the signatures of virulence evolution to be visible over this time scale. It is important to note that in recent years drug treatment has reduced HIV replication rates, which could impact evolutionary rates.

A major challenge for determining the virulence evolution trajectory for HIV has been defining and measuring virulence. Since mortality rates of HIV infected patients are 100%, this is an unsuitable metric because there is no variation between strains. The length of time from infection to mortality is the most obvious alternative but this definition is unsatisfying in practice because mortality occurs many years after infection, so more predictive measures were sought. These measures center around correlates of disease progression and the rate of onset of AIDS. Initially, the best predictor of disease progression was CD4 cell count in patients (Fahey *et al.,* 1990). However, this metric is only visible after significant disease has already occurred and predictive metrics earlier in infection were desired. It was shown that plasma viral loads after initial acute infection (referred to as the viral set-point) are a strong predictor of disease progression rates (Mellors *et al.,* 1996), and combining CD4 counts and set-point viral loads provided more predictive power (Mellors *et al.,* 1997). Recent studies have shown that set-point viral loads and CD4 counts tend to be correlated at a broad categorical level, but poorly correlated on an individual patient level, further reinforcing the importance of considering both metrics when inferring virulence (Rodriguez *et al.,* 2006). This variation in HIV virulence measures, and their associated caveats, is relevant for interpretation of the empirical studies and predictions of HIV virulence evolution.

Due to the avirulence of the ancestor of HIV, simian immunodeficiency virus, in its natural non-human primate hosts (reviewed in (Apetrei *et al.,* 2004)), it is appealing to hypothesize that HIV will also evolve towards avirulence as it adapts to humans. There is some empirical evidence that HIV has decreased in virulence since it emerged in humans. A study comparing HIV replicative fitness during competition in primary human cells found that field isolates from a single clinic in 1986-1989 had, in 75% of cases, higher *ex vivo* competitive fitness than those from 2002-2003 (Arien *et al.,* 2005b). There is also evidence that *ex vivo* competitive fitness is positively correlated with the rate of disease progression (i.e. virulence) in patients (Jason *et al.,* 2004; Quinones-Mateu *et al.,* 2000), thus supporting the conclusion of the authors in the previous study that HIV became attenuated over the time period examined. The authors suggested that HIV attenuation could be a function of bottlenecks during transmission, which is supported by *in vitro* studies (Yuste *et al.,* 1999). There is also evidence of the evolution of decreased HIV virulence from observational studies of viral genotypes. From the 1990s to the mid-2000s there was an increase in prevalence of HIV-1 subtype C, which became the dominant subtype overall worldwide (Arien et al., 2007). Studies indicate that subtype C has lower *ex vivo* replicative fitness but equal transmission fitness as other HIV-1 subtypes (Arien et al., 2005a), indicating it may have lower virulence but high overall fitness *in vivo.* However, the authors also found that the *ex vivo* competitive replication and transmission fitness of HIV groups O and HIV-2 were lower than subtype C. Furthermore, HIV group O and HIV-2 have been shown to have lower transmission rates (Gilbert *et al.,* 2003; Kanki *et al.,* 1994) and lower prevalence in the field than subtype C (Arien *et al.,* 2007). Thus, the postulated negative relationship between virulence and fitness may not be linear, but might be explained by a virulence trade-off, where intermediate levels of virulence provide optimal fitness.

Indication of a virulence trade-off for HIV also emerged from other studies. In two separate HIV cohorts in Amsterdam and Zambia, the most common viral load set-points observed are of intermediate values (Fraser *et al.,* 2007) that match mathematical model predictions of the values at which HIV will maximize R_0 at equilibrium, or during the initial epidemic growth period (Figure 6.10). Ultimately, viral set-point in the field was hypothesized to be driven by a transmission-virulence trade-off (Fraser *et al.,* 2014). As set-point increases, the probability of transmission given a contact event is believed to increase (Fideli *et al.,* 2001; Quinn *et al.,* 2000). However, increases in viral set-point also increase the probability of patients developing AIDS and suffering morbidity and mortality, thus decreasing the likelihood and duration of time that patients engage in contact events (deWolf *et al.,* 1997; Mellors *et al.,* 1996). This fits with data from a study of a Swiss HIV cohort that found no change in HIV virulence from 1984 to 2003, as measured by CD4 counts and viral set-points that fluctuated through time with no clear pattern (Muller et al., 2006). If such a virulence trade-off does exist for HIV it would suggest that, if the virus has evolved decreased virulence since it emerged

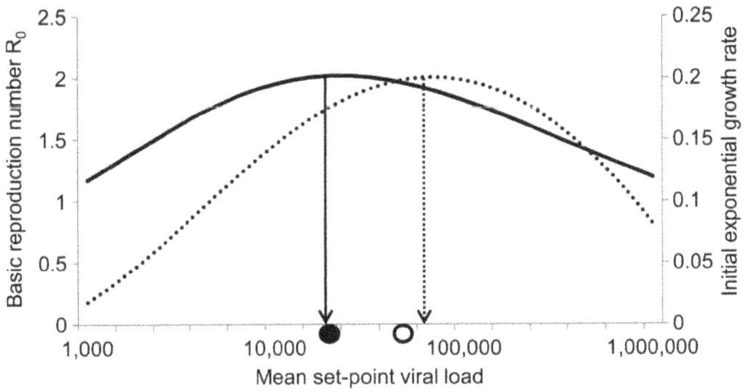

Figure 6.10. Optimal viral loads in a simple HIV transmission model. The solid vertical arrow shows the predicted viral set-point that maximizes R_0. The dotted vertical arrow shows the predicted set-point that maximizes the initial exponential growth rate. The observed mean viral loads for two HIV cohorts in Zambia (open circle) and Amsterdam (filled circle) are shown along the X axis. The model predictions closely match the field observations and demonstrate that intermediate virulence has optimal fitness in this system. Re-drawn with permission from Fraser *et al.*, 2007, copyright (2007) National Academy of Sciences, U.S.A.

in humans, it is unlikely to do so indefinitely, but rather should plateau at some level of intermediate virulence.

However, the increase in intervention strategies such as vaccination, drug treatment, or other transmission blocking methods, could change the shape of the virulence trade-off trajectory in this system, and drive virulence evolution in new directions. There is some evidence that such a phenomenon may be occurring. A study of two cohorts of HIV patients in Africa suggests that the evolution of host resistance may be driving the virulence evolution of HIV (Payne *et al.*, 2014). The authors found that in populations where the prevalence of host resistance genes was higher, set-point viral load was lower. It has previously been shown that the ability of HIV strains to evade host resistance genes comes with a cost to the virus of lower replication (Boutwell *et al.*, 2013; Crawford *et al.*, 2009; Martínez-Picado *et al.*, 2006). This led them to the conclusion that virulence, as measured by rate of disease progression, had also decreased in these viruses that were better able to evade the host response. However, these studies did not establish a relationship between rate of disease progression and viral set-point in these viral strains. This is notable because CD4 counts were lower in the host population with more resistance alleles, despite lower viral loads. In other words, it potentially took less viremia to cause more disease, suggesting the virus might actually be more virulent in the host population with higher resistance.

Ultimately it is difficult to say whether or not HIV has evolved decreased or increased virulence since it emerged in humans. Due to the inability to conduct *in vivo* studies and the ethical necessity to treat infected individuals, determining the answer to this question is challenging. There is now good evidence that HIV set-point viral load, and thus potentially virulence, is regulated in part by viral genetics (Fraser *et al.*, 2014), although host effects still account for the majority of HIV viral load variation among patients (Bonhoeffer *et al.*, 2003). This suggests that HIV virulence is a trait that could undergo evolution but host effects will have a major impact on its evolutionary trajectory. There is therefore no reason to assume that the direction of virulence evolution will be same for all HIV populations given the diversity in transmission and disease management worldwide (Arien *et al.,* 2007).

Virulence in hyperparasites and complex pathogen symbioses

Complex interactions involving viruses and sub-viral agents in multiple levels of parasitism are referred to as hyperparasitism. A common example is viruses that infect bacterial or cellular eukaryotic parasites that are themselves pathogens of humans or other hosts (Hartley *et al.,* 2012; Marquez and Roossinck, 2012). It has been demonstrated that phage infection of bacterial pathogens, especially upon integration, can modify bacterial phenotypes including virulence. In some cases phage infection has been shown to reduce bacterial virulence (Laanto *et al.,* 2012), but in many cases phages increase bacterial virulence by mechanisms such as encoding toxins (Fortier and Sekulovic, 2013; Hartley *et al.,* 2012). Although this is an important area of research, the virulence studied in these systems is not virulence of the virus itself, but virulence of its bacterial host in the hyper-host. Therefore impacts of viruses on the virulence of their hosts in complex symbioses will not be discussed further here.

A system demonstrating impacts of actual viral virulence within a complex hyperparasitic interaction is cryphonectria parasitica hypovirus 1 (CHV-1) in the fungal agent of chestnut blight, *Cryphonectria parasitica*. Chestnut blight is a serious fungal disease that has caused severe declines of chestnut trees throughout Europe and North America. CHV-1 is a naturally occurring virus of the fungus that causes "transmissible hypovirulence," in which virus infection reduces the virulence of the fungus, to the benefit of the chestnut tree hyperhost (Milgroom and Cortesi, 2004). Although most fungal viruses are persistent and asymptomatic on their hosts (Roossinck, 2010), CHV-1 is somewhat unique in being virulent on *C. parasitica*, causing measurable decreases in fungal growth and asexual sporulation (Bryner and Rigling, 2012). Experimental cross-infection studies comparing multiple CHV-1 and *C. parasitica* field isolates have demonstrated variation in both viral virulence and host tolerance, suggesting potential for virulence evolution (Peever *et al.,* 2000). A study exploring virulence differences in three CHV-1 strains revealed trade-offs in which the negative consequences of higher virulence in terms of reduced fungal host growth and sporulation, which the virus requires for vertical transmission, were counter-balanced by advantages in higher virus replication and higher horizontal transmission rates across vegetative incompatibility barriers (Bryner and Rigling, 2012). Figure 6.11 shows the correlation of CHV-1 virulence with increasing transmission in hosts of different vegetative incompatibility backgrounds. Thus a virulence-transmission trade-off functions within this hyperparasitic complex, based on a transmission mechanism unique to the fungal host.

Another hyperparasite system that has been studied in terms of virulence evolution involves the satellite RNAs of cucumber mosaic virus (CMV). CMV is a broad host range plant virus and satellite RNAs (satRNAs) are molecular parasites found associated with some CMV isolates. The satRNAs are small, non-coding RNAs that depend on CMV as a helper virus for their replication and transmission. When satRNAs are present they depress CMV replication and modify the

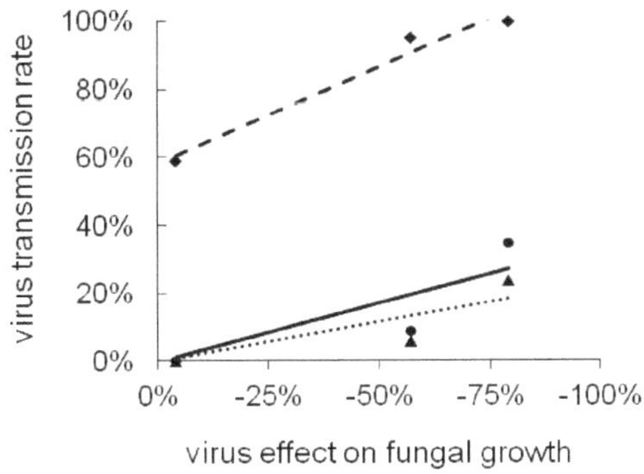

Figure 6.11. Positive correlation of virulence and transmission for the hyperparasite Cryphonectria hypovirus 1 (CHV-1) in its fungal parasite host, *Cryphonectria parasitica,* which causes the tree disease chestnut blight. Virus effect on fungal growth (virulence) is measured on dormant chestnut stems in comparison to a virus-free fungal strain. Virus transmission was assessed under different degrees of vegetative incompatibilty between the fungal donor and recipient, that is, when the donor and recipient were heteroallelic at the vegetative incompatibility locus *vic1* (diamonds, dashed trend line), *vic2* (triangles, dotted trend line), and *vic6* (circles, solid trend line), respectively. In all cases increased virulence was strongly correlated with increased transmissibility, potentially counterbalancing virulence costs. Redrawn with permission from Bryner and Rigling, 2012, copyright © 2012 The Society for the Study of Evolution.

symptoms induced by CMV in some plant hosts. In tomato, necrogenic genotypes of CMV satRNAs increase the moderate symptoms caused by CMV to cause systemic necrosis, while other satRNA genotypes attenuate the CMV symptoms. Studies of a natural epidemic of tomato necrosis involving satRNA hyperparasites in Spain between 1986-1992 comprise a well-documented observation of plant virus virulence evolution in the field (Escriu *et al.,* 2000, 2003). During the epidemic, mild disease caused by endemic CMV changed to necrotic disease that appeared and spread for the first few years, followed by a reduction in frequency and severity of disease, and eventually disappearance of necrosis by 1992. Characterization of virus isolates collected during the epidemic revealed that the initial emergence of necrotic disease was due to the appearance of necrogenic satRNAs associated with the endemic CMV. The reduced virulence observed later in the epidemic was due to a shift from necrogenic to attenuating satRNAs. Although experimental infections revealed that the necrogenic satRNAs had higher competitive fitness than attenuating satRNAs in mixed infections, single infection and co-infection models suggested a trade-off based on higher efficiency of aphid transmissibility of the attenuating satRNAs (Escriu *et al.,* 2003). These observations were broadened in a pair of complimentary studies testing the same strains of CMV and satRNAs in melon as an alternative host. These studies demonstrated host-specific virulence as there was no difference in symptoms caused by necrogenic or attenuating satRNAs in melon, but several other host-specific differences were found in parameters relevant to virulence evolution, facilitating modeling of transmission between different hosts in the field (Betancourt *et al.,* 2013, 2011).

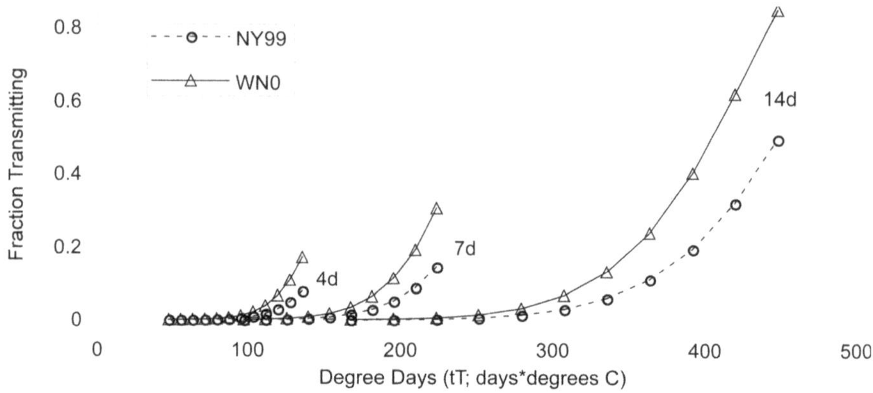

Figure 6.12. Impact of temperature on vector competence as a driver of West Nile Virus (WNV) evolution during its emergence and spread across North America. The original invading WNV strain NY99 was displaced by WN02, which is more efficiently transmitted by *Culex* mosquitos. Here each curve shows the fraction of mosquitoes transmitting NY99 (dotted lines) or WN02 (solid lines) at a fixed time period after feeding on WNV-infected blood (4, 7, or 14 days). Points on the lines show increasing temperatures (12°C to 32°C, symbol every 2°C). Results indicate that the advantage of WN02 over NY99 increases with warmer temperatures, which hypothetically facilitated the invasion of the WN02 strain. Reprinted with permission from Kilpatrick *et al.*, 2008, according to the PLoS Pathogens Creative Commons Attribution (CC BY) license.

Impact of environmental conditions on virulence evolution

Not surprisingly, there are numerous ways in which environmental conditions affect virulence evolution. The clear impact of temperature on insect vector transmission capacity has been demonstrated experimentally for West Nile virus, indicating temperature as an important factor in evolution of the virus during its spread across North America (Figure 6.12) (Kilpatrick *et al.,* 2008). The impact of the within-host chemical environment was illustrated by studies showing host nutritional status as a driving factor for evolution of more virulent Coxsacki B3 and influenza viruses in selenium-deficient hosts (Beck, 1997; Beck *et al.,* 2004). Similarly, the relative virulence and productivity of two isolates of nuclear polyhedrosis virus were shown to be determined by the tree species used as the food source by the larval moth host (Hodgson *et al.,* 2002). As a more general environmental feature, the theoretical prediction that spatial structure in terms of patchiness of host populations can constrain evolution of viral infectivity has been confirmed in one empirical study involving a baculovirus in larval moth hosts (Boots and Mealor, 2007).

In addition to virulence trade-offs based on quantitative viral traits, under certain environmental conditions viruses may provide qualitative benefits to their hosts that counter-balance the negative effects of their virulence and exploitation of host resources. Examples include CMV, which provides drought- or cold-tolerance to its plant hosts in stressful environments, and an aphid densovirus that promotes development of winged aphid morphs able to move to neighboring plants to access new food resources when aphid densities are high (Bao and Roossinck, 2013). The capacity of a pathogenic virus to become a conditional mutualist is a novel form of virulence trade-off that may be more prevalent than is currently appreciated (Bao and Roossinck, 2013; Marquez and Roossinck, 2012).

In general, it is widely recognized that permissive ecological conditions, often involving anthropogenic factors, play a major role in the occurrence of viral emergence and epidemics. This is particularly well described for arboviruses, where elimination of natural tropical environments creates a selection pressure for sylvatic viruses to adapt to humans and domestic animals as amplification and reservoir hosts, creating dengue-like urban transmission cycles with the potential for higher virulence in these hosts (Weaver, 2006). In fact, detailed analysis of a specific VEEV epidemic suggested that favorable environmental conditions were more likely to limit emergence than the occurrence of the original genetic mutation to higher virulence (Anishchenko *et al.*, 2006). An example of positive anthropogenic impact is effective infrastructure, such as mosquito-proof housing, that effectively prohibits arbovirus transmission to human hosts (Ewald, 2011).

Among anthropogenic impacts, the development of intensive agriculture of both plants and animals has created environments that may impact virulence evolution in many ways by providing dense populations of genetically uniform hosts [reviewed in Kennedy *et al.*, (in press)]. Surprisingly, the hypothesis that high density of hosts promotes evolution of higher virulence has not been demonstrated experimentally. Among observational studies, the varied patterns of rice cultivation in Africa have been described as drivers in the emergence and diversification of rice yellow mottle virus (Pinel-Galzi *et al.*, 2007). For animal viruses, avian influenza virus and infectious salmon anemia virus are well studied examples of viruses for which industrial-scale production of hosts, poultry and Atlantic salmon, respectively, has led to major increases in global disease burdens. In each case disease is caused by repeated evolution of high pathogenicity viral strains from low pathogenicity strains that are endemic in wild, sympatric hosts (Aamelfot *et al.*, 2014; Alexander, 2007). The genetic signatures associated with evolution of high pathogenicity strains are known for both viruses (Alexander, 2007; Markussen *et al.*, 2008; Mjaaland *et al.*, 2002) and pathogenic genotypes are found, with only rare exceptions, only in farmed host populations (Alexander, 2007; Christiansen *et al.*, 2011; Godoy *et al.*, 2008).

Genetics of virulence evolution

As sequencing technologies have advanced, increasing numbers of virulence evolution studies include genetic data for natural or experimental populations of viruses (Brusini *et al.*, 2013; Elde *et al.*, 2012; Kerr *et al.*, 2013; Snow and Cunningham, 2000). The numerous studies identifying virulence determinants for specific viruses are beyond the focus of this chapter. However, other relevant applications of sequence data include analyses for parallel mutations and inference of the number of mutational pathways to a change in virulence (Hajimorad *et al.*, 2011; Lafforgue *et al.*, 2011b; Pinel-Galzi *et al.*, 2007; Ping *et al.*, 2011; Taubenberger and Kash, 2010) , analyses of genetic diversity of virus populations (quasispecies) as an important virulence factor (Ciota and Kramer, 2010; Jerzak *et al.*, 2007; Korboukh *et al.*, 2014; Markussen *et al.*, 2013; Pfeiffer and Kirkegaard, 2005; Vignuzzi et al., 2006), and use of viral phylogenies to reveal heritability of virulence (Alizon *et al.*, 2010) and virulence as a selection pressure (Twiddy *et al.*, 2002). In general, genetic data provides powerful tools for confirming and understanding the phenotypic changes observed in virulence evolution studies.

Relevance of virulence evolution to disease management

Management of infectious diseases in humans, other animals, and plants, is a compelling goal that benefits in many ways from an understanding of virulence evolution. The first practical application of virulence evolution to disease management was the early development of highly successful vaccines for devastating diseases such as yellow fever and polio, by serial passaging of viruses in cultured cells to reduce virulence in humans. Although the original Sabin live-attenuated poliovirus vaccine has been in use for decades, research describing virulence thresholds has only recently provided an explanation for the rarity of vaccine-induced polio (Lancaster and Pfeiffer, 2011). Virulence evolution theory has proven to be a broad and useful framework for understanding and predicting consequences of our attempts to control disease, with the goal of avoiding human actions that unintentionally drive increases in pathogen virulence. Thus, understanding imperfect vaccines as a selective environment that may drive higher virulence with consequences in both vaccinated hosts (vaccine failure) and non-vaccinated hosts (extreme virulence) has motivated research into improved vaccines that avoid these impacts. Similarly, basic understanding of virus evolution in genetically resistant hosts, including RNAi-mediated resistance, has informed research to improve durability of resistant plant genotypes in agriculture (Hajimorad *et al.,* 2011; Lafforgue *et al.,* 2011b; Sacristan and Garcia-Arenal, 2008). Virulence evolution is also relevant to effective pest control efforts for viruses such as myxoma virus in rabbits and CHV-1 in the chestnut blight fungus, which coincidentally serve as field and laboratory models for testing and informing theory predictions.

Current research and development applications of virulence evolution include continuing use of serial passage strategies to develop vaccines, and as an alternative strategy for bioselection of cancer therapeutic viruses (Garijo *et al.,* 2014; Yan *et al.,* 2003). Practical disease management ideas derived from virulence evolution research include mixed plantings of susceptible and resistant crop cultivars to reduce risk of evolving resistance-breaking viruses (Montarry *et al.,* 2012), use of MHC typing in captive brood stock programs to maximize MHC diversity and avoid pathogen specialization that facilitates increased virulence (Kubinak *et al.,* 2013), use of phages to control pathogenic bacteria (Hartley et al., 2012), models to determine optimal culling criteria (Bolzoni and De Leo, 2013), support of infrastructure to prevent transmission of acute emerging viruses (Ewald, 2011), and alteration of agricultural conditions such as host density to avoid evolution of increased virulence (Kennedy *et al.,* In press). Importantly, the effectiveness of disease control measures in avoiding the evolution of high virulence in the field has been demonstrated for infectious salmon anemia virus in farmed Atlantic salmon. Strict disease management actions have been successful in eliminating high pathogenicity virus strains from the Atlantic salmon farming industries in both the Faroe Islands and Chile, and continued surveillance shows widespread persistence of only low pathogenicity strains (Christiansen et al., 2011; Godoy *et al.,* 2013). This is a welcome example of human intervention reducing the frequency and impact of virulence evolution and supporting maintenance of reduced virulence in the field (Figure 6.13). Collectively the potential value of the

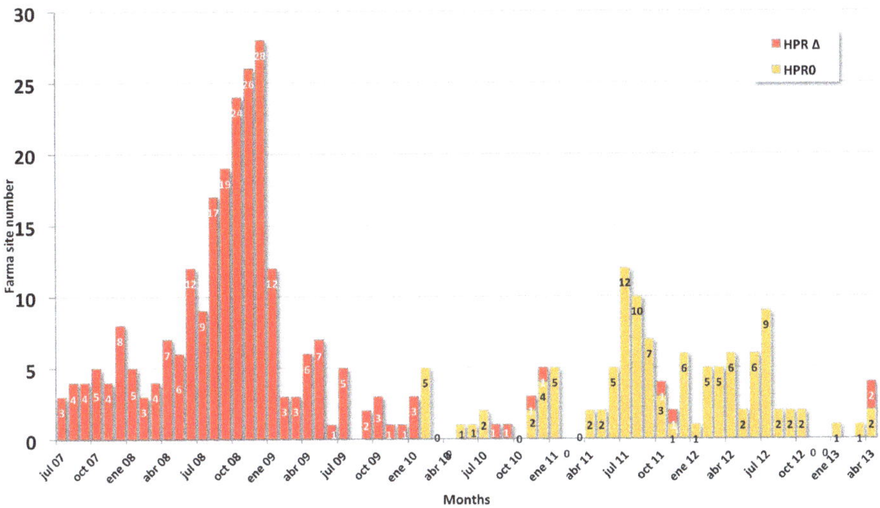

Figure 6.13. Reduced virulence of infectious salmon anaemia virus (ISAV) in farmed Atlantic salmon in Chile after successful disease control measures. A virulent genotype of ISAV, ISAV-HPRΔ (black), emerged in the Chilean salmon farming industry in 2007 and caused catastrophic disease.

contributions of virulence evolution theory has been recognized by increased research activity in ecology and evolution of infectious disease over the last 20 years.

Conclusions

Following the burst of creative modeling activity that generated the body of theory collectively known as Virulence Evolution Theory in the 1980s, later reviews criticized the lack of empirical testing of the various predictions of the theory (Bull, 1994; Froissart *et al.,* 2010). This chapter describes increased empirical research conducted within the framework of virulence evolution that has resulted in substantial progress in many areas. Importantly, the evolution of viral virulence in the field has been documented by observational studies for an increasing number of viruses (Table 6.4). As appropriate for the concept of selection of optimal levels of virulence, the observations include both virulence increases for influenza, MDV, and DENV-2, and virulence decreases for myxoma virus, CMV, and tobamoviruses, with other systems such as HIV lacking a consistent trend in either direction. In experimental studies we have seen viral virulence evolving in response to a variety of selection pressures including mode of transmission, enhanced transmission, enhanced vector competence, host changes, mixed infections, varied external environmental conditions, and replication in vaccinated or genetically resistant hosts. Research addressing the predicted transmission trade-off mechanism has shown empirical support in several cases, but it has also revealed that this mechanism does not apply in all host-viral systems. Therefore a broader range of alternative trade-offs, such as recovery-virulence trade-offs and durability-fecundity trade-offs, are being investigated. Overall the field of virulence evolution has matured from its purely theoretical origins to embrace empirical

Table 6.4. Observations of viral virulence evolution in the field

Virus-host pathosystem[a]	Event	Virulence measure	Virulence change	Hypothesized explanation	Refs[b]
Influenza-humans	1918 pandemic	severity of disease symptoms and mortality	transient increase	high frequency military transport of severely ill patients	50
AIV-chickens	24 primary outbreaks 1959-2004	severity of disease symptoms and mortality	increase	adaptation to new domesticated host, with favorable conditions in poultry farming	78
MDV-chickens	continuous disease since 1950s	severity of disease symptoms and mortality	sequential increases	response to poultry farming conditions and vaccination	54
MYXV-rabbits	Australian and European biocontrol releases 1950-1952	severity of disease symptoms and mortality	decrease to intermediate level, later increase	co-evolution of host resistance and viral virulence	37
HIV-humans	1986-2003 Belgium	*Ex vivo* replicative fitness as predictor of time to progression to mortality	decrease	natural result of adaptation to human host after zoonotic emergence	77
HIV-humans	2002-2005 South Africa 2007-2008 2012-2013 Botswana	Set-point viral load as predictor of time to progression to mortality	decrease or increase depending on metric used	adaptation to host resistance	84
CMV-tomato	1986-1992 tomato necrosis epidemic, Spain	change of plant symptoms, frequency of necrosis	decrease	evolution of associated satellite RNA population from necrogenic to attenuative	20
tobamovirus-pepper	1984-2004 pepper disease, Spain	ability to infect and cause disease on resistant pepper	decrease	response to reduced planting prevalence of resistant pepper cultivar	22
DENV2-human	1997 outbreak, Cuba	severity of disease symptoms and mortality	increase	virus mutation to increased viral fitness	85
ISAV-Atlantic salmon	multiple outbreaks since 1984	occurrence of disease and mortality	increase	virus evolution by specific mutations/deletions, with favorable conditions in intensive fish farming	80, 82, 83
ISAV-Atlantic salmon	surveillance after disease control, Faroe Islands, Chile	absence of disease, dominance of avirulent viral genotypes	decrease	hygeine and biosecurity prevent conditions that favor evolution of virulent virus genotypes	79, 81

[a]AIV, avian influenza; CMV, cucumber mosaic virus; DENV, Dengue virus; HIV, human immunodeficiency virus; ISAV, infectious salmon anemia virus; MDV, Marek's disease virus; MYXV, myxomavirus

[b] References numbered between 1-76 are listed in footnotes of Tables 1 and 2; 77, Arien *et al.*, 2005b; 78, Alexander *et al.*, 2007; 79, Christiansen *et al.*, 2011; 80, Godoy *et al.*, 2008; 81, Godoy *et al.*, 2013; 82, Markussen *et al.*, 2008; 83, Mjaaland *et al.*, 2002; 84, Payne *et al.*, 2014; 85, Rodriguez-Roche *et al.*, 2011.

testing, facilitating a more critical assessment of the strengths and weaknesses of the theory. Finally, although the potential of utilizing evolutionary theory to manage virulence evolution has been doubted by some, clear examples in some systems suggest promise for this approach.

References

Aamelfot, M., Dale, O., and Falk, K. (2014). Infectious salmon anaemia–pathogenesis and tropism. Journal of fish diseases *37*, 291-307.

Alexander, D.J. (2007). An overview of the epidemiology of avian influenza. Vaccine *25*, 5637-5644.

Alizon, S., Hurford, A., Mideo, N., and Van Baalen, M. (2009). Virulence evolution and the trade-off hypothesis: history, current state of affairs and the future. Journal of Evolutionary Biology *22*, 245-259.

Alizon, S., von Wyl, V., Stadler, T., Kouyos, R.D., Yerly, S., Hirschel, B., Boni, J., Shah, C., Klimkait, T., Furrer, H., *et al.* (2010). Phylogenetic approach reveals that virus genotype largely determines HIV set-point viral load. Plos Pathogens *6*.

Anderson, R.M. (1982). The population dynamics of infectious diseases: theory and applications (New York: Chapman and Hall Ltd).

Anderson, R.M., and May, R.M. (1979). Population biology of infectious diseases: Part I. Nature *280*, 361-367.

Anderson, R.M., and May, R.M. (1982a). Coevolution of host and pathogen. Parasitology *85*, 411-426.

Anderson, R.M., and May, R.M. (1982b). Population biology of infectious diseases (Berlin: Springer).

Anishchenko, M., Bowen, R.A., Paessler, S., Austgen, L., Greene, I.P., and Weaver, S.C. (2006). Venezuelan encephalitis emergence mediated by a phylogenetically predicted viral mutation. Proceedings of the National Academy of Sciences of the United States of America *103*, 4994-4999.

Apetrei, C., Robertson, D.L., and Marx, P.A. (2004). The history of SIVS and AIDS: epidemiology, phylogeny and biology of isoaltes from naturally SIV infected non-human primates (NHP) in Africa. Frontiers in Bioscience *9*, 225-254.

Arien, K.K., Abraha, A., Quinones-Mateu, M.E., Kestens, L., Vanham, G., and Arts, E.J. (2005a). The replicative fitness of primary human immunodeficiency virus type 1 (HIV-1) group M, HIV-1 group O, and HIV-2 isolates. Journal of Virology *79*, 8979-8990.

Arien, K.K., Troyer, R.A., Gali, Y., Colebunders, R.L., Arts, E.J., and Vanham, G. (2005b). Replicative fitness of historical and recent HIV-1 isolates suggests HIV-1 attenuation over time. Aids *19*, 1555-1564.

Arien, K.K., Vanham, G., and Arts, E.J. (2007). Is HIV-1 evolving to a less virulent form in humans? Nat Rev Microbiol *5*, 141-151.

Atkins, K.E., Read, A.F., Savill, N.J., Renz, K.G., Islam, A., Walkden-Brown, S.W., and Woolhouse, M.E.J. (2013a). Vaccination and reduced cohort duration can drive virulence evolution: Marek's disease virus and industrialized agriculture. Evolution *67*, 851-860.

Atkins, K.E., Read, A.F., Savill, N.J., Renz, K.G., Walkden-Brown, S.W., and Woolhouse, M.E.J. (2011). Modelling Marek's Disease Virus (MDV) infection: parameter estimates for mortality rate and infectiousness. Bmc Veterinary Research *7*.

Atkins, K.E., Read, A.F., Walkden-Brown, S.W., Savill, N.J., and Woolhouse, M.E.J. (2013b). The effectiveness of mass vaccination on Marek's disease virus (MDV) outbreaks and detection within a broiler barn: A modeling study. Epidemics *5*, 208-217.

Bangham, C.R.M., and Kirkwood, T.B.L. (1993). Defective interfering particles and virus evolution. Trends in Microbiology *1*, 260-264.

Bao, X.D., and Roossinck, M.J. (2013). A life history view of mutualistic viral symbioses: quantity or quality for cooperation? Current Opinion in Microbiology *16*, 514-518.

Beck, M.A. (1997). Increased virulence of coxsackievirus B3 in mice due to vitamin E or selenium deficiency. Journal of Nutrition *127*, S966-S970.

Beck, M.A., Handy, J., and Levander, O.A. (2004). Host nutritional status: the neglected virulence factor. Trends in Microbiology *12*, 417-423.

Bedhomme, S., Hillung, J., and Elena, S.F. (2015). Emerging viruses: why they are not jacks of all trades? Current opinion in virology *10*, 1-6.

Bedhomme, S., Lafforgue, G., and Elena, S.F. (2012). Multihost experimental evolution of a plant RNA virus reveals local adaptation and host-specific mutations. Molecular Biology and Evolution *29*, 1481-1492.

Berngruber, T.W., Froissart, R., Choisy, M., and Gandon, S. (2013). Evolution of virulence in emerging epidemics. Plos Pathogens *9*.

Betancourt, M., Escriu, F., Fraile, A., and Garcia-Arenal, F. (2013). Virulence evolution of a generalist plant virus in a heterogeneous host system. Evolutionary Applications *6*, 875-890.

Betancourt, M., Fraile, A., and Garcia-Arenal, F. (2011). Cucumber mosaic virus satellite RNAs that induce similar symptoms in melon plants show large differences in fitness. J Gen Virol *92*, 1930-1938.

Bolker, B.M., Nanda, A., and Shah, D. (2010). Transient virulence of emerging pathogens. J R Soc Interface *7*, 811-822.

Bolzoni, L., and De Leo, G.A. (2013). Unexpected consequences of culling on the eradication of wildlife diseases: The role of virulence evolution. American Naturalist *181*, 301-313.

Bonhoeffer, S., Funk, G.A., Gunthard, H.F., Fischer, M., and Muller, V. (2003). Glancing behind virus load variation in HIV-1 infection. Trends in Microbiology *11*, 499-504.

Bonhoeffer, S., Lenski, R.E., and Ebert, D. (1996). The curse of the pharaoh: The evolution of virulence in pathogens with long living propagules. Proceedings of the Royal Society of London Series B: Biological Sciences *263*, 715-721.

Boots, M., and Mealor, M. (2007). Local interactions select for lower pathogen infectivity. Science *315*, 1284-1286.

Boutwell, C.L., Carlson, J.M., Lin, T.-H., Seese, A., Power, K.A., Peng, J., Tang, Y., Brumme, Z.L., Heckerman, D., Schneidewind, A., *et al.* (2013). Frequent and Variable Cytotoxic-T-Lymphocyte Escape-Associated Fitness Costs in the Human Immunodeficiency Virus Type 1 Subtype B Gag Proteins. Journal of Virology *87*, 3952-3965.

Bremermann, H.J., and Pickering, J. (1983). A game-theoretical model of parasite virulence. Journal of Theoretical Biology *100*, 411-426.

Brown, S.P., Hochberg, M.E., and Grenfell, B.T. (2002). Does multiple infection select for raised virulence? Trends in Microbiology *10*, 401-405.

Brunner, J.L., and Collins, J.P. (2009). Testing assumptions of the trade-off theory of the evolution of parasite virulence. Evolutionary Ecology Research *11*, 1169-1188.

Brusini, J., Wang, Y., Matos, L.F., Sylvestre, L.S., Bolker, B.M., and Wayne, M.L. (2013). Virulence evolution in a host-parasite system in the absence of viral evolution. Evolutionary Ecology Research *15*, 883-901.

Bryner, S.F., and Rigling, D. (2012). Virulence not only costs but also benefits the transmission of a fungal virus. Evolution *66*, 2540-2550.

Bull, J.J. (1994). Perspective: Virulence. Evolution *48*, 1423-1437.

Bull, J.J., and Ebert, D. (2008). Invasion thresholds and the evolution of nonequilibrium virulence. Evolutionary Applications *1*, 172-182.

Bull, J.J., Molineux, I.J., and Rice, W.R. (1991). Selection of benevolence in a host-parasite system. Evolution *45*, 875-882.

Burnet, F.M., and White, D.O. (1972). Natural History of Infectious Disease (University Press).

Campbell, N.A. (1996). Biology, 4 edn (The Benjamin/Cummings Publishing Company, Inc.).

Cardenas, C., Carmona, M., Gallardo, A., Labra, A., and Marshall, S.H. (2014). Coexistence in field samples of two variants of the infectious salmon anemia virus: A putative shift to pathogenicity. Plos One *9*.
Carrasco, P., de la Iglesia, F., and Elena, S.F. (2007). Distribution of fitness and virulence effects caused by single-nucleotide substitutions in tobacco etch virus. Journal of Virology *81*, 12979-12984.
Carrillo, C., Lu, Z., Borca, M.V., Vagnozzi, A., Kutish, G.F., and Rock, D.L. (2007). Genetic and phenotypic variation of foot-and-mouth disease virus during serial passages in a natural host. Journal of Virology *81*, 11341-11351.
Casadevall, A., and Pirofski, L.A. (1999). Host-pathogen interactions: Redefining the basic concepts of virulence and pathogenicity. Infection and Immunity *67*, 3703-3713.
Chao, L., Hanley, K.A., Burch, C.L., Dahlberg, C., and Turner, P.E. (2000). Kin selection and parasite evolution: higher and lower virulence with hard and soft selection. Quarterly Review of Biology *75*, 261-275.
Chen, W., and Baric, R.S. (1996). Molecular anatomy of mouse hepatitis virus persistence: Coevolution of increased host cell resistance and virus virulence. Journal of Virology *70*, 3947-3960.
Christiansen, D.H., Ostergaard, P.S., Snow, M., Dale, O.B., and Falk, K. (2011). A low-pathogenic variant of infectious salmon anemia virus (ISAV-HPR0) is highly prevalent and causes a non-clinical transient infection in farmed Atlantic salmon (Salmo salar L.) in the Faroe Islands. J Gen Virol *92*, 909-918.
Ciota, A.T., Ehrbar, D.J., Matacchiero, A.C., Van Slyke, G.A., and Kramer, L.D. (2013). The evolution of virulence of West Nile virus in a mosquito vector: implications for arbovirus adaptation and evolution. Bmc Evolutionary Biology *13*.
Ciota, A.T., and Kramer, L.D. (2010). Insights into arbovirus evolution and adaptation from experimental studies. Viruses-Basel *2*, 2594-2617.
Coffey, L.L., Vasilakis, N., Brault, A.C., Powers, A.M., Tripet, F., and Weaver, S.C. (2008). Arbovirus evolution *in vivo* is constrained by host alternation. Proceedings of the National Academy of Sciences of the United States of America *105*, 6970-6975.
Coffey, L.L., and Vignuzzi, M. (2011). Host alternation of chikungunya virus increases fitness while restricting population diversity and adaptability to novel selective pressures. Journal of Virology *85*, 1025-1035.
Collinge, S.K., and Ray, C. (2006). Disease Ecology community structure and pathogen dynamics (New York: Oxford University Press).
Cooper, V.S., Reiskind, M.H., Miller, J.A., Shelton, K.A., Walther, B.A., Elkinton, J.S., and Ewald, P.W. (2002). Timing of transmission and the evolution of virulence of an insect virus. Proceedings of the Royal Society of London Series B-Biological Sciences *269*, 1161-1165.
Coutts, A.J., Dawson, S., Willoughby, K., and Gaskell, R.M. (1994). Isolation of feline respiratory viruses from clinically healthy cats at UK cat shows. Veterinary Record *135*, 555-556.
Crawford, H., Lumm, W., Leslie, A., Schaefer, M., Boeras, D., Prado, J.G., Tang, J., Farmer, P., Ndung'u, T., Lakhi, S., *et al.* (2009). Evolution of HLA-B*5703 HIV-1 escape mutations in HLA-B*5703-positive individuals and their transmission recipients. J Exp Med *206*, 909-921.
Dawson, S., Smyth, N.R., Bennett, M., Gaskell, R.M., McCracken, C.M., Brown, A., and Gaskell, C.J. (1991). Effect of primary-stage feline immunodeficiency virus-infection on subsequent feline calcivirus vaccination and challenge in cats. Aids *5*, 747-750.
Day, T. (2003). Virulence evolution and the timing of disease life-history events. Trends in Ecology and Evolution *18*, 113-118.
Day, T., and Proulx, S.R. (2004). A general theory for the evolutionary dynamics of virulence. American Naturalist *163*, E40-E63.
De Paepe, M., and Taddei, F. (2006). Viruses' life history: Towards a mechanistic basis of a trade-off between survival and reproduction among phages. PLoS Biol *4*, 1248-1256.

deWolf, F., Spijkerman, I., Schellekens, P.T., Langendam, M., Kuiken, C., Bakker, M., Roos, M., Coutinho, R., Miedema, F., and Goudsmit, J. (1997). AIDS prognosis based on HIV-1 RNA, CD4+ T-cell count and function: Markers with reciprocal predictive value over time after seroconversion. Aids *11*, 1799-1806.
Diekmann, O., Heesterbeek, J.A.P., and Metz, J.A.J. (1990). On the definition and the computation of the basic reproduction ratio R_0 in models for infectious-diseases in heterogeneous populations. Journal of Mathematical Biology *28*, 365-382.
Domingo, E., Escarmís, C., Sevilla, N., Moya, A., Elena, S.F., Quer, J., Novella, I.S., and Holland, J.J. (1996). Basic concepts in RNA virus evolution. The FASEB Journal *10*, 859-864.
Doumayrou, J., Avellan, A., Froissart, R., and Michalakis, Y. (2013). An experimental test of the transmission-virulence trade-off hypothesis in a plant virus. Evolution *67*, 477-486.
Drake, J.W. (1993). Rates of spontaneous mutation among RNA viruses. Proceedings of the National Academy of Sciences *90*, 4171-4175.
Duffy, S., Shackelton, L.A., and Holmes, E.C. (2008). Rates of evolutionary change in viruses: patterns and determinants. Nature Reviews Genetics *9*, 267-276.
Dunn, J.R., Silva, R.F., Lee, L.F., and Witter, R.L. (2012). Competition between two virulent Marek's disease virus strains *in vivo*. Avian Pathology *41*, 267-275.
Dwyer, G., Levin, S.A., and Buttel, L. (1990). A simulation-model of the population-dynamics and evolution of myxomatosis. Ecological Monographs *60*, 423-447.
Ebert, D. (1998). Evolution - Experimental evolution of parasites. Science *282*, 1432-1435.
Elde, N.C., Child, S.J., Eickbush, M.T., Kitzman, J.O., Rogers, K.S., Shendure, J., Geballe, A.P., and Malik, H.S. (2012). Poxviruses deploy genomic accordions to adapt rapidly against host antiviral defenses. Cell *150*, 831-841.
Elena, S.F. (2001). Evolutionary history conditions the timing of transmission in vesicular stomatitis virus. Infection Genetics and Evolution *1*, 151-159.
Elena, S.F., Agudelo-Romero, P., Carrasco, P., Codoner, F.M., Martin, S., Torres-Barcelo, C., and Sanjuán, R. (2008). Experimental evolution of plant RNA viruses. Heredity *100*, 478-483.
Elliot, S.L., Adler, F.R., and Sabelis, M.W. (2003). How virulent should a parasite be to its vector? Ecology *84*, 2568-2574.
Elsworth, P., Cooke, B.D., Kovaliski, J., Sinclair, R., Holmes, E.C., and Strive, T. (2014). Increased virulence of rabbit haemorrhagic disease virus associated with genetic resistance in wild Australian rabbits (Oryctolagus cuniculus). Virology *464-465*.
Escriu, F., Fraile, A., and Garcia-Arenal, F. (2000). Evolution of virulence in natural populations of the satellite RNA of Cucumber mosaic virus. Phytopathology *90*, 480-485.
Escriu, F., Fraile, A., and Garcia-Arenal, F. (2003). The evolution of virulence in a plant virus. Evolution *57*, 755-765.
Ewald, P. (1994). Evolution of infectious disease (Oxford: Oxford University Press).
Ewald, P.W. (1983). Host-parasite relations, vectors, and the evolution of disease severity. Annual Review of Ecology and Systematics *14*, 465-485.
Ewald, P.W. (2011). Evolution of virulence, environmental change, and the threat posed by emerging and chronic diseases. Ecological Research *26*, 1017-1026.
Fahey, J.L., Taylor, J.M.G., Detels, R., Hofmann, B., Melmed, R., Nishanian, P., and Giorgi, J.V. (1990). The prognostic value of cellular and serologic markers in infection with human immunodeficiency virus type 1. New England Journal of Medicine *322*, 166-172.
Fenner, F. (1983). The Florey Lecture, 1983 - Biological-control, as exemplified by smallpox eradication and myxomatosis. Proceedings of the Royal Society Series B-Biological Sciences *218*, 259-285.
Fenner, F., and Ratcliffe, R. (1965). Myxomatosis (London: Cambridge University Press).

Fideli, O.S., Allen, S.A., Musonda, R., Trask, S., Hahn, B.H., Weiss, H., Mulenga, J., Kasolo, F., Vermund, S.H., and Aldrovandi, G.M. (2001). Virologic and immunologic determinants of heterosexual transmission of human immunodeficiency virus type 1 in Africa. Aids Research and Human Retroviruses *17*, 901-910.
Fortier, L.C., and Sekulovic, O. (2013). Importance of prophages to evolution and virulence of bacterial pathogens. Virulence *4*, 354-365.
Fraile, A., Hily, J.M., Pagan, I., Pacios, L.F., and Garcia-Arenal, F. (2014). Host resistance selects for traits unrelated to resistance-breaking that affect fitness in a plant virus. Molecular Biology and Evolution *31*, 928-939.
Fraile, A., Pagan, I., Anastasio, G., Saez, E., and Garcia-Arenal, F. (2011). Rapid genetic diversification and high fitness penalties associated with pathogenicity evolution in a plant virus. Molecular Biology and Evolution *28*, 1425-1437.
Frank, S.A. (1996). Models of parasite virulence. Quarterly Review of Biology *71*, 37-78.
Fraser, C., Hollingsworth, T.D., Chapman, R., de Wolf, F., and Hanage, W.P. (2007). Variation in HIV-1 set-point viral load: Epidemiological analysis and an evolutionary hypothesis. Proceedings of the National Academy of Sciences *104*, 17441-17446.
Fraser, C., Lythgoe, K., Leventhal, G.E., Shirreff, G., Hollingsworth, T.D., Alizon, S., and Bonhoeffer, S. (2014). Virulence and pathogenesis of HIV-1 infection: An evolutionary perspective. Science *343*, 1328-+.
Freeman, S., and Herron, J.C. (1998). Evolutionary Analysis (New Jersey: Prentice-Hall).
Froissart, R., Doumayrou, J., Vuillaume, F., Alizon, S., and Michalakis, Y. (2010). The virulence-transmission trade-off in vector-borne plant viruses: a review of (non-)existing studies. Philos Trans R Soc B-Biol Sci *365*, 1907-1918.
Furio, V., Garijo, R., Duran, M., Moya, A., Bell, J.C., and Sanjuán, R. (2012). Relationship between within-host fitness and virulence in the vesicular stomatitis virus: Correlation with partial decoupling. Journal of Virology *86*, 12228-12236.
Gandon, S., and Day, T. (2008). Evidences of parasite evolution after vaccination. Vaccine *26*, C4-C7.
Gandon, S., Mackinnon, M.J., Nee, S., and Read, A.F. (2001). Imperfect vaccines and the evolution of pathogen virulence. Nature *414*, 751-756.
Garijo, R., Hernandez-Alonso, P., Rivas, C., Diallo, J.S., and Sanjuán, R. (2014). Experimental evolution of an oncolytic vesicular stomatitis virus with increased selectivity for p53-deficient cells. Plos One *9*.
Garver, K., Batts, W., and Kurath, G. (2006). Virulence comparisons of infectious hematopoietic necrosis virus U and M genogroups in sockeye salmon and rainbow trout. Journal of Aquatic Animal Health *18*, 232-243.
Gilbert, P.B., McKeague, I.W., Eisen, G., Mullins, C., Guéye-Ndiaye, A., Mboup, S., and Kanki, P.J. (2003). Comparison of HIV-1 and HIV-2 infectivity from a prospective cohort study in Senegal. Statistics in Medicine *22*, 573-593.
Gimeno, I.M. (2008). Marek's disease vaccines: A solution for today but a worry for tomorrow? Vaccine *26*, C31-C41.
Godoy, M.G., Aedo, A., Kibenge, M.J.T., Groman, D.B., Yason, C.V., Grothusen, H., Lisperguer, A., Calbucura, M., Avendano, F., Imilan, M., *et al.* (2008). First detection, isolation and molecular characterization of infectious salmon anaemia virus associated with clinical disease in farmed Atlantic salmon (*Salmo salar*) in Chile. Bmc Veterinary Research *4*.
Godoy, M.G., Kibenge, M.J.T., Suarez, R., Lazo, E., Heisinger, A., Aguinaga, J., Bravo, D., Mendoza, J., Llegues, K.O., Avendano-Herrera, R., *et al.* (2013). Infectious salmon anaemia virus (ISAV) in Chilean Atlantic salmon (*Salmo salar*) aquaculture: emergence of low pathogenic ISAV-HPR0 and re-emergence of virulent ISAV-HPR Delta: HPR3 and HPR14. Virology Journal *10*.
Goldhill, D.H., and Turner, P.E. (2014). The evolution of life history trade-offs in viruses. Current opinion in virology *8*, 79-84.

Graham, A.L., Allen, J.E., and Read, A.F. (2005). Evolutionary causes and consequences of immunopathology. In Annu Rev Ecol Evol Syst (Palo Alto: Annual Reviews), pp. 373-397.

Greene, I.P., Wang, E.Y., Deardorff, E.R., Milleron, R., Domingo, E., and Weaver, S.C. (2005). Effect of alternating passage on adaptation of sindbis virus to vertebrate and invertebrate cells. Journal of Virology *79*, 14253-14260.

Grenfell, B.T., and Dobson, A.P. (1995). Ecology of infectious diseases in natural populations (Cambridge: Cambridge University Press).

Group, F.I.I.S. (2010). Four year efficacy of prophylactic human papillomavirus quadrivalent vaccine against low grade cervical, vulvar, and vaginal intraepithelial neoplasia and anogenital warts: randomised controlled trial. BMJ: British Medical Journal *341*.

Hajimorad, M.R., Wen, R.H., Eggenberger, A.L., Hill, J.H., and Maroof, M.A.S. (2011). Experimental adaptation of an RNA virus mimics natural evolution. Journal of Virology *85*, 2557-2564.

Haller, S.L., Peng, C., McFadden, G., and Rothenburg, S. (2014). Poxviruses and the evolution of host range and virulence. Infection Genetics and Evolution *21*, 15-40.

Handel, A., Lebarbenchon, C., Stallknecht, D., and Rohani, P. (2014). Trade-offs between and within scales: environmental persistence and within-host fitness of avian influenza viruses. Proceedings of the Royal Society B: Biological Sciences *281*, 20133051.

Hardin, G. (1968). Tragedy of Commons. Science *162*, 1243-&.

Hartley, M.-A., Ronet, C., and Fasel, N. (2012). Backseat drivers: the hidden influence of microbial viruses on disease. Current opinion in microbiology *15*, 538-545.

Heineman, R.H., and Brown, S.P. (2012). Experimental evolution of a bacteriophage virus reveals the trajectory of adaptation across a fecundity/longevity trade-off. Plos One *7*.

Herrera, M., Garcia-Arriaza, J., Pariente, N., Escarmis, C., and Domingo, E. (2007). Molecular basis for a lack of correlation between viral fitness and cell killing capacity. PLoS Pathogens *3*, 498-507.

Hillung, J., Cuevas, J.M., Valverde, S., and Elena, S.F. (2014). Experimental evolution of an emerging plant virus in host genotypes that differ in their susceptibility to infection. Evolution *68*, 2467-2480.

Hodgson, D.J., Hitchman, R.B., Vanbergen, A.J., Hails, R.S., Possee, R.D., and Cory, J.S. (2004). Host ecology determines the relative fitness of virus genotypes in mixed-genotype nucleopolyhedrovirus infections. Journal of Evolutionary Biology *17*, 1018-1025.

Hodgson, D.J., Vanbergen, A.J., Hartley, S.E., Hails, R.S., and Cory, J.S. (2002). Differential selection of baculovirus genotypes mediated by different species of host food plant. Ecology Letters *5*, 512-518.

Hoelzer, K., and Parrish, C.R. (2010). The emergence of parvoviruses of carnivores. Vet Res *41*, 39.

Hudson, P.J., Rizzoli, A., Grenfell, B.T., Heesterbeek, H., and Dobson, A.P. (2001). The Ecology of Wildlife Diseases (New York: Oxford University Press).

Jason, D.B., Hecht, F.M., Wrin, T., Segal, M.R., Ramstead, C.A., Liegler, T.J., Busch, M.P., Petropoulos, C.J., Hellmann, N.S., Kahn, J.O., *et al.* (2004). Higher $CD4^+$ T cell counts associated with low viral "pol" replication capacity among treatment naive adults in early HIV-1 infection. The Journal of Infectious Diseases *190*, 251-256.

Jerzak, G.V., Bernard, K., Kramer, L.D., Shi, P.-Y., and Ebel, G.D. (2007). The West Nile virus mutant spectrum is host-dependant and a determinant of mortality in mice. Virology *360*, 469-476.

Kanki, P.J., Travers, K.U., Mboup, S., Hsieh, C.C., Marlink, R.G., Gueyendiaye, A., Siby, T., Thior, I., Hernandezavila, M., Sankale, J.L., *et al.* (1994). Slower heterosexual spread of HIV-2 than HIV-1. Lancet *343*, 943-946.

Kashiwagi, A., and Yomo, T. (2011). Ongoing phenotypic and genomic changes in experimental coevolution of RNA bacteriophage Q beta and Escherichia coli. Plos Genetics *7*.

Keeling, M.J., and Rohani, P. (2008). Modeling Infectious Diseases in Humans and Animals (New Jersey: Princeton University Press).

Kell, A.M., Wargo, A.R., and Kurath, G. (2013). The role of virulence in *in vivo* superinfection fitness of the vertebrate RNA virus infectious hematopoietic necrosis virus. Journal of Virology *87*, 8145-8157.

Kell, A.M., Wargo, A.R., and Kurath, G. (2014). Viral fitness does not correlate with three genotype displacement events involving infectious hematopoietic necrosis virus. Virology *464*, 146-155.

Kennedy, D.A., Read, A.F., Kurath, G., Wargo, A.R., Brito, I., Purcell, M.K., and Winton, J. (In press). Potential drivers of virulence evolution in aquaculture.

Kermack, W.O., and McKendrick, A.G. (1927). A contribution to the mathematical theory of epidemics. Proceedings of the Royal Society of London Series B-Biological Sciences *115*, 700-721.

Kerr, P.J. (2012). Myxomatosis in Australia and Europe: A model for emerging infectious diseases. Antiviral Research *93*, 387-415.

Kerr, P.J., Ghedin, E., DePasse, J.V., Fitch, A., Cattadori, I.M., Hudson, P.J., Tscharke, D.C., Read, A.F., and Holmes, E.C. (2012). Evolutionary history and attenuation of myxoma virus on two continents. Plos Pathogens *8*.

Kerr, P.J., Rogers, M.B., Fitch, A., DePasse, J.V., Cattadori, I.M., Twaddle, A.C., Hudson, P.J., Tscharke, D.C., Read, A.F., Holmes, E.C., *et al.* (2013). Genome scale evolution of myxoma virus reveals host-pathogen adaptation and rapid geographic spread. Journal of Virology *87*, 12900-12915.

Kilpatrick, A.M., Meola, M.A., Moudy, R.M., and Kramer, L.D. (2008). Temperature, viral genetics, and the transmission of West Nile virus by *Culex pipiens* mosquitoes. Plos Pathogens *4*.

Kochin, B.F., Bull, J.J., and Antia, R. (2010). Parasite evolution and life history theory. PLoS Biol *8*.

Korber, B., Muldoon, M., Theiler, J., Gao, F., Gupta, R., Lapedes, A., Hahn, B.H., Wolinsky, S., and Bhattacharya, T. (2000). Timing the ancestor of the HIV-1 pandemic strains. Science *288*, 1789-1796.

Korboukh, V.K., Lee, C.A., Acevedo, A., Vignuzzi, M., Xiao, Y., Arnold, J.J., Hemperly, S., Graci, J.D., August, A., and Andino, R. (2014). RNA virus population diversity, an optimum for maximal fitness and virulence. Journal of Biological Chemistry *289*, 29531-29544.

Kovaliski, J., Sinclair, R., Mutze, G., Peacock, D., Strive, T., Abrantes, J., Esteves, P.J., and Holmes, E.C. (2014). Molecular epidemiology of rabbit haemorrhagic disease virus in Australia: when one became many. Mol Ecol *23*, 408-420.

Kubinak, J.L., and Potts, W.K. (2013). Host resistance influences patterns of experimental viral adaptation and virulence evolution. Virulence *4*, 410-418.

Kubinak, J.L., Ruff, J.S., Cornwall, D.H., Middlebrook, E.A., Hasenkrug, K.J., and Potts, W.K. (2013). Experimental viral evolution reveals major histocompatibility complex polymorphisms as the primary host factors controlling pathogen adaptation and virulence. Genes and Immunity *14*, 365-372.

Kubinak, J.L., Ruff, J.S., Hyzer, C.W., Slev, P.R., and Potts, W.K. (2012). Experimental viral evolution to specific host MHC genotypes reveals fitness and virulence trade-offs in alternative MHC types. Proceedings of the National Academy of Sciences - PNAS *109*, 3422-3427.

Laanto, E., Bamford, J.K.H., Laakso, J., and Sundberg, L.R. (2012). Phage-driven loss of virulence in a fish pathogenic bacterium. Plos One *7*.

Lafforgue, G., Martínez, F., Sardanyés, J., De la Iglesia, F., Niu, Q.-W., Lin, S.-S., Solé, R.V., Chua, N.-H., Daròs, J.-A., and Elena, S.F. (2011a). Tempo and mode of plant RNA virus escape from RNA interference-mediated resistance. Journal of virology *85*, 9686-9695.

Lafforgue, G., Sardanyés, J., and Elena, S.F. (2011b). Differences in accumulation and virulence determine the outcome of competition during tobacco etch virus coinfection. Plos One *6*.

Lambrechts, L., and Scott, T.W. (2009). Mode of transmission and the evolution of arbovirus virulence in mosquito vectors. Proceedings of the Royal Society B-Biological Sciences *276*, 1369-1378.

Lancaster, K.Z., and Pfeiffer, J.K. (2011). Mechanisms controlling virulence thresholds of mixed viral populations. Journal of Virology *85*, 9778-9788.

Leggett, H.C., Benmayor, R., Hodgson, D.J., and Buckling, A. (2013). Experimental evolution of adaptive phenotypic plasticity in a parasite. Curr Biol *23*, 139-142.

Levin, B.R. (1996). The evolution and maintenance of virulence in microparasites. Emerging Infectious Diseases *2*, 93 -102.

Levin, B.R., and Bull, J.J. (1994). Short-sighted evolution and the virulence of pathogenic microbes. Trends in Microbiology *2*, 76-81.

Levy, J.A. (1993). Pathogenesis of human immunodeficiency virus infection. Microbiological Reviews *57*, 183-289.

Li, Y., Carroll, D.S., Gardner, S.N., Walsh, M.C., Vitalis, E.A., and Damon, I.K. (2007). On the origin of smallpox: Correlating variola phylogenics with historical smallpox records. Proceedings of the National Academy of Sciences *104*, 15787-15792.

Lion, S. (2013). Multiple infections, kin selection and the evolutionary epidemiology of parasite traits. Journal of Evolutionary Biology *26*, 2107-2122.

Lipsitch, M., and Moxon, E.R. (1997). Virulence and transmissibility of pathogens: What is the relationship? Trends in Microbiology *5*, 31-37.

Lofquist, J.M., Weimert, N.A., and Hayney, M.S. (2003). Smallpox: A review of clinical disease and vaccination. American Journal of Health-System Pharmacy *60*, 749.

Mackinnon, M.J., Gandon, S., and Read, A.F. (2008). Virulence evolution in response to vaccination: The case of malaria. Vaccine *26*, C42-C52.

Markussen, T., Jonassen, C.M., Numanovic, S., Braaen, S., Hjortaas, M., Nilsen, H., and Mjaaland, S. (2008). Evolutionary mechanisms involved in the virulence of infectious salmon anaemia virus (ISAV), a piscine orthomyxovirus. Virology *374*, 515-527.

Markussen, T., Sindre, H., Jonassen, C.M., Tengs, T., Kristoffersen, A.B., Ramsell, J., Numanovic, S., Hjortaas, M.J., Christiansen, D.H., and Dale, O.B. (2013). Ultra-deep pyrosequencing of partial surface protein genes from infectious salmon anaemia virus (ISAV) suggest novel mechanisms involved in transition to virulence. PloS one *8*, e81571.

Marquez, L.M., and Roossinck, M.J. (2012). Do persistent RNA viruses fit the trade-off hypothesis of virulence evolution? Current Opinion in Virology *2*, 556-560.

Marriott, A.C., and Dimmock, N.J. (2010). Defective interfering viruses and their potential as antiviral agents. Reviews in Medical Virology *20*, 51-62.

Marshall, I.D., and Fenner, F. (1958). Studies in the epidemiology of infectious Myxomatosis of rabbits: V. Changes in the innate resistance of Australian wild rabbits exposed to Myxomatosis. The Journal of Hygiene *56*, 288-302.

Martínez-Picado, J., Prado, J.G., Fry, E.E., Pfafferott, K., Leslie, A., Chetty, S., Thobakgale, C., Honeyborne, I., Crawford, H., Matthews, P., *et al.* (2006). Fitness cost of escape mutations in p24 Gag in association with control of human immunodeficiency virus type 1. Journal of Virology *80*, 3617-3623.

Massad, E. (1987). Transmission rates and the evolution of pathogenicity. Evolution *41*, 1127-1130.

May, R.M., and Anderson, R.M. (1979). Population biology of infectious diseases : Part II. Nature *280*, 455-461.

May, R.M., and Anderson, R.M. (1983). Epidemiology and genetics in the coevolution of parasites and hosts. Proceedings of the Royal Society of London Series B, Biological Sciences *219*, 281-313.

McGee, L.W., and Remold, S.K. (2013). Pleiotropic interactions involving virulence and replication rate in experimentally evolved vesicular stomatitis virus. Evolutionary Ecology Research *15*, 567-578.

Mead-Briggs, A., and Vaughan, J. (1975). The differential transmissibility of myxoma virus strains of differing virulence grades by the rabbit flea *Spilopsyllus cuniculi* (Dale). The Journal of Hygiene *75*, 237-247.

Mellors, J.W., Munoz, A., Giorgi, J.V., Margolick, J.B., Tassoni, C.J., Gupta, P., Kingsley, L.A., Todd, J.A., Saah, A.J., Detels, R., *et al.* (1997). Plasma viral load and CD4(+) lymphocytes as prognostic markers of HIV-1 infection. Annals of Internal Medicine *126*, 946-954.

Mellors, J.W., Rinaldo, C.R., Gupta, P., White, R.M., Todd, J.A., and Kingsley, L.A. (1996). Prognosis in HIV-1 infection predicted by the quantity of virus in plasma. Science *272*, 1167-1170.

Messenger, S.L., Molineux, I.J., and Bull, J.J. (1999). Virulence evolution in a virus obeys a trade-off. Proceedings of the Royal Society B-Biological Sciences *266*, 397-404.

Methot, P.O. (2012). Why do parasites harm their host? On the origin and legacy of Theobald Smith's "Law of Declining Virulence". History and Philosophy of the Life Sciences *34*, 561-601.

Milgroom, M.G., and Cortesi, P. (2004). Biological control of chestnut blight with hypovirulence: a critical analysis. Annu Rev Phytopathol *42*, 311-338.

Mjaaland, S., Hungnes, O., Teig, A., Dannevig, B.H., Thorud, K., and Rimstad, E. (2002). Polymorphism in the infectious salmon anemia virus hemagglutinin gene: importance and possible implications for evolution and ecology of infectious salmon anemia disease. Virology *304*, 379-391.

Montarry, J., Cartier, E., Jacquemond, M., Palloix, A., and Moury, B. (2012). Virus adaptation to quantitative plant resistance: erosion or breakdown? Journal of Evolutionary Biology *25*, 2242-2252.

Morisse, J., Le Gall, G., and Boilletot, E. (1991). Hepatitis of viral origin in Leporidae: introduction and aetiological hypotheses. OIE Revue Scientifique et Technique *10*, 269-310.

Moudy, R.M., Meola, M.A., Morin, L.L.L., Ebel, G.D., and Kramer, L.D. (2007). A newly emergent genotype of West Nile virus is transmitted earlier and more efficiently by Culex mosquitoes. American Journal of Tropical Medicine and Hygiene *77*, 365-370.

Muller, V.a.i., Ledergerber, B.b., Perrin, L.c., Klimkait, T.d., Furrer, H.e., Telenti, A.f., Bernasconi, E.g., Vernazza, P.h., Gunthard, H.F.b., Bonhoeffer, S.i., *et al.* (2006). Stable virulence levels in the HIV epidemic of Switzerland over two decades. AIDS *20*, 889-894.

Murall, C.L., Bauch, C.T., and Day, T. (2015). Could the human papillomavirus vaccines drive virulence evolution? Proceedings of the Royal Society B-Biological Sciences *282*.

Nair, V. (2005). Evolution of Marek's disease - A paradigm for incessant race between the pathogen and the host. Veterinary Journal *170*, 175-183.

Nations, U. (2014). HIV 2013 estimates with uncertainty bounds 1990-2013, U.G.R.-. 2014, ed.

Novella, I., Clarke, D., Quer, J., Duarte, E., Lee, C., Weaver, S., Elena, S., Moya, A., Domingo, E., and Holland, J. (1995). Extreme fitness differences in mammalian and insect hosts after continuous replication of vesicular stomatitis virus in sandfly cells. J Virol *69*, 6805-6809.

Novella, I.S., Hershey, C.L., Escarmis, C., Domingo, E., and Holland, J.J. (1999). Lack of evolutionary stasis during alternating replication of an arbovirus in insect and mammalian cells. Journal of Molecular Biology *287*, 459-465.

Nowak, M.A., and May, R.M. (1994). Superinfection and the evolution of parasite virulence. Proceedings of the Royal Society of London Series B, Biological Sciences *255*, 81-89.

Ogbunugafor, C.B., Alto, B.W., Overton, T.M., Bhushan, A., Morales, N.M., and Turner, P.E. (2013). Evolution of increased survival in RNA viruses specialized on cancer-derived cells. American Naturalist *181*, 585-595.
Ojosnegros, S., Beerenwinkel, N., Antal, T., Nowak, M.A., and Escarmis, C. (2010). Competition-colonization dynamics in an RNA virus. Proceedings of the National Academy of Sciences *107*, 2108-2112.
Olmsted, R.A., Baric, R.S., Sawyer, B.A., and Johnston, R.E. (1984). Sindbis virus mutants selected for rapid growth in cell culture display attenuated virulence in animals. Science *225*, 424-427.
Pagan, I., Alonso-Blanco, C., and Garcia-Arenal, F. (2007). The relationship of within-host multiplication and virulence in a plant-virus system. Plos One *2*.
Pagan, I., Montes, N., Milgroom, M.G., and Garcia-Arenal, F. (2014). Vertical transmission selects for reduced virulence in a plant virus and for increased resistance in the host. Plos Pathogens *10*.
Pascua, P.N.Q., Song, M.S., Lee, J.H., Baek, Y.H., Kwon, H.I., Park, S.J., Choi, E.H., Lim, G.J., Lee, O.J., Kim, S.W., *et al.* (2012). Virulence and transmissibility of H1N2 influenza virus in ferrets imply the continuing threat of triple-reassortant swine viruses. Proceedings of the National Academy of Sciences of the United States of America *109*, 15900-15905.
Pasteur, L., Chamberlain, C.E., and Roux, E. (1881). Compte rendu sommaire des expe´riences faites a` Pouilly-le-Fort, pre` s Melun, sur la vaccination charbonneuse. Comptes-rendus des se´ances de l'Acade´mie des Sciences *92*, 1378-1383.
Payne, R., Muenchhoff, M., Mann, J., Roberts, H.E., Matthews, P., Adland, E., Hempenstall, A., Huang, K.-H., Brockman, M., Brumme, Z., *et al.* (2014). Impact of HLA-driven HIV adaptation on virulence in populations of high HIV seroprevalence. Proceedings of the National Academy of Sciences *111*, E5393-E5400.
Peever, T.L., Liu, Y.C., Cortesi, P., and Milgroom, M.G. (2000). Variation in tolerance and virulence in the chestnut blight fungus-hypovirus interaction. Applied and Environmental Microbiology *66*, 4863-+.
Peleg, J. (1971). Growth of viruses in arthropod cell cultures: applications. I. Attenuation of semliki forest (SF) virus in continuously cultured *Aedes aegypti* mosquito cells (Peleg) as a step in production of vaccines. Current topics in microbiology and immunology *55*, 155.
Peñaranda, M.M.D., Wargo, A.R., and Kurath, G. (2011). *In vivo* fitness correlates with host-specific virulence of infectious hematopoietic necrosis virus (IHNV) in sockeye salmon and rainbow trout. Virology *417*, 312-319.
Pfeiffer, J.K., and Kirkegaard, K. (2005). Increased fidelity reduces poliovirus fitness and virulence under selective pressure in mice. PLoS pathogens *1*, e11.
Pileri, E., Gibert, E., Soldevila, F., Garcia-Saenz, A., Pujols, J., Diaz, I., Darwich, L., Casal, J., Martin, M., and Mateu, E. (2015). Vaccination with a genotype 1 modified live vaccine against porcine reproductive and respiratory syndrome virus significantly reduces viremia, viral shedding and transmission of the virus in a quasi-natural experimental model. Veterinary Microbiology *175*, 7-16.
Pinel-Galzi, A., Rakotomalala, M., Sangu, E., Sorho, F., Kanyeka, Z., Traore, O., Sereme, D., Poulicard, N., Rabenantoandro, Y., Sere, Y., *et al.* (2007). Theme and variations in the evolutionary pathways to virulence of an RNA plant virus species. Plos Pathogens *3*, 1761-1770.
Ping, J.H., Keleta, L., Forbes, N.E., Dankar, S., Stecho, W., Tyler, S., Zhou, Y., Babiuk, L., Weingartl, H., Halpin, R.A., *et al.* (2011). Genomic and protein structural maps of adaptive evolution of human influenza A virus to increased virulence in the mouse. Plos One *6*.
Power, A.G. (1992). Patterns of virulence and benevolence in inset-borne pathogens of plants. Critical Reviews in Plant Sciences *11*, 351-372.
Quinn, T.C., Wawer, M.J., Sewankambo, N., Serwadda, D., Li, C.J., Wabwire-Mangen, F., Meehan, M.O., Lutalo, T., Gray, R.H., and Rakai Project Study, G. (2000). Viral

load and heterosexual transmission of human immunodeficiency virus type 1. New England Journal of Medicine *342*, 921-929.
Quinones-Mateu, M.E., Ball, S.C., Marozsan, A.J., Torre, V.S., Albright, J.L., Vanham, G., van der Groen, G., Colebunders, R.L., and Arts, E.J. (2000). A dual infection/ competition assay shows a correlation between *ex vivo* human immunodeficiency virus type 1 fitness and disease progression. Journal of Virology *74*, 9222-9233.
Radford, A.D., Dawson, S., Coyne, K.P., Porter, C.J., and Gaskell, R.M. (2006). The challenge for the next generation of feline calicivirus vaccines. Veterinary Microbiology *117*, 14-18.
Ransome, A. (1880). On epidemic cycles. Proc Manchester Lit Phil Soc *19*, 75-96.
Ransome, A. (1881). On the form of the epidemic wave and some of its probable causes. Trans Epidem Soc Lond *1*, 96.
Read, A.F. (1994). The evolution of virulence. Trends in Microbiology *2*, 73-76.
Read, A.F., and Taylor, L.H. (2001). The ecology of genetically diverse infections. Science *292*, 1099-1102.
Remold, S. (2012). Understanding specialism when the jack of all trades can be the master of all. Proceedings of the Royal Society B: Biological Sciences, rspb20121990.
Rico-Hesse, R. (2007). Dengue virus evolution and virulence models. Clinical Infectious Diseases *44*, 1462-1466.
Roche, B., Drake, J.M., and Rohani, P. (2011). The curse of the Pharaoh revisited: evolutionary bi-stability in environmentally transmitted pathogens. Ecology Letters *14*, 569-575.
Rodriguez-Roche, R., Sanchez, L., Burgher, Y., Rosario, D., Alvarez, M., Kouri, G., Halstead, S.B., Gould, E.A., and Guzman, M.G. (2011). Virus role during intraepidemic increase in dengue disease severity. Vector-Borne and Zoonotic Diseases *11*, 675-681.
Rodriguez, B., Sethi, A.K., Cheruvu, V.K., Mackay, W., Bosch, R.J., Kitahata, M., Boswell, S.L., Mathews, W.C., Bangsberg, D.R., Martin, J., *et al.* (2006). Predictive value of plasma HIV RNA level on rate of CD4 T-cell decline in untreated HIV infection. Jama-Journal of the American Medical Association *296*, 1498-1506.
Roossinck, M.J. (2010). Lifestyles of plant viruses. Philosophical Transactions of the Royal Society B: Biological Sciences *365*, 1899-1905.
Ross, J., and Sanders, M.F. (1977). Innate resistance to Myxomatosis in wild rabbits in England. The Journal of Hygiene *79*, 411-415.
Sacristán, S., Fraile, A., Malpica, J.M., and García-Arenal, F. (2005). An analysis of host adaptation and its relationship with virulence in Cucumber mosaic virus. Phytopathology *95*, 827-833.
Sacristan, S., and Garcia-Arenal, F. (2008). The evolution of virulence and pathogenicity in plant pathogen populations. Molecular Plant Pathology *9*, 369-384.
Saiz, J., and Domingo, E. (1996). Virulence as a positive trait in viral persistence. J Virol *70*, 6410-6413.
Schmid-Hempel, P. (2011). Virulence. In Evolutionary Parasitology, O.U. Presss, ed. (New York: Oxford University Press).
Simmons, H.E., Dunham, J.P., Zinn, K.E., Munkvold, G.P., Holmes, E.C., and Stephenson, A.G. (2013). Zucchini yellow mosaic virus (ZYMV, Potyvirus): Vertical transmission, seed infection and cryptic infections. Virus Research *176*, 259-264.
Singh, K.R.P. (1971). Propagation of arboviruses in Singh's *Aedes* cells. Current Topics in Microbiology and Immunology *55*, 127-133.
Smith, T. (1904). Some problems in the life-hostory of pathogeneic microorganisms. Science *20*, 817-832.
Snow, M., and Cunningham, C.O. (2000). Virulence and nucleotide sequence analysis of marine viral haemorrhagic septicaemia virus following *in vivo* passage in rainbow trout *Oncorhynchus mykiss*. Diseases of Aquatic Organisms *42*, 17-26.

Spackman, E., and Swayne, D.E. (2013). Vaccination of gallinaceous poultry for H5N1 highly pathogenic avian influenza: Current questions and new technology. Virus Research *178*, 121-132.

Stewart, A., Logsdon, J.J., and Kelly, S. (2005). An empirical study of the evolution of virulence under both horizontal and vertical transmission. Evolution *59*, 730-739.

Suzuki, K., Okada, H., Itoh, T., Tada, T., Mase, M., Nakamura, K., Kubo, M., and Tsukamoto, K. (2009). Association of increased pathogenicity of Asian H5N1 highly pathogenic avian influenza viruses in chickens with highly efficient viral replication accompanied by early destruction of innate immune responses. Journal of Virology *83*, 7475-7486.

Taubenberger, J.K., and Kash, J.C. (2010). Influenza virus evolution, host adaptation, and pandemic formation. Cell Host and Microbe *7*, 440-451.

Taylor, W.P., and Marshall, I.D. (1975a). Adaptation studies with ross river virus - laboratory mice and cell-cultures. J Gen Virol *28*, 59-72.

Taylor, W.P., and Marshall, I.D. (1975b). Adaptation studies with ross river virus - retention of field level virulence. J Gen Virol *28*, 73-83.

Ten Haaft, P., Verstrepen, B., Uberla, K., Rosenwirth, B., and Heeney, J. (1998). A pathogenic threshold of virus load defined in simian immunodeficiency virus- or simian-human immunodeficiency virus-infected macaques. Journal of Virology *72*, 10281-10285.

Thomas, F., Guegan, J.F., and Renaud, F. (2009). Ecology and Evolution of Parasitism (New York: Oxford University Press).

Thomas, S., and Elkinton, J.S. (2004). Pathogenicity and virulence. Journal of Invertebrate Pathology *85*, 146-151.

Toft, C.A., and Karter, A.J. (1990). Parasite-host coevolution. Trends in Ecology and Evolution *5*, 326-329.

Truyen, U., Everman, J.F., Vieler, E., and Parrish, C.R. (1996). Evolution of canine parvovirus involved loss and gain of feline host range. Virology *215*, 186-189.

Turner, P.E., and Chao, L. (1999). Prisoner's dilemma in an RNA virus. Nature *398*, 441-443.

Turner, P.E., and Elena, S.F. (2000). Cost of host radiation in an RNA virus. Genetics *156*, 1465-1470.

Twiddy, S.S., Farrar, J.J., Chau, N.V., Wills, B., Gould, E.A., Gritsun, T., Lloyd, G., and Holmes, E.C. (2002). Phylogenetic relationships and differential selection pressures among genotypes of dengue-2 virus. Virology *298*, 63-72.

Van der Goot, J.A., De Jong, M.C.M., Koch, G., and Van Boven, M. (2003). Comparison of the transmission characteristics of low and high pathogenicity avian influenza A virus (H5N2). Epidemiology and infection *131*, 1003-1013.

Vasilakis, N., Deardorff, E.R., Kenney, J.L., Rossi, S.L., Hanley, K.A., and Weaver, S.C. (2009). Mosquitoes put the brake on arbovirus evolution: experimental evolution reveals slower mutation accumulation in mosquito than vertebrate cells. PLoS pathogens *5*, e1000467.

Vignuzzi, M., Stone, J.K., Arnold, J.J., Cameron, C.E., and Andino, R. (2006). Quasispecies diversity determines pathogenesis through cooperative interactions in a viral population. Nature *439*, 344-348.

Vignuzzi, M., Wendt, E., and Andino, R. (2008). Engineering attenuated virus vaccines by controlling replication fidelity. Nat Med *14*, 154-161.

Villarreal, L.P. (2009). Persistence pays: how viruses promote host group survival. Current opinion in microbiology *12*, 467-472.

Wallis, C.M., Stone, A.L., Sherman, D.J., Damsteegt, V.D., Gildow, F.E., and Schneider, W.L. (2007). Adaptation of plum pox virus to a herbaceous host (*Pisum sativum*) following serial passages. J Gen Virol *88*, 2839-2845.

Walther, B.A., and Ewald, P.W. (2004). Pathogen survival in the external environment and the evolution of virulence. Biological Reviews *79*, 849-869.

Wargo, A.R., Garver, K.A., and Kurath, G. (2010). Virulence correlates with fitness *in vivo* for two M group genotypes of infectious hematopoietic necrosis virus (IHNV). Virology *404*, 51-58.

Wargo, A.R., and Kurath, G. (2011). *In vivo* fitness associated with high virulence in a vertebrate virus is a complex trait regulated by host entry, replication, and shedding. Journal of Virology *85*, 3959-3967.

Wargo, A.R., and Kurath, G. (2012). Viral fitness: definitions, measurement, and current insights. Current Opinion in Virology *2*, 538-545.

Weaver, S.C. (2006). Evolutionary influences in arboviral disease. Quasispecies: Concept and Implications for Virology *299*, 285-314.

Weaver, S.C., Brault, A.C., Kang, W.L., and Holland, J.J. (1999). Genetic and fitness changes accompanying adaptation of an arbovirus to vertebrate and invertebrate cells. Journal of Virology *73*, 4316-4326.

West, S.A., and Buckling, A. (2003). Cooperation, virulence and siderophore production in bacterial parasites. Proceedings of the Royal Society B-Biological Sciences *270*, 37-44.

Witter, R.L. (1997). Increased virulence of Marek's disease virus field isolates. Avian Diseases *41*, 149-163.

Woolhouse, M.E.J., Taylor, L.H., and Haydon, D.T. (2001). Population biology of multihost pathogens. Science *292*, 1109-1112.

Xu, L.L., Bao, L.L., Zhou, J.F., Wang, D.Y., Deng, W., Lv, Q., Ma, Y.L., Li, F.D., Sun, H.H., Zhan, L.J., *et al.* (2011). Genomic polymorphism of the pandemic A (H1N1) influenza viruses correlates with viral replication, virulence, and pathogenicity *in vitro* and *in vivo.* Plos One *6*.

Yan, W., Kitzes, G., Dormishian, F., Hawkins, L., Sampson-Johannes, A., Watanabe, J., Holt, J., Lee, V., Dubensky, T., and Fattaey, A. (2003). Developing novel oncolytic adenoviruses through bioselection. Journal of virology *77*, 2640-2650.

Yuste, E., Sánchez-Palomino, S., Casado, C., Domingo, E., and López-Galíndez, C. (1999). Drastic fitness loss in human immunodeficiency virus type 1 upon serial bottleneck events. Journal of Virology *73*, 2745-2751.

Zarate, S., and Novella, I.S. (2004). Vesicular stomatitis virus evolution during alternation between persistent infection in insect cells and acute infection in mammalian cells is dominated by the persistence phase. Journal of Virology *78*, 12236-12242.

7

Taxonomy Advancement and Genome Size Change: Two Perspectives on RNA Virus Genetic Diversity

Chris Lauber[1,2] and Alexander E. Gorbalenya[1,3]*

[1] *Department of Medical Microbiology, Leiden University Medical Center, Leiden, The Netherlands;* [2] *Institute for Medical Informatics and Biometry, Technische Universität Dresden, Dresden, Germany;* [3] *Faculty of Bioengineering and Bioinformatics, Lomonosov Moscow State University, Moscow, Russia*

**Corresponding author: a.e.gorbalenya@lumc.nl*

Abstract

Relatively frequent insertion and deletion events, ranging from single nucleotides to full-length genes, and extreme point mutation rates make macroevolutionary studies of RNA viruses an exquisite challenge. Here we review recent advancements in the field concerning the study of both of these aspects and with regard to two specific applications - virus taxonomy and the analysis of genome size evolution. For the former we observe a dominant trend towards utilizing results from genetics-based virus classification during recent years. We also briefly discuss parallel developments for taxonomic studies of cellular organisms where very similar techniques are applied, but observe little cross-talk with virus taxonomy. For RNA virus genome size evolution we outline an emerging general pattern of genome (segment) enlargement being associated with a host- and mutation-constrained acquisition of enzymes that seemingly improved the rudimentary RNA virus replication machinery and promoted further expansion. We discuss possible consequences for these findings for the upper limit on the size of the RNA virus genome segment currently observed for single-segment positive stranded nidoviruses.

Introduction

Some time ago it was recognized that RNA viruses are evolving extremely fast because of error-prone genome replication (Holland *et al.*, 1982). While our view of this relation is more nuanced currently (Bouvet *et al.*, 2012; Minskaia *et al.*, 2006; Smith *et al.*, 2014c; Snijder *et al.*, 2003) (see also below), high mutation rate remains a major attribute of RNA viruses. It along with large along with large population sizes and a short replication cycle made studies of microevolution of RNA viruses feasible, assisted by different models of molecular evolution and computational tools developed over recent decades (Bhatt *et al.,* 2010; Bouckaert *et al.,* 2014; Drummond *et al.,* 2006). These studies are concerned with quantitative analysis of genetic heterogeneity accumulated over relatively small time periods, generally less than the most recent 100 years or so when virus samples were collected. As we have learned, synonymous and occasional non-synonymous substitutions at individual sites dominate during the evolution of different virus lineages that have diverged from a common ancestor within this limited time frame and formed subsets of species. Their patterns can be used to deduce phylodynamics of these lineages in individual and multiple hosts under changing selection pressure (Pybus and Rambaut, 2009; Volz *et al.,* 2013).

Things get more complicated when researchers try to understand the evolution of more diverged virus lineages that are recognized as forming taxa of higher ranks in virus taxonomy - from genus to order and beyond (King *et al.,* 2012). As the scale of divergence increases dramatically, the rapid rate of mutation becomes a major obstacle for reliable deduction of parameters of macroevolution. At this scale, most sites, even in the most conserved proteins, may be mutated repeatedly to the extent that reconstruction of the full chain of evolutionary events becomes highly uncertain if not outright impossible, a phenomenon known as mutation or substitution saturation (Holmes, 2003b). Fortunately, homology signals may be retained even during large scale divergence of RNA viruses (Gorbalenya *et al.,* 2010; Koonin *et al.,* 2015; Kuchibhatla *et al.,* 2014) (see also below), possibly because of constraints of the molecular environment that impose strong purifying selection and restrict accumulation of non-synonymous substitutions to few sites renewed slowly in time (Koonin and Gorbalenya, 1989). Still, because of the lack of virus fossils, an independent verification and calibration of evolutionary scenarios produced with the help of phylogenetic analysis of contemporary sequences may not be feasible. It is possible to alleviate this challenge to some extent for viruses that have co-evolved with their hosts [chief examples are from DNA viruses, e.g. (McGeoch *et al.,* 1995)] or whose ancestral genomes have been integrated into the host genome (Benveniste and Todaro, 1974; Cui and Holmes, 2012; Gilbert and Feschotte, 2010; Jaenisch, 1976; Katzourakis and Gifford, 2010). Despite these notable exceptions, dating of macroevolution of RNA viruses remain technically extremely challenging (Holmes, 2011).

At the genetic level, mutations may be classified into two major classes: substitutions and insertions/deletions (often called indels). Analysis of substitutions informs advancement of virus taxonomy using only genomic sequences, which is particularly relevant in the era of metagenomics studies. Accumulation of indels leads to genome size evolution which is most constrained in RNA viruses but remains the primary source of major functional innovation. Below, we review recent progress of studies of virus evolution from the perspective of taxonomy and genome size change.

Uncovering barriers to genetic divergence of RNA viruses

When searching the public sequence database GenBank/RefSeq (Benson *et al.,* 2010) using the keyword 'virus' roughly 3.4 million nucleotide entries (April 2015) show up. Perhaps not surprisingly, a large fraction is contributed by human immunodeficiency virus 1, influenza A virus and hepatitis C virus due to their high medical relevance and outstanding efforts to study these pathogens. However, this wealth of genetic data also involves a great many other viruses, both established and tentative, the latter mostly represented by so-called genomic survey sequences from metagenomics studies. On the one hand, this explosive growth of sequence information allows us to study the genetic diversity and evolution of viruses in unprecedented detail. On the other hand, it increasingly challenges virus taxonomy since other types of virus characterization normally cannot keep up with the pace set by genome sequencing. Hence, it is tempting to consider genome sequences as the ultimate source of information to be utilized in virus taxonomy, a notion which is supported by additional observations. First, they bring highly accurate knowledge due to the low sequencing error rates which generally fall well below 1% (Wesche *et al.,* 2004). Second, sequencing is fast and relatively cheap nowadays (Lander, 2011) as ever more efficient techniques are being developed (Eid *et al.,* 2009; Rothberg *et al.,* 2011). Third, genome sequences are easy to digitize and compare, readily enabling quantitative analyses. And last but not least, the use of nucleotide sequences as carriers of genetic information and heredity presents a universal property common to all biological entities on earth.

Utilization of this wealth of genetic data for virus classification can be traced back as far as 1988 when Shukla and Ward conducted a groundbreaking study in which they classified various potyvirus strains using only coat protein sequences (Shukla and Ward, 1988). The approach was adopted later in taxonomic studies to classify viruses of various families and genome types including picornaviruses (Chan *et al.,* 2010; Lauber and Gorbalenya, 2012c; Oberste *et al.,* 1999), caliciviruses (Schuffenecker *et al.,* 2001; Zheng *et al.,* 2006), geminiviruses (Fauquet and Stanley, 2003), coronaviruses (de Groot *et al.,* 2012; Gonzaalez *et al.,* 2003), potyviruses (Adams *et al.,* 2005), flexiviruses (Adams *et al.,* 2004), papillomaviruses (Bernard *et al.,* 2010; de Villiers *et al.,* 2004), poxviruses (Lefkowitz *et al.,* 2006), rotaviruses (Matthijnssens *et al.,* 2008), hantaviruses (Maes *et al.,* 2009), mesoniviruses (Lauber *et al.,* 2012), hepaciviruses (Smith *et al.,* 2014a), hepeviruses (Smith *et al.,* 2014b), and filoviruses (Bao *et al.,* 2012; Lauber and Gorbalenya, 2012a). In each of these studies it was assumed, either explicitly or implicitly, that the genetic divergence within a taxon is generally lower than between different taxa and that this information can be used to classify the viruses at hand (Figure 7.1). This utilization of divergence thresholds of genes or sets of genes for assisting virus classification is becoming popular in virus taxonomy (King *et al.,* 2012). There is, however, no consensus on key parameters of the method among the different studies, which includes (i) what genome regions to include, (ii) what sequence type – nt or aa – to use, (iii) what type of alignments – pairwise or multiple – to compile, and (iv) what measure of sequence similarity – uncorrected percentage identities or distances that correct for multiple substitutions at the same site – to calculate. A classification of picornaviruses basing on a multiple alignment of all proteins conserved across the family and using evolutionary distances, that can account for the high mutation rates of RNA viruses resulting in multiple substitutions at the same sequence site (Drake and Holland, 1999; Duffy *et al.,* 2008), was found to be most consistent and stable (Lauber and Gorbalenya, 2012b), although this finding needs to be verified for more virus families.

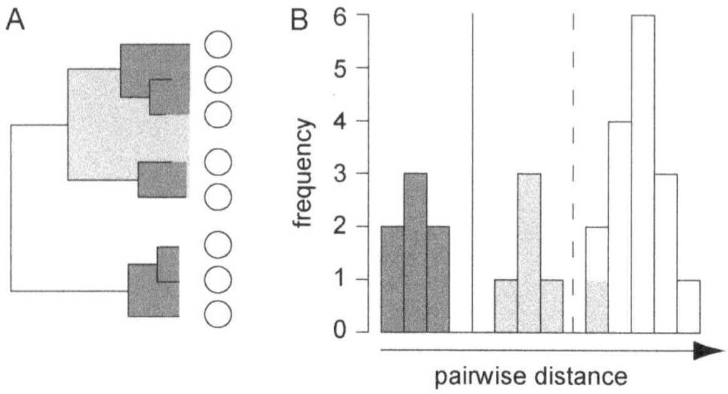

Figure 7.1. Grouping viruses based on thresholds in the distribution of pairwise genetic divergence. Shown is a cartoon example involving eight viruses (white circles) that illustrates the relationship between phylogeny (A), selection of thresholds in the distribution of intervirus genetic divergence (B), and the accompanied virus classification (C). The phylogeny of the viruses encompasses three species and two genera (A). Intra-species (dark-grey) and intra-genus (light-grey) intervirus genetic distances are highlighted. According to the tree, alternative species groupings seem possible, by setting different thresholds on genetic divergence. The frequency distribution of all pairwise distances extracted from the tree shows two regions of distance discontinuity, partitioning the pairwise distance range into intra-species (dark-grey), intra-genus (light-grey), and inter-genus (white), with a single exception for the genus level (B). Using these distance discontinuities as thresholds on genetic divergence (solid and dashed vertical lines) results in a hierarchical classification of the viruses into the three species and two genera, respectively (C). The single intra-genus distance violating the threshold is shown in white dashed line. Ideas adapted from (Lauber and Gorbalenya, 2012bc).

As sequenced viral genomes are accumulating with ever-increasing speed, efficient computational tools are needed to assist virus taxonomy. PAirwise Sequence Comparison (PASC) (Bao *et al.,* 2008), promoted by NCBI, has emerged as the de facto standard for pairwise-distance-based virus classification in this respect and was utilized in several of the studies mentioned above. A main objective of PASC is the speedy classification of newly identified viruses with sequenced genomes. To do so, PASC depends on pre-established classifications, usually brought by the ICTV taxonomy, for roughly 60 out of 110 families or floating genera established currently (Adams *et al.,* 2015). A new virus is classified using thresholds on its similarity to taxa in the respective pre-established classification. Importantly, PASC calculates these demarcation thresholds have been defined a priori as the lowest intra-level (for instance intra-species) similarity observed across virus pairs in the pre-established classification. Exactly this approach, however, presents a potential pitfall of the method since no golden standard classification is available for any virus family, due to the intrinsic lack of fossil data in virology. Hence, PASC is bound to propagate and may aggravate inaccuracies of the initial classification with the incorporation of newly identified viruses. Only a few studies have tried to approach the problem of threshold determination objectively, independent of any pre-existing classification. Matthijnssens *et al.* (2008) selected the threshold at which the ratio of intergroup to intragroup sequence identities dropped below one. Unfortunately, the basis for this choice was neither explained in detail nor evaluated rigorously. To address these and other deficiencies, we developed *D*ivErsity p*A*rtitioning by hie*R*archical

Clustering (DEmARC), an evolutionary-based framework for genetics-based virus classification. It enables the user to measure the support for both the demarcation thresholds and the inferred taxa in a quantitative manner, thus allowing for an objective selection of thresholds and resulting virus groupings (Lauber and Gorbalenya, 2012b).

Notably, DEmARC can serve not only pure classification purposes but also the prediction of biological properties of the analyzed viruses and the inferred virus taxa (Lauber and Gorbalenya, 2012c). Intuitively, a newly identified virus classified as belonging to a known virus species is expected to show phenotypic properties similar to that of other viruses in the species, a prediction which is also available through traditional virus taxonomy. Yet, a DEmARC-based classification can offer additional insights, due to the fact that virus taxa of the same level are delimited objectively by applying the same criterion (a demarcation threshold on genetic divergence) universally to all viruses at hand. Specifically, this concerns the natural genetic diversity of a taxon that might be heavily underestimated by the current virus sampling, but can be predicted by utilizing information from other, well-sampled taxa of the same level. Hence, DEmARC offers the means to identify those taxa on which to focus future virus discovery efforts in order to obtain a comprehensive picture of the natural genetic diversity of the virus family/genus under consideration (Lauber and Gorbalenya, 2012a). Notably, the predicted genetic diversity of a taxon presents only an upper limit of the actual natural diversity. For instance, the moderately sampled picornavirus species *Hepatitis A virus* (Feinstone *et al.,* 1973; Knowles *et al.,* 2012) with its few dozen available complete genome sequences, shows a relatively low genetic diversity (Lauber and Gorbalenya, 2012c), which may be due to either sampling specifics or the unusually low evolutionary rate exhibited by these viruses (Hanada *et al.,* 2004; Moratorio *et al.,* 2007; Sanchez *et al.,* 2003).

From a more fundamental perspective, the presence of peaks and valleys in the pairwise distance distribution, like those seen for the well-sampled family *Picornaviridae* (Lauber and Gorbalenya, 2012c), may provide an insight into commonalities across viral lineages during evolution. We proposed (Lauber and Gorbalenya, 2012c) that the observed distance discontinuities at and above the genus level and the distance peaks separated by these discontinuities could be explained by periods of, respectively, mass extinction and mass speciation of viral lineages, possibly reflecting large-scale changes in the environment that had a bearing on their hosts. In contrast, distance discontinuity associated with the species rank may be linked to peculiarities of virus replication as discussed below.

Virus species: constructs made to build taxonomy versus biological entities sharing key properties

There is a long-lasting dispute on the suitability of pairwise distances as a single criterion for classifying viruses (Van Regenmortel, 2007). This debate is largely linked to the question whether virus species, forming the basic level in virus taxonomy (Pringle, 1991), are real biological entities (Bishop, 1985; Kingsbury, 1985; Milne, 1984) or simply constructs in our mind (Van Regenmortel, 2003) developed for the convenience of biologists. If the former is true then certain biological properties, which could be used to discriminate between virus species, are expected to exist. In the case of eukaryotic organisms, for which it is generally accepted that species are evolving biological entities (Coyne and Orr, 2004), such properties usually include genetic incompatibilities between species that result in separated gene pools. It is tempting to apply this biological species concept, originally introduced by Dobzhansky in 1937 (Dobzhansky, 1937), to viruses, although its validity was questioned for organisms that reproduce asexually (Hendry, 2009). Nevertheless, virus clusters resembling species can be delimited genetically at the family level (Lauber and Gorbalenya, 2012c) or even across related families (Lauber *et al.,* 2012) through distance discontinuity in the conserved proteins. This distance discontinuity, which is the result of inter-virus distances being generally lower inside a cluster than between clusters (Figure 7.1), is nontrivial and could be explained by only two causes (assuming the calculated distances to be an adequate estimate of the real genetic distances). First, the distance discontinuity could be due to insufficient sampling of both the number and the diversity of the analyzed sequences. If this is not the case, as presumably for the well-sampled family *Picornaviridae*, the observed distance discontinuity is likely due to biological factors enforcing constraints that limit the divergence of viruses of the same but not of different clusters. In the latter scenario, it could be argued that the delineated virus clusters correspond to biological species. A plausible factor of speciation could be the action of homologous recombination if restricted to viruses of the same species, and this was already suggested for picornaviruses (Lukashev, 2010). In this way, homologous recombination could resemble the exchange of genetic material during (ordinary) sexual reproduction, thereby setting a barrier to genetic divergence and, consequently, to speciation, whereas viruses from different species continue to diverge by mutation. If true, not only would it provide a biological foundation for the recognition of virus species as real, evolving entities but also should it pin medical relevance to species (Jiang *et al.,* 2007; Tibayrenc, 2006). This would have great implications for many branches of virology including virus diagnostics, antiviral research, and epidemiological studies.

Future research efforts should also involve scrutinizing other classification approaches that rely on genetic sequences. Among them are phylogeny-based techniques like the branching index (Hraber *et al.,* 2008), which can be used to infer statistically whether a query sequence clusters with a known clade in the tree, and a method that determines an increase of the branching rate in the tree to define the species boundary (Pons *et al.,* 2006). The latter study is from the field of DNA barcoding, a recently emerged line of research that aims at genetics-based taxonomy of cellular organisms (Casiraghi *et al.,* 2010; Hebert *et al.,* 2003). Most of these methods (Hebert *et al.,* 2003; Lefebure *et al.,* 2006; Rach *et al.,* 2008; Sarkar *et al.,* 2002) face similar limitations like their counterparts in virus classification, including the dependency on a golden standard (Meyer and Paulay, 2005), as well as additional challenges owing to the large sizes of cellular genomes (Casiraghi *et al.,* 2010). Nevertheless, genetics-based (RNA) virus classification should not continue to ignore such parallel developments.

Uncovering barriers to gene length in RNA viruses

Comparative sequence analysis in virology is usually concerned with genetic variation (nucleotide or amino acid differences among the compared sequences) and its utilization for making biological inferences. These include structural and functional predictions for a sequence or specific sequence residues, or

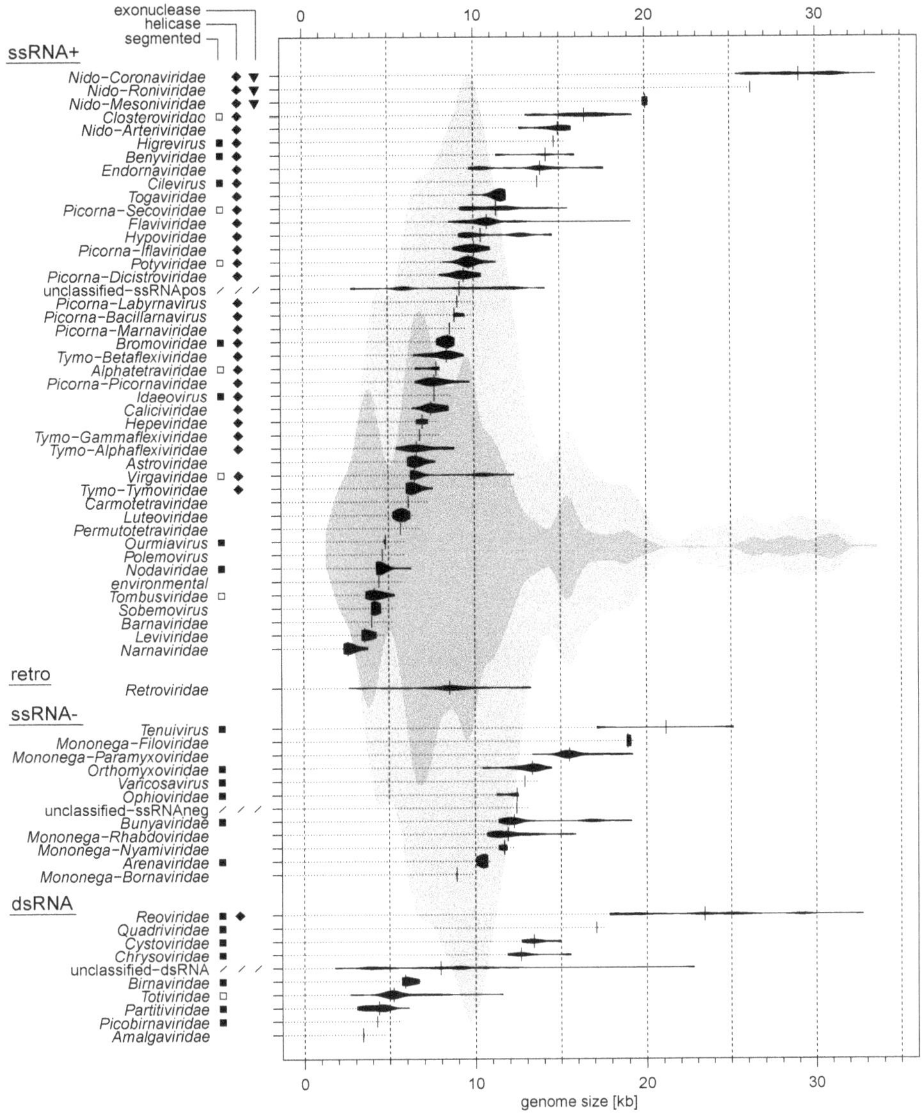

Figure 7.2. Genome sizes of RNA viruses and the relation to genome segmentation, and helicase and exoribonuclease expression. Shown are beanplots (Kampstra, 2008) of genome sizes for all known families or floating genera of viruses with ssRNA+, ssRNA-, and dsRNA genomes, and retroviruses. Sizes were extracted from the Viral Genomes Resource at NCBI (Bao *et al.*, 2004) (April 2015). A bean (black shape) shows the density distribution of individual genome sizes for a virus group. The median genome size per group is indicated by black vertical bars and was used for sorting. Additional information is shown next to virus group names: some (open square) or all (filled square) viruses have segmented genomes; expression of a helicase (diamond); expression of an exoribonuclease (triangle); / not applicable. Note that retroviruses don't encode a helicase due to specifics in their replication cycle (Kadare and Haenni, 1997). The joint distribution of all RNA virus genome sizes and all sizes of the largest genome segment are shown, respectively, as a light-gray and a dark-gray bean in the background. The average RNA virus genome size is 10.4 kb. Ideas adapted from (Gorbalenya *et al.*, 2006) and (Gorbalenya and Koonin, 1989).

reconstructing the evolutionary history of the viruses under study. There is, however, a second dimension that has received relatively little attention so far: the length of genetic sequences (in number of nt or aa). This includes both the total size of a viral genome and the size of genome regions encoding functional elements, for instance proteins. It is generally acknowledged that the genome size of RNA viruses is strongly constrained as a result of (i) the low fidelity of their polymerases (Holmes, 2003a) which would drive larger genomes into an 'error catastrophe' (Bull *et al.,* 2007; Eigen, 2002), (ii) the selection for high replication speed (Belshaw *et al.,* 2008; Elena and Sanjuán, 2005), and (iii) the relative inflexibility in expanding virus particles (virions) with icosahedral capsids to accommodate larger genomes (Chirico *et al.,* 2010). As a consequence, RNA virus genome sizes are in the range from two to 34 kb with an average of about 10 kb (Figure 7.2). Currently, the largest known RNA genome is employed by a single-segment python nidovirus (Bodewes *et al.,* 2014; Stenglein *et al.,* 2014) followed by an insect reovirus with its 16 dsRNA genome segments (Zhou *et al.,* 2014). When counting only the size of the largest genome segment (single RNA molecule) nidoviruses with genomes above 20 kb, which comprise coronaviruses, toro-/bafiniviruses, roniviruses, and mesoniviruses, outrank all other known RNA viruses. Still, all genes of an RNA virus are confined to severely constrained genomic space. This leads often to the encoding of multiple functions within the same genomic region, for instance through overlapping open reading frames (ORFs) (Belshaw *et al.,* 2007; Firth, 2014; Firth and Brown, 2006; Gorbalenya *et al.,* 2006; Keese and Gibbs, 1992; Krakauer, 2000; Sabath *et al.,* 2012) or RNA regulatory elements being situated inside a protein-coding gene (McKnight and Lemon, 1998). As would be expected, ORF overlap was found to be largest for viruses with the smallest genomes and vice versa (Belshaw *et al.,* 2007).

The exceptionally large genomes of nidoviruses that exceed about two- to three-fold the average size of a ssRNA+ genome, uniquely encode a 3'-5' exoribonuclease (Figure 7.2). It was proposed that this enzyme improves the otherwise low fidelity of genome replication (Snijder *et al.,* 2003), and allowed a subset of nidoviruses to overcome a genome size threshold of about 20 kb (Nga *et al.,* 2011). Furthermore, a second threshold of about 8 kb set by astroviruses seems to be associated with the expression of a protein with an NTP-binding motif indicative of a helicase (Figure 7.2). This strong correlation was already noticed more than 25 years ago (Gorbalenya and Koonin, 1989). Since then the helicase acquisition was implicated in promoting the enlargement of RNA virus genomes above the observed threshold (Gorbalenya *et al.,* 2006). At least in nidoviruses, the acquired helicase might have improved quality control over the virus RNA progeny by stimulating selective decay of aberrant RNA genomes (Deng *et al.,* 2014; Lehmann *et al.,* 2015). The helicase expression seems to also be associated with the size of the largest genome segment of dsRNA viruses but not with retrovirus and ssRNA- virus genomes (Figure 7.2). The latter observation indicates that the unwinding of long stretches of duplex RNA (for instance during replication and transcription), which is the expected primary function of a viral helicase, could be achieved through other (unconventional) mechanisms (Kadare and Haenni, 1997), or, alternatively, the formation of such dsRNA structures would need to be prevented in the first place. Notably, a few other protein domains in addition to the exoribonuclease and helicase, including a methyltransferase, and proteases with chymotrypsin- and papain-like folds are frequently employed by ssRNA+ viruses but none shows such a universal association with genome size (Gorbalenya and Koonin, 1993).

These findings show, at least in the case of dsRNA and ssRNA+ viruses, that the expression of RNA-processing enzymes with specific functions allows some RNA viruses to employ larger genomes than those that lack these enzymes, resulting in a considerable variety of genomic sizes. This is contrary to the prevailing perception that RNA viruses are simply limited in genome size with essentially no variability (concerning both the whole genome and distinct genetic elements within the genome), which would be worth a detailed analysis. In fact, different genome regions (e.g. genes) might be constrained differently depending on the encoded function. It was shown, for instance, that the gene length of polymerases, but not of nucleocapsids (both defined very broadly) increases with increasing genome size among RNA viruses, and the authors linked this size increase of the protein to an improved replication fidelity (Belshaw *et al.,* 2008). For the *Rhabdoviridae*, a family of the order *Mononegavirales* with negative-sense single-stranded RNA genomes, it was shown that genome expansion primarily happened through the insertion of additional transcriptional units (e.g. ORFs) between the five genes conserved across the family, although length variation of non-protein-coding regions was also accompanied with genome size change (Walker *et al.,* 2015). ORF insertions were found to be clade-specific and restricted to few or often a single genomic locus. The authors link this enormous genome plasticity of rhabdoviruses to their discontinuous transcription strategy (Walker *et al.,* 2015). Comparable size plasticity was described for the two genomic regions encompassing the 5'-proximal ORF1a and 3'-proximal ORFs, respectively, of corona- and other nidoviruses. The latter are expressed through subgenomic RNAs and also show frequent insertion of new genes (Gorbalenya *et al.,* 2006; Liu *et al.,* 2014; Snijder *et al.,* 2003).

For picornaviruses a striking negative correlation between sequence conservation and size variation of viral proteins was observed (Gorbalenya and Lauber, 2010). Specifically, the sizes of the six proteins conserved in sequence across picornaviruses (three capsid proteins plus putative helicase, proteinase, and polymerase proteins) proved to be constrained most strongly. On the one hand, this is not surprising since the most conserved proteins are essential to the virus and, thus, their function must be retained during evolution. These six picornavirus proteins control the three main steps in the virus life cycle – genome replication, genome expression, and encapsidation. On the other hand, the apparent barrier to size variation of the conserved proteins is nontrivial given that these proteins accepted as much as three replacements per residue (on average) when comparing the most distant picornavirus pair using maximum likelihood-based genetic distances (Lauber and Gorbalenya, 2012b). This shows that evolutionaryconstraints on picornavirus proteins act not only in the dimension of amino acid substitutions but also in a second one, defined by insertions and deletions.

Thesee striking observations were supported by a similar analysis of nidoviruses, in which the combined sizes of key replicative enzymes (those encoded in ORF1b) within a family or even several families were found to be most strongly constrained as well (Lauber *et al.,* 2013). Moreover, the combined set of these ORF1b-encoded proteins expanded first in the transition from small nidovirus genomes of at most 16 kb (represented by contemporary arteriviruses) to large genomes of at least 20 kb (mesoni-, roni-, toro-, and coronaviruses). Specifically, the above mentioned exonuclease and two methyltransferases have been inserted in ORF1b (Chen *et al.,* 2009; Snijder *et al.,* 2003). This supports the dominant role of replicative proteins in the control of gene/genome length during RNA virus evolution. However, a second stage of expansion (from 20 to 26 kb) predominantly involved the ORF1a, indicating that another, still illusive factor was acquired that allowed some nidoviruses to expand their genomes even further (Lauber *et al.*, 2013). We proposed that the ancestor of the *Coronaviridae* might have acquired an ADP-ribose-1"-phosphatase (ADRP) that was universally present in these but not other nidoviruses. However, the ADRP is apparently missing in a new python virus, which was tentatively placed into the *Torovirinae* subfamily of the *Coronaviridae* family (Stenglein *et al.,* 2014). Only after this second stage, the third main genome region – the 3'-proximal ORFs (3'ORFs) – expanded through the acquisition of genes with diverse, often unknown functions that may vary even between closely related nidoviruses (Gorbalenya *et al.,* 2006). In summary, these findings suggest a functional hierarchy of the three genome regions (ORF1b, ORF1a, and 3'ORFs) in the control of gene and genome size during evolution. Importantly, the three regions are characterized not only by different expression mechanisms which results in unequal molecular amounts of protein products, but also by different degrees of genetic divergence (proteins encoded in ORF1b and 3'ORFs show the highest and lowest sequence conservation, respectively). Additionally , the above hierarchical model defines universal constraints that have acted independently and simultaneously on each nidovirus lineage during evolution, while recognizing that contemporary nidoviruses may have reached different points on this trajectory of genome expansion (Lauber et al., 2013). Arteriviruses, for example, seem to be unable to overcome a barrier to genome size at around 16 kb due to missing factors in ORF1b that includes the exonuclease, whereas mesoniviruses, frozen in a stage of intermediate genome size of around 20 kb, are lacking a different factor predicted to be located in ORF1a.

Besides bringing important fundamental insights, the relationship of sequence conservation and size variation has immediate practical implications. For instance, it could provide guidance for key decisions in genetic engineering experiments, as it predicts where the insertion of the gene of interest will likely compromise the virus (namely in ORF1b) and where it will not (3'ORFs and, possibly, ORF1a). Nidoviruses at different points of the genome expansion trajectory may differ in this respect. An analogous reasoning can be applied to the L and 2A regions of the picornavirus genome, which show a large diversity of encoded proteins and, consequently, the largest tolerance of size variation (Gorbalenya and Lauber, 2010). Moreover, this relation and the resulting practical implications can be extended to the level of a single gene/protein, as shown in a recent study of poliovirus, where it was found that the insertion of short nucleotide stretches is only tolerated at gene regions of low sequence conservation (Teterina *et al.,* 2011). These insertions could be used for tagging specific proteins in order to purify the protein or associated protein complexes from infected cells, or to visualize the location and dynamics of a protein over time.

Future studies should be devoted to the analysis of additional RNA virus families in order to verify the observed correlation genetic size variation and sequence conservation beyond picorna- and nidoviruses. Moreover, further insights into the emergence of the largest known RNA genomes employed by nidoviruses might be gained by including viruses not belonging to the order. In this respect it would be natural to consider barna-, sobemo-, luteo- and astroviruses (Gorbalenya *et al.*, 2006) since they show the same genomic organization (ORF1a, ORF1b, 3'ORFs) but have much smaller genomes than nidoviruses (Figure 7.2). Equally important would be a broader coverage of the natural diversity within the order *Nidovirales*, especially in the genome size range between arteri- and roniviruses. This may also help to determine the additional factor(s), predicted to be located in ORF1a, which allowed roni-, toro- and coronaviruses to expand their genomes beyond that of mesoniviruses. Such analyses could finally help to determine the ultimate upper limit of the RNA (virus) genome size, which, in turn, would contribute to our understanding of fundamental evolutionary processes like the proposed transition from RNA- to DNA-based life forms (Forterre, 2006; Gilbert, 1986).

Acknowledgement

This review is based on a chapter of the PhD thesis of CL prepared during his studies in the group of AEG. The authors are indebted to many colleagues who helped to conduct their research cited in this review. The financial support for this research was received from the Netherlands Bioinformatics Centre, the European Union Framework Programs 6 and 7; the Collaborative Agreement in Bioinformatics between Leiden University Medical Center and Moscow State University (MoBiLe project), and the Leiden University Fund. AEG is member of the Netherlands Bioinformatics Center (NBIC) Faculty.

References

Adams, M.J., Antoniw, J.F., Bar-Joseph, M., Brunt, A.A., Candresse, T., Foster, G.D., Martelli, G.P., Milne, R.G., and Fauquet, C.M. (2004). The new plant virus family Flexiviridae and assessment of molecular criteria for species demarcation. Arch Virol *149*, 1045-1060.

Adams, M.J., Antoniw, J.F., and Fauquet, C.M. (2005). Molecular criteria for genus and species discrimination within the family Potyviridae. Arch Virol *150*, 459-479.

Adams, M.J., Hendrickson, R.C., Dempsey, D.M., and Lefkowitz, E.J. (2015). Tracking the changes in virus taxonomy. Arch Virol *160*, 1375-1383.

Bao, Y., Chetvernin, V., and Tatusova, T. (2012). PAirwise Sequence Comparison (PASC) and its application in the classification of filoviruses. Viruses *4*, 1318-1327.

Bao, Y., Kapustin, Y., and Tatusova, T. (2008). Virus Classification by Pairwise Sequence Comparison (PASC). In Encyclopedia of Virology, 5 vols, B.W.J. Mahy, and M.H.V. Van Regenmortel, eds. (Oxford, Elsevier. Vol. 5), pp. 342-348.

Bao, Y.M., Federhen, S., Leipe, D., Pham, V., Resenchuk, S., Rozanov, M., Tatusov, R., and Tatusova, T. (2004). National Center for Biotechnology Information Viral Genomes Project. J Virol *78*, 7291-7298.

Belshaw, R., Gardner, A., RarnbaUt, A., and Pybus, O.G. (2008). Pacing a small cage: mutation and RNA viruses. Trends in Ecology & Evolution *23*, 188-193.

Belshaw, R., Pybus, O.G., and Rambaut, A. (2007). The evolution of genome compression and genomic novelty in RNA viruses. Genome Research *17*, 1496-1504.

Benson, D.A., Karsch-Mizrachi, I., Lipman, D.J., Ostell, J., and Sayers, E.W. (2010). GenBank. Nucleic Acids Res *38*, D46-D51.

Benveniste, R.E., and Todaro, G.J. (1974). Evolution of C-type viral genes: inheritance of exogenously acquired viral genes. Nature *252*, 456-459.

Bernard, H.U., Burk, R.D., Chen, Z.G., van Doorslaer, K., zur Hausen, H., and de Villiers, E.M. (2010). Classification of papillomaviruses (PVs) based on 189 PV types and proposal of taxonomic amendments. Virology *401*, 70-79.

Bhatt, S., Katzourakis, A., and Pybus, O.G. (2010). Detecting natural selection in RNA virus populations using sequence summary statistics. Infection, genetics and evolution : journal of molecular epidemiology and evolutionary genetics in infectious diseases *10*, 421-430.

Bishop, D.H.L. (1985). The Genetic-Basis for Describing Viruses As Species. Intervirology *24*, 79-93.

Bodewes, R., Lempp, C., Schurch, A.C., Habierski, A., Hahn, K., Lamers, M., von Dornberg, K., Wohlsein, P., Drexler, J.F., Haagmans, B.L., *et al.* (2014). Novel divergent nidovirus in a python with pneumonia. The Journal of general virology *95*, 2480-2485.

Bouckaert, R., Heled, J., Kuhnert, D., Vaughan, T., Wu, C.H., Xie, D., Suchard, M.A., Rambaut, A., and Drummond, A.J. (2014). BEAST 2: a software platform for Bayesian evolutionary analysis. PLoS Comput Biol *10*, e1003537.

Bouvet, M., Imbert, I., Subissi, L., Gluasis, L., Canard, B., and Decroly, E. (2012). RNA 3'-end mismatch excision by the severe acute respiratory syndrome coronavirus nonstructural protein nsp10/nsp14 exoribonuclease complex. Proc Natl Acad Sci U S A *109*, 9372-9377.

Bull, J.J., Sanjuán, R., and Wilke, C.O. (2007). Theory of lethal mutagenesis for viruses. J Virol *81*, 2930-2939.

Casiraghi, M., Labra, M., Ferri, E., Galimberti, A., and De Mattia, F. (2010). DNA barcoding: a six-question tour to improve users' awareness about the method. Briefings in Bioinformatics *11*, 440-453.

Chan, Y.F., Sam, I.C., and Abubakar, S. (2010). Phylogenetic designation of enterovirus 71 genotypes and subgenotypes using complete genome sequences. Infect Genet Evol *10*, 404-412.

Chen, Y., Cai, H., Pan, J., Xiang, N., Tien, P., Ahola, T., and Guo, D.Y. (2009). Functional screen reveals SARS coronavirus nonstructural protein nsp14 as a novel cap N7 methyltransferase. Proc Natl Acad Sci U S A *106*, 3484-3489.

Chirico, N., Vianelli, A., and Belshaw, R. (2010). Why genes overlap in viruses. Proceedings of the Royal Society B-Biological Sciences *277*, 3809-3817.

Coyne, J.A., and Orr, H.A. (2004). Speciation (Sunderland, Massachusetts, U.S.A., Sinauer Associates, Inc.).

Cui, J., and Holmes, E.C. (2012). Endogenous RNA viruses of plants in insect genomes. Virology *427*, 77-79.

de Groot, R.J., Baker, S.C., Baric, R., Enjuanes, L., Gorbalenya, A.E., Holmes, K.V., Perlman, S., Poon, L.L., Rottier, P.J.M., Talbot, P.J., *et al.* (2012). Family Coronaviridae In Virus Taxonomy, Ninth Report of the International Committee on Taxonomy of Viruses, A.M.Q. King, M.J. Adams, E.B. Carstens, and E.J. Lefkowitz, eds. (Amsterdam, Elsevier Academic Press), pp. 806-828.

de Villiers, E.M., Fauquet, C., Broker, T.R., Bernard, H.U., and zur Hausen, H. (2004). Classification of papillomaviruses. Virology *324*, 17-27.

Deng, Z., Lehmann, K.C., Li, X., Feng, C., Wang, G., Zhang, Q., Qi, X., Yu, L., Zhang, X., Feng, W., *et al.* (2014). Structural basis for the regulatory function of a complex zinc-binding domain in a replicative arterivirus helicase resembling a nonsense-mediated mRNA decay helicase. Nucleic Acids Res *42*, 3464-3477.

Dobzhansky, T. (1937). Genetics and the Origin of Species (Columbia University Press).

Drake, J.W., and Holland, J.J. (1999). Mutation rates among RNA viruses. Proc Natl Acad Sci U S A *96*, 13910-13913.

Drummond, A.J., Ho, S.Y.W., Phillips, M.J., and Rambaut, A. (2006). Relaxed phylogenetics and dating with confidence. Plos Biology *4*, 699-710.

Duffy, S., Shackelton, L.A., and Holmes, E.C. (2008). Rates of evolutionary change in viruses: patterns and determinants. Nature Reviews Genetics *9*, 267-276.
Eid, J., Fehr, A., Gray, J., Luong, K., Lyle, J., Otto, G., Peluso, P., Rank, D., Baybayan, P., Bettman, B., *et al.* (2009). Real-time DNA sequencing from single polymerase molecules. Science *323*, 133-138.
Eigen, M. (2002). Error catastrophe and antiviral strategy. Proc Natl Acad Sci U S A *99*, 13374-13376.
Elena, S.F., and Sanjuán, R. (2005). Adaptive value of high mutation rates of RNA viruses: Separating causes from consequences. J Virol *79*, 11555-11558.
Fauquet, C.M., and Stanley, J. (2003). Geminivirus classification and nomenclature: progress and problems. Ann Appl Biol *142*, 165-189.
Feinstone, S.M., Kapikian, A.Z., and Purcell, R.H. (1973). Hepatitis A - Detection by Immune Electron-Microscopy of A Viruslike Antigen Associated with Acute Illness. Science *182*, 1026-1028.
Firth, A.E. (2014). Mapping overlapping functional elements embedded within the protein-coding regions of RNA viruses. Nucleic Acids Res *42*, 12425-12439.
Firth, A.E., and Brown, C.M. (2006). Detecting overlapping coding sequences in virus genomes. BMC Bioinformatics *7*, 75.
Forterre, P. (2006). The origin of viruses and their possible roles in major evolutionary transitions. Virus Res *117*, 5-16.
Gilbert, C., and Feschotte, C. (2010). Genomic fossils calibrate the long-term evolution of hepadnaviruses. Plos Biology *8*.
Gilbert, W. (1986). Origin of Life - the RNA World. Nature *319*, 618-618.
Gonzaalez, J.M., Gómez-Puertas, P., Cavanagh, D., Gorbalenya, A.E., and Enjuanes, L. (2003). A comparative sequence analysis to revise the current taxonomy of the family Coronaviridae. Arch Virol *148*, 2207-2235.
Gorbalenya, A.E., Enjuanes, L., Ziebuhr, J., and Snijder, E.J. (2006). Nidovirales: Evolving the largest RNA virus genome. Virus Res *117*, 17-37.
Gorbalenya, A.E., and Koonin, E. (1993). Comparative analysis of amino-acid sequences of key enzymes of replication and expression of positive-strand RNA viruses: validity of approach and functional and evolutionary implications. SovSciRevD PhysicochemBiol *11*, 1-84.
Gorbalenya, A.E., and Koonin, E.V. (1989). Viral-Proteins Containing the Purine Ntp-Binding Sequence Pattern. Nucleic Acids Res *17*, 8413-8440.
Gorbalenya, A.E., and Lauber, C. (2010). Origin and Evolution of the Picornaviridae Proteome. In The Picornaviruses, E. Ehrenfeld, E. Domingo, and R.P. Roos, eds. (Washington, ASM Press), pp. 253-270.
Gorbalenya, A.E., Lieutaud, P., Harris, M.R., Coutard, B., Canard, B., Kleywegt, G.J., Kravchenko, A.A., Samborskiy, D.V., Sidorov, I.A., Leontovich, A.M., et al. (2010). Practical application of bioinformatics by the multidisciplinary VIZIER consortium. Antivir Res *87*, 95-110.
Hanada, K., Suzuki, Y., and Gojobori, T. (2004). A large variation in the rates of synonymous substitution for RNA viruses and its relationship to a diversity of viral infection and transmission modes. Molecular Biology and Evolution *21*, 1074-1080.
Hebert, P.D.N., Ratnasingham, S., and deWaard, J.R. (2003). Barcoding animal life: cytochrome c oxidase subunit 1 divergences among closely related species. Proceedings of the Royal Society of London Series B-Biological Sciences *270*, S96-S99.
Hendry, A.P. (2009). Evolutionary Biology Speciation. Nature *458*, 162-164.
Holland, J., Spindler, K., Horodyski, F., Grabau, E., Nichol, S., and VandePol, S. (1982). Rapid evolution of RNA genomes. Science *215*, 1577-1585.
Holmes, E.C. (2003a). Error thresholds and the constraints to RNA virus evolution. Trends in Microbiology *11*, 543-546.
Holmes, E.C. (2003b). Molecular clocks and the puzzle of RNA virus origins. J Virol *77*, 3893-3897.

Holmes, E.C. (2011). What Does Virus Evolution Tell Us about Virus Origins? J Virol *85*, 5247-5251.

Hraber, P., Kuiken, C., Waugh, M., Geer, S., Bruno, W.J., and Leitner, T. (2008). Classification of hepatitis C virus and human immunodeficiency virus-1 sequences with the branching index. J Gen Virol *89*, 2098-2107.

Jaenisch, R. (1976). Germ line integration and Mendelian transmission of the exogenous Moloney leukemia virus. Proc Natl Acad Sci U S A *73*, 1260-1264.

Jiang, P., Faase, J.A.J., Toyoda, H., Paul, A., Wimmer, E., and Gorbalenya, A.E. (2007). Evidence for emergence of diverse polioviruses from C-cluster coxsackie A viruses and implications for global poliovirus eradication. Proc Natl Acad Sci U S A *104*, 9457-9462.

Kadare, G., and Haenni, A.L. (1997). Virus-encoded RNA helicases. J Virol *71*, 2583-2590.

Kampstra, P. (2008). Beanplot: A Boxplot Alternative for Visual Comparison of Distributions. Journal of Statistical Software, Code Snippets *28*, 1-9.

Katzourakis, A., and Gifford, R.J. (2010). Endogenous viral elements in animal genomes. PLoS Genet *6*, e1001191.

Keese, P.K., and Gibbs, A. (1992). Origins of Genes - Big-Bang Or Continuous Creation. Proc Natl Acad Sci U S A *89*, 9489-9493.

King, A.M.Q., Adams, M.J., Carstens, E.B., and Lefkowitz, E.J. (2012). Virus Taxonomy, Ninth Report of the International Committee on Taxonomy of Viruses (Academic Press).

Kingsbury, D.W. (1985). Species Classification Problems in Virus Taxonomy. Intervirology *24*, 62-70.

Knowles, N.J., Hovi, T., Hyypia, T., King, A.M.Q., Lindberg, A.M., Pallansch, M.A., Palmenberg, A.C., Simmonds, P., Skern, T., Stanway, G.*, et al.* (2012). Family Picornaviridae In Virus Taxonomy, Ninth Report of the International Committee on Taxonomy of Viruses, A.M.Q. King, M.J. Adams, E.B. Carstens, and E.J. Lefkowitz, eds. (Academic Press), pp. 855-880.

Koonin, E.V., Dolja, V.V., and Krupovic, M. (2015). Origins and evolution of viruses of eukaryotes: The ultimate modularity. Virology, *479-480*, 2-25.

Koonin, E.V., and Gorbalenya, A.E. (1989). Evolution of RNA genomes: does the high mutation rate necessitate high rate of evolution of viral proteins? Journal of Molecular Evolution *28*, 524-527.

Krakauer, D.C. (2000). Stability and evolution of overlapping genes. Evolution *54*, 731-739.

Kuchibhatla, D.B., Sherman, W.A., Chung, B.Y., Cook, S., Schneider, G., Eisenhaber, B., and Karlin, D.G. (2014). Powerful sequence similarity search methods and in-depth manual analyses can identify remote homologs in many apparently "orphan" viral proteins. J Virol *88*, 10-20.

Lander, E.S. (2011). Initial impact of the sequencing of the human genome. Nature *470*, 187-197.

Lauber, C., Goeman, J.J., Parquet, M.D., Nga, P.T., Snijder, E.J., Morita, K., and Gorbalenya, A.E. (2013). The Footprint of Genome Architecture in the Largest Genome Expansion in RNA Viruses. PLoS Pathog *9*.

Lauber, C., and Gorbalenya, A.E. (2012a). Genetics-Based Classification of Filoviruses Calls for Expanded Sampling of Genomic Sequences. Viruses-Basel *4*, 1425-1437.

Lauber, C., and Gorbalenya, A.E. (2012b). Partitioning the Genetic Diversity of a Virus Family: Approach and Evaluation through a Case Study of Picornaviruses. J Virol *86*, 3890-3904.

Lauber, C., and Gorbalenya, A.E. (2012c). Toward Genetics-Based Virus Taxonomy: Comparative Analysis of a Genetics-Based Classification and the Taxonomy of Picornaviruses. J Virol *86*, 3905-3915.

Lauber, C., Ziebuhr, J., Junglen, S., Drosten, C., Zirkel, F., Nga, P.T., Morita, K., Snijder, E.J., and Gorbalenya, A.E. (2012). Mesoniviridae: a proposed new family in the order

Nidovirales formed by a single species of mosquito-borne viruses. Arch Virol, *157*, 1623-1628

Lefebure, T., Douady, C.J., Gouy, M., and Gibert, J. (2006). Relationship between morphological taxonomy and molecular divergence within Crustacea: Proposal of a molecular threshold to help species delimitation. Molecular Phylogenetics and Evolution *40*, 435-447.

Lefkowitz, E.J., Wang, C., and Upton, C. (2006). Poxviruses: past, present and future. Virus Res *117*, 105-118.

Lehmann, K.C., Snijder, E.J., Posthuma, C.C., and Gorbalenya, A.E. (2015). What we know but do not understand about nidovirus helicases. Virus Res, 202, 12-32.

Liu, D.X., Fung, T.S., Chong, K.K., Shukla, A., and Hilgenfeld, R. (2014). Accessory proteins of SARS-CoV and other coronaviruses. Antivir Res *109*, 97-109.

Lukashev, A.N. (2010). Recombination among picornaviruses. Reviews in Medical Virology *20*, 327-337.

Maes, P., Klempa, B., Clement, J., Matthijnssens, J., Gajdusek, D.C., Kruger, D.H., and Van Ranst, M. (2009). A proposal for new criteria for the classification of hantaviruses, based on S and M segment protein sequences. Infect Genet Evol *9*, 813-820.

Matthijnssens, J., Ciarlet, M., Heiman, E., Arijs, I., Delbeke, T., McDonald, S.M., Palombo, E.A., Iturriza-Gomara, M., Maes, P., Patton, J.T., *et al.* (2008). Full genome-based classification of rotaviruses reveals a common origin between human Wa-like and porcine rotavirus strains and human DS-1-like and bovine rotavirus strains. J Virol *82*, 3204-3219.

McGeoch, D.J., Cook, S., Dolan, A., Jamieson, F.E., and Telford, E.A. (1995). Molecular phylogeny and evolutionary timescale for the family of mammalian herpesviruses. J Mol Biol *247*, 443-458.

McKnight, K.L., and Lemon, S.M. (1998). The rhinovirus type 14 genome contains an internally located RNA structure that is required for viral replication. Rna *4*, 1569-1584.

Meyer, C.P., and Paulay, G. (2005). DNA barcoding: Error rates based on comprehensive sampling. Plos Biology *3*, 2229-2238.

Milne, R.G. (1984). The Species Problem in Plant Virology. Microbiological Sciences *1*, 113-117.

Minskaia, E., Hertzig, T., Gorbalenya, A.E., Campanacci, V., Cambillau, C., Canard, B., and Ziebuhr, J. (2006). Discovery of an RNA virus 3 '-> 5 ' exoribonuclease that is critically involved in coronavirus RNA synthesis. Proc Natl Acad Sci U S A *103*, 5108-5113.

Moratorio, G., Costa-Mattioli, M., Piovani, R., Romero, H., Musto, H., and Cristina, J. (2007). Bayesian coalescent inference of hepatitis A virus populations: evolutionary rates and patterns. J Gen Virol *88*, 3039-3042.

Nga, P.T., Parquet, M.D., Lauber, C., Parida, M., Nabeshima, T., Yu, F.X., Thuy, N.T., Inoue, S., Ito, T., Okamoto, K., *et al.* (2011). Discovery of the First Insect Nidovirus, a Missing Evolutionary Link in the Emergence of the Largest RNA Virus Genomes. PLoS Pathog *7*, e1002215.

Oberste, M.S., Maher, K., Kilpatrick, D.R., and Pallansch, M.A. (1999). Molecular evolution of the human enteroviruses: Correlation of serotype with VP1 sequence and application to picornavirus classification. J Virol *73*, 1941-1948.

Pons, J., Barraclough, T.G., Gómez-Zurita, J., Cardoso, A., Duran, D.P., Hazell, S., Kamoun, S., Sumlin, W.D., and Vogler, A.P. (2006). Sequence-based species delimitation for the DNA taxonomy of undescribed insects. Systematic Biology *55*, 595-609.

Pringle, C.R. (1991). The 20Th Meeting of the Executive-Committee of the International-Committee-On-Virus-Taxonomy - Virus Species, Higher Taxa, A Universal Virus Database, and Other Matters. Arch Virol *119*, 303-304.

Pybus, O.G., and Rambaut, A. (2009). Evolutionary analysis of the dynamics of viral infectious disease. Nat Rev Genet *10*, 540-550.

Rach, J., DeSalle, R., Sarkar, I.N., Schierwater, B., and Hadrys, H. (2008). Character-based DNA barcoding allows discrimination of genera, species and populations in Odonata. Proceedings of the Royal Society B-Biological Sciences *275*, 237-247.

Rothberg, J.M., Hinz, W., Rearick, T.M., Schultz, J., Mileski, W., Davey, M., Leamon, J.H., Johnson, K., Milgrew, M.J., Edwards, M., *et al.* (2011). An integrated semiconductor device enabling non-optical genome sequencing. Nature *475*, 348-352.

Sabath, N., Wagner, A., and Karlin, D. (2012). Evolution of viral proteins originated de novo by overprinting. Molecular Biology and Evolution *29*, 3767-3780.

Sanchez, G., Bosch, A., Gómez-Mariano, G., Domingo, E., and Pinto, R.M. (2003). Evidence for quasispecies distributions in the human hepatitis A virus genome. Virology *315*, 34-42.

Sarkar, I.N., Thornton, J.W., Planet, P.J., Figurski, D.H., Schierwater, B., and DeSalle, R. (2002). An automated phylogenetic key for classifying homeoboxes. Molecular Phylogenetics and Evolution *24*, 388-399.

Schuffenecker, I., Ando, T., Thouvenot, D., Lina, B., and Aymard, M. (2001). Genetic classification of "Sapporo-like viruses". Arch Virol *146*, 2115-2132.

Shukla, D.D., and Ward, C.W. (1988). Amino-Acid Sequence Homology of Coat Proteins As A Basis for Identification and Classification of the Potyvirus Group. J Gen Virol *69*, 2703-2710.

Smith, D.B., Bukh, J., Kuiken, C., Muerhoff, A.S., Rice, C.M., Stapleton, J.T., and Simmonds, P. (2014a). Expanded classification of hepatitis C virus into 7 genotypes and 67 subtypes: updated criteria and genotype assignment web resource. Hepatology *59*, 318-327.

Smith, D.B., Simmonds, P., Jameel, S., Emerson, S.U., Harrison, T.J., Meng, X.J., Okamoto, H., Van der Poel, W.H., and Purdy, M.A. (2014b). Consensus proposals for classification of the family Hepeviridae. The Journal of general virology *95*, 2223-2232.

Smith, E.C., Sexton, N.R., and Denison, M.R. (2014c). Thinking Outside the Triangle: Replication Fidelity of the Largest RNA Viruses. Ann Rev Virol *1*, 111-132.

Snijder, E.J., Bredenbeek, P.J., Dobbe, J.C., Thiel, V., Ziebuhr, J., Poon, L.L.M., Guan, Y., Rozanov, M., Spaan, W.J.M., and Gorbalenya, A.E. (2003). Unique and conserved features of genome and proteome of SARS-coronavirus, an early split-off from the coronavirus group 2 lineage. J Mol Biol *331*, 991-1004.

Stenglein, M.D., Jacobson, E.R., Wozniak, E.J., Wellehan, J.F., Kincaid, A., Gordon, M., Porter, B.F., Baumgartner, W., Stahl, S., Kelley, K., *et al.* (2014). Ball python nidovirus: a candidate etiologic agent for severe respiratory disease in Python regius. mBio *5*, e01484-01414.

Teterina, N.L., Lauber, C., Jensen, K.S., Levenson, E.A., Gorbalenya, A.E., and Ehrenfeld, E. (2011). Identification of tolerated insertion sites in poliovirus non-structural proteins. Virology *409*, 1-11.

Tibayrenc, M. (2006). The species concept in parasites and other pathogens: a pragmatic approach? Trends in Parasitology *22*, 66-70.

Van Regenmortel, M.H.V. (2003). Viruses are real, virus species are man-made, taxonomic constructions. Arch Virol *148*, 2481-2488.

Van Regenmortel, M.H.V. (2007). Virus species and virus identification: Past and current controversies. Infect Genet Evol *7*, 133-144.

Volz, E.M., Koelle, K., and Bedford, T. (2013). Viral phylodynamics. PLoS Comput Biol *9*, e1002947.

Walker, P.J., Firth, C., Widen, S.G., Blasdell, K.R., Guzman, H., Wood, T.G., Paradkar, P.N., Holmes, E.C., Tesh, R.B., and Vasilakis, N. (2015). Evolution of genome size and complexity in the rhabdoviridae. PLoS Pathog *11*, e1004664.

Wesche, P.L., Gaffney, D.J., and Keightley, P.D. (2004). DNA sequence error rates in Genbank records estimated using the mouse genome as a reference. Dna Sequence *15*, 362-364.

Zheng, D.P., Ando, T., Fankhauser, R.L., Beard, R.S., Glass, R.I., and Monroe, S.S. (2006). Norovirus classification and proposed strain nomenclature. Virology *346*, 312-323.

Zhou, Y., Qin, T., Xiao, Y., Qin, F., Lei, C., and Sun, X. (2014). Genomic and biological characterization of a new cypovirus isolated from Dendrolimus punctatus. Plos One *9*, e113201.

8

Understanding Adaptation Through Experimental Evolution with Viruses: From Simple to Complex Environments

Valerie J. Morley and Paul E. Turner*

Department of Ecology & Evolutionary Biology, Yale University, 165 Prospect St, New Haven, CT 06520, USA

**corresponding author: paul.turner@yale.edu*

Abstract

A central goal of evolutionary biology is to better understand the processes that promote versus constrain adaptation in evolving populations; this aim is crucial for refining predictions on whether or not populations will successfully adapt to environmental change. Experimental evolution of viruses in controlled laboratory settings provides a powerful method for investigating generalized theories of adaptive evolution, and a vital approach for testing particular predictions on virus adaptive potential. Examples include the use of viruses to elucidate the dynamics of clonal interference between beneficial mutations vying to undergo fixation, the prevalence and character of epistatic interactions between genes, and the relative potencies of natural selection versus genetic drift in dictating evolutionary outcomes. While these efforts have helped refine predictions on adaptation occurring in constant environments, adaptation in complex or heterogeneous environments remains relatively seldom studied. Although organisms commonly encounter environments that are spatially and/or temporally complex, investigating the effects of environmental heterogeneity on evolution proves challenging in most study systems. Recent studies demonstrate that experimental evolution in viruses is a tractable and rigorous approach for resolving how evolution proceeds in heterogeneous environments. In this chapter, we review virus experiments that illuminate how basic evolutionary forces – especially selection, drift and migration – operate to dictate the fate of alleles and populations in simple and complex environments. We also suggest many exciting and open questions that remain to be explored using virus experimental models, such as empirical mapping of fitness landscapes, and the role of heterogeneous environments in promoting versus constraining virus adaptation.

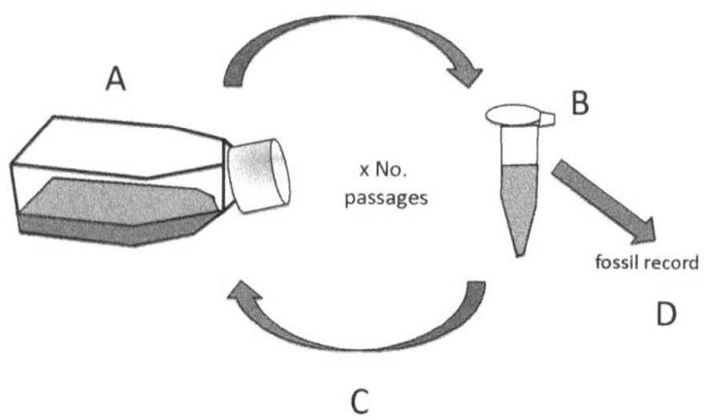

Figure 8.1. Experimental evolution by serial passage. **A.** Cells are infected from a virus stock or an infectious virus clone, and the infected cells are incubated for a set period of time (e.g., 24 or 48 hours). **B.** Virus progeny is harvested from the flask. **C.** A subset of the harvested virus is used to infect a fresh flask of cells, thereby initiating the next passage. **D.** The remainder of the harvested virus can be frozen at each passage to create a "fossil record" through time, allowing direct comparisons between an ancestral virus and its evolved derivatives.

Introduction

With his publication of *The Origin of Species* in 1859, Darwin became the first to convincingly propose a mechanism of evolution—natural selection—that could explain the magnificent diversity of life. Darwin's theory engendered the notion that by describing the basic mechanisms of evolution, one could explain the origination of diversity and how organisms are shaped over generations by their environments. Moreover, evolutionary biology holds the tantalizing if elusive promise of growing into a predictive field, through which we might foresee how a bacterial population would respond to antibiotic use, or how algal populations might adapt in response to a changing climate. Historically, scientists have relied on comparative studies to investigate evolutionary questions. In this approach, living species and populations are compared to one another, and sometimes to fossils, to give insight into how populations have evolved over time. The advent of DNA sequencing technology revolutionized the study of evolutionary biology. Sequencing allowed populations and species to be compared at the genetic level, and phylogenetics emerged as a way to resolve evolutionary relationships. Most of our empirical knowledge of evolution in natural populations stems from these comparative and molecular approaches.

An alternative and complimentary approach to study evolution is to observe populations evolving in real time. Some studies have successfully documented contemporary evolution in natural populations of multicellular organisms, such as the famous and ongoing research on Darwin's finches in the Galapagos Islands (Grant and Grant, 2014). While these studies have provided invaluable insights, they are typically time and labor intensive. Moreover, it is simply impossible to observe a large number of generations for natural populations of many multicellular species within a researcher's lifetime. Experimental microbial populations, on the other hand, allow us to observe evolution happening in real time in replicated environments in the laboratory. Experimental evolution with microbial populations is a powerful tool for testing theoretical evolutionary predictions. Researchers have embraced bacteria, viruses, and single-celled eukaryotes such as yeasts and algae as models to study evolution in the laboratory, and these experimental systems have provided great insight into the basic mechanisms of evolution.

Viruses are extremely powerful models for experimental evolution (Figure 8.1). Both bacteriophages and viruses of eukaryotes have been used as experimental models, and these virus systems provide many compelling advantages. First of all, very large viral populations can be maintained in the laboratory. The arthropod-borne virus (arbovirus) *Vesicular stomatitis virus* (VSV), for example, reaches titers of about 10^9 virus particles per mL within 24 hours when grown on cells in laboratory tissue culture. Secondly, many model viruses have short generation times in the laboratory. A virus population may go through five generations in a 24-hour incubation period, a rate which allows 20 days of serial passaging to equal 100 virus generations (Turner and Chao, 1998). Third, RNA viruses have famously high mutation rates on the order of 10^{-3} to 10^{-5} misincorporations per nucleotide per replication (Drake, 1993; Holland *et al.,* 1982). This means that beneficial mutations, the raw material for adaptation by natural selection, enter the population more frequently than they do for DNA-based organisms. Fourth, viruses generally have extremely small genomes relative to cellular organisms. For known RNA viruses, the mean genome size is approximately 10,000 nucleotides. All described single-stranded DNA viruses also have small genomes, with none larger than 11,000 nucleotides. Double-stranded DNA viruses have a much wider range of genome sizes than RNA viruses, with the largest exceeding one-million nucleotides (Holmes, 2009). Working with small genomes makes it easily affordable to sequence whole genomes of replicate experimental strains. Many viruses can also be manipulated through site-directed mutagenesis, which allows researchers to engineer mutations of interest onto different genetic backgrounds to confirm fitness effects and interactions among mutations. Finally, viruses can be stored frozen and revived by thawing, permitting direct comparisons between evolved populations and their ancestors, and providing a means to "replay" evolution by founding new lines from a frozen sample to test the repeatability of adaptation.

Given these advantages, it is unsurprising that researchers have been extremely successful in using experimental viral systems to test theoretical predictions pertaining to basic evolutionary mechanisms. Experiments in virus evolution have led to insights into the dynamics of adaptation, the distribution of mutational fitness effects, the process of clonal interference, selective sweeps, the prevalence and character of epistatic interactions, and genetic drift. Some of the main questions motivating these studies are: Can adaptation continue indefinitely even in a constant environment? What is the distribution of mutational fitness effects? How reproducible is evolution? What types of molecular events are involved in adaptation, and what types of genes are affected? How common is genetic parallelism in populations evolved in replicate environments? How frequent are interactions between mutations, and how do these interactions shape adaptive trajectories? How rugged are fitness landscapes?

Both theoretical and empirical studies of adaptation have primarily focused on populations evolving in novel but constant environments. After an environmental change, an evolving population may suffer a decrease in fitness. The population is then expected to adapt to the new environment by fixing a series of beneficial mutations that move it toward a new fitness optimum. Thanks to innovative theoretical and experimental work, we have reached a point where we can make strong predictions about what an adaptive walk in a novel but constant environment should look like. Populations adapting to constant environments are

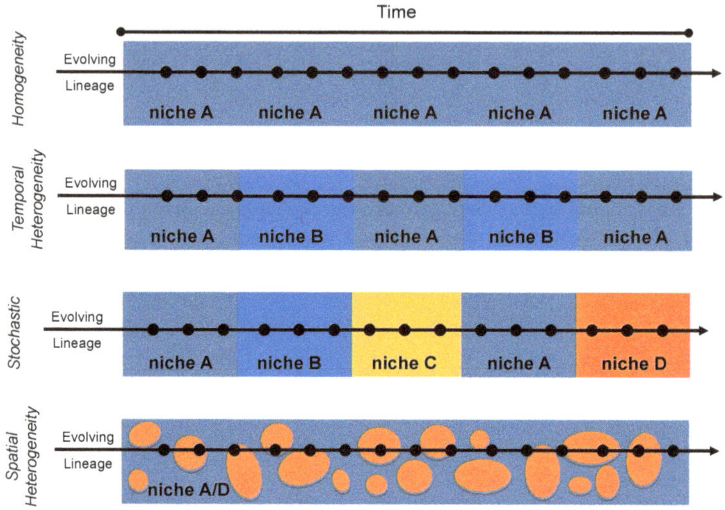

Figure 8.2. Evolution in constant versus heterogeneous environments. The pace of environmental turnover can fundamentally differ for populations with rapid generation times, experiencing homogeneous (constant) versus various types of heterogeneous environments. Filled circles represent generation times experienced by the evolving lineages. Whereas lineages evolving in homogeneous environments experience selection in only a single niche, those encountering temporally or spatially heterogeneous environments can experience predictable (e.g., seasonal) or stochastic niche turnover.

expected to first fix beneficial mutations of large effect, followed by beneficial mutations of decreasing effect size. This pattern of diminishing returns results in a rapid initial increase in fitness, followed by an approach toward a fitness plateau.

While adaptation to constant environments is well characterized, the effects of environmental complexity on adaptive dynamics remain seldom studied. The impact of environmental complexity is of great interest, because most organisms in nature navigate environments that are heterogeneous in space and time (Figure 8.2). Temporal heterogeneity may be predictable, such as the changing of the seasons and the shortening and lengthening of days, or it may be unpredictable, such as the day-to-day changes in weather conditions. Adaptive responses to temporal environmental heterogeneity are likely also influenced by the grain of these changes relative to the generation time of the organism. Organisms may respond very differently to environmental fluctuations that occur over the course of their lifetimes as opposed to fluctuations that span many generations. Organisms also face spatial structure in their environments. When a malarial parasite infects a human host, it enters through the bloodstream, must relocate to the liver, replicate in the liver, and then relocate to the bloodstream. After replicating in the bloodstream, the parasite is ingested by a mosquito vector, in which it must replicate while navigating through the diverse tissue types of the midgut and salivary glands. At a larger scale, the parasite deals with the spatial structure of the landscape, including different climatic conditions and host genotypes across its range. Despite the importance of environmental heterogeneity, it has received relatively little theoretical or experimental attention in the study of adaptation.

Environmental complexity is the norm, and evolutionary biologists are becoming increasingly interested in studying adaptation in spatially and temporally heterogeneous environments. Theoreticians have made some headway in modeling the effects of heterogeneity, but their predictions are challenging to test empirically. Experimental evolution with viruses provides a tractable way to investigate the complexities of environmental heterogeneity in controlled laboratory settings.

In this chapter, we focus on how experimental work with viruses has been used to elucidate basic principles of evolutionary biology. We discuss insights from experimental evolution with viruses concerning the dynamics of adaptation, clonal interference, and evolution in very small populations. We also discuss the use of viruses to experimentally characterize epistasis and to estimate empirical fitness landscapes. Then, we shift our focus to experimental evolution in complex environments, encompassing spatial and temporal heterogeneity and migration between variable environmental patches. Finally, we conclude by highlighting exciting open questions and future directions in the experimental study of virus evolution.

Dynamics of adaptation

An adaptive walk is a process through which a population increases its fitness in a given environment by fixing a series of beneficial mutations. Understanding the dynamics of adaptive walks is a central challenge in evolutionary biology and can elucidate the process by which populations respond to novel environmental challenges. Moreover, an ability to describe and predict how populations respond to environmental changes has important implications for applications as diverse as vaccine design, the use of antimicrobials, wildlife conservation, monitoring disease emergence, and predicting how populations might respond to climate change.

Advances in the theory of adaptation allow us to make strong predictions about the dynamics of an adaptive walk. For decades, scientific discussion of adaptive genetics was dominated by the infinitesimal model. The infinitesimal model posits that populations adapt to novel environments by fixing a multitude of beneficial alleles of very small effect. Although Darwin had no understanding of genes or genetics, he essentially was the first to make this argument when he hypothesized that natural selection acts upon very small phenotypic differences in a population, and thus adaptation must happen very gradually. In 1930 R. A. Fisher published *The Genetical Theory of Natural Selection*, an influential work that promoted the infinitesimal model (Fisher, 1930). Fisher showed that while mutations of very small phenotypic effect have a 50% probability of being beneficial, larger mutations have a rapidly decreasing probability of being favorable. Therefore, Fisher argued, adaptation must usually be based on the accumulation of mutations of very small effect. Fisher's model seemingly settled the debate over the genetics of adaptation, and the infinitesimal model became generally accepted.

While Fisher's argument was extremely influential, it was unfortunately incomplete in ways that would not be recognized for half a century. In 1983, Kimura pointed out a major oversight in Fisher's reasoning (Kimura, 1983). Many favorable mutations never go to fixation because they are lost early on due to stochastic processes. Although mutations of large phenotypic effect are less likely to be beneficial, they are more likely than smaller-effect mutations to sweep to fixation. Based on this observation, Kimura argued that mutations of intermediate phenotypic effect would be most important in adaptation. In 1992 Orr and Coyne published a paper arguing that there was little evidence, either theoretical or empirical, to support the infinitesimal view of adaptation (Orr and Coyne, 1992). Their call to action inspired a flurry of theoretical work that fundamentally changed our understanding of adaptive genetics. This newer theory has overturned the infinitesimal model by emphasizing the importance of beneficial mutations of large phenotypic effect in adaptation. This theory predicts that the distribution of mutational effects fixed over the course of an adaptive walk is roughly exponential, with a few mutations of large beneficial effect fixing early in the adaptive walk, followed by fixation of a larger number of mutations of smaller effect (Orr, 1998, 1999, 2002).

Theorists have provided compelling predictions for how adaptation should proceed, but empirical evidence is needed to support these claims. Experimental evolution with viruses provides a tractable way to test the predictions of adaptive theory. Here, experimental work on virus adaptation has focused on two questions: What is the distribution of mutational fitness effects? What are the dynamics of adaptation for a population in a novel environment?

The first step towards understanding adaptive walks is estimating the distribution of phenotypic effects of the factors fixed during an adaptive walk. Theory predicts that most mutations are deleterious or neutral, and that beneficial mutations of large effect are greatly outnumbered by beneficial mutations of small effect (Fisher, 1930). This prediction has been supported by experimental work in virus systems. Empiricists working with viruses have sought to quantify the proportion of mutations that are lethal, deleterious, and neutral, and to describe the distribution of beneficial fitness effects. These experiments have generally taken two approaches: mutation accumulation experiments and site-directed mutagenesis. In a mutation accumulation experiment, individual virus clones are repeatedly selected and propagated. In this extreme bottlenecking regime, genetic drift overwhelms selection, and non-lethal deleterious substitutions that might otherwise be lost to selection are observed to fix. This approach allows the experimenter to observe non-lethal mutations, while minimizing the effects of natural selection. Site-directed mutagenesis is a more comprehensive and rigorous approach. In this type of experiment, point mutations are introduced to randomly chosen sites in the virus genome, and the fitness effects of these mutations are recorded. This is an unbiased approach in which the proportion of lethal mutations can also be observed. From these experiments, a comprehensive picture of the virus mutational landscape is beginning to emerge.

For RNA and ssDNA viruses, it seems clear that the majority of mutations are lethal or deleterious. Site-directed mutagenesis studies have consistently found 20-40% of substitutions to be lethal (Carrasco *et al.*, 2007; Domingo-Calap *et al.*, 2009; Sanjuán, 2010; Sanjuán *et al.*, 2004b). Next generation sequencing analysis of the poliovirus mutant spectrum suggests that this number might be lower for synonymous substitutions, for which research found only 10% to be lethal (Acevedo *et al.*, 2014). Non-lethal mutations have been shown to reduce fitness by about 10-13% on average (Domingo-Calap *et al.*, 2009; Sanjuán, 2010). For VSV 30% of observed non-lethal mutations were deleterious and 27% were neutral (Sanjuán *et al.*, 2004b). These numbers are similar for *Tobacco etch virus* (TEV) for which 36% of viable mutations are deleterious and 27% are neutral (Carrasco *et al.*, 2007). These experimental results support theoretical predictions that most mutations are deleterious or neutral.

Despite their importance for adaptation, beneficial mutations are relatively rare. Only 4% of substitutions produced by site-directed mutagenesis in VSV were beneficial (Sanjuán *et al.*, 2004b). None of the experimental nucleotide substitutions in the TEV genome had a measurable beneficial effect *in vivo* (Carrasco *et al.*, 2007). As predicted by theory, experiments typically show that beneficial mutations of small effect are more common than those of large effect [although see (Rokyta *et al.*, 2008; Burch and Chao, 1999; Sanjuán, 2010; Sanjuán *et al.*, 2004b)]. It should be noted that similar studies in cellular organisms have found a much lower rate of deleterious and lethal mutations, suggesting that ssDNA and RNA viruses are less genetically robust (i.e., less capable of maintaining a constant phenotype when mutations enter the genome) (Domingo-Calap *et al.*, 2009; Elena *et al.*, 2006; Sanjuán, 2010). This sensitivity to mutation may result from viruses having small genomes that lack gene redundancy, and that often contain overlapping reading frames; these genetic architectures tend to prevent alternate genetic solutions and pathways as mechanisms for fostering robustness (Elena *et al.*, 2006).

Describing the distribution of mutational effects allows us to understand the raw material available to adapting populations. The next step is mapping the dynamics of adaptive walks observed in the laboratory. To accomplish this, replicate experimental virus populations are evolved in a novel environment, such as on a novel host type or at a challenging temperature, and changes in fitness are mapped through time (in generations). Results from such experiments allow us to make generalizations about virus adaptation in constant environments. First, adaptation to new environments initially occurs very rapidly, and then decelerates with time (Betancourt, 2009; Bull *et al.*, 1997; Elena *et al.*, 1998; Holder and Bull, 2001; Novella *et al.*, 1995a; Rokyta *et al.*, 2009; Wichman *et al.*, 1999). This finding supports theoretical models of adaptive walks, in which mutations of large beneficial effect are the first to be substituted, followed by mutations of smaller beneficial effect (Orr, 1998, 1999, 2002). Experimental populations consistently show large gains in fitness early on, with diminishing returns as the experiment progresses. However, data from one evolution experiment only supported Orr's model of adaptation after being adjusted for mutational bias (Rokyta *et al.*, 2005). This dataset suggests that the mutations of largest benefit may not always be the first to reach fixation, because not all mutations are equally likely to occur. Overall, virus populations have been shown to adapt through an initial rapid improvement, followed by diminishing returns as the population approaches a fitness plateau.

A surprising amount of genetic parallelism has been observed among replicate evolving virus lineages (Bull *et al.*, 1997; Cuevas *et al.*, 2002; Rokyta *et al.*, 2009; Wichman *et al.*, 1999). Genetic parallelism (convergence) occurs when the same substitutions in the nucleic acid sequence are fixed in multiple replicate lineages evolving under the same environmental conditions. In an experiment with multiple adapting lines of ssDNA phage, half of the substitutions observed occurred in more than one replicate lineage (Bull *et al.,* 1997). High levels of parallelism have even been reported across genotypes in adapting ssDNA phage lineages (Rokyta *et al.,* 2009). Parallel genetic changes provide strong evidence that the mutations of interest were adaptive.

Experimental evolution of RNA viruses has been invaluable in testing the predictions of adaptive theory. The distribution of mutational effects in viruses has shown to be dominated by lethal, deleterious, and neutral effects. Beneficial mutations are rare, and beneficial mutations of small effect outnumber those of large effect. Despite their rarity, beneficial mutations of large effect are expected to reach fixation early in adaptive walks due to their large selection coefficients. This leads to a characteristic pattern of adaptation in constant environmental conditions, in which populations show an initial rapid increase in fitness, followed by decelerating fitness increases.

Clonal interference

Populations adapt through the fixation of beneficial mutations. In large populations with high mutation rates, it is possible that multiple beneficial mutations will arise concurrently in different genetic backgrounds within the population. In an asexual population, these beneficial mutations cannot be recombined onto the same genetic background. Therefore, different beneficial mutations will compete with one another until one sweeps to fixation. This process is known as clonal interference, and it is invoked to explain how competition between beneficial mutations can slow the rate of adaptation. Gerrish and Lenski's theoretical model of clonal interference establishes a number of predictions for adapting asexual populations (Gerrish and Lenski, 1998): (i) The probability that a given mutation will go to fixation decreases with increasing population size and increasing mutation rate (ii) Beneficial mutations that reach fixation will have greater effect sizes with increasing population size and increasing mutation rate (iii) The rate of adaptation is an increasing but decelerating function of population size and mutation rate (iv) There should be numerous mutations that become transiently common but fail to eventually fix because they are outcompeted (v) Transient polymorphisms may result in the "leapfrog" effect, where the most common genotype at a given moment is less closely related to the immediately preceding one than to an earlier genotype.

Viruses provide a powerful model to test these predictions in the laboratory. Particularly useful are negative-sense RNA viruses, which rarely recombine, and therefore are near to truly asexually reproducing lineages. Several early experiments tested the above predictions by evolving populations of viruses at differing effective population sizes. Experiments with the dsRNA phage Φ6 and negative-sense ssRNA *Vesicular stomatitis virus* (VSV) both supported the prediction that increasing effective population size correlated positively with the effect size of fixed beneficial mutations (Burch and Chao, 1999; Miralles *et al.,* 1999a, 2000). Observations from VSV also support the prediction that larger population sizes correlate with increasing but decelerating rates of adaptation (Miralles *et al.,* 1999a, 2000).

While the above experimental work supports theoretical predictions for clonal interference, other experiments have also identified nuances that were less obvious in theoretical treatments. Models of clonal interference suggest that increasing the number of beneficial mutations simultaneously present in a population, either through increasing population size or mutation rate, should intensify interference and therefore retard adaptation. Experimental work with the RNA phage MS2 suggests that this may not always be the case. The Gerrish and Lenski model does not allow for multiple beneficial mutations to arise on a single genetic background before that lineage sweeps to fixation, although it should be noted that some models allow this possibility (Kim and Orr, 2005). RNA viruses have very high mutation rates and can exist in very large populations. If population size and mutation rate are high enough, beneficial mutations may arise often enough to occur on the same genetic background and sweep to fixation together. Bollback and Huelsenbeck (2007) find evidence for this in their experimental work with phage MS2, and argue that the combination of large population size and high mutation rate may ameliorate the effects of clonal interference. If this is the case, clonal interference may be most intense at intermediate population sizes and mutation rates.

While population size and mutation rate are known to be important determinants of the intensity of clonal interference, experimental work with the phage ΦX174 suggests that the selective environment may also be influential. Increasing population size and mutation rate both increase the number of beneficial mutations occurring simultaneously in a population, and therefore intensify competition among lineages. Pepin and Wichman (2008) demonstrate that the selective environment may also influence the number of beneficial mutations occurring concurrently in an adapting population. They evolved populations of phage ΦX174 in either a benign or harsh environment, based on the abundance of $CaCl_2$. They found that populations evolved in the harsh environment only fixed mutations in one portion of one gene and showed no evidence of clonal interference. Populations adapting to the benign environment, on the other hand, had a variety of substitutions in many genes and showed evidence of clonal interference. The authors surmise that fewer beneficial mutations may have been available in the harsh environment. This paucity of available beneficial mutations may have acted similarly to small population size or low mutation rate in decreasing the number of competing beneficial mutations and ameliorating the effects of clonal interference. This body of work is a good example of how experimental evolution with viruses can be used to test theoretical predictions for adaptation, and also to inform new directions and modifications to existing theory.

Epistasis and empirical fitness landscapes

Epistasis, or the dependence of mutational effects on genetic background, has implications for evolutionary theories including the origin and maintenance of sex (Kondrashov, 1988, 1993; Otto, 2009; Otto and Feldman, 1997), the predictability and repeatability of evolution (Blount *et al.*, 2008; Bridgham *et al.*, 2009), and the topography of fitness landscapes (Poelwijk *et al.*, 2011). Epistasis occurs when the effect of a mutation changes in magnitude or sign depending on its genetic context. For example, two mutations (A and B) that are each beneficial when they appear singly in a genome might be less beneficial in combination than their additive individual benefits would suggest ($AB < Ab + aB$).

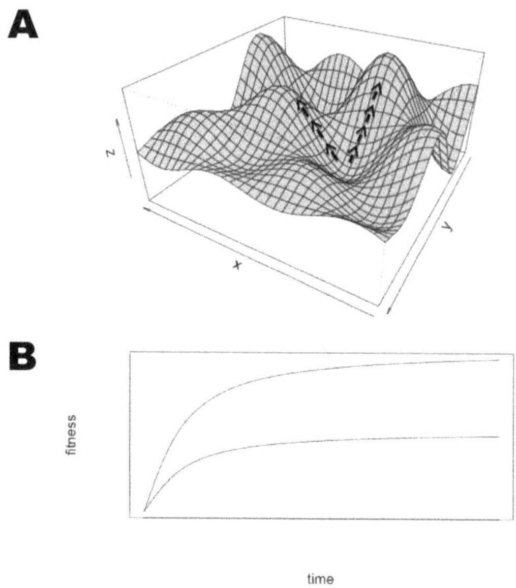

Figure 8.3. Fitness landscapes. **A.** A typical representation of a hypothetical fitness landscape, in which the x-y plane represents genotypic space, and fitness is height on the landscape mapped on the z-axis. The arrows represent two alternate adaptive trajectories, ending at different fitness peaks. **B.** Fitness trajectories of the two imagined populations in 2A. Both populations increase in fitness through time, but one population evolves to a lower fitness optimum (peak) than the other.

To visualize the effects of epistasis on adaptive trajectories, the concept of the fitness landscape is useful. Introduced by Sewall Wright (Wright, 1932a), the fitness landscape is a means to visualize the genotype-fitness map. The fitness landscape is classically illustrated as a three-dimensional mountainous landscape, in which genotypic space is represented on the x-y plane and fitness is represented on the z-axis (Figure 8.3). In this model, adapting populations "walk" through the landscape, moving to areas of genotypic space associated with higher fitness values. In the absence of epistasis, fitness landscapes are expected to be unimodal, with one fitness peak. In this scenario, a beneficial mutation is equally beneficial in any background, and so all populations should adapt towards the same peak even if they have different starting points. Wright recognized, however, that pervasive epistasis would make the fitness landscape rugged. A rugged landscape has multiple peaks. Because adapting populations can only be selected to move towards higher points on the fitness landscape, populations with different starting genotypes might end up at different local optima.

This idea was further developed with the formal introduction of the concept of sign epistasis (Weinreich *et al.,* 2005). Sign epistasis is an epistatic interaction that changes the sign of mutational effects on fitness such that a given mutation can be either deleterious or beneficial depending on its genetic background. Sign epistasis can limit the number of adaptive trajectories available to an evolving population (DePristo *et al.,* 2007; Weinreich *et al.,* 2006; Weinreich *et al.,* 2005). For example, certain genotypes can become stuck on local fitness optima because they are surrounded by genotypes of relatively low fitness. This leads to the exciting suggestion that by understanding how epistasis constrains adaptation, evolution could become more predictable (de Visser and Krug, 2014).

Despite the importance of epistasis and fitness landscapes, their implications remained largely theoretical for decades due to a lack of empirical data on the prevalence and character of epistatic interactions. While the fitness landscape is an intuitively appealing concept, it is extraordinarily challenging to gather large-scale data on the topography of empirical fitness landscapes. Finally, data from experimental work with viruses and other organisms are beginning to shed light on the nature of epistatic interactions and their implications for fitness landscapes. These data are crucial to inform theoretical work and simulated landscapes. Viruses, particularly RNA viruses, have compact genomes that commonly encode multifunctional proteins and overlapping coding regions (Holmes, 2009). This architecture may make them particularly subject to epistatic constraint. Experimental work on epistatic interactions in viruses has led to three main conclusions: (i) epistatic interactions are common and variable (ii) empirical landscapes are rugged, and epistasis limits the number of pathways available to adapting populations and (iii) the nature of epistatic interactions depends on the environment.

The most simple but rigorous way to test for epistasis is to select a pair of substitutions and then engineer them into a genome singly and in combination through site-directed mutagenesis. If the effect of the pair-wise interaction significantly deviates from the additive effects of the two mutations, the interaction is epistatic. These experiments can test for three key types of epistasis: positive (synergistic) epistasis, negative (antagonistic) epistasis, and sign epistasis. In a positive epistatic interaction, the effect of a double mutation is greater than expected based on the mutations' additive effects. In negative epistasis, the two mutations have a smaller effect in combination than their combined single effects. Sign epistasis, as described above, occurs when a mutation can be either beneficial or deleterious depending on its genetic background. Reciprocal sign epistasis occurs when two deleterious mutations are beneficial in combination, or when two beneficial mutations are deleterious in combination. Experimental work using site-directed mutagenesis to analyze pair-wise interactions shows that epistatic interactions are common in virus genomes (Caudle *et al.,* 2014; da Silva *et al.,* 2010; Parera and Martínez, 2014; Pepin and Wichman, 2007; Rokyta *et al.,* 2011; Sanjuán *et al.,* 2005, 2004a). Within a study, both antagonistic and synergistic interactions are typically observed. Sign epistasis has also been demonstrated (Rokyta *et al.,* 2011). Both intergenic (Sanjuán *et al.,* 2005; 2004a) and intragenic (da Silva *et al.,* 2010; Pepin and Wichman, 2007) epistasis have been reported. These experiments are informative for determining the prevalence and character of pair-wise epistatic interactions, but to truly understand the effects of epistasis on adaptation, larger data sets are required to move towards an understanding of the topography of fitness landscapes.

Two primary methods have been used to empirically map virus fitness landscapes. In the first method, a set of mutations is selected, and the fitness effects of these mutations are measured singly and in combination. This approach allows for systematic mapping of epistatic interactions for a limited set of mutations, and enables observation of pair-wise as well as higher-order interactions. This approach was employed to study a set of seven amino acid changes in the envelope glycoprotein of human immunodeficiency virus 1 (HIV-1) (da Silva *et al.*, 2010). A total of 53 mutants were tested, including single mutants and combinations of mutations. Fifty-six percent of interactions tested were significantly epistatic, and epistatic interactions were demonstrated for higher-order interactions. The highest-level significant net interaction occurred among five amino acids. This study suggests a considerably rugged landscape for the HIV-1 envelope glycoprotein, characterized by both pair-wise and higher-order epistatic interactions.

The second approach to mapping fitness landscapes combines high-throughput measurements of mutational effects with computational analyses to map genotype to phenotype. Hinkley *et al.* (2011) analyzed a dataset measuring the *in vitro* fitness of 70,081 virus samples isolated from HIV-1 infected individuals. From this large dataset, they estimated the fitness effects and pair-wise epistatic interactions of 1,859 single amino acid variants found within the HIV-1 protease and reverse transcriptase sequences. Fitness effects were measured across sixteen environments *in vitro*, consisting of a drug-free treatment and treatments including fifteen individual antiviral drugs. This approach allows the reconstruction of approximate fitness landscapes for these proteins. Across all environments, the fitness landscapes for these viral proteins are highly rugged and characterized by strong epistasis (Hinkley *et al.*, 2011; Kouyos *et al.*, 2012). This study is perhaps the first to offer a quantitative description of a large realistic and biologically relevant fitness landscape.

The rugged landscapes revealed by systematic and large-scale analyses of HIV-1 suggest that viral adaptation is constrained by epistasis. Epistatic interactions can limit the adaptive pathways available to an evolving population. John Maynard Smith (1970) famously explained this concept through use of a metaphor in which one must change the word WORD to GENE by changing one letter at a time, where every intermediate state is also a word. This can be accomplished through the following pathway: WORD → WORE → GORE → GONE → GENE. Notice that in this analogy, some mutations are only tolerated after the appearance of others. The ancestral sequence does not tolerate three of the four eventual changes, as GORD, WERD, and WOND are not words. Through empirical mapping of epistatic interactions and fitness landscapes, we now have biological examples of this type of adaptive constraint.

To explore how epistatic interactions influence an adaptive walk, Gong *et al.* (2013) experimentally created all intermediate genotypes along a 39-mutation evolutionary trajectory of the influenza virus nucleoprotein. They also introduced each mutation individually in the ancestral genome. The mutational series making up the adaptive walk was reconstructed from naturally evolving influenza virus sequences. Upon reconstructing all of the natural intermediates, they demonstrated that all intermediates had high fitness, in accordance with Maynard Smith's analogy of evolution proceeding through functional sequences. However, three of the mutations were significantly deleterious in the ancestral genome, despite having fixed during natural evolution without a negative impact. Upon further analysis, they demonstrated that these mutations were all preceded by a permissive mutation that allowed them to fix by epistatically altering the mutational effect. One mutation also was followed by a subsequent compensatory mutation, which ablated any remaining deleterious effect. The epistatically constrained mutations were shown to contribute to viral immune escape (Gong and Bloom, 2014; Gong *et al.,* 2013). This experiment highlights the role of epistasis in constraining natural protein evolution, and underscores the importance of contingency in viral adaptation.

Other studies have shown similar roles for epistasis in constraining evolution. Da Silva *et al.* (2010) used their HIV-1 data to reconstruct minimum-length mutational trajectories between two allele states. These trajectories are defined as the most direct evolutionary paths between alleles involving only single-step mutations without reversals. Adaptation is constrained when sign epistasis makes some of these minimum-length mutational trajectories inaccessible. For this study, they reconstructed trajectories from an ancestral genotype to a genotype including seven mutations that together are putatively involved in adaptation to an alternative host-cell co-receptor. They found that all twenty-four of the minimum-length trajectories experimentally tested were selectively inaccessible due to containing at least one single-step mutation that does not increase fitness. Because these mutations fixed in natural evolving populations, they must be accessible from a pathway that was not experimentally tested. Once again, this suggests that adaptive pathways are highly constrained by epistasis, and that contingency places an important role in adaptation. Other studies have emphasized the role of compensatory mutation in adaptation. These studies have engineered deleterious mutations into viruses, and then evolved these mutants in replicated environments to determine the frequency of mutational reversions as compared to compensatory mutations. These studies have shown that compensatory mutation is much more common than reversion (Poon and Chao, 2005; Sanjuán *et al.,* 2005), which once again implies a strong role of epistasis in determining adaptive trajectories.

Finally, epistatic interactions have been shown to depend on their environment. This is perhaps unsurprising. The phenotypic effects of mutations depend on their environment, and hence their interactions also are environmentally dependent. In experiments with the ssDNA phage ID11, epistatic interactions have been shown to be more pronounced at higher temperatures (Caudle *et al.,* 2014; Rokyta *et al.,* 2011). Epistatic interactions also varied for HIV-1 across different types of host cells (da Silva *et al.,* 2010) and drug treatments (Hinkley *et al.,* 2011; Kouyos *et al.,* 2012), although all landscapes showed high ruggedness.

To truly understand a fitness landscape, one would need to measure the interactions between all possible mutations within an entire genome. While this is obviously extraordinarily challenging, work to measure epistasis in viral pair-wise interactions and larger systematic sets of mutations is beginning to elucidate the topography of real biological landscapes. This recent surge of empirical data has the potential to greatly inform evolutionary theories that depend on epistasis. Looking forward, increasing our understanding of epistasis has the potential to move evolutionary biology towards becoming more of a predictive science (de Visser and Krug, 2014).

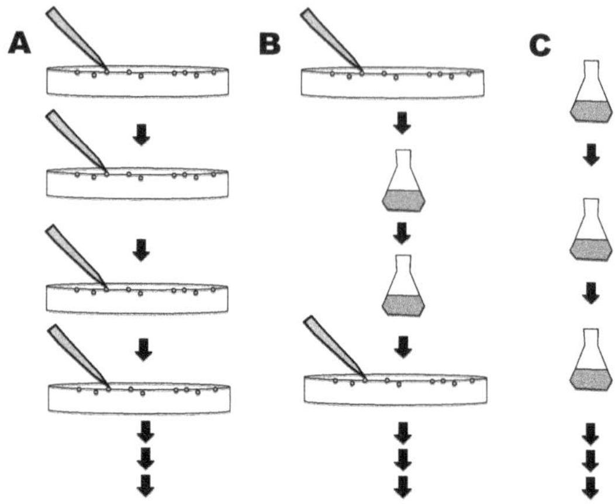

Figure 8.4. Evolution experiments with bottlenecking. **A.** Serial plaque-to-plaque transfers. One viral plaque is chosen randomly to initiate the next experimental passage. Under this regime, fitness is expected to decline through time due to the random fixation of mutations via genetic drift, and the expectation that most mutations are deleterious for fitness. **B.** Large population size transfers with intermittent bottlenecks. In this regime, virus lineages alternate between single-plaque transfers and two rounds of growth at large population sizes. Fitness is expected to decline, because infrequent extreme bottlenecks are sufficient for drift to overwhelm effects of natural selection. **C.** Serial passage at large population size. In this regime, virus populations are propagated at large population sizes, allowing evolutionary increases in fitness via natural selection.

Evolution in very small populations: drift, decay, and Muller's ratchet

Natural selection and genetic drift are the two primary forces of evolution. Adaptation by natural selection has been the primary focus of this chapter. During the process of adaptation, natural selection acts to increase the frequency of beneficial mutations in a population, and deleterious mutations are selected against. Natural selection operates most effectively at large population sizes. In very small populations, the effect of genetic drift can overwhelm natural selection. Genetic drift occurs when the frequency of alleles in a population changes stochastically over time due to random sampling. In large populations this random sampling primarily impacts the frequency of selectively neutral alleles, which over time will drift towards fixation or extinction. In very small populations, however, reproductive success becomes a matter of chance, and genetic drift overwhelms selection.

Studying systems in which genetic drift is the dominant evolutionary force is important for understanding the evolution of sex and the survival and evolution of small populations. Experiments with extreme bottlenecking have also been used to study the distribution and rate of deleterious mutations and epistatic interactions between mutations. To conduct these experiments, a virus lineage is propagated by randomly choosing a single virus plaque at each transfer to initiate the next experimental passage (Figure 8.4A). Because a plaque originates from a single virus clone, this process puts the virus population through a bottleneck of one individual at each passage. Under this experimental regime, there is minimal genetic variation on which natural selection can act, and genetic drift dominates.

Mutations enter the genome at their normal rate but are subject to minimal selection, so any non-lethal mutation that appears in a serially passaged clone instantly goes to fixation. These mutation accumulation experiments therefore reveal the distribution of non-lethal mutations minimally biased by selection.

These plaque-to-plaque mutation accumulation experiments have been central to providing evidence for the Muller's ratchet hypothesis (Felsenstein, 1974; Maynard-Smith, 1976; Muller, 1964). Muller's ratchet predicts that as mutations accumulate in small asexual populations, mutation-free individuals will decrease in frequency and eventually be lost to drift. If deleterious mutations outnumber beneficial or compensatory mutations, the fitness of the population will decline over time as mutations accumulate. In the absence of sexual recombination, genomes free of deleterious mutations cannot be restored, unless created through very rare back mutations. Viruses provide a powerful model to test Muller's ratchet, because viruses have a high mutation rate and can be propagated asexually at extremely small population sizes. In the first experiment to provide empirical support for Muller's ratchet using a viral system, Chao (1990) propagated lines of the dsRNA phage Φ6 in a plaque-to-plaque transfer regime. On average, Φ6 lines evolved in this regime of intensified genetic drift incurred a significant drop in fitness, with the relative fitness of some replicates decreasing by as much as 71%. This strongly suggests that the viruses in this experiment were subject to the effects of Muller's ratchet. Subsequent experiments have shown the same effect in the –ssRNA virus VSV (Clarke *et al.*, 1993; Duarte *et al.*, 1992) and the +ssRNA foot-and-mouth disease virus (FMDV) (Escarmis *et al.*, 1996). Fitness decay in serially plaque-to-plaque transferred VSV lines was shown to occur in a step-wise fashion, in accordance with the predictions of Muller's ratchet (Clarke *et al.*, 1993).

These experiments have demonstrated that virus populations undergoing repeated extreme bottlenecks experience fitness declines; but under what conditions is fitness recovery possible after fitness decay? When low fitness virus clones generated through mutation accumulation experiments are propagated at high population sizes, they can rapidly recover lost fitness (Clarke *et al.*, 1993). While serial passage at large population size can foster evolutionary rescue of a population, inserting periodic bottlenecks in this regime will erase the effect (Duarte *et al.*, 1993; Elena *et al.*, 1996). Several studies have employed a passaging regime in which a single plaque of VSV is transferred, followed by two transfers at high population sizes, followed by repetition of this cycle (Figure 8.4B). In both studies, the evolving virus lines incurred fitness losses similar to lines that underwent serial single plaque transfers. Two rounds of replication at high population size were not enough for selection to overcome the effect of drift in the intermittent bottlenecks. It is important to recall that when population size fluctuates over time, the effective population size equals the harmonic mean of passage bottlenecks – a population size predominantly dictated by the smallest bottlenecks experienced by the population. Fluctuation between bottlenecks and large population sizes is especially relevant for viruses. Viruses often go through a small bottleneck during interhost transmission, followed by amplification to extremely large intrahost populations. This may make them particularly susceptible to the effects of drift and Muller's ratchet despite large intrahost population sizes.

The effect of population size on evolutionary trajectories may vary depending on the initial fitness of the virus population. If the fitness of a virus clone is extremely low, it may be possible for fitness to recover even through propagation at low population sizes (Elena *et al.,* 1998). This may be due to more beneficial mutations of large effect size being available to a population with low fitness in a given environment. In other words, these populations have more room for improvement than populations with high starting fitness. It has also been demonstrated that VSV populations with high starting fitness require larger bottleneck sizes to maintain fitness than do lower fitness populations (Novella *et al.,* 1995b). Once again, this is likely due to a shift in the availability of beneficial versus deleterious mutations. If a population is nearer to a peak in the fitness landscape, there is little room for improvement, and mutations may be more likely to push the population toward a lower fitness value.

Population starting fitness influences the likelihood of fitness decay and recovery, but population size also influences where fitness approaches a plateau. One of the most consistently documented phenomena in experimental evolution is that populations adapting for long periods of time in a constant environment do not increase in fitness monotonically; rather, these populations show fitness plateaus, indicating waiting periods of fitness stasis until small-effect beneficial alleles sweep to fixation and fitness again measurably improves (Elena and Lenski, 2003; Elena and Sanjuán, 2007). Silander *et al.* (2007) showed that phage ΦX174 populations subjected to mutation accumulation also decline toward a low fitness plateau. When they evolved phage ΦX174 populations at a range of population sizes, they found that replicate populations converged at fitness values that depended on the size of the evolving population. Through a theoretical model, they predicted that both high and low fitness plateaus must result from either a change in the ratio of beneficial to deleterious mutations as populations approach the plateau, or a change in the size of mutational fitness effects. Through mutation accumulation experiments, they were able to show that as fitness declines, the ratio of beneficial to deleterious mutations dramatically increases. In contrast, the size of mutational fitness effects remains essentially constant as population fitness changes. These findings have significant implications for both adaptation and fitness declines, and provide insight into the dynamic relationship between population size and evolutionary trajectories.

Alternating host environments

The majority of theoretical and empirical work on adaptation has focused on populations evolving in simple and constant environments. Most organisms, however, live in spatially and temporally complex environments. Viruses may contend with particularly complicated environments involving sometimes heterogeneous biotic and abiotic environmental challenges (Wasik and Turner, 2013). While understanding how spatial and temporal environmental heterogeneity influences adaptation is of utmost importance to understanding adaptation in natural populations, there is very limited experimental work exploring this topic. An exception is the experimental evolution of arboviruses in fluctuating host environments. This work is motivated by the complex life histories of arboviruses in nature and a desire to understand the evolutionary pressures that shape them. In nature, arboviruses alternate obligately between vertebrate and invertebrate hosts (Figure 8.2). For example, endemic dengue virus circulates alternately between human hosts and the mosquito vector *Aedes aegypti*. Understanding how obligate host cycling impacts adaptation is a long-standing question in arbovirus evolution, with broad implications for the evolution of environmental specialization versus generalization and the importance of performance trade-offs across environments [for a recent review on life history trade-offs in viruses see (Goldhill and Turner, 2014)].

The dynamics of arbovirus evolution have long been viewed through the lens of the trade-off hypothesis. This hypothesis posits that arboviruses maintain the ability to cycle between two hosts at the expense of gaining higher fitness by specializing on one host. This type of hypothesis has long been used to explain the evolution of specialism. The reasoning is that there must be a cost to generalism, otherwise every organism would benefit from expanding its niche to take advantage of more resources. This viewpoint is often summarized with the phrase "a jack-of-all-trades is a master of none." Arboviruses have been a useful system for testing the predictions of the trade-off hypothesis. While conclusions from these experiments have direct relevance for understanding arbovirus evolution, they can also be more broadly extended to explain how populations adapt in response to cyclical environmental fluctuations.

The main relevant prediction of the trade-off hypothesis is that if a virus population is experimentally released from cycling between two hosts, it should evolve a higher level of fitness on the selected host. This gain in fitness is typically expected to come at a cost to fitness on the bypassed host relative to the ancestral virus. If observed, this trade-off between the selected and the bypassed host could have one of two origins. The first is antagonistic pleiotropy, in which mutations beneficial in one environment are deleterious in another environment. The second possibility is mutation accumulation, whereby a population accumulates substitutions that are neutral in the selected environment, but deleterious in an alternate environment. Experimental evolution with arboviruses has generally been inconsistent with the trade-off hypothesis. These experiments clearly show that viruses are capable of adapting to two hosts without suffering a cost. There are examples of trade-offs, but they do not seem to be a general rule. Three main generalizations can be taken away from this body of work: (i) Selection on one host type leads to increased fitness on that host type (ii) This often, but not always, is correlated with a reduction in fitness on the bypassed host (iii) Populations evolved in an alternating host regime gain fitness on both host types, often to the same or greater extent as populations serially passaged on a single host type.

Most alternating host experiments have been carried out *in vitro* using various cell lines as host types. Alphaviruses, a genus of positive-sense single-stranded RNA viruses, have been a popular model. Two *in vitro* experiments with Eastern equine encephalitis virus (EEEV) have shown mixed support for the trade-off hypothesis. The first of these experiments passaged EEEV populations serially on mosquito cells, mammalian cells, or alternating between the two cell types (Weaver *et al.*, 1999). EEEV lines passaged on a single host showed fitness increases on the selected host type relative to the ancestral virus, and suffered a decrease in fitness on the bypassed host type. EEEV lines evolved in the alternating regime showed increases on both host types, but these increases were of a lesser magnitude on each host type as compared to the respective single-host passaged lines. Single-host passaged lines also accumulated more mutations than the alternating lines, which may suggest that host alternation constrains the rate of evolution. A second similar experiment with EEEV used avian cells instead of mammalian cells, which is closer to the natural ecology of EEEV (Cooper and Scott, 2001). They found that serial passage on avian or mosquito cells led to improvements on the selected host type, but they saw no evidence of a cost on the bypassed host type. Virus lines passaged in an alternating regime showed improvements on both cell types, but to a lesser extent than lines serially passaged on a single host.

Experiments with other alphavirus systems have shown equally mixed support for trade-offs between host types. An experiment with Sindbis virus (SINV) evolved virus lines strictly on mammalian cells, strictly on mosquito cells, or using alternating passages between the two types (Greene *et al.,* 2005). SINV lines passaged on a single host gained fitness on the selected host and lost fitness on the bypassed host. Lines evolved with alternating passages gained fitness on both host types, but these gains were lesser than those for single-host passaged lines. These results generally align with predictions from the trade-off hypothesis. Single-host passaged lines in this experiment also accumulated more substitutions over the course of the experiment than alternately passaged lines, as was observed for EEEV (Weaver *et al.,* 1999). In a similar experiment with chikungunya virus (CHIKV), virus lines that alternated between mammalian and invertebrate cells actually showed greater gains on each host type than lines passaged strictly on either host (Coffey and Vignuzzi, 2011). This result directly contradicts the predictions of the trade-off hypothesis.

In vitro experiments with host alternation have also been performed for *Flaviviruses and Rhabdoviruses,* two other families containing arboviruses. In one experiment, dengue virus (DENV) was evolved on human cells, on mosquito cells, or in an alternating regime (Vasilakis *et al.,* 2009). In agreement with previously described studies, the authors found that single-host passaged lines showed fitness gains in the selected host type and fitness losses in the bypassed host type. Most DENV lines passaged in the alternating regime demonstrated fitness gains on both host types. Results from this experiment do not suggest that DENV evolution is constrained by host alternation. Several studies have used VSV as a model for host alternation. In an experiment in which VSV was evolved in single or alternating host passage on mammalian and sandfly cells, virus lines evolved in an alternating regime gained fitness on both cell types equal to that gained by single-host passaged lines (Novella *et al.,* 1999). Single-host passaged lines in this experiment also showed improvements on the bypassed host type, although this improvement was less than that for all other treatments. A second experiment evolved VSV on two novel types of mammalian cells, human and canine, with serial passage on alternating or single host types (Turner and Elena, 2000). VSV lines evolved in the alternating regime gained fitness on both host types equal to that gained by single-host passaged virus lines. Virus lines passaged on human cells lost fitness on their original laboratory host cell (hamster cells), and decreased in fitness on canine cells relative to their ancestor. Virus lines evolved on canine cells, however, lost fitness on human cells, but not on the original hamster host cell line. This reaffirms that while trade-offs certainly exist, they are specific to various virus and cell types and are not a universal rule.

Several studies have investigated the evolutionary consequences of host alternation *in vivo*. These experiments attempt to add a higher level of biological reality to our experimental knowledge of virus evolution by forcing viruses to navigate the complexities of a live organismal host. These systems incorporate characteristics such as variation in tissue types and host immune responses that are not present in experimental tissue culture systems. In one experiment, West Nile virus (WNV) was serially passaged in chicks (*Gallus gallus domesticus*) or the mosquito vector *Culex pipiens* (Deardorff *et al.*, 2011). Serial passage in chicks led to reduced fitness in one vector species, *Culex quinquefasciatus*, but fitness gains in *Cx. pipiens*. Serial passage in *Cx. pipiens* led to reduced viral fitness in chicks, but little change in virus fitness occurred on either mosquito vector species. Once again, this experiment shows that trade-offs can result from specialization on one host but are not universal. A second experiment passaged Venezuelan equine encephalitis virus (VEEV) *in vivo* serially in rodents or mosquitoes, or alternating between the two host types (Coffey *et al.*, 2008). Virus lines passaged serially in mosquitoes showed increased infectivity in mosquitoes, but lower viremia in mice. Lines passaged serially in mice showed higher viremia in mice and lower infectivity in mosquitoes. Lines passaged alternately between mosquitoes and mice did not show fitness increases in either host. The results from this experiment align with predictions from the trade-off hypothesis, from which we would expect fitness to be constrained in virus populations that cycle between two hosts.

From these experiments, it is clear that while the trade-off hypothesis may explain some instances of arbovirus evolution, it is not a universal explanation. Experiments in alternating host environments clearly show that selection on a single type of host *in vitro* or *in vivo* leads to fitness gains on the selected host. These fitness gains sometimes, but not always, come at a fitness cost on the bypassed host. Virus populations evolved in alternating host environments reach high levels of fitness on both hosts, and frequently reach fitness equal to that of populations evolved in single-host environments. These experimental results suggest a need for new and wide-reaching theoretical insights into the evolution of specialism and generalism (Remold, 2012).

Stochastic temporal heterogeneity

Experiments investigating fluctuating host environments have provided an excellent starting point for considering virus evolution in complex environments, but other types of temporal heterogeneity remain largely unexplored. Virus populations can clearly evolve high fitness when forced to cycle between two host types, but does this pattern hold for other types of selection pressures? Can virus populations still achieve high-fitness generalist phenotypes if temporal change in the environment is stochastic rather than predictably cyclical (Figure 8.2)? A recent study addressed these questions by evolving populations of *Vesicular stomatitis virus* (VSV) in environments where the temperature varied by treatment and over time (Alto *et al.*, 2013). Replicated virus populations were evolved at 29°C, 37°C, in a treatment that alternated between 29°C and 37°C at each passage, or in a treatment where a temperature between 29°C and 37°C was assigned randomly each day. After about 100 virus generations, the evolved populations were assayed for fitness at 29°C and 37°C relative to their common ancestor.

As expected, VSV populations evolved in constant environments at 29°C and 37°C improved in their selected environments. No trade-offs were observed when these populations were assayed in their respective unselected environments. Virus populations that evolved in the predictably alternating environments showed the most improvement at both temperatures. This is consistent with results from the previously described host alternation experiments, which show that selection in alternating environments leads to the evolution of generalist populations with high fitness in both environments. As in experiments with alternating hosts, the populations evolved in alternating environments performed as well or better than populations evolved in constant environments, which suggests that generalism did not come at a cost to high fitness in either environment. This experiment is unusual in that it explored virus evolution in a fluctuating environment where the variable was something other than host type, but what is truly novel about this experiment is the addition of a stochastically changing environment. Of all the experimental populations, virus lines evolved in the stochastic temperature regime showed the least improvement at both 29°C and 37°C. It is possible that the stochastically changing temperatures created a situation where alleles that were strongly favored one day might be less strongly favored the next. This may have prevented potentially beneficial alleles from reaching fixation, and left the population metaphorically spinning its wheels.

This experiment raises the intriguing possibility that even rapidly adapting RNA viruses have trouble adapting to stochastically changing environments. Climate change models predict that weather patterns will grow increasingly unpredictable in the near future, and it is unclear how populations of organisms will evolve to cope with such change. This experiment suggests that studying how populations respond to different types of environmental heterogeneity will be a critical piece of this puzzle, and future work on this topic is essential.

Adaptation to spatially structured environments

Viruses also need to replicate in spatially structured and often patchy environments (Figure 8.2). Spatial structure occurs at several levels: a virus population needs to navigate various tissue types within a host, a virus may encounter different host species or genotypes in a host community, and a virus may experience different ecological conditions over its geographical range. Little experimental work has dealt with how spatial structure and spatially heterogeneous environments influence the evolution of virus populations. This is a ripe area for further research, and holds promise for helping us to understand viral evolution across many spatial scales.

One recent experiment investigated the impact of spatial structure in a homogenous environment (Ally *et al.,* 2014). Populations of phage ID11 were evolved on confluent lawns of *Escherichia coli* bacteria. At each passage, the spatial structure on the plate was either preserved during transfer, or it was disrupted and the phage population from the plate was mixed. At the end of the experiment, populations that evolved with spatial structure showed higher levels of global biodiversity across the plate than populations that were mixed. This is likely because spatial structure effectively creates loose subpopulations, and mutations that arise do not compete with one another globally the way they do in a mixed environment. In this sense, spatial structure may exacerbate the effects of clonal

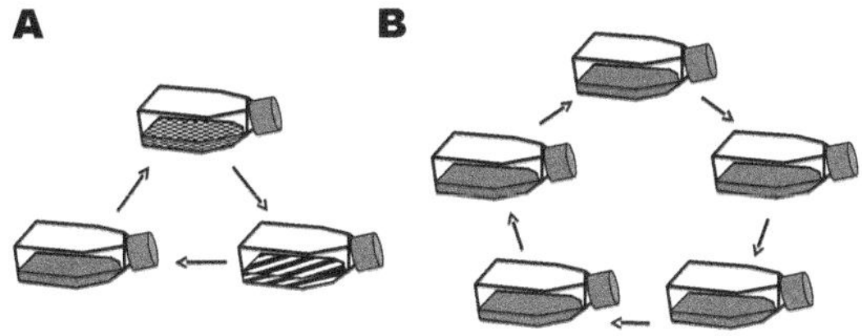

Figure 8.5. Experimental migration between environmental patches. **A.** Virus migration among three heterogeneous environmental patches as in Cuevas *et al.*, 2003. Each flask contains a different type of host cell, and a subset of the evolving virus populations are migrated clockwise to the neighboring flask (host type) at each passage. **B.** Migration among five homogenous patches as in Miralles *et al.*, 1999. Each flask contains the same type of host cell.

interference and result in favorable mutations taking longer to become globally fixed.

The effects of migration on adaptation in homogenous and heterogeneous landscapes

Migration is a powerful evolutionary force, but the effects of migration on adaptation can be challenging to predict. Some theoretical models predict that if populations are evolving in a heterogeneous landscape, gene flow among populations will inhibit local adaptation (Antonovics, 1976; Bolnick and Nosil, 2007; Ehrlich and Raven, 1969; Lenormand, 2002; Mayr, 1963). Populations in a variable landscape are subject to different local selective pressures, and alleles that are beneficial or neutral in one environment may be deleterious in another. Hence, when there is gene flow between locally adapted populations, adaptation in the receiving population may be swamped by the influx of maladapted genotypes. Other theoretical models emphasize the potential of migration to enhance adaptation (Gomulkiewicz *et al.*, 1999; Holt *et al.*, 2004; Wright, 1932b). Wright's shifting-balance hypothesis predicts that migration can enhance adaptation for populations situated at relatively low peaks in a fitness landscape (Wright, 1932b). Migrants may push the population away from its low fitness peak, thus enabling the population to cross a fitness valley and adapt toward a higher fitness optimum. Gene flow can also increase genetic variability in a population, thereby providing more raw material for natural selection to act upon and potentially enhancing adaptation. In this section we will address the impact of landscape heterogeneity on the effects of migration, the potential for migration to enhance adaptation in ecological sinks, and how migration affects the dynamics of antagonistic coevolution.

A key factor in predicting the effect of migration on adaptation is whether the populations are evolving in a homogenous or heterogeneous landscape. Migration among populations evolving in dissimilar environments has the potential to inhibit adaptation, while migration among populations evolving in similar environments can enhance adaptation. Cuevas *et al.* (2003) employed experimental evolution of VSV to explore the effects of migration among heterogeneous environmental patches on local adaptation. Each replicate system consisted of three flasks, each containing a different type of host cell, positioned on the vertices of an imaginary triangle. Virus populations were grown in each flask, and at each passage a portion of the virus population was transferred clockwise in a simulated migration (Figure 8.5A). This experimental design, consisting of unidirectional migration, is an extension of the one-dimensional stepping stone model. The rate of migration varied among treatments. As expected, in the absence of migration, each virus population had the highest fitness on its local cell type, although there were correlated improvements on other cell types. In other words, in the absence of migration, populations underwent local adaptation. With increasing migration, the fitness of viral populations on their local cell type decreased. This suggests that the increasing migration rate inhibits local adaptation. This is likely because migration among dissimilar patches introduces migrants that are less well adapted to the local environment.

While migration among heterogeneous environmental patches can hinder local adaptation, it is possible for gene flow among homogeneous patches to promote adaptation. In an experimental design similar to that described above, VSV populations were evolved in five flasks arranged in a circle (Miralles *et al.,* 1999b). Each flask contained the same type of host cell (Figure 8.5B). At each passage, a predetermined proportion of the virus population of each flask was transferred clockwise to the neighboring flask. After one hundred generations, the mean viral fitness averaged across patches was found to be positively correlated with migration rate. Meanwhile, variance for fitness across patches correlated negatively with migration rate. In this experiment, migration allowed beneficial mutations that arose in one patch to spread globally. Migration is therefore a source of adaptive alleles along with mutation.

Migration can facilitate evolutionary rescue in populations evolving in ecological sinks. An ecological sink is an environment in which death rate exceeds birth rate. Theory predicts that a sink population's local supply of beneficial mutations is likely to be small and dwindling, and migration from a source population has the potential to supply beneficial alleles (Gomulkiewicz *et al.,* 1999; Holt and Gomulkiewicz, 1997; Holt *et al.,* 2003; Holt *et al.,* 2004). For viruses, sinks can be designed experimentally by creating a treatment in which the bottleneck is greater than the basic reproductive rate ($R_0 < 1$). To test the hypothesis that migration can enhance evolution for populations evolving in sink habitats, dsRNA phage Φ6 was experimentally evolved under sink conditions (Ching *et al.,* 2013). Treatment populations received migrants at each passage from a frozen ancestral stock. The experimental populations evolved higher fitness in the sink environment, with the greatest fitness increases at intermediate levels of migration. Some populations reached a point where they would be able to survive in the sink conditions without any further migration, which suggests that migration may enhance virus emergence. The success of an intermediate level of migration suggests that an optimal balance can be struck between the addition of beneficial genetic variation and the inhibitory effects of migrant genotypes adapted to source conditions.

The potential of migration to rescue sink populations from extinction also depends on the degree of similarity of the selection pressures acting on the source and sink populations. Imagine that different environmental patches contain different hosts. In a range-expansion scenario, the source and sink populations are evolving on the same host type, but conditions in the sink are unfavorable to viral growth. In a host-shift scenario, the population in the sink environment is evolving on a novel host, while the source population evolves on the native host. To explore the dynamics of these two scenarios, phage Φ6 was evolved in either a range-expansion scenario, a host-shift scenario, or with migrants from a frozen stock which did not evolve over the course of the experiment (Dennehy *et al.*, 2010). Source and sink populations in the range-expansion treatment are likely to experience analogous selection pressures, and therefore migration is predicted to enhance adaptation to the sink. Observations from evolved Φ6 populations support this hypothesis, with populations in the range-expansion treatment reaching the highest fitness in the sink.

In the host-shift scenario, source and sink populations are experiencing potentially disruptive selection pressures, and migrants are expected to be less fit in the sink environment as the experiment proceeds. Sink populations in the host-shift treatment also increased in fitness, but their adaptive trajectories were more punctuated, and the final fitness was lower than that of the range-expansion populations. Populations receiving migrants from the ancestral stock did not adapt to sink conditions. This experiment confirms that migrants can enable adaptation in sink environments by supplying additional genetic variation. The degree to which migration can enable adaptation in sink environments depends, however, on whether source and sink populations are evolving under congruent selection pressures. Migration is more likely to enhance adaptation in an ecological sink if source and sink environments are similar, because similarity between selection regimes increases the likelihood that successful genotypes in the source population will also be fit in the sink population.

Patterns of migration may also influence the dynamics of antagonistic coevolution. Antagonistic coevolution between hosts and parasites can lead to local adaptation in parasite populations, meaning that parasites have higher infectivity or fitness on their local host population than on nonlocal host populations. Theory suggests that an intermediate level of migration between parasite populations might increase local adaptation (Gandon, 2002; Gandon *et al.*, 1996; Gandon and Michalakis, 2002; Lively, 1999), although high levels of migration could homogenize populations and swamp local adaptation (Gandon, 2002). To test the hypothesis that migration can bolster local adaptation of parasites in coevolving host-parasite systems, *Pseudomonas fluorescens* SBW25 bacteria and phage SBW25Φ2 were coevolved in the laboratory (Morgan *et al.*, 2005). Low levels of phage migration between populations increased local adaptation for the phage, while migration of just the bacteria had no effect on bacterial local adaptation compared to non-migrated controls. One possible explanation for this asymmetry is that the phages in this system are less evolvable than the bacteria due to smaller genomes and lower standing genetic diversity. Hence, they may have more to gain from the additional genetic variation supplied by migration. Indeed, when *P. fluorescens* SBW25 and phage SBW25Φ2 are coevolved with no migration, the bacteria seem to have the evolutionary advantage, and evolve high levels of resistance against the phage. However, when the same experiment is conducted with simultaneous migration of phage and bacteria, resistance levels drop dramatically (Morgan *et al.*, 2007). These experiments show that simultaneous or differential migration rates can profoundly alter the dynamics of antagonistic coevolution. In coevolving systems, migration seems to have the strongest advantage for the organism that is less evolvable due to genome constraints, low genetic diversity, or low population size.

Experimental work on virus migration has important implications for virus emergence and local adaptation. Experiments examining evolution in ecological sinks strongly suggest that migration can facilitate virus range expansion and potentially evolutionary shifts onto new types of hosts. Migration can also influence which player gains the upper hand in antagonistic coevolution and determine how well viruses can adapt to their local host populations. Importantly, principles from these experiments are broadly applicable to all types of organisms on many scales. Understanding the dynamics of local adaptation and how it is influenced by gene flow among populations is critical to understanding the origin, maintenance, and geographical distribution of biodiversity. Experimental evolution with viruses allows us to begin exploring the dynamics of local adaptation in homogenous and heterogeneous landscapes at a small scale. Coupled with studies from natural systems, these experiments have the potential to provide great insight into the effects of migration on local adaptation, range expansion, and coevolutionary processes.

Conclusion

Experimental work with viruses has greatly shaped the scientific understanding of basic evolutionary principles. These experiments have helped to characterize the dynamics of adaptation in simple environments, to test the predictions associated with genetic drift, and to explore the prevalence and evolutionary implications of epistasis. With this foundation in place, researchers are increasingly focusing on investigating adaptation in complex environments. Organisms face both spatial and temporal heterogeneity in their environments at many scales, and it remains unclear how generalizations concerning adaptation in simple environments translate to adaptation in complex environments. In the coming years, evolution experiments incorporating environmental heterogeneity could greatly increase our understanding of adaptation in natural populations.

Future Directions

1. Further investigate the effects of spatial structure on adaptation at different environmental scales
2. Explore the effects of temporal environmental heterogeneity on adaptation, incorporating different patterns of temporal change, the impact of the predictability of temporal variation, and different grains of temporal heterogeneity (i.e. the frequency of environmental fluctuations as compared to the generation time of the organism)
3. Explore adaptive responses in populations subject to multiple simultaneous selection pressures
4. Develop more wide-reaching theory to explain the evolution of specialism and generalism
5. Systematically map empirical fitness landscapes for defined sets of mutations
6. Investigate the prevalence of higher-order epistasis (interactions involving three or more mutations)
7. Incorporate empirical data on the prevalence and character of epistatic interactions into theories of epistasis, adaptive walks, and fitness landscapes to build more biologically realistic models

Acknowledgements

V.J.M. was supported by the U.S. National Science Foundation (NSF) Graduate Research Fellowship grant #DGE-1122492, and P.E.T by the NSF BEACON Center for Study of Evolution in Action.

References

Acevedo, A., Brodsky, L., and Andino, R. (2014). Mutational and fitness landscapes of an RNA virus revealed through population sequencing. Nature *505*, 686-690.

Ally, D., Wiss, V.R., Deckert, G.E., Green, D., Roychoudhury, P., Wichman, H.A., Brown, C.J., and Krone, S.M. (2014). The impact of spatial structure on viral genomic diversity generated during adaptation to thermal stress. PLoS ONE *9*.

Alto, B.W., Wasik, B.R., Morales, N.M., and Turner, P.E. (2013). Stochastic temperatures impede RNA virus adaptation. Evolution *67*, 969-979.

Antonovics, J. (1976). The nature of limits to natural selection. Annals of the Missouri Botanical Garden *63*, 224-247.

Betancourt, A.J. (2009). Genomewide patterns of substitution in adaptively evolving popultions of the RNA bacteriophage MS2. Genetics *181*, 1535-1544.

Blount, Z.D., Borland, C.Z., and Lenski, R.E. (2008). Historical contingency and the evolution of a key innovation in an experimetnal population of *Escherichia coli*. Proceedings of the National Academy of Sciences *105*, 7899-7906.

Bollback, J.P., and Huelsenbeck, J.P. (2007). Clonal interference is alleviated by high mutation rates in large populations. Molecular Biology and Evolution *24*, 1397-1406.

Bolnick, D.I., and Nosil, P. (2007). Natural selection in populations subject to a migration load. Evolution *61*, 2229-2243.

Bridgham, J.T., Ortlund, E.A., and Thornton, J.W. (2009). An epistatic ratchet constrains the direction of glucocorticoid receptor evolution. Nature *461*, 515-519.

Bull, J.J., Badgett, M.R., Wichman, H.A., Huelsenbeck, J.P., Hillis, D.M., Gulati, A., Ho, C., and Molineux, I.J. (1997). Exceptional convergent evolution in a virus. Genetics *147*, 1497-1507.

Burch, C.L., and Chao, L. (1999). Evolution by small steps and rugged landscapes in the RNA virus ϕ6. Genetics *151*, 921-927.

Carrasco, P., de la Iglesia, F., and Elena, S.F. (2007). Distribution of fitness and virulence effects caused by single-nucleotide substitutions in *tobacco etch virus*. Journal of Virology *81*, 12979-12984.

Caudle, S.B., Miller, C.R., and Rokyta, D.R. (2014). Environment determines epistatic patterns for a ssDNA virus. Genetics *196*, 267-279.

Chao, L. (1990). Fitness of RNA virus decreased by Muller's ratchet. Nature *348*, 454 - 455.

Ching, J., Musheyev, S.A., Chowdhury, D., Kim, J.A., Choi, Y., and Dennehy, J.J. (2013). Migration enhances adaptation in bacteriophage populations evolving in ecological sinks. Evolution *67*, 10-17.

Clarke, D.K., Duarte, E.A., Moya, A., Elena, S.F., Domingo, E., and Holland, J.J. (1993). Genetic bottlenecks and population passages cause profound fitness differences in RNA viruses. Journal of Virology *67*, 222-228.

Coffey, L.L., Vasilakis, N., Brault, A.C., Powers, A.M., Tripet, F., and Weaver, S.C. (2008). Arbovirus evolution *in vivo* is constrained by host alternation. Proceedings of the National Academy of Sciences *105*, 6970-6975.

Coffey, L.L., and Vignuzzi, M. (2011). Host alternation of Chikungunya virus increases fitness while restricting population diversity and adaptability to novel selective pressures. Journal of Virology *85*, 1025-1035.

Cooper, L.A., and Scott, T.W. (2001). Differential evolution of eastern equine encephalitis virus populations in response to host cell type. Genetics *175*, 1403-1412.

Cuevas, J.M., Elena, S.F., and Moya, A. (2002). Molecular basis of adaptive convergence in experimental populations of RNA viruses. Genetics *612*, 533-542.
Cuevas, J.M., Moya, A., and Elena, S.F. (2003). Evolution of RNA virus in spatially structured heterogeneous environments. Journal of Evolutionary Biology *16*, 456-466.
da Silva, J., Coetzer, M., Nedellec, R., Pastore, C., and Mosier, D.E. (2010). Fitness epistasis and constraints on adaptation in human immunodeficiency virus type 1 protein region. Genetics *185*, 293-303.
de Visser, J.A.G.M., and Krug, J. (2014). Empirical fitness landscapes and the predictability of evolution. Nature Reviews Genetics *15*, 480-490.
Deardorff, E.R., Fitzpatrick, K.A., Jerzak, G.V.S., Shi, P.Y., Kramer, L.D., and Ebel, G.D. (2011). West Nile virus experimental evolution *in vivo* and the trade-off hypothesis. PLoS Pathogens *7*.
Dennehy, J.J., Friedenberg, N.A., McBride, R.C., Holt, R.D., and Turner, P.E. (2010). Experimental evidence that source genetic variation drives pathogen emergence. Proceedings of the Royal Society B *277*, 3113-3121.
DePristo, M.A., Hartl, D.L., and Weinreich, D.M. (2007). Mutational reversions during adaptive protein evolution. Molecular Biology and Evolution *24*, 1608-1610.
Domingo-Calap, P., Cuevas, J.M., and Sanjuán, R. (2009). The fitness effects of random mutations in single-stranded DNA and RNA bacteriophages. PLoS Genetics *5*.
Drake, J.W. (1993). Rates of spontaneous mutation among RNA viruses. Proceedings of the National Academy of Sciences *90*, 4141-4175.
Duarte, E.A., Clarke, D.K., Moya, A., Domingo, E., and Holland, J.J. (1992). Rapid fitness losses in mammalian RNA virus clones due to Muller's ratchet. Proceedings of the National Academy of Sciences *89*, 6015-6019.
Duarte, E.A., Clarke, D.K., Moya, A., Elena, S.F., and Holland, J.J. (1993). Many-trillionfold amplification of single RNA virus particles fails to overcome the Muller's ratchet effect. Journal of Virology *67*, 3620-3623.
Ehrlich, P.R., and Raven, P.H. (1969). Differention of populations. Science *165*, 1228-1232.
Elena, S.F., Carrasco, P., Daròs, J., and Sanjuán, R. (2006). Mechanisms of genetic robustness in RNA viruses. EMBO Reports *7*, 168-173.
Elena, S.F., Davila, M., Novella, I.S., Holland, J.J., Domingo, E., and Moya, A. (1998). Evolutionary dynamics of fitness recovery from debilitating effects of Muller's ratchet. Evolution *52*, 309-314.
Elena, S.F., Gonzalez-Candelas, F., Novella, I.S., Duarte, E.A., Clarke, D.K., Domingo, E., Holland, J.J., and Moya, A. (1996). Evolution of fitness in experimental populations of vesicular stomatitis virus. Genetics *142*, 673-679.
Elena, S.F., and Lenski, R.E. (2003). Microbial genetics: evolution experiments with microorganisms: the dynamics and genetic bases of adaptation. Nature Reviews Genetics *4*, 457-469.
Elena, S.F., and Sanjuán, R. (2007). Virus evolution: Insights from an experimental approach. Annual Review of Ecology, Evolution, and Systematics *38*, 27-52.
Escarmis, C., Davila, M., Charpentier, N., Bracho, A., Moya, A., and Domingo, E. (1996). Genetic lessons associated with Muller's ratchet in an RNA virus. Journal of Molecular Biology *264*, 255-267.
Felsenstein, J. (1974). The evolutionary advantage of recombination. Genetics *78*, 737-756.
Fisher, R.A. (1930). The genetical theory of natural selection (Oxford: Oxford University Press).
Gandon, S. (2002). Local adaptation and the geometry of host-parasite coevolution. Ecology Letters *5*, 246-256.
Gandon, S., Capowiez, Y., Dubois, Y., Michalakis, Y., and Olivieri, I. (1996). Local adaptation and gene for gene coevolution in a metapopulation model. Proceedings of the Royal Society B *263*, 1003-1009.

Gandon, S., and Michalakis, Y. (2002). Local adaptation, evolutionary potential and host-parasite coevolution: interactions between migration, mutation, population size and generation time. Journal of Evolutionary Biology *15*.
Gerrish, P.J., and Lenski, R.E. (1998). The fate of competing beneficial mutations in an asexual population. Genetica *102*, 127-144.
Goldhill, D.H., and Turner, P.E. (2014). The evolution of life history trade-offs in viruses. Current Opinion in Virology *8*, 79-84.
Gomulkiewicz, R., Holt, R.D., and Barfield, M. (1999). The effects of density dependence and immigration on local adaptation and niche evolution in a black-hole sink environment. Theoretical Population Biology *55*, 283-296.
Gong, L.I., and Bloom, J.D. (2014). Epistatically interacting substitutions are enriched during adaptive protein evolution. PLoS Genetics *10*.
Gong, L.I., Suchard, M.A., and Bloom, J.D. (2013). Stability-mediated epistasis constrains the evolution of an influenza protein. eLife *2*.
Grant, P.R., and Grant, B.R. (2014). 40 Years of Evolution: Darwin's Finches on Daphne Major Island (Princeton, New Jersey: Princeton University Press).
Greene, I.P., Wang, E., Deardorff, E.R., Milleron, R., Domingo, E., and Weaver, S.C. (2005). Effect of alternating passage on adaptation of Sindbis virus to vertebrate and invertebrate cells. Journal of Virology *79*, 14253-14260.
Hinkley, T., Martins, J., Chappey, C., Haddad, M., Stawiski, E., Whitcomb, J.M., Petropoulos, C.J., and Bonhoeffer, S. (2011). A systems analysis of mutation effects in HIV-1 protease and reverse transcriptase. Nature Genetics *43*, 487-489.
Holder, K.K., and Bull, J.J. (2001). Profiles of adaptation in two similar viruses. Genetics *159*, 1393-1404.
Holland, J., Spindler, K., Horodyski, F., Grabau, E., Nichol, S., and VandePol, S. (1982). Rapid evolution of RNA genomes. Science *215*, 1577-1585.
Holmes, E.C. (2009). The Evolution and Emergence of RNA Viruses (Oxford: Oxford University Press).
Holt, R.D., and Gomulkiewicz, R. (1997). How does immigration influence local adaptation? A reexamination of a familiar paradigm. The American Naturalist *149*, 563-572.
Holt, R.D., Gomulkiewicz, R., and Barfield, M. (2003). The phenomenology of niche evolution via quantitative traits in a 'black-hole' sink. Proceedings of the Royal Society B *270*, 215-224.
Holt, R.D., Knight, T.M., and Barfield, M. (2004). Allee effects, immigration, and the evolution of species' niches. The American Naturalist *163*, 253-262.
Kim, Y., and Orr, H.A. (2005). Adaptation in sexuals vs. asexuals: clonal interference and the Fisher-Muller model. Genetics *171*, 1377-1386.
Kimura, M. (1983). The neutral theory of molecular evolution (Cambridge, U. K.: Cambridge University Press).
Kondrashov, A.S. (1988). Deleterious mutations and the evolution of sexual reproduction. Nature *336*, 435-440.
Kondrashov, A.S. (1993). Classification of hypotheses on the advantage of amphimixis. Journal of Heredity *84*, 372-387.
Kouyos, R.D., Leventhal, G.E., Hinkley, T., Haddad, M., Whitcomb, J.M., Petropoulos, C.J., and Bonhoeffer, S. (2012). Exploring the complexity of the HIV-1 fitness landscape. PLoS Genetics.
Lenormand, T. (2002). Gene flow and the limits to natural selection. Trends in Ecology and Evolution *17*, 183-189.
Lively, C. (1999). Migration, virulence and the geographic mosaic of adaptation by parasites. The American Naturalist *153*.
Maynard-Smith, J. (1976). The Evolution of Sex (Cambridge: Cambridge University Press).
Mayr, E. (1963). Animal species and evolution (Cambridge: Harvard University Press).
Miralles, R., Gerrish, P.J., Moya, A., and Elena, S.F. (1999a). Clonal interference and evolution of RNA viruses. Science *285*, 1745-1747.

Miralles, R., Moya, A., and Elena, S.F. (1999b). Effect of population patchiness and migration rates on adaptation and divergence of vesicular stomatitis virus quasispecies populations. Journal of General Virology *80*, 2051-2059.

Miralles, R., Moya, A., and Elena, S.F. (2000). Diminishing returns of population size in the rate of RNA virus adaptation. Journal of Virology *74*, 3566-3571.

Morgan, A.D., Brockhurst, M.A., Lopez-Pascua, L.D.C., Pai, C., and Buckling, A. (2007). Differential impact of simultaneous migration on coevolving hosts and parasites. BMC Evolutionary Biology *7*.

Morgan, A.D., Gandon, S., and Buckling, A. (2005). The effect of migration on local adaptation in a coevolving host-parasite system. Nature *437*, 253-256.

Muller, H.J. (1964). The relation of recombination to mutational advance. Mutation Research *1*, 2-9.

Novella, I.S., Duarte, E.A., Elena, S.F., Moya, A., Domingo, E., and Holland, J.J. (1995a). Exponential increases of RNA virus fitness during large population transmissions. Proceedings of the National Academy of Sciences *92*, 5841-5844.

Novella, I.S., Elena, S.F., Moya, A., Domingo, E., and Holland, J.J. (1995b). Size of genetic bottlenecks leads to virus fitness loss is determined by mean initial population fitness. Journal of Virology *69*, 2869-2872.

Novella, I.S., Hershey, C.L., Escarmis, C., Domingo, E., and Holland, J.J. (1999). Lack of evolutionary stasis during alternating replication of an arbovirus in insect and mammalian cells. Journal of Molecular Biology *287*, 459-465.

Orr, H.A. (1998). The Population Genetics of Adaptation: The Distribution of Factors Fixed during Adaptive Evolution. Evolution; international journal of organic evolution *52*, 935-949.

Orr, H.A. (1999). The evolutionary genetics of adaptation: a simulation study. Genetics Research *74*, 207-214.

Orr, H.A. (2002). The population genetics of adaptation: The adaptation of DNA sequences. Evolution; international journal of organic evolution *56*, 1317-1330.

Orr, H.A., and Coyne, J.A. (1992). The genetics of adaptation: A reassessment. The American Naturalist *140*, 725-742.

Otto, S.P. (2009). The evolutionary enigma of sex. The American Naturalist *174*, S1-S14.

Otto, S.P., and Feldman, M.W. (1997). Deleterius mutations, variable epistatic interactions, and the evolution of recombination. Theoretical Population Biology *51*, 134-147.

Parera, M., and Martínez, M.A. (2014). Strong epistatic interactions within a single protein. Molecular Biology and Evolution *31*, 1546-1553.

Pepin, K.M., and Wichman, H.A. (2007). Variable epistatic effects between mutations at host recognition site in φX174 bacteriophage. Evolution *61*, 1710-1724.

Pepin, K.M., and Wichman, H.A. (2008). Experimental evolution and genome sequencing reveal variation in levels of clonal interference in large populations of bacteriophage φX174. BMC Evolutionary Biology *8*.

Poelwijk, F.J., Tănase-Nicola, S., Kiviet, D.J., and Tans, S.J. (2011). Reciprocal sign epistasis is a necessary condition for multi-peaked fitness landscapes. Journal of Theoretical Biology *272*, 141-144.

Poon, A., and Chao, L. (2005). The rate of compensatory mutation in the DNA bacteriophage φX174. Genetics *170*, 989-999.

Remold, S. (2012). Understanding specialism when the jack of all trades can be the master of all. Proceedings of the Royal Society B *279*, 4861-4869.

Rokyta, D.R., Abdo, Z., and Wichman, H.A. (2009). The genetics of adaptation for eight microvirid bacteriophages. Journal of Molecular Evolution *69*, 229-239.

Rokyta, D.R., Beisel, C.J., Joyce, P., Ferris, M.T., Burch, C.L., and Wichman, H.A. (2008). Beneficial fitness effects are not exponential for two viruses. Journal of Molecular Evolution *67*, 368-376.

Rokyta, D.R., Joyce, P., Caudle, S.B., Miller, C.R., Beisel, C.J., and Wichman, H.A. (2011). Epistasis between beneficial mutations and the phenotype-to-fitness map for a ssDNA virus. PLoS Genetics *7*.

Rokyta, D.R., Joyce, P., Caudle, S.B., and Wichman, H.A. (2005). An empirical test of the mutational landscape model of adaptation using a single stranded DNA virus. Nature Genetics *37*, 441-444.

Sanjuán, R. (2010). Mutational fitness effects in RNA and single-stranded DNA viruses: common patterns revealed by site-directed mutagenesis studies. Philosophical Transactions of the Royal Society B *365*, 1975-1982.

Sanjuán, R., Cuevas, J.M., Moya, A., and Elena, S.F. (2005). Epistasis and the adaptability of an RNA virus. Genetics *170*, 1001-1008.

Sanjuán, R., Moya, A., and Elena, S.F. (2004a). The contribution of epistasis to the architecture of fitness in an RNA virus. Proceedings of the National Academy of Sciences *101*, 15376-15379.

Sanjuán, R., Moya, A., and Elena, S.F. (2004b). The distribution of fitness effects caused by single-nucleotide substitutions in an RNA virus. Proceedings of the National Academy of Sciences *101*, 8396-8401.

Silander, O.K., Tenaillon, O., and Chao, L. (2007). Understanding the evolutionary fate of finite populations: The dynamics of mutational effects. PLoS Biology *5*, 922-931.

Smith, J.M. (1970). Natural selection and the concept of a protein space. Nature *225*, 563-564.

Turner, P.E., and Chao, L. (1998). Sex and the evolution of intrahost competition in RNA virus φ6. Genetics *150*, 523-532.

Turner, P.E., and Elena, S.F. (2000). Cost of host radiation in an RNA virus. Genetics *156*, 1465-1470.

Vasilakis, N., Deardorff, E.R., Kenney, J.L., Rossi, S.L., Hanley, K.A., and Weaver, S.C. (2009). Mosquitoes put the brake on arbovirus evolution: experimental evolution reveals slower mutation accumulation in mosquito than vertebrate cells. PLoS Pathogens.

Wasik, B.R., and Turner, P.E. (2013). On the biological success of viruses. Annual Review of Microbiology *67*, 519-541.

Weaver, S.C., Brault, A.C., Kang, W., and Holland, J.J. (1999). Genetic and fitness changes accompanying adaptation of an arbovirus to vertebrate and invertebrate cells. Journal of Virology *73*, 4316-4326.

Weinreich, D.M., Delaney, N.F., DePristo, M.A., and Hartl, D.L. (2006). Darwinian evolution can follow only very few mutational paths to fitter proteins. Science *312*, 111-114.

Weinreich, D.M., Watson, R.A., and Chao, L. (2005). Sign epistasis and genetic constraint on evolutionary trajectories. Evolution *59*, 1165-1174.

Wichman, H.A., Badgett, M.R., Scott, L.A., Boulianne, C.M., and Bull, J.J. (1999). Different trajectories of parallel evolution during viral adaptation. Science *285*, 422-424.

Wright, S. (1932a). The roles of mutation, inbreeding, crossbreeding and selection in evolution. Proceedings of the 6th International Congress of Genetics, 356-366.

Wright, S. (1932b). The roles of mutation, inbreeding, crossbreeding and selection in evolution. Proceedings of the 6th International Congress of Genetics *1*, 356-366.

9

Evolution of Persistent Viruses in Plants

Marilyn J. Roossinck

Center for Infectious Disease Dynamics, Pennsylvania State University, University Park, PA 16802, USA. Email: mjr25@psu.edu

Abstract
Persistent plant viruses have a unique natural history. They are cytoplasmic RNA viruses found in many plants, including many crop and ornamental plants, and are the most common type of virus found in wild plants. Persistent plant viruses have very long associations with their hosts. They do not move between plant cells, are found in every cell, and are only vertically transmitted. Their evolutionary history remains obscure but many aspects of their lifestyles imply mutualistic relationships with their plant hosts. Many are related to viruses of fungi, including endophytic fungi, and in some cases there is evidence for transmission between plants and fungi.

Introduction

Plant viruses were the first recognized and studied viruses, launching the field of virology. *Tobacco mosaic virus*, the first described virus, was found because it caused a transmissible disease in tobacco that was not removed by the fine filters that remove bacteria (Beijerinck, 1898). Most of plant virology over the past century has focused on the acute disease-causing viruses of crops. Many serious diseases of crop plants have been caused by virus infections, and control efforts have generally had only short-term success, so acute viruses continue to be a focus of research and interest. However, plants are also infected by a group of viruses known as persistent viruses (Roossinck, 2010, 2012a). These viruses do not cause any apparent symptoms. They do not move between plant cells, but are found in the cytoplasm of every cell. Persistent viruses are not readily transmitted horizontally, but have nearly 100% vertical transmission rates. They are very poorly studied, even though they are common in crop and ornamental plants, and virus biodiversity studies indicate that they are the most common viruses found in wild plants (Roossinck, 2012b, 2014). Sequences of some members of the persistent virus families are integrated into the genomes of plants (Chiba *et al.*, 2011; Liu *et al.*, 2010), making them part of the emerging field of paleovirology. Plant genomes have many retrovirus- and pararetrovirus-like paleovirus elements as well. Another chapter in this book describes paleoviruses; in this chapter I will describe what is known about the evolution of persistent viruses in plants, an area of virus evolution that is rich in potential new research opportunities.

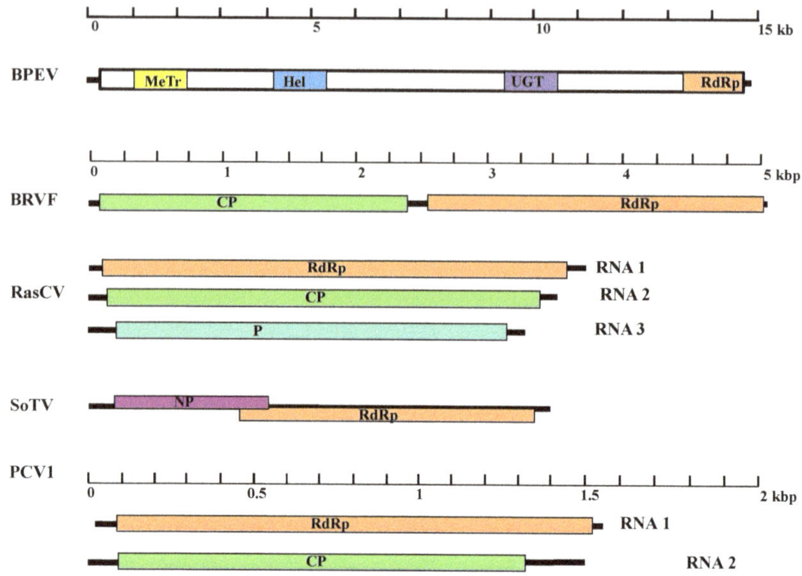

Figure 9.1. Genome organization of representative persistent plant viruses. BPEV, Bell pepper endornavirus. The virus encodes a large polyprotein, four domains are identified and the remainder are unknown. BRVF, Black raspberry associated virus F (unpublished totivirus found in black raspberry but assumed to be a fungal virus). RasCV, Raphanus sativus chrysovirus. SoTV, Southern tomato virus. PCV1, Pepper cryptic virus 1. MeTr, methyl transferase; Hel, helicase; UGT, glycosyltransferase; RdRp, RNA dependent RNA polymerase; CP, coat protein; P, putative protease, NP, putative nucleoprotein.

Persistent plant viruses

Persistent viruses of plants are cytoplasmic viruses characterized by a lack of cell-to-cell movement, strictly vertical transmission, low titers, and asymptomatic infections. They are disseminated throughout the plants only by cell division, and are found in all cells including the meristem. They are members of three families recognized by the International Committee for the Taxonomy of Viruses (ICTV), *Endornaviridae* and *Partitiviridae* (King *et al.,* 2012), and the recently named *Amalgaviridae*, and two families not yet recognized as infecting plants, *Chrysoviridae* and *Totiviridae (Roossinck, 2010)*. Members of many of these families also infect fungi, including endophytic fungi [i.e. fungi that live inside plants (Bao and Roossinck, 2013)]. The *Amalgaviridae*, *Chrysoviridae*, *Partitiviridae* and *Totiviridae* have double-stranded (ds) RNA genomes, while the *Endornaviridae* probably have single-stranded (ss) genomes, although they are usually found as dsRNA replicative intermediates (Roossinck *et al.,* 2011). Representative members of the five plant persistent virus familes are listed in Table 9.1, and their genome organizations are shown in Figure 9.1.

Amalgaviridae

The first member of this family was found in tomato (Sabanadzovic *et al.,* 2009), followed soon after by descriptions of related viruses in bean (Liu and Chen, 2009), rhododendron (Sabanadzovic *et al.,* 2010), and blueberry (Martin *et al.,*

Table 9.1. Representative viruses for each family of persistent plant viruses

Virus name	Family	RNA 1[a]	RNA 2	RNA 3
Bell pepper endornavirus	*Endornaviridae*	JN019858	n.a	n.a.
Black raspberry virus F	*Totiviridae*	NC_009890	n.a.	n.a.
Raphanus sativus chrysovirus	*Chrysoviridae*	JQ045335	JQ045336	JQ045337
Southern tomato virus	*Amalgaviridae*	EF442780	n.a.	n.a.
Pepper cryptic virus 1	*Partitiviridae*	JN117276	JN117277	n.a.

2011). The putative RNA dependent RNA polymerase (RdRp) of these viruses is distantly related to the RdRp of members of the family *Partitiviridae*. However, unlike the partitiviruses the amalgaviruses have a single genome segment encoding two putative proteins. The open reading frame (ORF) upstream of the RdRp has been dubbed a coat protein (CP) but there is no evidence of virus particles, nor is there any nucleotide or amino acid similarity between these sequences and any known proteins. The family name derives from the idea that because these virus genomes are monopartite they are related to the *Totiviridae* (Martin *et al.,* 2011), but there is no evidence of this relationship, and recently an analysis based on structural similarities suggest that the putative coat protein may be related to the nucleocapsid protein of the *Bunyaviridae* (Krupovic *et al.,* 2015). The *Bunyaviridae* are (-) sense single-stranded RNA viruses that infect many hosts including plants. In the plant bunyavirus genus *Tenuivirus*, the nucleocapsid protein associates with the viral RNA to form long filaments, rather than true virus particles (Falk and Tsai, 1998). Since no particles have been seen for any of the amalgaviruses, Krupovic *et al.* (2015) propose this structure for these viruses as well, and suggest that the amalgaviruses arose via a recombination between a partiti-like virus and a bunya-like virus. If this virus family is indeed an unencapsidated ds RNA virus it is the first described. All other characterized dsRNA viruses encapsidate their genomes along with at least one copy of their RdRp, and do not truly uncoat in the infected cells, but instead extrude their ss mRNA and pregenomes from the capsid (Safari and Roossinck, 2014). This may be a strategy to avoid the problems of dsRNAs that can ellicit a number of antiviral responses.

Partitiviridae

Virus particles without association with disease were first noticed in beet plants in the late 1960s. These particles were later purified and characterized as *Beet cryptic virus* (Kassanis *et al.,* 1977) a member of the *Alphacryptovirus* genus in the family *Partitiviridae*. The partitiviruses have two genomic ds RNAs, each containing one ORF, and encoding an RdRp and CP. For most of the partitiviruses the larger RNA encodes the RdRp, and the RdRp RNA is called RNA 1 even if it is not the larger of the two RNAs. Numerous members of this family have been described, many of them from fungi. Recently the division of species into various genera within the family has been re-examined comparing the RNA dependent RNA polymerase (Nibert *et al.,* 2014). Phylogenetic evidence shows that while some clades are specific to plants or to fungi, other clades have members that infect a mixture of hosts, implying that there has been transmission between

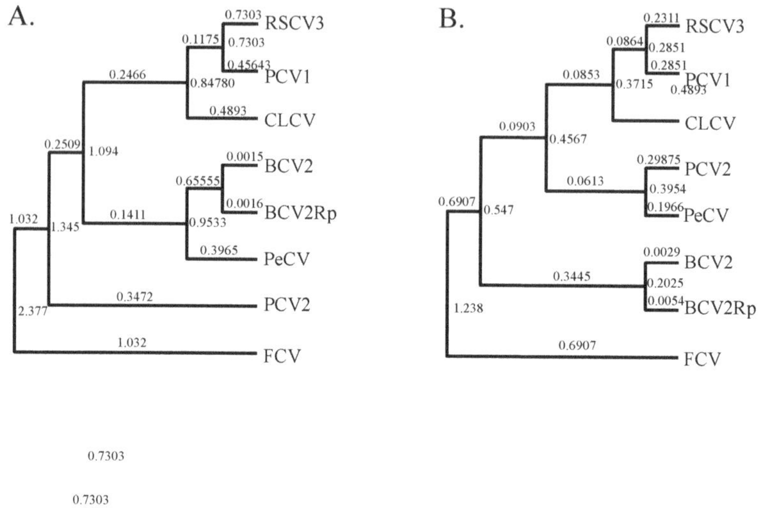

Figure 9.2. Comparison of phylogenetic trees for the Deltapartitiviruses. Amino acid sequences of the CP and RdRp were aligned in MAFFT and manually edited in Geneious 8. Baysian analysis was completed using the Mr Bayes (Huelsenbeck and Ronquist, 2001) plug-in in Geneious 8. Rate matrix was set to a poisson distribution with a gamma rate variation. Burn-in was 100,000 and total chain length was 1,100,000. Branch lengths were unconstrained. Relative branch lengths are shown above the lines; support for branches is shown at the nodes. Fig cryptic virus, which is included in the Deltapartitiviruses but is very distant from the others, was used as an outgroup. Virus names, abbreviations and accession numbers are listed in Table 9.2. A. Coat protein phylogeny. B. RdRp phylogeny.

Table 9.2. Deltapartitivirus names, abbreviations and protein accession numbers

Virus	Abbreviation	RdRp	CP
Beet cryptic virus 2	BCV2	ADP24757	ADP24756
Beet cryptic virus 2 Rp	BCV2Rp	AGQ49482	AGQ49466
Citrullus lanatus cryptic virus	CLCV	AGH12859	AGH12860
Fig cryptic virus	FCV	YP_004429258	CBW77437
Pepper cryptic virus 1	PCV1	AEJ07890	AYJ07891
Pepper cryptic virus 2	PCV2	AEJ07892	AEJ97893
Persimmon cryptic virus	PeCV	YP_006390091	YP_006390090
Raphanus sativus cryptic virus 3	RSCV3	YP_002364401	YP_002364402

plants and fungi (Li *et al.*, 2009; Nibert *et al.*, 2014; Roossinck, 2010). There is some direct evidence of transmission of partitiviruses among fungi, but not in plants (Nibert *et al.*, 2014), although there is one somewhat obscure report of a fungal partitivirus that has very close similarity to a plant partitivirus (Melnychuk *et al.*, 2005; Szegö *et al.*, 2010).

Plant virus evolution is often modular (Roossinck, 2005), and it seems plausible that reassortment has played a role in the evolution of the partitiviruses (Nibert *et al.*, 2014). A number of pepper cultivars are infected with both *Pepper cryptic virus* 1 (PCV1) and *Pepper cryptic virus* 2 (PCV2) (Sabanadzovic and Valverde, 2011), providing the opportunity for reassortment to occur. There are undoubtedly other examples of mixed infections in plants, and they are also quite common in fungi. A phylogenetic analysis of the RdRp and CP amino acid sequences of seven of the plant infecting viruses in the proposed Deltacryptovirus genus with complete genome information available, shows some incongruence between the two trees (Figure 9.2), providing support for reassortment in their evolutionary history. However, more data and deeper analyses are needed to confirm this. While the RdRp sequences of all of the Deltacryptoviruses appear to be homologous, this is much less certain for the coat proteins. Indeed, the diversity of coat protein sequences for the partitiviruses is extreme, and it seems unlikely that they are monophyletic.

Endornaviruses

During the 1980s large double-stranded RNAs (dsRNAs) were found in bean plants (Wakarchuk and Hamilton, 1985). These were not associated with any detectable particles or any observable symptoms, and they were not initially classified as viruses. Similar dsRNAs have been found in numerous plants since then and these make up the virus family *Endornaviridae*. Their evolutionary history is puzzling, because they contain genes that seem to have a wide variety of origins from plants to bacteria, and across the family only the RdRp could be considered homologous (Roossinck *et al.*, 2011). Like members of the *Partitiviridae*, the endornaviruses also infect fungi, and do not segregate in phylogenetic analyses by host type. Although classified as dsRNA viruses, their RNA dependent RNA polymerase (RdRp) more closely resembles a single stranded (ss) RNA RdRp, and they are in fact closest to the plant *Closteroviridae* (Roossinck *et al.*, 2011). It seems likely that they are in fact ssRNA viruses that, like the hypoviruses of *Cryphonectria*, are isolated as replicative intermediates.

Oryza sativa endornavirus is found in Japonica cultivars of rice, but not in the Indica cultivars (Horiuchi *et al.*, 2003). A related virus, *Oryza rufipogen endornavirus* is found in the wild progenitor of both Japonica and Indica rice. *Oryza sativa endornavirus* and *Oryza rufipogen endornavirus* share about 76% identity at the nucleotide level. The hosts of these two viruses diverged with the development of cultivated rice, about 10,000 years ago (Molina *et al.*, 2011). This gives us a probable timeline on the evolutionary rate of endornaviruses of about 0.34 mutations per genome per year, a surprising slow mutation rate given that the virus probably undergoes 3,000-4,000 generations per year (using a conservative estimate of ten generations/day), and dramatically lower than estimates for acute plant viruses (Pagán and Holmes, 2010).

Bell pepper endornavirus has been isolated and the complete sequence determined from five cultivars. Divergence within this group ranges from 0.5% to 12.2%. Unfortunately very little is currently known about the breeding history for these cultivars, and it isn't clear if the phylogeny of the pepper hosts would be congruent with that of the viruses, although the US isolate Yolo Wonder appears to be ancestral to the other isolates, which are all from Asia (Figure 9.3A). Recent

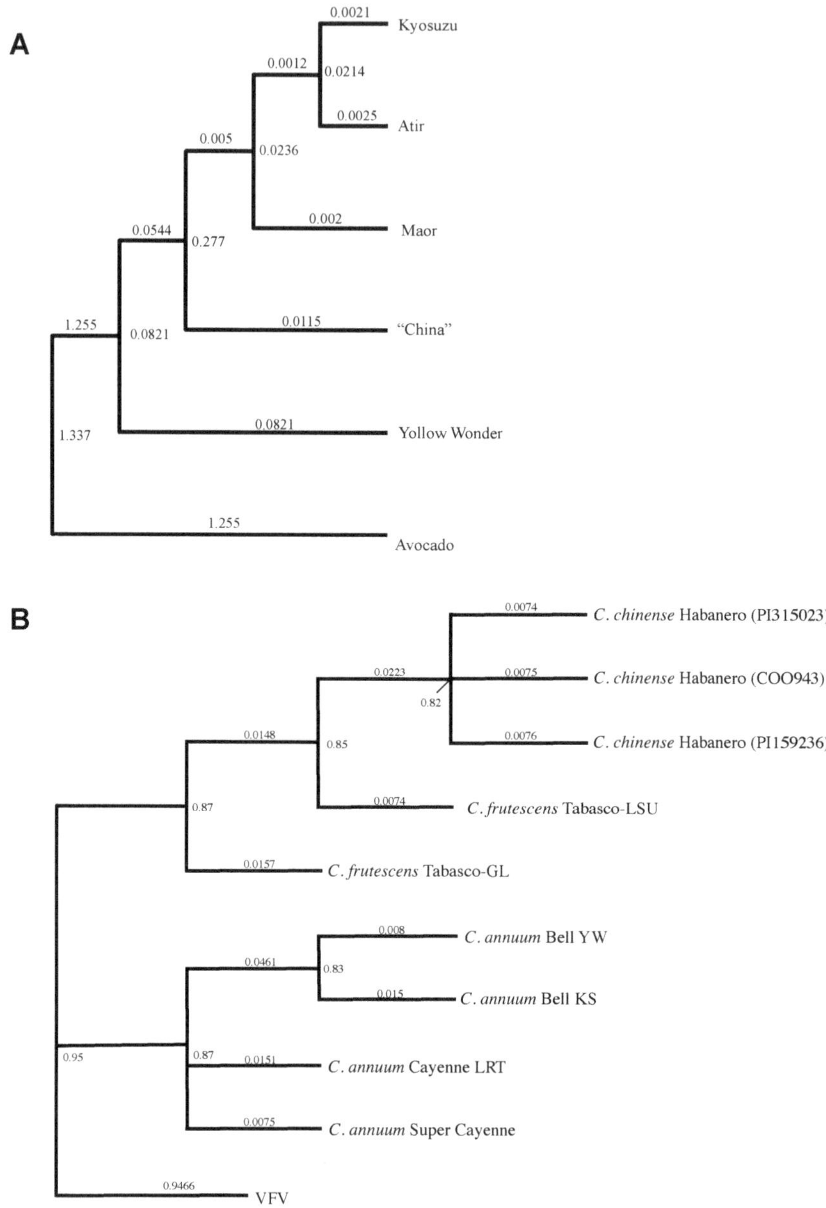

Figure 9.3. Phylogenetic analysis of Bell pepper endornavirus. A. Relationships among isolates from Bell pepper cultivars. Complete genomes (nucleotide sequences) were aligned using MAFFT, and manually edited in Geneious 8. Baysian analysis was completed using the Mr Bayes (Huelsenbeck and Ronquist, 2001) plug-in in Geneious 8. Rate matrix was set to a poisson distribution with a gamma rate variation. Burn-in was 100,000 and total chain length was 1,100,000. Branch lengths were unconstrained. Avocado endornavirus was used as an outgroup. B. Phylogenetic analyses of endornaviruses from different species and cultivars of pepper (reprinted with permission from (Okada *et al.*, 2011). Virus names, abbreviations and accession numbers are listed in Table 9.3.

Table 9.3. Endornavirus names, strains, and accession numbers

Virus	Host	Cultivar	Accession
Bell pepper endornavirus	*Capsicum annuum*	Kyosuzu	AB597230
Bell pepper endornavirus	*Capsicum annuum*	Yolo wonder	JN019858
Bell pepper endornavirus	*Capsicum annuum*	Atir	JQ951943
Bell pepper endornavirus	*Capsicum annuum*	"China"	KF709944
Bell pepper endornavirus	*Capsicum annuum*	Maor	KP455654
Avocado endornavirus	*Persea americana*	na	JQ951943

work on molecular markers in pepper will probably make it possible to clarify relationships among the pepper (Hill *et al.,* 2013). An analysis of only the RdRp of additional endornaviruses in the genus *Capsicum* suggests that their relationships are congruent with the hosts [(Okada *et al.,* 2011); Figure 9.3B]. An analysis of the time frame when congruence between virus and host is lost would help establish the length of time that plants have been infected with these viruses.

Chrysoviridae and Totiviridae

Deep sequence analysis of dsRNAs enriched from about 9000 individual plants shows many sequences that are related to members of the families *Totiviridae* and *Chrysoviridae* (Roossinck, 2012b). These two virus families are also common in fungi, and it seems likely that they also have persistent lifestyles, as there is no evidence of any movement proteins. Recently a tripartite chrysovirus was described in radish, and the sequence of a related virus from Anthirium, an ornamental Hawaiian plant has been reported (Li *et al.,* 2013). These two plant chrysoviruses differ from their fungal counterparts in the number of segments (three vs. four), conservation of the 3' non-translated regions, and size of the virions, but the similarity of the three putative proteins to fungal chrysoviruses puts them squarely in the same family.

Effects of persistent plant viruses on host evolution

Very high rates of vertical transmission in parasites is indicative of a mutualistic relationship. The persistent viruses approach transmission rates of 100%, and clearly have extremely long associations with their hosts. In most cases no function has been attributed to these viruses, although in one case, white clover, a persistent virus plays an important role in the host plant. In legumes the symbiotic relationship between bacteria and plant results in the formation of nodules that are able to fix atmospheric nitrogen. This allows legumes to grow under low nitrogen conditions, but when adequate nitrogen is available it is better for the plant not to make the costly nodules. In white clover the coat protein gene of *White clover cryptic virus* suppresses nodulation under high nitrogen conditions. The coat protein could be expressed in another legume, lotus, and it functioned in the same way (Nakatsukasa-Akune *et al.,* 2005). Hence the plant apparently has co-opted the coat protein of the virus for a new function. Since the diversity of viruses is extreme, it is not surprising that these types of co-incidental mutualism can occur between plants and viruses, and indeed mutualistic relationships are probably much more common than previously thought (Roossinck, 2011).

In beans (*Vicia faba*) an endornavirus is associated with cytoplasmic male sterility (Pfeiffer, 1999). Cytoplasmic male sterility is probably a negative trait for a plant, although it may be advantageous in some circumstances because it forces outcrossing. Cytoplasmic male sterility has been used extensively by the crop breeding industry to produce hybrids. Overall the cell biology of persistent virus infections is poorly studied, but a recent report about two distinct endornaviruses that can co-infect common bean (*Phaseolis vulgaris*) shows that the two viruses have different subcellular locations (Okada *et al.,* 2013).

There has been little evidence for host immune responses for most persistent plant viruses. Some viral metagenomic studies have used siRNAs, the plants adaptive immune response to many virus infections, to detect novel plant viruses (Roossinck *et al.,* 2015), but so far there has been only one report of reassembly of a complete endornavirus genome using siRNA sequences (Sela *et al.,* 2012). This may be related to the nature of the endornaviruses, that do not seem to encapsidate their RNAs and hence probably leave their replicative form of dsRNA exposed to the cell. It is also possible that most persistent viruses become recognized by the plant as "self", and are protected from immune response by other means.

Experimental evidence of host effects are complicated with persistent viruses. In most cases there are no uninfected hosts of the same cultivar to compare with infected hosts. However, vertical transmission is not 100%; Jalapeño pepper plants free of *Pepper cryptic virus 1* were obtained by screening about 50 plants grown from a single seed lot (Valverde and Gutierrez, 2008), and virus-free bell pepper plants lacking *Bell pepper endornavirus* were obtained in a similar way by screening over 100 plants (Pita and Roossinck, 2013). This provides a system for more study, but it will be necessary to use several individual lines obtained from single virus-free seeds to obtain reliable results.

Conclusions

The persistent plant viruses are widespread in plants but have remained very poorly studied. There are many intriguing aspects of their evolution. Why do the host plants maintain them for thousands of years? Why do they encapsidate, when they do not have any apparent external phase? If they do indeed escape the host immune response, how is this accomplished? How do some of them become incorporated into the host genomes? In the era of deep sequence analysis, more and more of these viruses are being discovered, and discovery will hopefully lead to in-depth characterization that may answer some of these important questions.

References

Bao, X., and Roossinck, M.J. (2013). Multiplexed interactions: viruses of endophytic fungi. In Fungal Viruses, S. Ghabrial, ed. (Elsevier), pp. 37-57.

Beijerinck, M.W. (1898). Concerning a contagium vivum fluidum as cause of the spot disease of tobacco leaves. In Phylopathological Classics, No 7, J. Johnson, ed. (St. Paul: American Phytopathological Society), pp. 33-52.

Chiba, S., Kondo, H., Tani, A., Saisho, D., Sakamoto, W., Kanematsu, S., and Suzuki, N. (2011). Widespread endogenization of genome sequences of non-retroviral RNA viruses into plant genomes. PLoS Pathogens 7, e1002146.

Falk, B.W., and Tsai, J.H. (1998). Biology and molecular biology of viruses in the genus *Tenuivirus*. Annu. Rev. Phytopathol. *36*, 139-163.

Hill, T.A., Ashrafi, H., Reyes-Chin-Wo, S., Yao, J., Stoffel, K., Truco, M.-J., Kozik, A., Michelmore, R.W., and VanDeynze, A. (2013). Characterization of *Capsicum annuum* genetic diversity and population structure based on parallel polymorphism discovery with a 30K unigene pepper genechip. PLoS ONE *8*, e56200.

Horiuchi, H., Moriyama, H., and Fukuhara, T. (2003). Inheritance of *Oryza* sativa endornavirus in F1 and F2 hybrids betwween *japonica* and *indica* rice. Genes and Genetics Systems *78*, 229-234.

Huelsenbeck, J.P., and Ronquist, F. (2001). MRBAYES: Bayesian inference of phylogenetic trees. Bioinformatics *17*, 754-755.

Kassanis, B., White, R.F., and Woods, R.D. (1977). Beet cryptic virus. Phytopath Z *90*, 350-360.

King, A.M.Q., Adams, M.J., Carstens, E.B., and Lefkowitz, E.J., eds. (2012). Virus Taxonomy Ninth Report of the International Committee On Taxonomy of Viruses (San Diego, CA: Elsevier Academic Press).

Krupovic, M., Dolja, V.V., and Koonin, E.V. (2015). Plant viruses of the *Almagaviridae* family evolved via recombination between virusese with double-stranded and negative-strand RNA genomes. Biology Direct *10*:12.

Li, L., Liu, J., Xu, A., Want, T., Chen, J., and Zhu, X. (2013). Molecular characterization of a trisegmented chrysovirus isolated from the radish *Raphanus sativus*. Vir Res *176*, 169-178.

Li, L., Tiam, Q., Du, Z., Duns, G.J., and Chen, J. (2009). A novel double stranded RNA virus detected in Primula malacoides is a plant-isolated partitivirus closely related to partivirus infecting fungal species. Arch. Virol. *154*, 565-572.

Liu, H., Fu, Y., Jiang, D., Li, G., Xie, J., Cheng, J., Pend, Y., Ghabriel, S.A., and Yi, X. (2010). Widespread horizontal gene transfer from double-stranded RNA viruses to eukaryotic nuclear genomes. J. Virol. *84*, 11879-11887.

Liu, W., and Chen, J. (2009). A double-stranded RNA as the genome of a potential virus infecting Vicia faba. Virus Genes *39*, 126-131.

Martin, R.R., Zhou, J., and Tzanetakis, I.E. (2011). Blueberry latent virus: an amalgam of the *Partitiviridae* and *Totiviridae*. Virus Res. *155*, 175-180.

Melnychuk, M.D., V, G., Sotrtdibiv, and Oleksienko, I.P. (2005). Features of virus double-stranded RNA, isolated from microscopic fungi parasitizing on sugar beet. Mikrobio Z *67*, 52-58.

Molina, J., Sikora, M., Garud, N., Flowers, J.M., Rubinstein, S., REynolds, A., Huang, P., Jackson, S., Schaal, B.A., Bustamante, C.D., *et al.* (2011). Molecular evidence for a single evolutionary origin of domesticated rice. Proc Natl Acad Sci USA *108*, 8351-8356.

Nakatsukasa-Akune, M., Yamashita, K., Shimoda, Y., Uchiumi, T., Abe, M., Aoki, T., Kamizawa, A., Ayabe, S.-i., Higashi, S., and Suzuki, A. (2005). Suppression of root nodule formation by artificial expression of the *TrEnodDR1* (coat protein of *White clover cryptic virus 2*) gene in *Lotus japonicus*. Mol. Plant-Microbe Inteact *18*, 1069-1080.

Nibert, M.L., Ghabrial, S.A., Maiss, E., Lesker, T., Vainio, E.J., Jiang, D., and Suzuki, N. (2014). Taxonomic reorganization of family *Partitviridae* and other recent progress in partitivirus research. Vir Res *188*, 128-141.

Okada, R., Kiyota, E., Sabanadzovic, S., Moriyama, H., Fukuhara, T., Saha, P., Roossinck, M.J., Severin, A., and Valverde, R.A. (2011). Bell pepper endornavirus: molecular and biological properties, and occurrence in the genus *Capsicum*. J Gen Virol *92*, 2664-2673.

Okada, R., Yong, C.K., Valverde, R.A., Sabanadzovic, S., Aoki, N., Hotate, S., Kiyota, E., Moriyama, H., and Fukuhara, T. (2013). Molecular characterization of two evolutionarily distinct endornaviruses co-infecting common bean (*Phaseolus vulgaris*). J Gen Virol *94*, 220-229.

Pagán, I., and Holmes, E.C. (2010). Long-term evolution of the *Luteoviridae*: time scale and mode of virus speciation. J Virol *84*, 6177-6187.

Pfeiffer, P. (1999). Possible Mechanisms of the "447" Cytoplasmic Male Sterility in *Vicia faba*, and unconventional CMS associated with a double-stranded RNA replicon. In Anther and Pollen From Biology to Biotechnology, C. Clément, E. Pacini, and J.-C. Audran, eds. (Heidelberg: Springer-Verlag), pp. 101-112.

Pita, J.S., and Roossinck, M.J. (2013). Mapping viral functional domains for genetic diversity in plants. J Virol *87*, 790-797.

Roossinck, M.J. (2005). Symbiosis *versus* competition in the evolution of plant RNA viruses. Nature Rev Microbiol *3*, 917-924.Roossinck, M.J. (2010). Lifestyles of plant viruses. Phil Trans Royal Soc B *365*, 1899-1905.

Roossinck, M.J. (2011). The good viruses: Viral mutualistic symbioses. Nature Rev Microbiol *9*, 99-108.

Roossinck, M.J. (2012a). Persistent Plant Viruses: Molecular Hitchhikers or Epigenetic Elements? In Viruses: Essential Agents of Life, G. Witzany, ed. (Dordrecht: Springer), pp. 177-186.

Roossinck, M.J. (2012b). Plant virus metagenomics: Biodiversity and ecology. Annu Rev of Genet *46*, 357-367.

Roossinck, M.J. (2014). Metagenomics of plant and fungal viruses reveals an abundance of persistent lifestyles. Frontiers in Microbiol *5*, 767.

Roossinck, M.J., Martin, D.P., and Roumagnac, P. (2015). Plant virus metagenomics: advances in virus discovery. Phytopathology. *105,* 716-727.

Roossinck, M.J., Sabanadzovic, S., Okada, R., and Valverde, R.A. (2011). The remarkable evoluntionary history of endornaviruses. J Gen Virol *92*, 2674-2678.

Sabanadzovic, S., Ghanem-Sabanadzovic, N.A., and Valverde, R.A. (2010). A novel monopartite dsRNA virus from rhododendron. Arch Virol *155*, 1859-1863.

Sabanadzovic, S., Valverde, R., Brown, J.K., Martin, R.R., and Tzanetakis, I.E. (2009). Southern tomato virus: the link between the families *Totiviridae* and *Partitiviridae*. Virus Res *140*, 130-137.Sabanadzovic, S., and Valverde, R.A. (2011). Properties and detection of two cryptoviruses from pepper (*Capsicum annuum*). Virus Genes *43*, 307-312.

Safari, M., and Roossinck, M.J. (2014). How does the genome structure and lifestyle of a virus affect its population variation? Curr Op Virol *9*, 39-44.

Sela, N., Luria, N., and Dombrovsky, A. (2012). Genome assembly of Bell pepper *endornavirus* from small RNA. J Virol *86*, 7721.

Szegö, A., Enünlü, N., Deshmukh, S.D., Veliceasa, D., Hunyadi-Gulyás, E.v., Kühne, T., Ilyés, P., Potyondi, L., Medzihradszky, K., and Lukács, N. (2010). The genome of *Beet cryptic virus* 1 shows high homology to certain cryptoviruses present in phylogenetically distant hosts. Virus Genes *40*, 267-276.

Valverde, R.A., and Gutierrez, D.L. (2008). Molecular and biological properties of a putative partitivirus from jalapeño pepper (*Capsicum annuum* L.). Rev Mex Fitopat *26*, 1-6.

Wakarchuk, D.A., and Hamilton, R.I. (1985). Cellular double-stranded RNA in *Phaseolus vulgaris*. Plant Mol Biol *5*, 55-63.

10

Paleovirology: The Study of Endogenous Viral Elements

Amr Aswad and Aris Katzourakis*

Department of Zoology, University of Oxford, South Parks Road, Oxford OX1 3PS, UK
**Corresponding author: aris.katzourakis@zoo.ox.ac.uk*

Abstract

Viruses sometimes heritably integrate into the genomes of their hosts, resulting in genomic features known as endogenous viral elements (EVEs). Using EVEs, the field of paleovirology investigates the long-term evolution of viruses and their impact on hosts. One of the fruitful outcomes of high throughput genomics is the widespread availability of whole genome data, offering the unprecedented opportunity to investigate EVEs at a large scale. In this chapter, we review the consequent surge in paleovirology research that can be traced to landmark work performed over half a century ago. We describe general principles of EVE biology and the main methodological techniques used to study them. We show how EVEs can only be understood within an evolutionary framework and we outline a generalized workflow for conducting paleovirology studies. We review exemplar paleovirological discoveries from each of the main viral groups, highlighting a range of approaches to paleovirology. We also discuss the major implications that certain discoveries and insights from paleovirology could have on our understanding of both virus and genome evolution. Finally, we consider the current limitations and potential pitfalls in paleovirology, and anticipate the possible future directions of this exciting and rapidly growing field.

Introduction

Usage of the term paleovirology to denote the study of ancient viruses has risen in the last decade, reflecting the establishment of a distinct specialty of evolutionary virology. The precise definition continues to expand since it has emerged from infancy, but predominantly involves the analysis of viral remnants in host genomes. Under the general term endogenous viral elements (EVEs), these 'viral fossils' are the consequence of a germline infection that results in the heritable genome integration of a virus (or part thereof) (Katzourakis and Gifford, 2010). Alternatively, ancient viruses can be studied by examining the lasting effect they have had on the evolution of antiviral genes, a method described as indirect paleovirology in contrast to the direct approach of investigating EVEs (Patel *et al.,* 2011).

The first EVEs were endogenous retroviruses identified in chickens and mice in the late 60's and early 70's, even before retroviruses were formally named (Weiss, 2006). Endogenous retroviruses (ERVs) are by far the most abundant group of EVEs due to an obligatory genome integration step in their replication cycle. Retroviruses posses the reverse transcriptase enzyme that is responsible for converting their RNA genomes into cDNA, which is inserted into the host genome by means of an integrase. This is in contrast to other EVEs, for which a route to endogenisation is not obvious; many viruses do not ordinarily integrate into the host genome or even enter the nucleus during replication. Moreover, other RNA viruses do not have enzymes that can generate cDNA or integrate the viral genome, and most DNA viruses lack an integration mechanism. Nonetheless, it was recently revealed that there are endogenous representatives of viruses from all seven main genome types and replication strategies (Katzourakis and Gifford, 2010). This finding was a significant milestone in the emergence of paleovirology, and was followed by a series of comprehensive reviews in quick succession that synthesized research from different fields under a new paleovirological umbrella. This included the decades of established ERV research, advances in genomics and virology, discoveries of non-retroviral EVEs and relevant studies of the evolution of virus-host interactions (Aswad and Katzourakis, 2012; Feschotte and Gilbert, 2012; Holmes, 2011; Patel *et al.,* 2011).

This chapter will offer a synthesis of paleovirology research, with a focus on the most recent non-retroviral EVE work. We will describe a set of techniques that encompasses the latest methods and approaches in paleovirology, which are intimately linked to the nature of the data and the interpretation of results. Viruses are classified according to the guidelines of the International Committee on Taxonomy of Viruses (ICTV), which has a hierarchal structure similar to that of zoological taxonomy. Additionally, viruses can be categorized according to their method of mRNA synthesis under the Baltimore Classification system, which recognizes seven different viral groups (Baltimore, 1971).

EVE biology

The integration of ancient viruses that became EVEs cannot be observed and there are no clear sources of evidence for how it could have occurred (Holmes, 2011). We can, however, speculate on the plausible mechanisms of integration by drawing from our knowledge of virology, cell biology and genomics. Understanding how the germline integration of a virus becomes an EVE requires us to consider consequences to the progeny within the context of host population genetics. Despite the remarkably diverse replication strategies employed by viruses, they share a common fundamental life cycle: a viral particle or virion infects a host cell in order to parasitize the molecular machinery needed for reproduction. Ordinarily, a virus is specific to a host species (or closely related species), and is usually adapted to target a particular cell type; a phenomenon known as host/cellular tropism. Interestingly, EVEs that have been described are not necessarily related to viruses with germline-specific tropism. Moreover, it is not clear where a germline invasion would be most likely to occur; possible targets include either of the two gametes or their precursor cells, the zygote or an early-stage embryo. Once inside the cell the number of steps preceding integration for viruses will vary depending on their genome type. While dsDNA viruses are already in a state capable of integrating into a host genome, single-stranded DNA viruses first require the synthesis of a complementary strand. For RNA viruses, the only means of producing complementary dsDNA genomes is through the catalytic action of reverse transcriptase.

Once it becomes endogenous, an EVE will conceivably reach fixation in the population rapidly if it offers a beneficial function to the host. Several such cases of co-opted EVEs have been documented in the literature, and they vary widely in terms of the how they benefit the host, as well as the mechanism in which this is achieved. In the case of ERVs, the genes of most loci are inactive relics that cannot encode proteins and many are in fact removed through recombination of their long terminal repeats (LTRs) that leaves behind 'solo-LTRs'. However, both full-length ERVs and solo LTRs can affect transcriptional regulation in the genome by introducing promoters and alternative splice sites. While this can be detrimental to the host, it has also shown to be beneficial in some cases. For example, an LTR is responsible for the beneficial elevation of transcription of the mouse cytidine deaminase APOBEC3 (Sanville *et al.,* 2010). Another mechanism by which EVEs can benefit the host is through coopted genes, such as the retrovirus-derived *Syncytin* gene in mammals. Multiple *Syncytins* have been identified in mammals, and they are responsible for the cell-to-cell fusion required in placental formation (Lavialle *et al.,* 2013). EVE-derived immunity genes (EDIs) have also been described in a number of mammals, where an EVE has been co-opted to function against the exogenous viral counterpart that they originate from (Aswad and Katzourakis, 2012).

Identifying & Authenticating Endogenous Viruses

Identifying EVEs involves a combination of sequence similarity searches and evolutionary analyses. A region of a host genome that is similar to a known viral sequence is likely to be an EVE, except in the case of viral genes that originate from host-gene capture. In order to identify novel EVEs, sequence data gathered from extant exogenous viruses are compared against a host genome using a sequence similarity-searching algorithm such as BLAST (Camacho *et al.,* 2009).

Step 1: Virus-host sequence comparison

Most EVEs are inactivated relics that are no longer subject to the sequence-preserving effect of purifying selection. This means that an EVE sequence is likely to only share low levels of similarity to a query sequence from an extant virus, particularly at the nucleotide level. However, due to the redundancy of the genetic code, the protein sequence takes longer to diverge, and so amino acid comparison is a more sensitive means of identifying EVEs (Figure 10.1a). The strategy for choosing a set of query sequences depends on the EVE being sought, the candidate host genome(s), and the amount of available processing power. The simplest and fastest method would involve using a single viral protein sequence to query a six-frame translation of a single host genome using tBLASTn (Figure 10.1a). A more comprehensive search of the same genome for EVEs related to the chosen virus type would use protein sequences from multiple virus species.

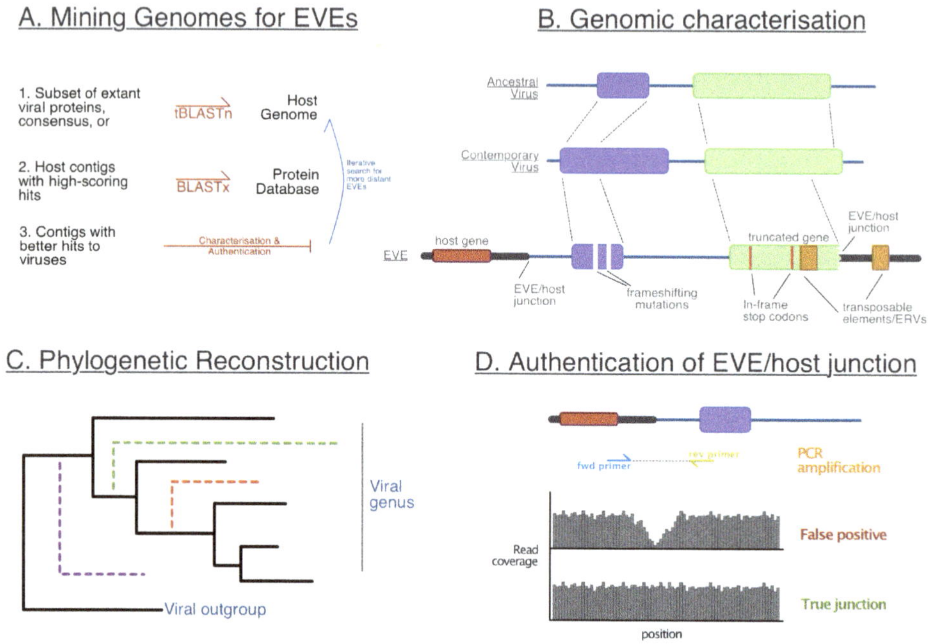

Figure 10.1. Finding and authenticating EVEs involves a multifaceted bioinformatic strategy that is informed by the nature of the data. **Panel A**: A strategic subset of extant viral proteins is used to query a genomic database of one or more hosts using the tBLASTn algorithm that searches a 6-frame translation of the target genome. From the resulting list of hits, a reciprocal BLASTx of the nucleotide sequences obtained from the genome(s) in question is performed against a general protein database to identify and roughly annotate each contig. Contigs from this search that contain hits to viruses, which out-score those to anything else, are shortlisted as putative EVEs for further analysis. **Panel B**: The extant viral query used to search host genomes for an EVE may not be identical to the ancestral virus that became the EVE. For instance in this hypothetical example, one of the genes was significantly shorter in the ancestral virus compared to the contemporary relative. For any contig suspected to contain an EVE, there are a number of genomic features we would expect to observe. Neutral genetic drift will often result in a number of in-frame stop codons and frame-shifting mutations. Recombination or errors at the point of integration could have caused the virus to become truncated, and the prolonged existence in a host genome will often result in the insertion of transposable elements such as SINES or LINEs into the EVE. In the most ideal scenario, the EVE/host junction will be clearly identifiable, with unambiguous host and viral genes on either side. **Panel C**: Phylogenetic reconstruction of the EVE is an essential part of an evolutionary analysis into the natural history of the virus in question. Various possible placements of an EVE in a tree of viral relatives are shown as dashed lines. The resulting topology that would be most easily interpretable is shown in red, where an EVE resolves deep within an existing viral group. In some cases however, long branch attraction can lead to the potentially incorrect placement of an EVE (shown here in green). The purple line shows another common result in paleovirology, where an EVE resolves outside known viral groups; this could either mean that it belongs to a previously unrecognised group, or the results could be used as justification for the expansion of an existing group to include the new diversity introduced by the EVE. **Panel D**: The authentication of the EVE/host junction can be achieved by confirmatory PCR and sequencing of a fragment amplified using primers designed on either side of the junction. Alternatively, or in addition, we can consult the raw sequence data of the locus, which will reveal whether or not the putative EVE is genuinely inserted into the genome or if the junction is just an assembly artifact.

The best strategy here would be to maximize diversity by choosing representatives from across the (extant) viral tree, or better yet, using all available proteins.

Rather than searching a single genome, multiple hosts can be targeted in a single search, as can multiple sets of viral queries. With unlimited computational resources, one can conceivably search every available host genome using every known viral protein. However, there will be an upper limit to what can be achieved by BLAST and similar programs, since sequence divergence due to genetic drift will eventually render the EVE unrecognizable (Kuchibhatla *et al.,* 2013). This sets a limit on the age of EVEs that can be found, with the exception of those that are co-opted by the host for at least some time after endogenisation (Aswad and Katzourakis, 2012). In such cases, the functional sequence in an EVE is maintained by purifying selection and therefore 'preserved' in the genomic fossil record for a longer time period. The alternative to using viral proteins to directly compare against host genomes is to use consensus sequences or sequence profiles that capture a lot of the viral diversity in a single query (Figure 10.1a). For example, programs such as HMMER (Finn *et al.,* 2011) use a hidden Markov model approach to compare individual sequences to sequence-profiles that capture both the sequence itself and also account for variability at each site. This contains information about how a set of sequences can evolve, such as the position of common insertions or deletions or the probability of an amino acid given the identity of the adjacent site. This allows us to detect faint patterns of similarity to viral queries and search for EVEs beyond the hypothetical age/divergence limit of BLAST. The efficacy of these tools has been demonstrated in viral genome annotation, where distant homologs of apparently unique genes were identified (Kuchibhatla *et al.,* 2013).

Step 2: Discriminating between true EVEs and false positives and determining the fixation in the population

A virus is considered an EVE once inherited vertically by the host from parent to a viable offspring. By this definition however, many more EVEs will have existed briefly, having never reached fixation in the population to be maintained and observed today. A similar set of analyses can be used to investigate the authenticity of an EVE as well as whether it is fixed in the host population. The most straightforward way to achieve both of these objectives is to identify homologous EVE loci. An orthologous EVE in another species or individual indicates that the loci are derived from a common ancestor, which indicates that they were vertically inherited. For EVEs that have not been co-opted, this also allows us to estimate the date of the integration by measuring the genetic distance between loci and dividing half of this by the host neutral rate of evolution. The probability that it is fixed is higher for older EVEs, if the individual being sampled is not part of a geographically isolated sub-population. This technique offers a conservative minimum bound for the age of an EVE, which can be compared to the expected time it would need to reach fixation according to the Hardy-Weinberg principle. Alternatively, the same dating technique can be applied to date the duplication event of paralogous EVEs in the same individual, which is a conservative minimum age of the EVE. This approach was used to date the integration of a lentivirus in the genome of the European rabbit at over 10 million years (Katzourakis *et al.,* 2007).

These methods depend on inferring either authenticity or fixation from the age of a putative EVE. If it is not possible to directly obtain an estimate of the integration date, we can also look for indirect evidence for the age of an EVE. A characteristic feature of most EVEs is the degradation of their sequence, either in the form of gene loss or nonsense mutations. This is because the population frequency of an EVE that is either neutral or only slightly detrimental to the host will be governed by genetic drift (Aswad and Katzourakis, 2012). This can be used as evidence to distinguish an EVE from an exogenous virus, particularly if it is very old and has had time to accumulate neutral mutations. However, evidence of degradation alone cannot exclusively prove that a viral sequence is an EVE, because the low fidelity of most viral polymerases means that a proportion of an exogenous viral population will exhibit similar patterns. Furthermore, host restriction factors are also known to hypermutate exogenous viruses (Zhang *et al.,* 2003), and so the coding capacity of a putative EVE can only act to supplement other sources of evidence. For example, an important step in the authentication process is the identification and characterization of the EVE/host junction, where one can discriminate the integration site by recognizing the difference between a host region and an unambiguously viral sequence (Figure 10.1b). This can be verified by the targeted amplification and sequencing of the junction, and additional assurance can be sought by interrogating the genome assembly for evidence of high coverage at the junction (Figure 10.1d). This is because viruses often have repetitive sequences that can be falsely connected to a repetitive region of the host genome by automatic contig assembly software. For example, an endogenous herpesvirus in the tarsier genome is integrated into the repetitive telomeric region, and a combination of high assembly coverage and targeted sequencing was used to confirm the authenticity of the integration site (Aswad and Katzourakis, 2014).

Step 3: Phylogenetic reconstruction and evolutionary analysis

In addition to characterizing the genomic locus, the authenticity of an EVE can be achieved as part of an evolutionary analysis that involves phylogenetic reconstruction. Although an EVE with remote similarity can still be characterized at the genomic level, phylogenetic analysis is a crucial step in the confident authentication of an EVE, as well as the key to understanding its natural history. Broadly speaking, EVEs that are undetectable by BLAST are also unlikely to yield robust phylogenetic results, since it will be difficult to align them with confidence to other viral sequences. In a phylogeny composed of viral sequences related to the query used to find the EVE as well as a number of more distantly related viruses, a true EVE should exhibit more sequence similarity to some viruses than to others (Figure 10.1c). Although not always the case (particularly for young insertions), the accumulation of mutations in an EVE as a result of neutral genetic drift can cause an EVE to have a longer terminal phylogenetic tree branches compared to viral sequences. This can sometimes return misleading results, because such long branches could resolve incorrectly in the tree (using some methods long branches tend to converge), making it difficult to precisely pinpoint the relationship of the EVE to extant viruses. In some cases, an EVE will resolve distantly to known viral groups, rather than within them, which can either mean that the phylogenetic range of the group needs to be expanded, or can be interpreted as evidence for a novel group of either extinct or unsampled viruses (Figure 10.1c).

EVEs from viruses of all known genome types

A number of EVEs are known that are related to each of the seven main groups of viruses as classified by the Baltimore system. Examples of EVEs from each of the groups have revealed aspects of viral evolutionary history, as has their host distributions and abundances. For example, the fact that ERVs are the most abundant type of EVE reveals that there might be a causal relationship between a viral replication strategy and its propensity to endogenise. For each of the seven groups, we will highlight examples of EVEs as a means of discussing the range of insights that have been gained in paleovirology, as well as methodological challenges that are often unique to a particular group.

Group I: Double-stranded DNA viruses

The largest viruses we know of possess dsDNA genomes, probably due to the biochemical stability of dsDNA compared to other nucleic acids. Thirty-three families of dsDNA viruses are currently recognized by the ICTV (King *et al.,* 2012), but only a small number of EVEs have been recognized. Their size is probably a barrier to endogenisation in that an insertion is more likely to be detrimental/fatal to the host the lager it is, because of the simultaneous introduction of a number of genes, regulatory elements and potential recombination sites. The scarcity of EVEs related to dsDNA viruses is also compounded by the bioinformatic challenges associated with searching for them. For example, while the number of host genomes in public databases continues to increase, the majority of them are in draft assembly states, meaning that large EVEs will be split over multiple short contigs. Moreover, dsDNA viruses are known to frequently co-opt host genes, which further complicates the identification of EVEs.

The *Polydnaviridae* (PDV) are a family of dsDNA virus EVEs that are associated with wasps in the *Ichneumonoidea* or *braconidae*, and are accordingly classified into the genus *Bracovirus* or *Ichnovirus,* respectively (Espagne *et al.,* 2004; Herniou *et al.,* 2013). The endogenisation of PDVs into wasp genomes has occurred multiple times in the evolutionary history of these wasps, leading to functional, vertically inherited EVEs that produce and package viral genes. These wasps parasitize lepidopteran larvae by injecting their eggs along with infectious PDV particles that 'protect' them from the caterpillar immune system (Herniou *et al.,* 2013). These PDVs were probably once viruses that became domesticated in the wasp genome, but they no longer replicate independently. The PDV 'macrolocus' is distributed across the wasp genome, and is differentially packaged into particles that do not individually constitute a complete polydnavirus genome. Interestingly, the PDV has undergone multiple duplication events that have resulted in the emergence of gene families, and the insertion of wasp genes into the PDV locus has been shown to have occurred over 86 million years ago (Herniou *et al.,* 2013). These characteristics highlight the ancient host-specific adaptations undergone by PDVs that characterize them as components of the genome, driven by selective pressure on the host rather than the virus.

A well-studied group of dsDNA viruses, in the order *Herpesvirales*, has a wide host range with particular host groups infected by each of the three constituent families. The *Alloherpesviridae* infect amphibians and fish, while mammals, birds and reptiles are susceptible to members of the *Herpesviridae*. The third family, *Malacoherpesviridae,* is populated by an oyster Herpesvirus (Davison *et al.,* 2009), but partially described malacoherpesvirus-like sequences were identified amidst the genome data of the lancelet (*Branchiostoma lanceolatum*). This could be evidence of a herpesvirus EVE in amphioxous, or else represents an exogenous virus that contaminated the sequencing sample (Savin *et al.,* 2010). Similarly, a herpesvirus related to roseoloviruses was identified in the genome of the Tarsier, a basal primate found in the Philippines, Malaysia, Indonesia and Brunei (Aswad and Katzourakis, 2014). In this study, the host-EVE junction was confirmed using experimental data and the sequences exhibited the nonsense mutations that are typical of anciently integrated EVEs. Although less ancient, the closely related human Roseolovirus has been shown to integrate into the germline and is inherited in ~1% of people (Arbuckle *et al.,* 2010).

A fascinating group of dsDNA viruses are the *Megavirales* (pending ICTV ratification) or nucleo-cytoplasmic large DNA viruses (NCLDVs) (Colson *et al.,* 2013). Among the megaviruses are the largest known viruses, which can exceed the size of the smallest cells, with genomes up to 2.5 MB in the case of the pandoravirus (Philippe *et al.,* 2013). Shortly following their discovery, it was found that some giant viruses are themselves susceptible to infection by viruses (known as virophages) that specifically parasitize the replication machinery of the host that was recruited by the giant virus known as a 'viral factory'(La Scola *et al.,* 2008). Unlike satellite viruses that require co-infection of a cell with a helper virus, virophages negatively affect the production of giant virus particles by ~80% (Campos and Boratto, 2014). EVEs known as mavericks, described in the genome of the giant virus' host amoeba, likely originate from the recurring integration of virophages (Fischer and Suttle, 2011; Katzourakis and Aswad, 2014). Interestingly, infection of a giant virus-infected amoeba with a virophage results in the partial recovery of the amoeba (Campos and Boratto, 2014). This could explain the repeated endogenisation of virophages in amoeba genomes, which could be co-opting them to defend against infection giant viruses (Fischer and Suttle, 2011; Katzourakis and Aswad, 2014).

Group II: Single-stranded DNA viruses

EVEs related to group II viruses are among the most abundant and widely represented non-retroviral EVEs in animal genomes, particularly those related to the family *Parvoviridae*. Well over a hundred different *Parvovirinae* and *Densovirinae* (the two subfamilies) have been identified in placental and non-placental mammals, fish, birds, a tunicate, crustaceans, insects and a flatworm (Belyi *et al.,* 2010a; Katzourakis and Gifford, 2010; Liu *et al.,* 2011a; Thézé *et al.,* 2014). An orthologous pair of parvovirus EVEs in mice and rats suggests that the integration event occurred at least 30 million years ago, offering a minimum age for the virus group for the first time (Katzourakis and Gifford, 2010). While nearly all of the vertebrate EVEs exhibit signs of degradation, an intact open reading frame of the NS1 gene was identified in a *Dependovirus*-like EVE in elephants (Katzourakis and Gifford, 2010), as well as a number of different invertebrate *Parvoviridae*-EVEs (Liu *et al.,* 2011a).

EVEs have also been described from a number of other families of ssDNA viruses. The exogenous viruses in the family *Circoviridae* have a host range limited to vertebrates. In addition to finding EVEs in species previously not known to be circovirus hosts, a diverse range of non-vertebrate genomes has been shown to harbor EVEs with circovirus similarity, including mites, nematodes, gastropods, hydrozoans and giardia (Belyi *et al.,* 2010b; Liu *et al.,* 2011b). Additionally, EVEs related to the plant-infecting *Geminiviridae* have been identified in several *Nicotania* species including many tobacco cultivars (*Nicotiana tabacum*), the black cottonwood (*Populus trichocarpa*) and green algae (Ashby *et al.*, 1997; Bejarano *et al.*, 1996; Liu *et al.*, 2011b).

Group III: Double-stranded RNA viruses

Group three viruses possess linear dsRNA genomes with between 1-4 segments or 10-12 in the case of the *Reoviridae*. The first dsRNA EVE to be identified, in the genome of the disease vector mosquito *Aedes aegypti,* is almost identical to segment 5 of the exogenous Liaoning virus isolated from another *Aedes* species (Katzourakis and Gifford, 2010). The presence of a host-flanking sequence and a large deletion excludes the possibility that the element is a contaminant, but the extremely high similarity to a currently circulating virus suggests that the integration occurred very recently. Indeed, it maybe that not all *A. aegypti* genomes contain the EVE, raising the possibility that this is an unfixed genome invasion.

Since the discovery of endogenous Liaoning virus, a number of studies revealed the widespread presence of dsRNA-related EVEs, particularly those related to the fungi- and plant-infecting *Partitiviridae* and *Totiviridae (Chu et al., 2014).* In addition to elements identified in four plants and four fungi, EVEs were identified in genomes that are not natural partitivirus/totivirus hosts. A partitivirus EVE was found in the genome of a gastropod, and a number of totuvirus-like EVEs we found in a nematode and protozoan (Liu *et al.*, 2010).

Group IV: Positive single-stranded RNA viruses

EVEs related to RNA viruses with positive-sense, single-stranded genomes have been identified in both animals and plants (Belyi *et al.*, 2010c; Chiba *et al.*, 2011; Crochu *et al.*, 2004; Katzourakis and Gifford, 2010; Roiz *et al.*, 2009; Tanne and Sela, 2005). Much of the diversity of group IV viruses is distributed among those that infect plants, but eight different families of mainly human and non-human vertebrate viruses are known. In the case of plants, EVEs with similarity to the plant-specific potyviruses were described in the genome of the common grapevine (*Vitis vinifera*) (Tanne and Sela, 2005). In the Western honeybee genome (*Apis melifera*), an EVE with similarity to the *Dicistoviridae* was identified and named the Israeli acute paralysis virus (IAPV) (Maori *et al.*, 2007). Interestingly, the IAPV is one of the extremely rare demonstrable examples of a viral endogenisation event that has yet to reach fixation in the population, having been identified in only 30% of bee populations (Maori *et al.*, 2007). In addition to the ~1% of humans known to carry a heritable integration of human herpesvirus 6 (Arbuckle *et al.*, 2010, 2013), we also know of an ERV present in some koala populations (Tarlinton *et al.*, 2006), which invaded the genome less than 50,000 years ago and continues to proliferate (Ishida *et al.*, 2014). Similarly, retroviruses that are still capable of intragenomic proliferation can give rise to new EVE paralogs that are polymorphic in the host population, such as the human ERV family HERV-K (HML2) (Belshaw *et al.*, 2005).

The *Flaviviridae* includes a variety of human disease-causing viruses that are transmitted by arthropod vectors, and in the case of two different mosquito species (*A. aegypti* and *A. albopictus*) EVEs were identified that are phylogenetically similar to the *Aedes* flaviviruses cell fusing agent and Kamiti River virus (Crochu *et al.*, 2004; Katzourakis and Gifford, 2010; Roiz *et al.*, 2009). It was also found that in the medaka genome, there is a putative flavivirus EVE with similarity to *Tamana Bat Virus*, although the authors could not rule out the possibility of contamination (Belyi *et al.*, 2010c).

Group V: Negative single-stranded RNA viruses

Among the group V viruses is the order *Mononegavirales*, which is composed of five families that mainly infect humans and other vertebrates, all but one of which are represented in the viral fossil record. *Bornaviridae*-like EVEs are especially abundant and are found in numerous and diverse hosts, including mammals, marsupials, reptiles, insects, fish and spiders (Belyi *et al.*, 2010c; Horie *et al.*, 2010, 2013; Katzourakis and Gifford, 2010). Many of these EVEs are only represented by the bornavirus nucleoprotein gene, and are referred to as endogenous borna-like N (EBLN). While there have been multiple independent insertions of bornavirus EVEs, orthologous loci in different species allow us to obtain a minimum bound of their integration date from estimates of their hosts' speciation dates. For example, a set of orthologous EBLNs in afrotherians can be dated to at least 65 million years old based on the divergence date of their hosts (Horie *et al.*, 2013). The identification of orthologous EBLNs in strepsirrhine primates indicated that a particular EBLN integration occurred over 54 million years ago and was selectively maintained for some time in all the extant host lineages, with coding capacity maintained in baboon and all great apes except orangutans (Horie *et al.*, 2013; Katzourakis and Gifford, 2010). Although a specific function has yet to be ascribed to any of these potentially functional ORFs, a recent study in the thirteen-lined ground squirrel demonstrated that EBLNs can act as a restriction factor of Bornavirus replication (Fujino *et al.*, 2014). This pioneering work has paved a route for translational research into *Borna Disease Virus*, which is a known zoonotic agent that can infect humans.

EVEs related to *Filoviridae* have also been identified in a range of mammals (Katzourakis and Gifford, 2010; Taylor *et al.*, 2010, 2011), which is surprising given that their exogenous counterparts are limited to primates and bats in central Africa (Holmes, 2011). The phylogeographic pattern of exogenous and endogenous filoviruses has been considered as evidence of large gaps in the sampling of exogenous filoviruses, because known exogenous viruses are more closely related to marsupials than to endogenous filoviruses of placental mammals (Katzourakis and Gifford, 2010; Taylor *et al.*, 2010). *Rhabdovirus*-like EVEs were initially identified in the genomes of the black-legged tick (*Ixodes scapularis*) and mosquitoes (*A. aegypti*) (Katzourakis and Gifford, 2010), and there is some evidence for rhabdovirus EVEs in other arthropods (Fort *et al.*, 2011). In addition to the *Mononegavirales,* there are a number of EVEs related to ssRNA negative strand viruses that are not assigned to a viral order. A short sequence in the black-legged tick genome resembling viruses from the genus *Quarjavirus* are the only orthomyxovirus-like EVEs described to date (Katzourakis and Gifford, 2010). This EVE is particularly interesting since the influenza viruses are also contained within the *Orthomyxoviridae*, and insights into their long-term evolution will be informative to intervention strategies.

Group VI: Retrotranscribing RNA viruses

Retroviruses represent the vast majority of the genomic fossil record. For example, while ~3% of the Human genome is composed of protein-coding genes, it is estimated that ERVs represent up to 8% (Tristem, 2000). Part of this abundance can be explained by their replication cycle that involves a genome integration step. Additionally however, retroviruses can continue to proliferate in the genome after the initial endogenisation event. The consequent evolutionary dynamics of ERVs in the genome are therefore distinct from those resulting from the endogenisation of other viral groups.

The repertoire of ERVs in mammalian genomes is structured into 'families' that originate from a single insertion. An ERV can proliferate by re-infecting the germline cell, but also intra-genomically if it is incapable of producing virions because of the loss/damage of some of their genes (Jern and Coffin, 2008). In the latter case, this means that we can track the natural history of a retroviral lineage that has only ever existed within the confines of a genome. Interestingly, the number of ERVs in the genomes of different mammals varies widely and the mechanisms behind this non-uniformity are not fully understood. An increased capacity to proliferate intragenomically in some species could explain the differences observed in some cases. It has been shown that the loss of the envelope gene (*env*) in the IAP family of murine ERVs has resulted in an increased rate of proliferation despite an inability to re-infect (Magiorkinis *et al.,* 2012).

The fact that the loss of *env* has occurred multiple times, independently, suggests that the process might be adaptive (Magiorkinis *et al.,* 2012). Evolving towards intragenomic replication might be favorable to an ERV due to the harmful effects of *env* expression to the host cell, as well as the consequent harmful effect of viral infection to the host in general (Magiorkinis *et al.,* 2012). Similarly, another observation in the abundance and distribution of ERVs could be explained as an adaptive response to the harmful effects of ERVs. In mammals, the number of ERVs in a species correlates negatively with larger body size (Katzourakis *et al.,* 2014). This could be explained as an adaptive response by larger hosts that are under a higher selective pressure to control the mutagenic effects of ERVs. This could be due to the increases in risk associated with having a higher number of cells that also undergo a greater number of cell divisions since larger mammals live longer (Katzourakis *et al.,* 2014).

Group VII: Retro-transcribing double-stranded DNA viruses

The two families of group VII viruses are the *Caulimoviridae* that infect plants and the vertebrate-infecting *Hepadnaviridae*, both of which possess dsDNA genomes that replicate via an RNA intermediate. This step in the replication cycle has also earned them the term pararetroviruses in reference to their reverse transcription, but unlike retroviruses they do not require a genome integration step.

A hepadnavirus EVE was identified in the genome of the black-legged tick (Katzourakis and Gifford, 2010), despite the fact that hepadnaviruses are not known to infect insects. This EVE could belong to an unknown vertebrate lineage of hepadnavirus that crossed species facilitated by the parasitic nature of ticks. Another set of hepadnavirus EVEs also expanded the previously known host range of hepadnaviruses to include reptiles. Two independent insertions of hepadnavirus-like EVEs were identified in the genome of speckled rattlesnakes (*Crotalus mitchellii* species complex), one of which is also shared by the king cobra (*Ophiophagus hannah*).

An EVE related to avian hepadnaviruses was identified the genome of the zebra finch by comparison to the duck hepatitis B virus (DHBV). The EVE is spread over multiple chromosomes but altogether represents almost 80% of the hypothesized ancestral virus (Gilbert and Feschotte, 2010a; Katzourakis and Gifford, 2010). The endogenous zebra finch HBV (eZHBV) is distributed across four different chromosomes, which means that the ancestral insertion could have been fragmented by rearrangements. Moreover, one of the fragments described was shown to have duplicated post-insertion, allowing the estimation of a minimum bound for the date of the integration. By dividing the molecular divergence of the duplicate fragments by an estimate of the host neutral evolutionary rate, we can estimate a minimum date of the duplication between 3.8 and 7.5 million years ago (Gilbert and Feschotte, 2010a).

The date of the hepadnavirus genome invasion was estimated to be between 19 and 40 million years ago by finding an ortholog of eZHBV in the distantly related dark-eyed junco (*Junco hyemalis*), thereby inferring the date from estimates of the birds' speciation date (Gilbert and Feschotte, 2010a). In addition to the Dark-eyed junco, fragments of endogenous avian hepadnaviruses were identified in the genomes of the Black-throated finch (*Poephila cincta*), Scaly-breasted monia (*Lonchura punctulata*) and Gouldian finch (*Chloebia gouldiae*), representing other orthologs of two distinct eZHBV integrations (Gilbert and Feschotte, 2010a). While these EVEs were all fragmented, two full-length avian hepadnavirus EVE were identified in the budgerigar (*Melopsittacus undulatus*), one of which also underwent a partial duplication that increased its length by ~30% (Liu *et al.,* 2012). Further details into the natural history of these EVEs emerged as the number of identified eZHBV accumulated. The combined strategic sampling for new loci and an analysis of their integration dates revealed that the oldest eZHBV must have endogenised before the divergence of neoavian birds, suggesting that the hepadnaviruses are at least ~74-94 million years old (Suh *et al.,* 2013).

Four of the seven genera of *Caulimoviridae* have been identified in plant genomes (Staginnus and Richert-Pöggeler, 2006). Multiple EVEs related to *Banana streak virus* (*Badnavirus*) were identified in the genome of banana (*Musa sp.*) (Iskra-Caruana *et al.,* 2014). In rice (*Oryza sp.*) a *Tungrovirus* EVE was identified and a *Petuvirus*-like EVE was found in the genome of petunia (*Petunia hybrida*) and the trifoliate orange (*Poncirus trifoliata*), which is closely related to citruses (Kunii *et al.,* 2004; Richert-Pöggeler *et al.,* 2003). *Cavemovirus*-like EVEs have also been identified in several species of tobacco (*Nicotiana sp.*), potato (*Solanum sp.*) and tomato (*Lycopersicon sp.*) (Hansen *et al.,* 2005; Jakowitsch *et al.,* 1999; Lockhart *et al.,* 2000).

Paleovirology: State of the art and future directions

The causes and consequences of EVEs

It has been less then a decade since non-retroviral EVEs have been studied under the same evolutionary framework – i.e. paleovirology – that also links them to the large body of research into endogenous retroviruses. The presence of ERVs in host genomes is relatively well understood in that we know the biological mechanisms responsible for their integration and subsequent intragenomic proliferation. EVE formation involves two steps; integration into the germline followed by vertical transmission to the progeny. For some EVEs we have been able to infer the mechanism of germline integration, either by extrapolating from the known properties of related exogenous viruses, or observations that indicate a plausible mechanism of integration. For example, we know that a number of herpesviruses have been shown to somatically integrate into the genome, including human roseoloviruses (Morissette and Flamand, 2010). Given the close genomic and phylogenetic similarity between roseoloviruses and the endogenous tarsier herpesvirus, we can infer that its integration involved homologous recombination with host telomeric repeats, which have been identified in the tarsier herpesvirus EVE (Aswad and Katzourakis, 2014). In the case of EVEs related to RNA viruses, it has been speculated that their conversion to dsDNA (a prerequisite of integration) could be achieved by reverse-transcriptase that is abundant in host genomes due to the presence of ERVs and other retroelements (Holmes, 2011). Indeed it has been shown that cDNA of bornaviruses can be detected in infected cell, and endogenous bornavirus EVEs exhibit features such as poly-A tails and target sited duplications that suggest LINE-mediated reverse transcription and integration (Horie *et al.,* 2010).

The herpesvirus and bornavirus examples are instances where we have inferred integration events by observing the outcome, but integration itself has also been experimentally demonstrated. Cell culture experiments showed that while the majority of detectable bornaviruses were in the cytosol, a proportion of integrated elements can be detected ~30 days after infection (Horie *et al.,* 2010). The group V *Lymphocytic choriomeningitis virus* can also become integrated into the genome after reverse transcription and recombination with a mouse ERV (Geuking *et al.,* 2009). Interestingly, the ERVs in this case belonged to the IAP family of *env*-less ERVs that are particularly prolific in the mouse genome (Magiorkinis *et al.,* 2012), which might positively influence the probability of recombination. We have also observed the process of vertical transmission of an integrated virus, which is the second step in the formation of an EVE. Both *Human Herpesvirus 6* and the koala retroviral EVE KoRV are unfixed in the host population, and the inheritance of the same EVE locus from parent to offspring has been demonstrated in genealogical studies (Arbuckle *et al.,* 2010; Ishida *et al.,* 2014). While none of these examples demonstrates the *in vivo* integration followed by proof of vertical inheritance, synthesizing the evidence for both steps in the endogenisation process allows us to support hypotheses about EVE formation.

The harmful effect of EVE integrations that disrupt genes and important regulatory regions are the same for both retroviral and non-retroviral EVEs. Active ERVs impose an additional fitness cost to the host because of the pathogenicity of the virus itself, as well as the potential harm of new integrations in the genome. Furthermore, inactivated ERV families contribute to non-homologous recombination that can result in major genomic rearrangements. On the other hand, we know of multiple instances where ERVs have played a role beneficial to the host. Syncitins are a group of related genes in mammalian genomes that are derived from the *env* gene of ERVs, and are a crucial component of placental development. A number of ERV-derived immunity genes (EDIs) have also been described, as well as co-opted ERV regulatory sequences that contribute to host genome function (Aswad and Katzourakis, 2012). To date, only one non-retroviral EVE has been shown to be beneficial to its host, but this is sure to change given the numerous EVEs with intact open reading frames and/or evidence of expression. At their initial discovery, evidence of purifying selection in EBLNs suggested that they might have been providing the host with a beneficial function, perhaps as an EDI (Aswad and Katzourakis, 2012; Horie *et al.*, 2013; Katzourakis and Gifford, 2010). It was later shown that, *in vitro,* an EBLN in the genome of the thirteen-lined ground squirrel (*Ictidomys tridecemlineatus*) was capable of both reducing *Borna disease virus* replication as well as inhibiting new infections (Fujino *et al.,* 2014). Other EVEs are on a similar route that will potentially lead to the characterization of novel functional genes. For example, the dependovirus identified in the genome of the degu (*Octodon degus*) is differentially expressed, which indicates a potentially tissue-specific function (Arriagada and Gifford, 2014).

Current and foreseeable challenges in paleovirology

While the precise definition and remit of paleovirology continues to adjust in these early years, at its core is the study of virus evolution within a framework that considers the deep history of EVEs. Across all of the different viral groups, the recurring finding is that viral groups are a lot older than has been inferred by extrapolation from contemporary sequences. This discrepancy between short and long term viral evolutionary rates is an unusual phenomenon that is an area of ongoing research [see (Aiewsakun and Katzourakis, 2015) for a review]. However, despite the prefix that implies a limitation of the scope to ancient EVEs, the methodology and objectives to studying recently integrated viruses largely overlap. In addition to ERVs, a number of non-retroviral EVEs had been described before they were recognized as part of a broader genomic fossil record, but it was not until the advent of cheap, rapid genome sequencing that they could be studied systematically. Efficient data management is one of the primary difficulties that paleovirologists currently face, a challenge that will grow alongside the rapid and voluminous generation of genomic data.

Another issue that is symptomatic of new disciplines is a general lack of standardization across several aspects of EVE research. At the most basic level, nomenclature is often inconsistent, with various studies referring to endogenous viruses differently. Although 'EVE' has been featured in a number of high impact publications since its inception, EVE sub-groups have also been designated as descriptors that sometimes overlap, such as 'non-retroviral integrated RNA viruses' (NIRVs) (Taylor *et al.,* 2011), 'non-retroviral RNA virus sequences' (NRVSs) (Chiba *et al.*, 2011) or 'endogenous plant pararetroviruses' (EPRVs) (Staginnus and Richert-Pöggeler, 2006). Moreover, classification within the ICTV requires a virus species to exhibit multiple distinguishable features or phenotypes that set it apart from other viruses (King *et al.,* 2012), which can be difficult if an EVE is only represented by a sequence. Similarly, viruses at higher taxonomic levels (e.g. genus) are typically grouped according to a set of distinguishing phenotypic criteria, such as morphology or pathogenicity. This means that even if phylogenetic evidence strongly supports the inclusion of an EVE within an established group, the rules that define that group cannot be evaluated for an EVE.

More importantly, there is a lot of variation in the standards and criteria for authenticating and describing an EVE. The evidence used to support the identity of a viral sequence as genuinely endogenous range from as little as sequence similarity that is not scrutinized further, to fully characterized genomic loci with robust phylogenetic reconstruction and additional sequencing. Furthermore, there are variations in what we think of as a single EVE, which becomes problematic in cases where a single integration is distributed across different loci as in the case of PDVs (Herniou *et al.,* 2013) or hepadnaviruses (Gilbert and Feschotte, 2010b; Katzourakis and Gifford, 2010; Liu *et al.,* 2012).

The size of genomic databases has already reached a point where the routine and efficient searching of all known viral proteins against a six-frame translation of all genomes is not feasible on standard desktop computers, requiring instead a computer cluster or super-computer. Even using a high-performance workstation, the search for EVEs is usually limited in scope, either by reducing the number of query sequences or target hosts, or both. This means that our knowledge of the EVE repertoire of host genomes is not usually complete, nor is the full distribution and abundance of EVEs related to particular viral groups. In addition to the taxonomic and nomenclature inconsistencies and variation in methodological rigor, this incompleteness of the genomic fossil record impedes empirical lines of enquiry that will help us understand aspects of viral evolution and history of host interaction. For example in the case of *Parvoviridae*-like EVEs, there have already been over a hundred described loci, the vast majority of which were identified nearly five years before the writing this chapter (Katzourakis and Gifford, 2010; Liu *et al.,* 2011a). The number of publically available host genomes has more than doubled since then, suggesting that a comparable number of new *Parvoviridae* EVEs are yet to be described. This means that it will be possible to investigate factors that influence their distribution in host species. Unlike ERVs, the fact that EVEs do not proliferate will allow us for the first time to dissociate the effect of endogenisation on host genomics, ecology and evolution from subsequent intra-genomic proliferation. The answers to such questions as well as new questions will emerge from paleovirology as it establishes norms and standard practices, as well as community resources such as curated, publically available non-redundant databases.

References

Aiewsakun, P., and Katzourakis, A. (2015). Endogenous viruses: Connecting recent and ancient viral evolution. Virology.

Arbuckle, J.H., Medveczky, M.M., Luka, J., Hadley, S.H., Luegmayr, A., Ablashi, D., Lund, T.C., Tolar, J., De Meirleir, K., Montoya, J.G., et al. (2010). The latent human herpesvirus-6A genome specifically integrates in telomeres of human chromosomes *in vivo* and *in vitro*. PNAS *107*, 5563–5568.

Arbuckle, J.H., Pantry, S.N., Medveczky, M.M., Prichett, J., Loomis, K.S., Ablashi, D., and Medveczky, P.G. (2013). Mapping the telomere integrated genome of human herpesvirus 6A and 6B. Virology *442*, 3–11.

Arriagada, G., and Gifford, R.J. (2014). Parvovirus-derived endogenous viral elements in two South American rodent genomes. J. Virol. *88*, 12158–12162.

Ashby, M.K., Warry, A., Bejarano, E.R., Khashoggi, A., Burrell, M., and Lichtenstein, C.P. (1997). Analysis of multiple copies of geminiviral DNA in the genome of four closely related Nicotiana species suggest a unique integration event. Plant Mol. Biol. *35*, 313–321.

Aswad, A., and Katzourakis, A. (2012). Paleovirology and virally derived immunity. Trends Ecol. Evol. *27*, 627–636.

Aswad, A., and Katzourakis, A. (2014). The first endogenous herpesvirus, identified in the tarsier genome, and novel sequences from primate rhadinoviruses and lymphocryptoviruses. PLoS Genet. *10*, e1004332.

Baltimore, D. (1971). Expression of animal virus genomes. Bacteriol. Rev. *35*, 235–241.

Bejarano, E.R., Khashoggi, A., Witty, M., and Lichtenstein, C. (1996). Integration of multiple repeats of geminiviral DNA into the nuclear genome of tobacco during evolution. PNAS *93*, 759–764.

Belshaw, R., Dawson, A.L.A., Woolven-Allen, J., Redding, J., Burt, A., and Tristem, M. (2005). Genomewide screening reveals high levels of insertional polymorphism in the human endogenous retrovirus family HERV-K(HML2): implications for present-day activity. J. Virol. *79*, 12507–12514.

Belyi, V.A., Levine, A.J., and Skalka, A.M. (2010a). Sequences from ancestral single-stranded DNA viruses in vertebrate genomes: the parvoviridae and circoviridae are more than 40 to 50 million years old. J. Virol. *84*, 12458–12462.

Belyi, V.A., Levine, A.J., and Skalka, A.M. (2010b). Sequences from ancestral single-stranded DNA viruses in vertebrate genomes: the parvoviridae and circoviridae are more than 40 to 50 million years old. J. Virol. *84*, 12458–12462.

Belyi, V.A., Levine, A.J., and Skalka, A.M. (2010c). Unexpected inheritance: multiple integrations of ancient bornavirus and ebolavirus/marburgvirus sequences in vertebrate genomes. PLoS Pathog. *6*, e1001030.

Camacho, C., Coulouris, G., Avagyan, V., Ma, N., Papadopoulos, J., Bealer, K., and Madden, T.L. (2009). BLAST+: architecture and applications. BMC Bioinformatics *10*, 421.

Campos, R., and Boratto, P. (2014). Samba virus: a novel mimivirus from a giant rain forest, the Brazilian Amazon. Virol. J.

Chiba, S., Kondo, H., Tani, A., Saisho, D., Sakamoto, W., Kanematsu, S., and Suzuki, N. (2011). Widespread endogenization of genome sequences of non-retroviral RNA viruses into plant genomes. PLoS Pathog. *7*, e1002146.

Chu, H., Jo, Y., and Cho, W.K. (2014). Evolution of endogenous non-retroviral genes integrated into plant genomes. Curr. Plant Biol.

Colson, P., De Lamballerie, X., Yutin, N., Asgari, S., Bigot, Y., Bideshi, D.K., Cheng, X.-W., Federici, B.A., Van Etten, J.L., Koonin, E. V, et al. (2013). "Megavirales", a proposed new order for eukaryotic nucleocytoplasmic large DNA viruses. Arch. Virol. *158*, 2517–2521.

Crochu, S., Cook, S., Attoui, H., Charrel, R.N., De Chesse, R., Belhouchet, M., Lemasson, J.-J., de Micco, P., and de Lamballerie, X. (2004). Sequences of flavivirus-related RNA viruses persist in DNA form integrated in the genome of Aedes spp. mosquitoes. J. Gen. Virol. *85*, 1971–1980.

Davison, A.J., Eberle, R., Ehlers, B., Hayward, G.S., McGeoch, D.J., Minson, A.C., Pellett, P.E., Roizman, B., Studdert, M.J., and Thiry, E. (2009). The order Herpesvirales. Arch. Virol. *154*, 171–177.

Espagne, E., Dupuy, C., Huguet, E., Cattolico, L., Provost, B., Martins, N., Poirié, M., Periquet, G., and Drezen, J.M. (2004). Genome sequence of a polydnavirus: insights into symbiotic virus evolution. Science *306*, 286–289.

Feschotte, C., and Gilbert, C. (2012). Endogenous viruses: insights into viral evolution and impact on host biology. Nat. Rev. Genet. *13*, 283–296.

Finn, R.D., Clements, J., and Eddy, S.R. (2011). HMMER web server: interactive sequence similarity searching. Nucleic Acids Res. *39*, W29–W37.

Fischer, M., and Suttle, C. (2011). A virophage at the origin of large DNA transposons. Science (80-.). *332*, 231–234.

Fort, P., Albertini, A., Van-Hua, A., Berthomieu, A., Roche, S., Delsuc, F., Pasteur, N., Capy, P., Gaudin, Y., and Weill, M. (2011). Fossil Rhabdoviral Sequences Integrated into Arthropod Genomes: Ontogeny, Evolution, and Potential Functionality. Mol. Biol. Evol. *29*, 381–390.

Fujino, K., Horie, M., Honda, T., Merriman, D.K., and Tomonaga, K. (2014). Inhibition of Borna disease virus replication by an endogenous bornavirus-like element in the ground squirrel genome. PNAS *111*, 13175–13180.
Geuking, M.B., Weber, J., Dewannieux, M., Gorelik, E., Heidmann, T., Hengartner, H., Zinkernagel, R.M., and Hangartner, L. (2009). Recombination of retrotransposon and exogenous RNA virus results in nonretroviral cDNA integration. Science *323*, 393–396.
Gilbert, C., and Feschotte, C. (2010a). Genomic fossils calibrate the long-term evolution of hepadnaviruses. PLoS Biol. *8*, 12.
Gilbert, C., and Feschotte, C. (2010b). Genomic fossils calibrate the long-term evolution of hepadnaviruses. PLoS Biol.
Hansen, C.N., Harper, G., and Heslop-Harrison, J.S. (2005). Characterisation of pararetrovirus-like sequences in the genome of potato (Solanum tuberosum). Cytogenet. Genome Res. *110*, 559–565.
Herniou, E.A., Huguet, E., Thézé, J., Bézier, A., Periquet, G., and Drezen, J.-M. (2013). When parasitic wasps hijacked viruses: genomic and functional evolution of polydnaviruses. Philos. Trans. R. Soc. Lond. B. Biol. Sci. *368*, 20130051.
Holmes, E.C. (2011). The Evolution of Endogenous Viral Elements. Cell Host Microbe *10*, 368–377.
Horie, M., Honda, T., Suzuki, Y., Kobayashi, Y., Daito, T., Oshida, T., Ikuta, K., Jern, P., Gojobori, T., Coffin, J.M., et al. (2010). Endogenous non-retroviral RNA virus elements in mammalian genomes. Nature *463*, 84–87.
Horie, M., Kobayashi, Y., Suzuki, Y., and Tomonaga, K. (2013). Comprehensive analysis of endogenous bornavirus-like elements in eukaryote genomes. Philos. Trans. R. Soc. Lond. B. Biol. Sci. *368*, 20120499.
Ishida, Y., Zhao, K., Greenwood, A.D., and Roca, A.L. (2014). Proliferation of Endogenous Retroviruses in the Early Stages of a Host Germ Line Invasion. Mol. Biol. Evol. msu275 – .
Iskra-Caruana, M.-L., Duroy, P.-O., Chabannes, M., and Muller, E. (2014). The common evolutionary history of badnaviruses and banana. Infect. Genet. Evol. *21*, 83–89.
Jakowitsch, J., Mette, M.F., van Der Winden, J., Matzke, M.A., and Matzke, A.J. (1999). Integrated pararetroviral sequences define a unique class of dispersed repetitive DNA in plants. PNAS *96*, 13241–13246.
Jern, P., and Coffin, J.M. (2008). Effects of retroviruses on host genome function. Annu. Rev. Genet. *42*, 709–732.
Katzourakis, A., and Aswad, A. (2014). The origins of giant viruses, virophages and their relatives in host genomes. BMC Biol. *12*, 51.
Katzourakis, A., and Gifford, R.R.J. (2010). Endogenous viral elements in animal genomes. PLoS Genet. *6*, e1001191.
Katzourakis, A., Tristem, M., Pybus, O.G., and Gifford, R.J. (2007). Discovery and analysis of the first endogenous lentivirus. PNAS *104*, 6261–6265.
Katzourakis, A., Magiorkinis, G., Lim, A.G., Gupta, S., Belshaw, R., and Gifford, R. (2014). Larger mammalian body size leads to lower retroviral activity. PLoS Pathog. *10*, e1004214.
King, A.M.Q., Adams, M.J., Carstens, E.B. and Lefkowitz, E.J. (2012). Virus Taxonomy: Ninth Report of the International Committee on Taxonomy of Viruses (San Diego, USA: Elsevier).
Kuchibhatla, D.B., Sherman, W.A., Chung, B.Y.W., Cook, S., Schneider, G., Eisenhaber, B., and Karlin, D.G. (2013). Powerful Sequence Similarity Search Methods and In-Depth Manual Analyses Can Identify Remote Homologs in Many Apparently "Orphan" Viral Proteins. J. Virol. *88*, 10–20.
Kunii, M., Kanda, M., Nagano, H., Uyeda, I., Kishima, Y., and Sano, Y. (2004). Reconstruction of putative DNA virus from endogenous rice tungro bacilliform virus-like sequences in the rice genome: implications for integration and evolution. BMC Genomics *5*, 80.

Lavialle, C., Cornelis, G., Dupressoir, A., Esnault, C., Heidmann, O., Vernochet, C., and Heidmann, T. (2013). Paleovirology of "syncytins", retroviral env genes exapted for a role in placentation. Philos. Trans. R. Soc. Lond. B. Biol. Sci. *368*, 20120507.

Liu, H., Fu, Y., Jiang, D., Li, G., Xie, J., Cheng, J., Peng, Y., Ghabrial, S.A., and Yi, X. (2010). Widespread horizontal gene transfer from double-stranded RNA viruses to eukaryotic nuclear genomes. J. Virol. *84*, 11876–11887.

Liu, H., Fu, Y., Xie, J., Cheng, J., Ghabrial, S.A., Li, G., Peng, Y., Yi, X., and Jiang, D. (2011a). Widespread endogenization of densoviruses and parvoviruses in animal and human genomes. J. Virol. *85*, 9863–9876.

Liu, H., Fu, Y., Li, B., Yu, X., Xie, J., Cheng, J., Ghabrial, S.A., Li, G., Yi, X., and Jiang, D. (2011b). Widespread horizontal gene transfer from circular single-stranded DNA viruses to eukaryotic genomes. BMC Evol. Biol. *11*, 276.

Liu, W., Pan, S., Yang, H., Bai, W., Shen, Z., Liu, J., and Xie, Y. (2012). The first full-length endogenous hepadnaviruses: identification and analysis. J. Virol. *86*, 9510–9513.

Lockhart, B.E., Menke, J., Dahal, G., and Olszewski, N.E. (2000). Characterization and genomic analysis of tobacco vein clearing virus, a plant pararetrovirus that is transmitted vertically and related to sequences integrated in the host genome. J. Gen. Virol. *81*, 1579–1585.

Magiorkinis, G., Gifford, R.J., Katzourakis, A., De Ranter, J., and Belshaw, R. (2012). Env-less endogenous retroviruses are genomic superspreaders. PNAS *109*, 7385–7390.

Maori, E., Tanne, E., and Sela, I. (2007). Reciprocal sequence exchange between non-retro viruses and hosts leading to the appearance of new host phenotypes. Virology *362*, 342–349.

Morissette, G., and Flamand, L. (2010). Herpesviruses and chromosomal integration. J. Virol. *84*, 12100–12109.

Patel, M.R., Emerman, M., and Malik, H.S. (2011). Paleovirology—ghosts and gifts of viruses past. Curr. Opin. Virol. *1*, 304–309.

Philippe, N., Legendre, M., Doutre, G., and Couté, Y. (2013). Pandoraviruses: amoeba viruses with genomes up to 2.5 Mb reaching that of parasitic eukaryotes. Science (80-.).

Richert-Pöggeler, K.R., Noreen, F., Schwarzacher, T., Harper, G., and Hohn, T. (2003). Induction of infectious petunia vein clearing (pararetro) virus from endogenous provirus in petunia. EMBO J. *22*, 4836–4845.

Roiz, D., Vázquez, A., Seco, M.P.S., Tenorio, A., and Rizzoli, A. (2009). Detection of novel insect flavivirus sequences integrated in Aedes albopictus (Diptera: Culicidae) in Northern Italy. Virol. J. *6*, 93.

Sanville, B., Dolan, M.A., Wollenberg, K., Yan, Y., Martin, C., Yeung, M.L., Strebel, K., Buckler-White, A., and Kozak, C.A. (2010). Adaptive evolution of Mus Apobec3 includes retroviral insertion and positive selection at two clusters of residues flanking the substrate groove. PLoS Pathog. *6*, e1000974.

Savin, K.W., Cocks, B.G., Wong, F., Sawbridge, T., Cogan, N., Savage, D., and Warner, S. (2010). A neurotropic herpesvirus infecting the gastropod, abalone, shares ancestry with oyster herpesvirus and a herpesvirus associated with the amphioxus genome. Virol. J. *7*, 308.

La Scola, B., Desnues, C., Pagnier, I., Robert, C., Barrassi, L., Fournous, G., Merchat, M., Suzan-Monti, M., Forterre, P., Koonin, E., et al. (2008). The virophage as a unique parasite of the giant mimivirus. Nature *455*, 100–104.

Staginnus, C., and Richert-Pöggeler, K.R. (2006). Endogenous pararetroviruses: two-faced travelers in the plant genome. Trends Plant Sci. *11*, 485–491.

Suh, A., Brosius, J., Schmitz, J., and Kriegs, J.O. (2013). The genome of a Mesozoic paleovirus reveals the evolution of hepatitis B viruses. Nat. Commun. *4*, 1791.

Tanne, E., and Sela, I. (2005). Occurrence of a DNA sequence of a non-retro RNA virus in a host plant genome and its expression: evidence for recombination between viral and host RNAs. Virology *332*, 614–622.

Tarlinton, R.E., Meers, J., and Young, P.R. (2006). Retroviral invasion of the koala genome. Nature *442*, 79–81.
Taylor, D., Leach, R., and Bruenn, J. (2010). Filoviruses are ancient and integrated into mammalian genomes. BMC Evol. Biol. *10*, 193.
Taylor, D.J., Dittmar, K., Ballinger, M.J., and Bruenn, J. (2011). Evolutionary maintenance of filovirus-like genes in bat genomes. BMC Evol. Biol. *11*, 336.
Thézé, J., Leclercq, S., Moumen, B., Cordaux, R., and Gilbert, C. (2014). Remarkable diversity of endogenous viruses in a crustacean genome. Genome Biol. Evol. *6*, 2129–2140.
Tristem, M. (2000). Identification and Characterization of Novel Human Endogenous Retrovirus Families by Phylogenetic Screening of the Human Genome Mapping Project Database. J. Virol. *74*, 3715–3730.
Weiss, R.A. (2006). The discovery of endogenous retroviruses. Retrovirology *3*, 67.
Zhang, H., Yang, B., Pomerantz, R.J., Zhang, C., Arunachalam, S.C., and Gao, L. (2003). The cytidine deaminase CEM15 induces hypermutation in newly synthesized HIV-1 DNA. Nature *424*, 94–98.

11

Population Genetic Modeling of Viruses

Jayna Raghwani[1]*, Oliver G. Pybus[1] and Christopher J.R. Illingworth[2]

[1] University of Oxford, Depart of Zoology, South Parks Road, Oxford, OX1 3PS, UK;
[2]University of Cambridge, Department of Genetics, Downing Street, Cambridge, CB2 3EH, UK. Email: jayna.raghwani@zoo.ox.ac.uk; oliver.pybus@zoo.ox.ac.uk; chris.illingworth@gen.cam.ac.uk

**Corresponding Author: jayna.raghwani@zoo.ox.ac.uk*

Abstract

Population genetic models provide a powerful approach to the study of the evolutionary dynamics of viruses. The application of population genetic approaches to viruses has grown in recent years and has been facilitated by advances in sequencing technologies that make it feasible to obtain multiple 'genetic' snapshots of evolving viral populations through time. Populations of viruses are often shaped by a combination of high mutation rates, strong selection, large population sizes, and recurrent bottlenecks, and thus inhabit a different region of population genetic "parameter space" to cellular organisms. In this chapter, we discuss how traditional and novel population genetic models have provided insights into virus evolutionary processes and the means by which pathogenic viruses may be combatted. In particular, we illustrate how population genetics can be used to understand adaptive evolution in viruses and review recent work on whether viral evolution can be predicted.

Introduction

Population genetics concerns itself with the nature and dynamics of genetic variation in natural populations. While many biological questions can be adequately addressed by considering the genetic composition of just one or a few representative individuals from a species [as exemplified by the singular "human genome" project (Lander *et al.*, 2001; Venter *et al.*, 2001)], natural populations are typically genetically diverse. Population genetics aims to understand the natural processes that create, extinguish, and modify the genetic variation within populations and has, since its inception, sought to describe evolutionary processes using mathematical models (Fisher, 1930).

Several features of the biology of viruses make them particularly interesting as subjects of population genetic study. First, the genomes of most pathogenic viruses are comparatively small and simple in structure. As such, the number of viral gene and genome sequences available for analysis can be very large, exceeding 19,500 complete genomes for the influenza A virus. In addition the functions, structures and interactions of viral proteins are often well characterised at the molecular level, making it easier to understand the effects of viral mutations on important traits such as transmission and virulence. Second, many viruses, particularly RNA viruses, evolve exceptionally rapidly and have rates of mutation that are orders of magnitude faster than those observed in multicellular organisms. This means that significant evolutionary change can accrue in viral populations over timescales as short as weeks and months; this rapid evolution can have important consequences for how viral pathogens are treated and controlled.

In this chapter we outline the growing application of population genetic models to viral populations, and illustrate how they have provided insights into evolutionary processes and the means by which pathogenic viruses may be combatted. We begin by providing a brief introduction to the basic components of population genetic theory and their application to viral populations. In the second section, we describe how population genetics has been used to understand and measure the adaptive evolution of viral populations. Lastly, we review recent work on the important question of whether viral evolution can be predicted. Influenza A viruses feature repeatedly in our chapter and are of particular interest in viral population genetics, due in part to their importance for global human health and because of the wealth of influenza genomic data available for analysis.

Population genetic processes

Population genetic theory is well established and based on mathematical models of evolutionary processes that were first developed during the 'modern synthesis' of evolutionary biology during the 1930s and 1940s (Fisher, 1930; Wright, 1931; Dobzhansky, 1937; Huxley, 1942). Models are used to explain how processes such as mutation, natural selection and random genetic drift generate the genetic variation observed in natural populations. In this section, we introduce the major processes of population genetics and briefly outline the role of each in virus populations.

Natural selection

Natural selection acts via the reproduction, or lack of reproduction, of individuals in a population. The likelihood that a given individual will reproduce and the number of offspring that results depend upon a variety of factors. If the presence or absence of a mutation (or other type of heritable genetic change) is one such factor then, over time, the frequency of that genetic variant in the population will change. The process of natural selection is modelled in population genetics by assigning a *selection coefficient* (denoted σ) to each genetic variant. The selection coefficient represents the degree to which the variant is *advantageous* ($\sigma > 0$) or *deleterious* ($\sigma < 0$) to individuals that carry it. Population genetic models can predict the outcome of natural selection by making simplifying assumptions. For example, we can consider a single genetic locus in a hypothetical population of infinite size, at which two variants (or alleles), denoted A and a, are present. In the absence of any other evolutionary effects, the change in the frequency of allele *A* due to natural selection is given by

$$dx_A/dt = \sigma x_A(1-x_A) \quad [Eq.\ 1]$$

where x_A is the frequency of *A* in the population. Under positive selection ($\sigma > 0$), the frequency of *A* will increase, while under negative selection ($\sigma < 0$) its frequency will decrease. If x_A increases to 1.0 then the allele is said to have become *fixed*, or *substituted*, within the population. The rate at which such fixations accrue in a virus population is referred to as either the *substitution rate*, the *fixation rate*, or the *evolutionary rate* of the virus. This rate can be readily estimated from viral genomes that have been sampled at different time points (Drummond *et al.,* 2003). The term '*selective sweep*' is used to describe the process whereby a strongly-selected variant becomes rapidly fixed in a population. A detailed review of the determinants and patterns of viral substitution rates is provided elsewhere (Duffy *et al.,* 2008).

Identifying the action of natural selection is an important question in population genetics. The presence of selection within viral genes is often investigated by the measurement of dN/dS, the ratio of the rates of non-synonymous and synonymous fixation among sampled individuals (Yang and Bielawski, 2000). This approach is based on the assumption that synonymous substitutions (i.e. those that do not change the corresponding amino acid in a gene) are selectively neutral (i.e. are neither advantageous nor deleterious) and therefore provide a baseline rate of fixation. If non-synonymous mutations are mostly deleterious, then the average rate of fixation of non-synonymous mutations will be slower than that of neutral mutations, hence dN/dS < 1. Conversely, if a gene or set of codons has experienced strong positive selection, then non-synonymous fixation rates may exceed synonymous ones, giving dN/dS > 1. Thus, given a set of divergent gene sequences, the dN/dS method can be used to infer the nature of selective pressures on those sequences. The dN/dS method has been applied many times to virus species (e.g. Bennett *et al.,* 2003; Gaschen *et al.,* 2002) and seems well suited to identifying sites in viral proteins that are subjected to repeated immune selection. Furthermore, when protein structural data is explicitly incorporated into estimating dN/dS ratios, it can significantly improve the power to detect selection (Tusche *et al.,* 2012; Meyer and Wilke, 2012).

Some limitations apply to the application of dN/dS to viral populations. For example, accurate measurement requires a sufficient number of fixations to have occurred. Over very short time-scales, as may be relevant to some viral studies, estimates of dN/dS may be harder to interpret (Mugal *et al.,* 2013). Further, the assumption that synonymous mutations are selectively neutral is open to question. Synonymous mutations in the C2-V5 region of the HIV-1 *env* gene have been inferred to be mildly deleterious (Zanini and Neher, 2013) and influenza virus populations show evidence of adaptation for codon usage (Kryazhimskiy *et al.,* 2008a). Experiments conducted with polioviruses suggest that synonymous mutations can exhibit a spectrum of mutational effects, from lethally deleterious to advantageous (Acevedo *et al.,* 2014). The existence of secondary RNA structure in RNA virus species (Simmonds *et al.,* 2004) is one reason why synonymous sites may not be selectively neutral.

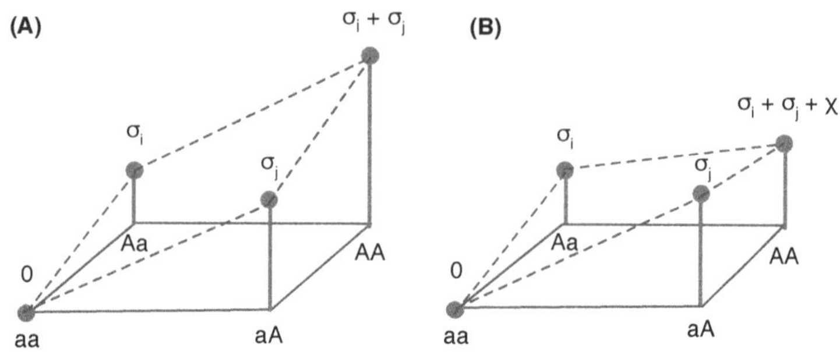

Figure 11.1. Illustration of a hypothetical fitness landscape. Regions of increased height correspond to genotypes with increased relative fitness. The arrow shows a path of adaptation towards the local fitness maximum at point A. This itself is separated from the global fitness maximum at B; crossing from A to B would require adaptation across a valley of lower fitness genotypes.

More direct methods for measuring selection are available when sequences are sampled from rapidly-evolving populations at different times. Specifically, Equation 1 can be directly fitted to observed allele frequencies, and parameters that describe the time of origin and strength of selection of a given mutation can be inferred. Such an approach was taken by Ganusov *et al.* (2011) to measure the fitnesses of differing CTL escape responses against HIV-1. Such approaches are straightforward to apply to large populations in which allele frequencies evolve independently from each another, and have been further adapted to consider populations of unknown size (Bollback *et al.,* 2008) and situations in which selected alleles interact with one another during asexual evolution (Illingworth and Mustonen, 2011).

The behaviour of a population in response to natural selection is often described using the concept of a *fitness landscape* (Figure 11.1). Each point on the landscape's surface represents a different genotype (i.e. a genome sequence) and the surface's height at each point represents the relative fitness of that genotype. Fitness landscapes can be helpful for describing the dynamics of adaptation in qualitative terms; for example, positive selection allows a population to '*climb*' the surface towards a '*fitness peak*', or cause it to become stuck at a local maximum, requiring it to cross a '*fitness valley*' in order to achieve further adaptation. The concept of a fitness landscape has been generalised to account for time-dependent fitness effects, giving rise to the notion of a '*fitness sea-scape'* (Mustonen and Lässig, 2009) upon which a population is evolving to climb a continually shifting fitness peak. Evolution in viral populations is further complicated by the simultaneous but independent action of natural selection at different levels of organisation]so called *multi-level* selection (Belshaw *et al.,* 2011)]. Mutations that are advantageous to the virus at one scale (e.g. within a patient taking anti-viral drugs) may be deleterious at another scale (e.g. reducing transmission between hosts).

Epistasis
So far we have only considered the evolution of single genetic variants in a population. However, the fitness effects of new mutations depend also on the genomic background on which they arise. *Epistasis* describes the presence of interactions between selective effects acting at different genetic loci. In the absence of epistasis, selective effects combine in an additive manner; if an allele A_i at locus i has selection coefficient σ_i, and allele A_j at locus j has selection coeffecient σ_j, then an individual with both alleles will exhibit the selection coefficient ($\sigma_i + \sigma_j$). Under epistasis, the combined effect of the alleles is different from this sum. Positive epistasis occurs if the combined effect of the two alleles is greater than their sum, and negative epistasis arises when the combined effect is less than their sum. Under sign epistasis, one allele changes the direction of the effect of another; a beneficial mutation becomes deleterious, or vice versa. In real populations epistasis is an important factor in understanding how natural selection modulates genotype frequencies and generates adaptation.

Viral populations provide some of the clearest examples of epistasis in action. For example, using sequences from HIV patients that underwent anti-viral drug therapy, Bonhoeffer *et al.* (2004) demonstrated that positive epistasis is prevalent in the HIV reverse transcriptase and protease genes. Similarly, epistatic interactions have been observed in other viral pathogens, such as influenza A virus and hepatitis C virus, indicating that it is a pervasive force in viral evolution (Gong *et al.,* 2013; Kryazhimskiy *et al.,* 2011a; Parera and Martínez, 2014; Shapiro *et al.,* 2006). Furthermore, there is growing evidence that epistatic effects can, to a large degree, be explained by protein structure, often leading to compensatory changes that restore protein stability (Gong *et al.,* 2013) . A recent analysis of the evolutionary trajectory of the nucleoprotein from influenza A virus revealed that destabilizing mutations typically occur together with stabilizing mutations, close to the time of mutation arising (Gong *et al.,* 2013). Other viral phenotypes are also likely to influence epistasis in the genome, such as those that arise via protein-to-protein interactions or RNA secondary structure (Sanjuán, 2006; Wilke *et al.,* 2003).

Mutation
Mutation is a random process of nucleotide sequence change, resulting in the creation of new alleles. Mutation is vital to adaptation, creating the genetic diversity upon which natural selection is able to act. The mutation rate of a virus has important consequences for its biology; for example, viruses with higher mutation rates might be expected to more readily evolve resistance to a particular drug or adapt to a new host species (e.g. Russell *et al.,* 2012). In the population genetic literature, the term mutation most commonly refers to a change in the allele (or nucleotide) present at a given locus (or site), although insertion and deletion changes in genetic sequences can also be modelled.

Some viruses, particularly those with small and/or RNA genomes, exhibit exceptionally high rates of genetic mutation. Measurements of the rate of mutation have been conducted for a range of viruses (Sanjuán *et al.,* 2010). Estimated rates vary substantially between DNA viruses (10^{-8} to 10^{-6} mutations per nucleotide per cellular infection) and RNA viruses (10^{-6} to 10^{-4} mutations per nucleotide per cellular infection). These mutation rates are consistent with a broadly inverse relationship between genome size and mutation rate (Drake, 1991; Linz *et al.*, 2014).

Although mutations may have a broad range of fitness effects, the majority of new mutations are likely to be disadvantageous and statistical analyses suggest that deleterious polymorphisms are abundant in most RNA virus species (Pybus *et al.*, 2007). The high rates of influenza virus mutation may be responsible for low viral survival rate; although more than 10,000 viruses may be produced by a single infected cell (Sidorenko and Reichl, 2004), the majority of these viruses are not able to initiate infections in other cells (Marcus *et al.*, 2009).

Population genetic models can help elucidate the evolutionary consequences of mutation acting in the absence of natural selection. Consider again an infinite population of individuals characterised by a single genetic locus with two-alleles. The change in the frequency of allele *A* due to mutation alone is given by the equation:

$$dxA/dt = \mu\ (1-2xA) \qquad [Eq.\ 2]$$

where μ is the rate of mutation, and mutation from any one allele to another is equally likely. In the absence of any other evolutionary effects, the frequency of variant *A* will eventually tend to one half. However, this process is so slow (because μ is so small) that mutation alone is unlikely to cause rapid or substantial shifts in allele frequencies, and mutation as a process is only significant when very rare genetic variants are being considered (see below).

Random genetic drift

The equations described above for natural selection and mutation are both based on the assumption of a population of infinite size. This deterministic approximation provides a valuable insight into the mechanisms of evolution, and for large populations under strong selection, evolving over short timescales, it can provide an adequate description of reality (e.g. Illingworth *et al.*, 2012). However, in smaller populations, and over longer timescales, stochastic effects become important for evolution. In a population of finite size, the number of offspring produced by any one individual is subject to random events. Thus even when a polymorphic allele is not subject to mutation or selection, its frequency will change from generation to generation as a result of these events, via a process known as *random genetic drift*. The importance of genetic drift in a population depends on the size of the population, *N*. In a randomly-mating population of constant size, for a polymorphic allele at frequency *x*, drift becomes more important than selection if $x<1/N\sigma$, where σ is the magnitude of selection on the allele. A second allele frequency threshold exists at $x<1/N\mu$, above which genetic drift becomes more important than mutation. A more thorough review on stochastic evolutionary effects in viral populations may be found elsewhere (Rouzine *et al.*, 2001).

In more complex evolutionary scenarios, the effective population size, N_e, is used to characterise the extent to which a population is subjected to random genetic drift (Wright, 1931). In a population of viruses N_e can be very different to the census number of individuals in the population. For example, if a population of cells is infected with a retrovirus, each cell may release a large number of viruses,

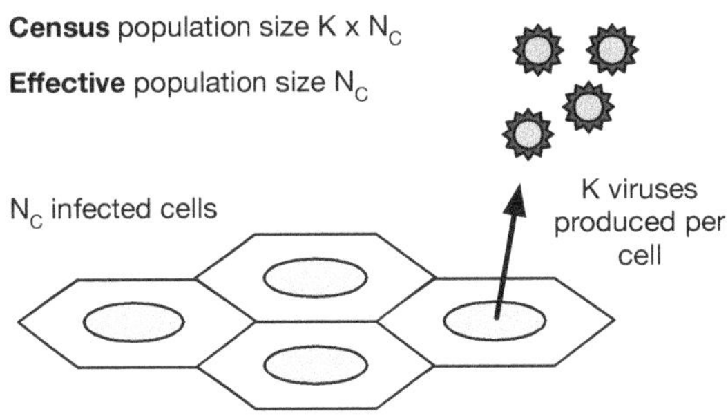

Figure 11.2. Census and effective population sizes. A toy model of retroviral infection is shown. Within a host, a constant number of NC cells are each infected by a single retroviral genotype. Each infected cell releases viruses into the bloodstream, maintaining a constant K copies of each viral genotype in the bloodstream. In this case, the census population, equal to the total number of viruses in the bloodstream, is equal to KNC. However, the dynamics of the viral population are dependent on the infection of new cells and the death of infected cells, such that the effective size of the viral population is NC.

but changes in the viral population will be driven by what happens in the smaller population of infected cells (Figure 11.2).

The relative importance of selection versus genetic drift for populations of HIV-1 within infected individuals has been a matter of controversy. Application of a range of population genetic tests to *env* gene sequences initially suggested that the effective population size of the virus was low, at close to 10^3 (Brown, 1997; Shriner *et al.,* 2004), implying that stochastic models would be required to model HIV evolution. More recently, however, the low genetic diversity identified by these tests has been explained by the occurrence of multiple selective sweeps in the population and a more thorough analysis suggests a much larger value for N_e (Pennings *et al.,* 2014), implying that selection, rather than drift, is the key driver of HIV evolution within infected individuals. A high N_e for HIV-1 seems logical, given the rapid and predictable evolution of HIV-1 anti-viral drug resistance.

Recombination

Recombination is the process by which genetic information is exchanged or combined between DNA or RNA molecules. *Recombination* is distinct from *reassortment*, a separate process of genetic exchange exhibited by viruses with segmented genomes, in which different viruses co-infecting a single host swap segments of genetic material. A necessary condition for both recombination and reassortment is the co-infection of a single host cell by two different viruses. There is wide variation in the ability of viruses to undergo recombination, largely determined by viral genome architecture. For example, in negative-sense RNA viruses, such as influenza viruses, recombination appears to be very rare, probably due to the viral genome being coated with ribonucleoproteins inside the host cell. Since the uncoating of viral RNA is not necessary for replication, the nucleoprotein-genomic complex may impede template switching by the RNA polymerase (Chare *et al.,* 2003; Conzelmann, 1998; Elena and Sanjuán, 2007; Worobey and Holmes, 1999).

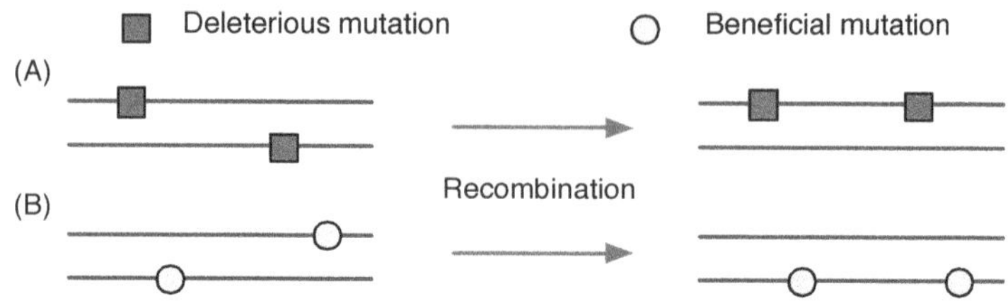

Figure 11.3. Importance of recombination for evolution (A) Recombination allows the removal of deleterious mutations from genotypes in the population. Although in this case, the doubly-deleterious genome may not be viable, an individual carrying neither deleterious mutation survives to the next generation. (B) Recombination allows for faster adaptation. Without recombination, genotypes carrying beneficial mutations would compete with one another for success. Recombination allows for the creation of a fitter double mutant faster than would occur via mutation alone.

From an evolutionary perspective, recombination and reassortment are important for at least two reasons (Figure 11.3). Firstly, they guard against the accumulation of deleterious mutations. In the absence of recombination, a single genome passed via inheritance from individual to individual would acquire an increasing number of mutations. As most mutations are deleterious this may cause a decrease over time in the fitness of each individual, via a process known as Muller's ratchet (Felsenstein, 1974), the existence of which in viral populations has been demonstrated experimentally (Chao, 1990). Recombination enables deleterious mutations to be removed from the population by joining together regions of the genome that lack deleterious mutations. Secondly, recombination has the potential to increase the rate of adaptation in a population. Consider a scenario in which two different genetic variants are required in combination for a genome to gain a fitness advantage. In an asexual population, the fitter genotype can only be generated if the two variants are fixed sequentially, or if they occur by mutation independently upon a single genome. By contrast, with recombination or reassortment, the double-variant can be created from individuals with one each of the two variants, more rapidly creating the fitter genotype.

The importance of recombination for adaptation has led to interest in measuring the rate of recombination in HIV. While recombination rates in other organisms can be estimated via the observation of specific meiotic events (Illingworth *et al.*, 2013; e.g. Mancera *et al.*, 2008), inferring recombination rates in a virus requires a more indirect population genetic approach. One approach to inferring the rate of recombination considered the acquisition of two-site variants in HIV populations (Neher and Leitner, 2010). The study examined sites in HIV *env* sequences that were collected from a patient. The researchers first noted pairs of sites at which variant alleles were observed, albeit not on the same sequence at the same time. When the virus population was observed at a later time point, double-variant sequences could be seen. The observation of a double-variant sequence can be explained by three processes; (i) the double variant existed at the first time point, but was not sampled, (ii) the double variant was created in the intervening time via mutation, or (iii) the double variant was created via the recombination of two single-variant sequences. In the former two cases, the probability that a double-variant sequence would be observed is independent of the distance along the genome between the two individual sites. However, the probability of a recombination event between two loci increases the further the two sites are apart. Searching for a dependence upon distance in the probability of observing double-variants gave an estimate for the rate of recombination of 1.4×10^{-5} per site per generation.

This concludes our very brief overview of population genetic processes acting within viral populations. In the next two sections we discuss how population genetic methods can be used to study viral evolution using sequence data. Specifically, we address two important empirical questions; (i) what are the factors underlying viral adaptation, and (ii) is the evolution of fast-evolving viruses predictable? These questions have been investigated in great detail for influenza A viruses, particularly human influenza, and we describe this work in further detail below.

Adaptive evolution in viruses: the influenza A virus as a case study

Influenza A virus is a highly versatile pathogen, native to birds, that regularly jumps between host species. The viruses that circulate among humans reflect a small subset of the diversity found in the virus' natural reservoir, wild waterfowl, and in other ecologically linked species such as poultry and pigs (Kida *et al.*, 1994; Scholtissek, 1987; Webster *et al.*, 1992). Within humans, influenza is characterised by a rapid process of adaptive evolution that enables the virus to evade herd immunity (Grenfell *et al.*, 2004). Viruses with pandemic potential typically emerge via reassortment, whereas seasonal influenza is associated with *antigenic drift*. Despite its name, antigenic drift is a selective process whereby novel antigenic variants arise in the population due to mutation at key sites involved in immune recognition (typically known as antigenic sites or viral epitopes). The capacity of influenza to constantly re-emerge and the difficulty in controlling influenza transmission distinguishes the virus as an exceptional global health concern.

The influenza A virus genome comprises eight segments. Two encode the surface antigens, haemagglutinin (HA) and neuraminidase (NA), while the remaining six encode proteins that are important for replication, transcription, cellular entry and viral assembly. Influenza A viruses are characterized by great antigenic diversity and are classified according to the specific subtypes of the two antigens, e.g. HxNy, where x and y range from 1-18 and 1-11, respectively. Despite this enormous diversity, only a restricted set of antigenic subtypes (H1N1, H2N2 and H3N2) have been seen to sucessfully emerge in humans (Palese, 2004; Webster *et al.*, 1992). Although this may simply reflect the limited window of observation, evidence from seroprevalence data prior to the 1918 pandemic suggests that circulating strains were of subtypes H3 and H1 (Dowdle, 1999; Palese, 2004; Worobey *et al.*, 2014), suggesting a pattern of subtype emergence and recycling that is predicted by some mathematical models (Recker *et al.*, 2007). The most recent pandemic was caused by the swine-origin H1N1 virus in 2009 that rapidly disseminated throughout the world, infecting many millions of people after its first appearance in Mexico.

Influenza A virus strains that circulate as epidemics in human populations originate from pandemic viral lineages (Webster *et al.*, 1992). Until 2009, the two main subtypes responsible for influenza A infections in humans were H3N2 and 'seasonal' H1N1; the latter has since been replaced by the swine-origin H1N1 2009 pandemic virus (Palese and Wang, 2011). Influenza transmission dynamics vary between tropical and temperate regions, which respectively act as source and sink populations for the virus (Bahl *et al.*, 2011; Bedford *et al.*, 2010; Nelson *et al.*, 2007; 2006; Rambaut *et al.*, 2008; Russell *et al.*, 2008). Human migration, especially air travel, maintains this global transmission pattern and also contributes significantly to seasonal viral diversity in temperate regions by exporting viruses from tropical or well-connected populations (Bahl *et al.*, 2011; Lemey *et al.*, 2014).

One of the striking features of seasonal influenza A virus evolution is the surprisingly low genetic diversity of viruses (belonging to the same subtype) present in humans at any one time, especially in the antigenic genes. This observation is non-intuitive given the very large antigenic repertoire of the virus. However, both population genetic and simulation studies have provided compelling evidence that this observed pattern is driven by frequency-dependent selection on immune-escape mutations, such that rare antigenic variants are positively selected due to their ability to evade the host immune system (Bush, 1999; Cobey and Koelle, 2008; Ferguson *et al.*, 2003; Koelle *et al.*, 2006; Recker *et al.*, 2007; Shih *et al.*, 2007; Wolf *et al.*, 2006). These evolutionary dynamics are likely to be complemented by genetic hitchhiking (a process that arises due to genetic linkage where the allele frequency at one locus is affected by the action of selection at another locus) and founder effects (which occur when a new population is established by a small number of individuals from a larger population) further reducing the overall genetic diversity of the genome (Rambaut *et al.*, 2008). Below, we discuss some of the major findings about the adaptive evolutionary process in seasonal influenza.

Determinants of adaptive evolution

It is widely accepted that the molecular adaptation of influenza A viruses is strongly coupled with antigenic evolution; i.e. sites in the immunogenic part of the viral surface antigens are found to be under constant immune selection and thus play a major role in viral adaptation (Bush, 1999; Cobey and Koelle, 2008; Ferguson *et al.*, 2003; Koelle *et al.*, 2006; Shih *et al.*, 2007; Wolf *et al.*, 2006). Although NA has an essential function in the viral life cycle, the main antibody (humoral) epitopes are located in HA, in particular in the HA1 domain. Consequently, most studies that have investigated adaptive evolution in influenza A have focused on this gene region. Furthermore, H3N2 viruses are better studied than H1N1 viruses, reflecting the past epidemiological dominance of the H3N2 strain and the enormous data set of geographically- and temporally-referenced gene sequences available for it.

The evolutionary and ecological processes described above give rise to a distinctive ladder-like phylogenetic structure for the influenza virus antigenic genes (Figure 11.4) (Bush, 1999; Ferguson *et al.*, 2003; Fitch *et al.*, 1991; 1997). This topology indicates that only one viral lineage, located on the 'trunk' of the tree, persists through time. Although there is evidence for genome-wide selective sweeps, the genetic diversity at segments other than HA and NA is less restricted, such that multiple viral lineages may co-exist over several years (Rambaut *et al.*, 2008). Thus the evolutionary dynamics operating in seasonal influenza are the

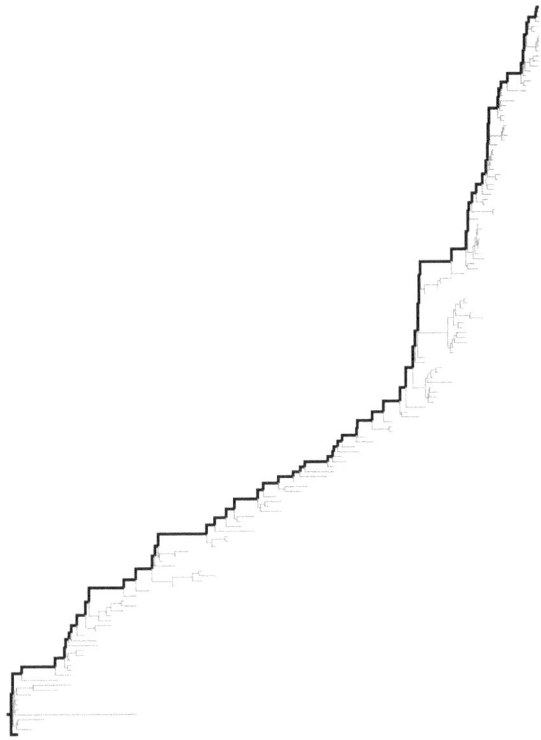

Figure 11.4. Time-scaled phylogeny of HA from seasonal influenza A/H3N2 viruses sampled between 1968 to 2008. The characteristic ladder-like pattern reflects the continuous turnover of viral lineages due to immune selection. As a consequence, only one lineage persists over time, which is indicated by the trunk of the tree (black line). All other viruses (light grey), despite descending from the surviving lineage, are doomed to extinction.

result of a complex interplay of mutation, selection, reassortment and gene flow (Holmes *et al.*, 2005; Rambaut *et al.*, 2008).

Early studies of influenza virus adaptive evolution have been largely motivated by the short-term efficacy of seasonal influenza vaccines, which need to be regularly updated when a novel antigenic variant appears in the viral population. To understand the mechanisms underlying viral adaptation, a common approach has been to identify, typically using dN/dS ratios, the molecular differences between the persistent trunk lineage and the ill-fated co-circulating viruses found on the phylogenetic side branches. These analyses confirm that the trunk lineage is indeed evolving under strong positive selection, driven by nonsynonymous substitutions in the epitope regions of HA (Bush, 1999; Fitch *et al.*, 1991; 1997; Wolf *et al.*, 2006). Conversely, the non-trunk lineages appear to be more characterized by purifying selection (Wolf *et al.*, 2006).

The discovery using dN/dS methods of only a small number of sites under selection implies that adaptive evolution in seasonal influenza is shaped by *episodic selection* (Koelle *et al.*, 2006; Wolf *et al.*, 2006). This type of natural selection emerges when selection is very strong and the rate of generation of beneficial mutations is low, such that adaptive mutations become fixed in the population via sequential selective sweeps (Sniegowski and Gerrish, 2010). The main evidence that episodic selection operates in seasonal influenza comes from Wolf *et al.* (2006). That study, based on dN/dS, suggested that rapid viral adaptation is confined to short periods, during which novel antigenic variants emerge. The longer intervening periods, in contrast, are characterized by little evolutionary change, indicating that purifying selection and neutral evolution dominate (Wolf *et al.,* 2006). Antigenic cluster transitions, in which a novel antigenic variant replaces the current antigenically dominant strain, are associated with multiple substitutions at antigenic sites, indicating that positive epistasis is likely to be present.

However, these results should be interpreted in light of the known limitations of the dN/dS method for detecting selection. It is well known that dN/dS is a conservative measure of selection; it is particularly sensitive to a certain mode of selection that involves repeated fixation of adaptive mutations at the same sites. Therefore, sites that do not undergo strong, diversifying selection may not be identified by this popular approach, e.g. adaptive substitutions that confer small selective advantages or that involve only single selective sweeps (Yang and Nielsen, 2000; Pybus and Shapiro 2009).

A subsequent study by Shih *et al.* (2007) found many more sites under positive selection (63 versus 15) by examining changes in the frequency of putative adaptive mutations (Shih *et al.,* 2007). Encouragingly, many of the sites identified by dN/dS ratios also belonged to the set of sites identified by Shih *et al.* The authors also observed multiple fixations occurring at the same time in epitope regions, supporting the idea that epistatic interactions are important in driving influenza virus adaptation. Importantly this study indicates that analyses based on dN/dS ratios alone are likely to provide an incomplete picture of influenza evolution, and provides strong evidence that adaptive evolution is an ongoing process involving mutations that confer both small and large fitness effects.

Mode and tempo of adaptive evolution

An alternative approach to detect and quantify adaptive evolution in viral populations sampled serially through time has been developed by Bhatt *et al.* (2011). This method builds upon the McDonald-Kreitman (MK) test (McDonald and Kreitman, 1991) and its subsequent extensions (Smith and Eyre-Walker, 2002; Williamson, 2003) and can directly estimate the rate of molecular adaptation (adaptive fixations per codon per year) from gene sequence data. The MK test is a simple and robust population genetic test of whether selection has acted upon a population or species. It relies on comparing different classes of mutations that contribute to diversity within populations and to divergence between populations (Figure 11.5). The standard MK test uses sequences from a closely related species to discriminate between fixed mutations (those that are carried by all members of the study population but are distinct from the sister species; dark grey in Figure 11.5) and polymorphic mutations (those carried by only some individuals in the study population; light grey in Figure 11.5). Mutations that arise after the divergence from the sister species are typically called ‘derived’ changes. Mutations are further classified into nonsynonymous changes (diamonds in Figure 11.5) and synonymous changes (circles in Figure 11.5).

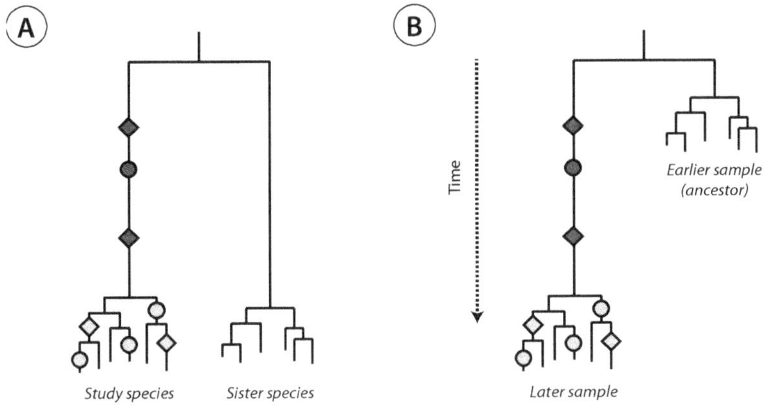

Figure 11.5. Estimating adaptive fixations in viral populations: A) In the standard McDonald Kreitman test, the different classes of mutations in the study species are inferred by comparing with a more distantly related sister species or population. Any mutations that arise on the internal branch leading to the study species represent fixations (dark grey), shared by all individuals in the population, while all other mutations correspond to polymorphisms (light grey). The general concept for inferring adaptive substitutions is centered on the expectation that the ratio of nonsynonymous (diamonds) and synonymous (circles) observed at fixed and polymorphic sites, respectively, should be equivalent (see main text for detail). Thus, any excess nonsynonymous fixations can be directly translated to the number of adaptive substitutions. B) For viral populations that are temporally sampled, sequences sampled from the earliest sample can be used to infer mutations in subsequent samples. Considering multiple time points generates a measure of the number of adaptive substitutions over time, allowing rates of molecular adaptation to be estimated using regression-fitting methods.

Under neutral evolution, the rate of divergence between species and the levels of polymorphism observed within species are simply functions of the mutation rate. Hence if synonymous changes are assumed to be non-selected then the following equality should hold true:

$$N_{fixed}/S_{fixed} = N_{polymorphic}/S_{polymorphic} \qquad [Eq.\ 3]$$

where N_{fixed} and S_{fixed}, respectively, refer to nonsynonymous and synonymous changes that are fixed in the population, while $N_{polymorphic}$ and $S_{polymorphic}$, respectively, refer to nonsynonymous and synonymous changes that are segregating at frequencies below 1.0 in the population. Positive selection may be inferred when $N_{fixed}/S_{fixed} > N_{polymorphic}/S_{polymorphic}$, indicating an excess of fixed nonsynonymous mutations detected compared to that expected under neutral evolution.

Bhatt *et al.* (2011, 2010) have made a number of key extensions to the standard MK test so that this framework can be applied to temporally-sampled viral populations. Instead of classifying fixed and polymorphic sites using a sister species or population, they use sequences sampled from the population at an earlier time (see Figure 11.5B). A sequence from the earlier time point represents an ancestral virus, which is used to identify the number of derived mutations in later samples.

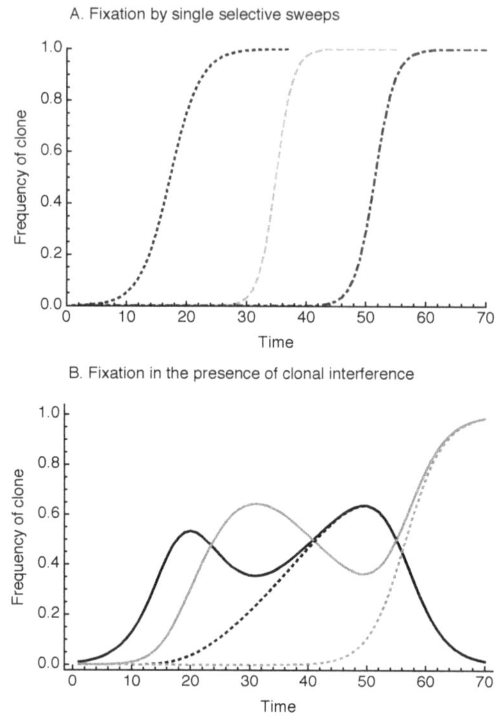

Figure 11.7. Adaptive fixation in the absence and presence of clonal interference. A) A standard view of adaptive fixation. Beneficial mutations that overcome genetic drift, i.e. by reaching frequencies substantially greater than 1/N, become fixed in the population by sequential selective sweeps. The speed at which the beneficial mutation fixes is proportional to the strength of selection conferred. Under this scenario it is assumed that the beneficial mutation rate, μb, is low, such that $N\mu b << 1$. This can arise if either μb, or the population size N are low. B) Clonal interference arises in the population as result of high mutation rate, low recombination rate and large population size. As a consequence, multiple beneficial mutations enter and sweep concurrently in the population. This leads to a complex dynamics where ongoing sweeps are helped, or impeded, by the arrival of subsequent beneficial mutations on other backgrounds (two genetic backgrounds depicted; grey and black). A more strongly beneficial mutant lineage that emerges later (solid grey line) can initially outcompete an earlier beneficial lineage (solid black line). A later beneficial mutation arising in the black lineage (dotted black line) sweeps to fixation in that lineage, and enables it to outcompete the grey lineage. However, another beneficial mutation in the gray lineage (dotted grey line) causes this lineage finally to reach fixation. When a beneficial mutant lineage eventually fixes, it can appear as if multiple mutations have emerged simultaneously.

Since the MK test was originally devised to detect selection in species with low rates of evolution (and typically small population sizes), many of the assumptions of the standard MK test are violated by viral sequence data. In particular, the MK test assumes that each site in the sequence can experience at most one mutation (this is called the 'infinite sites model' in the population genetics literature). This assumption is problematic for viral sequences, which, due to their high mutation rates and large population sizes, will often exhibit multiple changes in the same nucleotide site. Thus, observing more than two different nucleotides at a particular site is common in many viral data sets. Ignoring this phenomenon will lead to reduced statistical power to detect positive selection, as the number of mutations at polymorphic sites, especially those involving nonsynonymous changes, will be underestimated. To address this issue Bhatt *et al.* (2010) introduced a proportional counting method to partially account for the likelihood of multiple mutations.

Another assumption of the standard MK test that does not apply to fast-evolving viruses is that all polymorphisms are selectively neutral. The combined effect of large population sizes and very high mutation rates means that a significant number of deleterious polymorphisms will exist at low frequencies in viral populations (Pybus *et al.*, 2007). Figure 11.6 illustrates the distribution of mutation frequencies in an idealised population. Bhatt *et al.* (2011) explored the balance between the forces of mutation, selection and genetic drift using realistic values for viral mutation rates and viral selection coefficients. They found that mutations observed at low frequencies are most likely to be deleterious, while mutations under positive selection are most likely to observed at high frequencies. Selectively neutral (or near-neutral) mutations predominate among polymorphisms that are observed at intermediate frequencies. An important assumption of this model is that selection is sufficiently strong to rapidly drive selected sites to either high or low frequencies in the population and ignores complex interactions as result of genetic linkage. Under these assumptions, it is possible to estimate the number of adaptive substitutions in the study population using the following equation:

$$\alpha = N_{high}\ (1-(S_{high}/N_{high})(N_{mid}/S_{mid}) \qquad [Eq.\ 4]$$

where N_{high} and S_{high} respectively correspond to the number non-synonymous and synonymous mutations in the high site-frequency range, and N_{mid} and S_{mid} respectively correspond to the number of non-synonymous and synonymous mutations in the intermediate site-frequency range. Equation 4 is a generalization of the related methods of Smith and Eyre-Walker (2002) and Williamson (2003). Further, by estimating the number of adaptations that have accumulated in the viral population at each time point after the first sampling time a time-series is generated. The rate of molecular adaptation per year can then be obtained by fitting a regression model to this time series.

Using this framework Bhatt *et al.* (2011) estimated the rate of adaptive evolution across the genome of the human influenza A virus (Bhatt *et al.*, 2011). As expected they found that the main genomic regions of viral adaptation are the antigenic genes, in particular surface residues of the HA and NA genes, which adapt at around 0.0025-0.004 adaptive substitutions per codon per year (Bhatt *et al.*, 2011). In addition, they estimate a non-zero, although much lower, rate of adaptive evolution at internal residues in the antigenic genes and in other genomic segments. Interestingly, this approach also revealed that the seasonal H1N1 virus underwent significantly less adaptive evolution than H3N2 viruses, in line with epidemiological and serological evidence indicating that the seasonal H1N1 virus was subjected to stronger genetic drift than H3N2 viruses (Bedford *et al.*, 2014; Hay *et al.*, 2001; Rambaut *et al.*, 2008). A strength of Bhatt et al.'s approach lies in providing an estimate of adaptive evolution that is easy to interpret and readily comparable. This enables a direct evaluation of adaptation rates acting across different genes, populations and even species. In contrast, interpreting dN/dS ratios between different datasets can be difficult (Kryazhimskiy *et al.*, 2008b; Mugal *et al.*, 2013; Rocha *et al.*, 2006; Sheridan *et al.*, 2004).

Linkage effects on viral adaptation

The high adaptation rates estimated for influenza HA and NA genes and the apparent lack of intrasegment recombination suggest that beneficial mutations in influenza virus populations may be competing with each other for fixation (Neher and Hallatschek, 2013; Strelkowa and Lässig, 2012). New beneficial mutations that arise can interfere with ongoing selective sweeps and potentially outcompete earlier beneficial mutations if they confer a greater selective advantage (Figure 11.7). Consequently some beneficial mutations may be lost from the population even after rising to intermediate frequencies (Figure 11.7). This mode of evolution is known in population genetics as *clonal interference* and is characteristic of large asexual populations that evolve rapidly. A similar process of selective interference, called the Hill-Robertson effect, also occurs in sexual populations ((Hill and Robertson, 1966) when there is low recombination between sites under weak selection. An important consequence of clonal interference, and of genetic linkage in general, is that selection does not act independently on each mutation but rather operates on the genotype of an entire genome or genomic region (Neher and Shraiman, 2009). It has also been suggested that the presence of clonal interference can make times between adaptive fixations more regular (Gerrish, 2001), which may explain why influenza A virus HA sequences evolve in a notably 'clock-like' manner. Furthermore, in comparison to freely recombining genomic regions, under clonal interference it is expected that (i) adaptive fixation rates are reduced, and (ii) the mean selection coefficient of fixations is higher.

Detecting the presence of clonal interference using sampled sequence data is difficult if standard population genetics are used, due to the inherent non-independence among observed polymorphisms as a result of linkage and interference. However, two recent studies have investigated the role of clonal interference in the adaptation of seasonal influenza A virus using novel population genetic methods and viral sequences sampled serially through time (Illingworth and Mustonen, 2012; Strelkowa and Lässig, 2012). Interestingly, both studies provide strong evidence for clonal interference and background selection in the molecular evolution of influenza HA genes.

In the first study, it was noted that fixation events in the HA gene occur non-independently and are temporally clustered (Strelkowa and Lässig, 2012). Fixation events are accompanied by frequent dips in sequence diversity, strongly suggesting that the fixations occurred via selective sweeps. [These observations are line with previous studies that report multiple concurrent fixations in HA epitopes (Kryazhimskiy *et al.*, 2011b; Plotkin *et al.*, 2002; Shih *et al.*, 2007; Wolf *et al.*, 2006)]. This pattern was seen both for nonsynonymous and synonymous changes in HA, suggesting that hitchhiking and epistatic interactions are important. As a result, mutations that are not positively selected, but are genetically linked, will rise in frequency together with the selected mutation. The times taken for mutations to fix further supports this theory since they are similarly distributed for nonsynonymous and synonymous changes. If mutations were unlinked then the former, being under greater positive selection, should fix more rapidly than the latter.

Clonal interference was explored further using an elegant model that considers the probabilities that nonsynonymous and synonymous mutations reach a certain threshold frequency (Strelkowa and Lässig, 2012). The authors introduced the *propagator ratio,* $g(x)$, which is a measure of the ratio of nonsynonymous to synonymous mutations that reach a given frequency, x. A further variable, called the *loss propagator ratio*, $h(x)$, was introduced to describe the ratio of nonsynonymous to synonymous mutations that reach the frequency x before eventually being lost from the population. Fundamentally, this approach is related to the McDonald-Kreitman test, in that positive selection may be inferred if there is an increased likelihood of observing nonsynonymous mutations at a certain frequency in the population, i.e. $g(x) > 1$ or $h(x) > 1$. Similarly, negative selection may be inferred if $g(x) < 1$ or $h(x) < 1$.

Strelkowa and Lässig, (2012) used the propagator ratios $g(x)$ and $h(x)$ to investigate the mutations arising in the epitope and non-epitope regions of the HA gene from H3N2 influenza A viruses. Nonsynonymous mutations are approximately twice as likely to become fixed than synonymous mutations in the HA epitopes ($g(1) = 2.3$; a similar value was reported by Bhatt *et al.* (2011). Consequently, from the 80 amino acid substitutions observed in the HA epitopes over 39 years, 45 fixations were inferred to be adaptive. Since the mean time for fixation is approximately 3 years, we can infer that an average of at least three advantageous mutations are present at any given time in influenza virus population, supporting the hypothesis of clonal interference. The authors also inferred many more beneficial mutations among those alleles that were eventually lost from the viral population, which further supports the role of clonal interference in the molecular evolution of HA. Similar to findings from analyses based upon dN/dS, nonsynonymous mutations in non-epitope regions were found to be under negative selection (Strelkowa and Lässig, 2012).

The second study of clonal interference in influenza viruses (Illingworth and Mustonen, 2012) examined the importance of linkage effects on the adaptive evolution of HA. Using time-resolved allele frequency data obtained from HA sequences from H3N2 influenza A viruses sampled over 40 years, the authors estimated the level of selection acting upon mutant alleles by fitting two simple population genetic models, which either excluded or included the effects of selection arising from linkage between sites. Despite using a very simple model of additive selection, complex evolutionary dynamics were captured by the latter model (Illingworth and Mustonen, 2012). This model represented the selection acting upon a specific site as the sum of (i) the inherent benefit or disadvantage of that mutation, plus (ii) the effect of selection acting upon other sites in linkage disequilibrium with the first site. Summing these linkage effects over all sites gave a total, time-dependent 'effective selection' pressure acting on the site in question. Further, Illingworth and Mustonen, (2012) were able to decompose the total 'effective selection' into the inherent selection acting upon a mutation and the influence of interference from other sites. The relative contributions of both sources of selection were then examined for influenza virus mutations that were eventually fixed, and for those that were lost from the viral population. Both inherent selection and linkage effects were found to be important; fixed mutations generally exhibited more inherent positive selection, arose on more beneficial genetic backgrounds, and avoided the worst effects of negative interference from other alleles. However, selection arising from genetic linkage (i.e. clonal interference) was inferred to be more important in the eventual fate (fixation or loss) of a mutation than was the inherent selective advantage of that mutation (Illingworth and Mustonen, 2012). These results complement those of Strelkowa and Lässig, (2012) in showing that many mutations in the HA epitopes that do not become fixed are nevertheless under strong positive selection.

Although the discussion here has mostly focused on the HA gene, it will be interesting to see whether interference effects and epistatic interactions are significant in the adaptive evolution of other influenza virus genes. Positive epistasis has been recently observed in the NA and NP genes, indicating that it is pervasive force in seasonal influenza (Gong *et al.,* 2013; Kryazhimskiy *et al.,* 2011b; Shapiro *et al.,* 2006). Whether positive epistasis is a general feature in the molecular evolution of rapidly-evolving viruses is an ongoing question for future research. For example, strong epistatic interactions have been reported in the hepatitis C virus (HCV) and both negative and positive interactions were found in the HCV protease gene, NS3 (Parera and Martínez, 2014).

Reassortment may also be important in determining the adaptive evolution of seasonal influenza. There is some evidence that reassortment can facilitate antigenic evolution by (i) rescuing low-frequency novel antigenic variants from less-fit genetic backgrounds or (ii) generating substantially novel antigenic combinations (Holmes *et al.,* 2005; Nelson *et al.,* 2008; Rambaut *et al.,* 2008). Further, a recent study has provided evidence that reassortment also shapes adaptive evolution in influenza A genes other than HA (Neverov *et al.,* 2014). Clearly further studies are required to understand the exact role of reassortment in influenza virus adaptation and emergence, and the field would benefit greatly from methods that can quantify this evolutionary process.

The emerging view of adaptive evolution in seasonal influenza A viruses is that it is a complex and ongoing process in the antigenic genes. This phenomenon, at least in HA, has been shown to be driven by multiple adaptive mutations competing for fixation that results from a combination of strong genetic linkage, high mutation rates and large population sizes. In contrast to this continual change at the genetic level, the evolution of antigenic phenotypes is punctuated (Smith *et al.,* 2004; Wolf *et al.,* 2006) and a single amino acid substitution can underlie an antigenic cluster transition (Koel *et al.,* 2013). Clearly mutations with both small and large fitness effects contribute to the molecular adaptation of HA genes. The discrepancy between genotype and phenotype change, which has been a topic of long-standing interest to evolutionary biologists (Waddington, 1942), strongly implies that adaptive evolution, even in the antigenic genes, is by no means restricted to changes in the viral epitopes. Increasing replication efficiency and maintaining protein stability or protein-protein interactions are also expected to be crucial to viral fitness.

Predicting virus evolution

Evolutionary biologists have for a long time been fascinated by the question of whether evolution is a repeatable and therefore predictable natural process (Gould, 1990). Because natural selection can, through time, amplify the consequences of rare, random events (a phenomenon known as *historical contingency*), it is not immediately obvious that future evolution can be forecasted.

While predicting the future is no easy task, there are reasons to believe that, for virus populations, making useful predictions may be a possibility. First, viral populations are extremely large in number. As explained earlier, the effects of random genetic drift are much reduced when a population is large and it is more likely that the evolutionary behavior of the population is, in the short-term, close to being deterministic. Second, knowledge of viral population diversity is relatively easy to obtain. The small genomes of viruses, particularly RNA viruses, means that the number of sequences available from a population may exceed 10^4 or 10^5. Third, viruses have much simpler genomes and life cycles than other organisms, and fourth, they evolve so quickly that evolutionary trajectories can be replicated in the laboratory (Meyer *et al.,* 2012). Viruses therefore make attractive targets for predictions of future evolution and as a consequence, population genetic techniques are now being employed to predict how virus populations will evolve. We outline this work in this section, much of which is motivated by the problem of choosing contemporary influenza viruses for inclusion in future influenza vaccines.

Models of prediction

We here divide approaches to predicting evolution roughly into two categories according to their scope. First, we describe 'fitness-based' prediction methods. The evolution of a population may be predicted by characterising the relative fitnesses of different individuals, or of groups of individuals, that comprise the population. As selection acts through time, fitter individuals and groups will grow in frequency. Further, in the majority of cases, offspring arising in later generations will be descendants of a subset of current, higher-fitness individuals. Although nothing may be known about the effects of mutations that will arise in the population in the future, aspects of the genetic composition of future populations can be inferred from estimates of current fitnesses.

One disadvantage of fitness-based prediction methods is that they do not directly distinguish specific potential mutations from each other. A second form of prediction, here denoted 'fitness-landscape-based' prediction methods, does precisely this. In this case, an estimate is made of the fitness landscape upon which the population evolves (see Figure 11.1). Given this, and knowledge of a population's current genetic state, the probability that a specific mutation will emerge at some point in future can be estimated, because the probability that the new variant will establish itself is not just a function of its fitness (landscape height) but also of the genetic background on which it arises (landscape position).

Fitness-based prediction: seasonal influenza viruses

As discussed earlier, a key driver of seasonal influenza virus evolution is the antigenic phenotype of the virus, which is typically defined in the laboratory by measuring the way in which an immune response provoked by one viral strain affects the growth of a second strain (Hirst, 1943). The degree of antigenic divergence between pairs of strains can be projected using statistical methods onto a two-dimensional plane, or antigenic space, thereby describing visually the historic path of influenza antigenic evolution (Smith *et al.*, 2004). This method, called antigenic cartography, has been used to evaluate whether or not future virus evolution can be predicted. By constructing a model of viral evolution within antigenic space, Bedford *et al.* (2012) showed that host immunity to previous strains of the virus acquired by the human population restricts the diversity of the viral population and channels future change in antigenic phenotype, such that the evolution of the virus may be predictable over a one- or two-year timescale.

Other approaches have adopted more explicitly population genetic approaches and have estimated the fitnesses of observed influenza viruses to make a prediction of the future viral population. Characterising viruses by sequence rather than antigenic phenotype potentially gives a much more complete description of the virus, since the sequence encodes not only the antigenic properties of the virus (Steinbrück and McHardy, 2012) but also other phenotypes, such as the stability of protein structure (Gong *et al.*, 2013; Wylie and Shakhnovich, 2011), and protein electrostatics (Arinaminpathy and Grenfell, 2010). Indeed, the selective effect of any phenotype, known or unknown, that impacts upon viral fitness should be observable over time in the genetic structure of the viral population.

One of the earliest fitness-based prediction studies was undertaken by Bush *et al.* (1999). The authors estimated the phylogeny of the HA gene of the H3N2 subtype, and compared rates of synonymous and non-synonymous substitutions to identify a subset of 18 codon positions in HA that were under positive selection (Bush, 1999). By examining the viral population over time, they noted that sequences with a higher number of substitutions in these codons were, on average, more related to influenza isolates in future years. More recently, Steinbrück and McHardy (2011) investigated influenza virus allele frequencies and noted that variants that initially showed a rapid increase in frequency (which, in the absence of interference, is an indicator of positive selection), were more likely to reach fixation in the global population.

A more sophisticated attempt at fitness-based prediction was recently introduced by Łuszka and Lässig (2014), who developed and utilised an explicit parametric model of viral fitness. They first assigned viral sequences into clades of strains with a mutual common ancestor, according to an estimated phylogeny. Next, they applied their fitness model to viruses within each clade. Nonsynonymous mutations in viral epitopes were assigned a positive fitness effect, which is then reduced according to the extent to which viruses with each mutation have previously circulated in the global population. This reduction is intended to model the effect of host immunity on the transmission of each strain. Mutations outside of viral epitopes were counted as deleterious, on the basis that these may destabilise protein structure, or confer some other negative effect. Finally, the evolutionary future of the population is predicted, with clades evolving in competition with each other and each changing in frequency according to the fitness of its composite strains. Application of the model to historical data, in order to make predictions for one year ahead, gave a better prediction of the optimal vaccine strain than the actual vaccine used, as measured by sequence distances at viral epitope sites (Łuksza and Lässig, 2014).

Much effort in population genetics has gone towards understanding the dynamics of continually adapting populations that are able to access a continual supply of beneficial mutations (Desai and Fisher, 2007; Gerrish and Lenski, 1998; Good *et al.*, 2012; Rouzine *et al.*, 2003; Wilke, 2004). Such populations evolve according to a 'travelling wave' model; that is, there exists a stable distribution of individual fitness values that moves continually in the direction of increasing fitness (Figure 11.8). Recent work has considered the shape of phylogenies of sequences sampled from such populations; individuals at the leading edge of the wave are likely to have multiple offspring in subsequent populations, giving distinctive patterns in the phylogeny, whereby multiple related sequences coalesce within short spaces of evolutionary time (Neher and Hallatschek, 2013). These ideas have also been used to predict the outcome of viral evolution (Neher *et al.*, 2014). Neher *et al.* (2014) constructed a phylogenetic tree from influenza virus sequences and used the shape of the tree to infer the fitness of its internal nodes. These estimates of fitness were then used to predict the sequence in the current population that will be most similar to sequences in the population at some future date.

Fitness landscape-based prediction: drug resistance

Predicting viral evolution requires an understanding of the fitness landscape upon which a pathogen is currently evolving so that the current and future states of the viral population can be evaluated. Ongoing research into the problem of drug resistance evolution by pathogens suggests this may be possible (de Visser and Krug, 2014). Advances in this field are aided by the fact that drug resistance

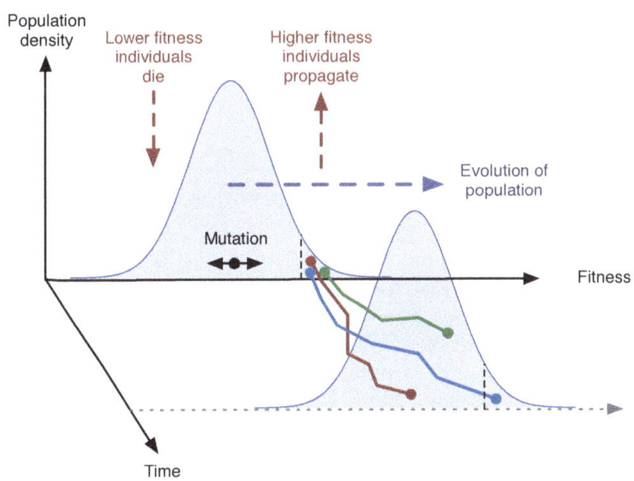

Figure 11.8. Travelling wave model of evolution A population comprises individuals with a distribution of fitnesses. A continual supply of beneficial mutations in the population creates individuals with ever-increasing fitness values at the leading (right hand) edge of the distribution. Individuals with increased fitnesses grow in relative frequency within the population, while less fit individuals (left hand edge of the bell curve) die out. Individuals in the high-fitness tail of the distribution go on to seed the majority of individuals in subsequent populations. Figure adapted from (Neher and Hallatschek, 2013).

evolution is often a comparatively constrained evolutionary scenario with a limited number of mutational trajectories to be explored. For example, a study of four amino acid replacements in the dihydrofolate reductase gene of *Plasmodium falciparum* characterised the different pathways by which these mutations may be gained (Lozovsky *et al.*, 2009), information that could be of use in tracking and predicting the evolution and emergence of anti-malarial drug resistance.

Fitness landscape studies have also generated insight into the optimisation of anti-viral combination therapies for treating infection with HIV (Lengauer and Sing, 2006). By characterising the genetic compositions of HIV populations before and after the acquisition of drug resistance, Beerenwinkel *et al.* (2005) derived networks that described the mutational pathways that must be traversed in order for the virus to develop resistance to a particular drug (Beerenwinkel *et al.*, 2005). This knowledge can subsequently be used to evaluate the value of a particular combination therapy. If a second drug decreases the fitness benefit to the virus of mutations in the resistance pathway to the first drug, then drug resistance may emerge more slowly when the drugs are given together and combinations of drugs that increase the barrier to resistance evolution can be identified.

In Beerenwinkel *et al.* (2005) evaluation of drug therapy outcomes, fitness landscapes were inferred indirectly: the appearance of a mutation in the population within a short period of time being a proxy for that mutation conferring increased fitness. More direct measurement of a fitness landscape has been achieved for the *gag* gene of HIV, with the aim of designing an optimal vaccine (Ferguson *et al.*, 2013). The authors reduced a multiple sequence alignment of *gag* sequences to binary sequences, representing the consensus and non-consensus amino acids in each position, from which a viral fitness landscape was estimated. This approach identifies regions of the HIV genome vulnerable to immune attack; priming of the immune system to target these regions may be a path to more effective therapies.

Fitness landscape-based prediction: cross-species transmission

While population genetics has been applied with good effect to established and newly emergent viral diseases, predicting the nature and behaviour of a pandemic that has yet to emerge is a substantially greater task. Generating a fitness landscape with predictive value requires sufficient data to evaluate both the current status of a population and the multiple directions in which it could evolve. The number of viruses that could potentially cross the species barrier and initiate transmission among humans is extremely large and reservoirs of viruses from many genera and families have been identified in bats and other species (Luis *et al.*, 2013; Wynne and Wang, 2013). Collecting sufficient data about all of these, though not inconceivable (Anthony *et al.*, 2013) would be a formidable task. The process of host adaptation poses a further challenge. Viruses exhibit evolutionary changes during the process of switching from one host to another (Campitelli *et al.*, 2004; Mehle and Doudna, 2009; Taubenberger and Kash, 2010; Wain *et al.*, 2007), implying that different fitness landscapes apply within different species; even if a virus were to be extensively characterised in its native host, this may be of little value for predicting its behavior within a human population.

Research into the question of whether or not H5N1 avian influenza A could develop airborne transmissibility among humans has given some insight into the importance of fitness landscapes for viral emergence (Russell *et al.*, 2012). Transmissibility is an important concern for H5N1 and the more recently identified H7N9 virus (Gao *et al.*, 2013; Zhu *et al.*, 2013). Both strains are highly pathogenic viruses that have made the first step in cross-species transmission, jumping the species barrier to cause sporadic infections in humans (Wolfe *et al.*, 2007). However, to cause a pandemic further viral evolution for increased transmissibility among humans would be required. In a series of controversial experiments, H5N1 and H7N9 viruses have been artificially constructed, or modified, to become transmissible between ferrets under lab conditions, thereby identifying mutations conferring airborne transmissibility (Herfst *et al.*, 2012; Imai *et al.*, 2012; Richard *et al.*, 2013). The fact that a different set of mutations evolved in each case to produce a similar conformational change in the HA protein structure, and the same phenotype of airborne transmissibility, illustrates that predicting phenotypic changes from sequence data is extremely challenging. Furthermore, adaptation to new hosts is likely to occur across the whole genome, not just in the antigenic genes (Bhatt *et al.*, 2013; Herfst *et al.*, 2012). Even given sufficient sequence data, evaluating the role of selection in the within-host evolution of influenza remains a theoretical challenge. Given the low rate of homologous recombination within flu segments (Boni *et al.*, 2010; Lam *et al.*, 2013) clonal interference and hitchhiking effects between mutations are likely to be pervasive, such that a naive inference of selection may give misleading results (Illingworth and Mustonen, 2011). One recent study (Illingworth *et al.*, 2014) utilised a hierarchical multi-locus fitness model to quantify the role of selection within individual influenza virus infected hosts, observed during experimental infections in swine influenza (Lloyd *et al.*, 2011; Murcia *et al.*, 2012). Evidence of selection was identified in the HA gene in six out of sixteen infections for which time-resolved data were available, including cases of simultaneous selection at more than one locus, and of time-dependent selection, in one case potentially due to an immune response against a specific clade of viruses.

This ability to infer fitness effects from within-host studies of viral evolution represents a first step in measuring fitness landscapes that can be used to predict viral pandemic potential. However, each evolutionary experiment in which evolution is observed represents only a single path across what may be a very complex evolutionary landscape; there will likely be much less constraint on the number of possible evolutionary trajectories than is observed in studies of drug resistance evolution (see above). Further, for influenza viruses, epistatic fitness effects are prevalent (Gong *et al.*, 2013; Kryazhimskiy *et al.*, 2011b; Lipsitch and Galvani, 2014). A single mutation, unobserved during a laboratory experiment, may significantly change the topology of the fitness landscape. Finally, despite many similarities between ferrets and humans (Belser *et al.*, 2011), the measurement of fitness effects in ferrets does not imply that the same effects pertain in a human population with varying age structure and immunological history.

Predicting the severity of viral outbreaks

Combatting the spread of pandemic viruses involves coordinated scientific and political campaigns on a global level, such as that which led to the eradication of

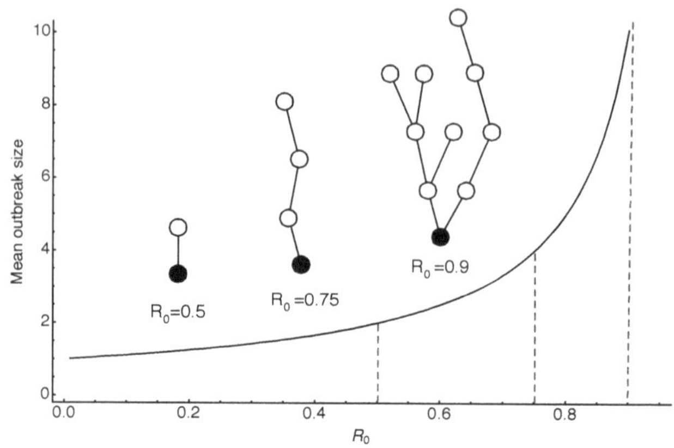

Figure 11.9. Mean outbreak size according to R_0. Examples are shown of typical outbreak patterns for R_0=0.5, R_0=0.75 and R_0=0.9. In each case, the first individual to be infected is shown in black; subsequently infected individuals are shown in white.

smallpox (Breman and Arita, 1980; Morse *et al.,* 2012). Population genetic models have begun to make a contribution to the surveillance, understanding and control of recent, and ongoing epidemics. As discussed below, phylogenetic analyses that incorporate a population genetic model called the coalescent (Lemey *et al.,* 2009), can be used to investigate and quantify the epidemiological spread of viral outbreaks.

Technologies for genome sequencing, combined with new tools for phylogenetic analysis, have enabled the rapid characterisation of viral outbreaks, and can be used to investigate the origin, transmission rate and epidemic potential of a newly emerged viral pathogen. For example, the MERS coronavirus (Middle Eastern Respiratory Syndrome) was first identified and isolated in 2012 (Zaki *et al.,* 2012) and has since infected more than 700 people. Within a year of MERS first being identified, phylogenetic analysis of the viral population had identified viral sequence diversity, with two distinct clades being observed. Examining the pattern of diversity in the collected genomes allowed the inference of transmission events between infected individuals. From these results, it was discovered that the virus had been introduced into the human population on multiple occasions, rather than a single introduction followed by extensive human-to-human transmission (Cotton *et al.,* 2013)

The importance of this conclusion for predicting and controlling MERS transmission is best understood in the context of some epidemiological theory. The potential size and transmissibility of an epidemic is characterised by epidemiologists using the parameter R_0, which represents the expected number of individuals in an immunologically-naive population to whom a single infected individual will transmit the disease. If R_0 is less than 1.0 then the number of people infected by a single zoonotic event will fit a statistical distribution that can be derived from the theory of branching processes (Figure 11.9). That is, each new introduction of the virus into humans will lead to an outbreak of limited size, and the total number of people infected will depend on the number of introductions. However, if R_0 is greater than one an epidemic is predicted to spread exponentially. Therefore the conclusion that observed MERS cases were caused by multiple independent cross species transmissions implied that the epidemic potential of the virus was likely to be low. A more formal analysis of the numbers of people infected by each zoonotic event supported this finding and gave an estimate for R_0 of between 0.5 and 0.92 (Breban *et al.*, 2013)

Measured in terms of the number of people affected, the MERS outbreak contrasts strongly with the 2009 H1N1 influenza A virus pandemic, which caused more than 200 000 estimated deaths worldwide in its first year (Dawood *et al.*, 2012). Soon after the H1N1 virus was initially reported, studies of the virus confirmed that it originated from a single source followed by sustained transmission among people, with the virus spreading initially from Mexico via air traffic routes (Fraser *et al.*, 2009). The 2009 H1N1 pandemic was notable in that it represented the first time that population genetic methods were used, alongside traditional approaches, to estimate the pandemic potential of a newly emerging virus. Specifically, coalescent models were applied to H1N1 virus sequences sampled early during the pandemic and used to estimate the outbreak's growth rate. The population genetic estimate of R_0 for H1N1 influenza was similar to that obtained using surveillance epidemiology (~1.6), implying that, similar to previous outbreaks of influenza, the virus was likely to spread globally (Fraser *et al.*, 2009). Further explanation of the application of coalescent models to viral epidemics is provided elsewhere (Pybus and Rambaut, 2009).

Phylogenetic analysis of the origin of the 2009 H1N1 virus suggested that it arose from multiple reassortment events, in which segments of the influenza virus genome from different sources combined to create a new viral strain (Smith *et al.*, 2009). One notable feature of the 2009 H1N1 strain was that the HA gene, which was derived from swine influenza viruses, had previously originated from human influenza virus, being related to the HA gene from the 1918 pandemic influenza virus that circulated among humans until 1957 (Taubenberger and Kash, 2010). Thus older people who had been previously exposed to related H1 viruses exhibited a degree of immunity against the new pandemic (Medina *et al.*, 2010; Xu *et al.*, 2010).

The analyses outlined above are valuable for predicting the potential danger posed by an epidemic, but are limited from a population genetic perspective in treating the virus as a homogeneous population; a single value of R_0 makes no distinction between viral genetic variants. However, in many cases genetics may underlie substantial differences between viruses, for example in virulence, transmissibility, or response to a particular therapy. Identifying these differences and their causes is an important task for targeting intervention, although during the time needed to prepare a response against a virus, the virus itself may change.

Conclusions

Population genetics has made a valuable contribution towards understanding the evolution of viral populations. A great deal of this progress has come about as a result of computational or theoretical researchers conducting analyses upon datasets collected by others. To this extent, the increasing public availability of pathogen sequence data should aid our understanding of viral disease.

The principle of open data, whereby data relevant to a published study are made publically available, has a long history. In the context of viruses, online databases containing genome sequences of HIV and influenza viruses have been of great value to the evolutionary and population genetic research communities (Bao *et al.*, 2008; Rhee, 2003). However, it is important that researchers on the front line of collecting data are adequately acknowledged and recognized for their work. For example, the GISAID influenza virus project allows researchers in developing countries to share data without fear that their work will not be credited (Bogner *et al.*, 2006).

We anticipate that future research in viral population genetics will progress in the direction of analyzing more data about a greater number of viruses, creating an increasingly complete global picture of viral evolution. Historically, population genetic studies have been biased towards a select few pathogens, such as influenza A virus and HIV-1. This is in part reflects the immediate public health threat posed by these viruses, which are associated with significant costs in terms of mortality, treatment and vaccine development. Where data collection has sometimes been sparse, the perceived urgency of studying these viruses has led to greater accumulation of sequence data, allowing for more sophisticated analyses of viral evolution. As sequencing technology becomes faster, cheaper, and more widely available, the range of viruses studied has grown, with population genetic approaches being applied to HCV, dengue viruses, herpes viruses and many others. However, substantial gaps in data collection exist in many resource-poor countries. For example, a 2010 study of the global migration of influenza neglects entirely the circulation of the virus in Africa, purely through a lack of available data (Bedford *et al.*, 2010). Completing the global picture is an important aim for future work.

One common tool used to investigate viral diversity is the Bayesian phylogenetic package, BEAST (Bouckaert *et al.*, 2014). This is particularly popular when considering the analysis of viral sequences that have been sampled serially through time. This package provides methods for coalescent-based inference of evolutionary parameters such as substitution rate and effective population size. Phylogenetic methods can be particularly powerful for analyzing novel viral outbreaks, where samples are typically limited. This has been recently demonstrated by the analysis of the 2014 Ebola outbreak, which revealed that West African viruses are descended from older human outbreaks from Central Africa (Gire *et al.*, 2014).

However, Bayesian phylogenetic methods are computationally demanding and difficult to apply to sets of viral sequences that number in the thousands or tens of thousands. Such data sets are increasingly common as sequence databases grow, and are being generated by ultra-deep sequencing methods. Hence alternative techniques to examine the evolution of viral populations should be considered. Population genetic techniques that examine the distribution and changes in allele and mutation frequencies through time in viral populations are an attractive option. If enough variation has accumulated then these population genetic approaches may also be suitable for slower evolving DNA viruses. For example, a recent investigation into the evolution of the varicella zoster virus (VZV), which causes chicken pox (Depledge *et al.*, 2014) was undertaken using deep sequencing of the viral genome. The live vaccination strain is heterogeneous, and occasionally causes rashes in inoculated patients. The authors examined viral populations from the vaccine and from skin lesions. Symptoms following vaccination were identified as being caused by selection acting on alleles already present in the vaccine population, rather than upon newly generated mutations (Depledge *et al.*, 2014) . The promise of increasingly large and refined sets of viral sequence data, together with a raft of novel population genetic techniques, heralds a promising future for the understanding, prediction, and treatment of infectious disease.

References

Acevedo, A., Brodsky, L., and Andino, R. (2014). Mutational and fitness landscapes of an RNA virus revealed through population sequencing. Nature *505*, 686–690.

Anthony, S.J., Epstein, J.H., Murray, K.A., Navarrete-Macias, I., Zambrana-Torrelio, C.M., Solovyov, A., Ojeda-Flores, R., Arrigo, N.C., Islam, A., Ali Khan, S., et al. (2013). A strategy to estimate unknown viral diversity in mammals. mBio *4*, e00598–13.

Arinaminpathy, N., and Grenfell, B. (2010). Dynamics of glycoprotein charge in the evolutionary history of human influenza. PLoS One *5*, e15674.

Bahl, J., Nelson, M.I., Chan, K.H., and Chen, R. (2011). Temporally structured metapopulation dynamics and persistence of influenza A H3N2 virus in humans. Proceedings of the National Academy of Sciences *108*, 19359-19364.

Bao, Y., Bolotov, P., Dernovoy, D., Kiryutin, B., Zaslavsky, L., Tatusova, T., Ostell, J., and Lipman, D. (2008). The Influenza Virus Resource at the National Center for Biotechnology Information. Journal of Virology *82*, 596–601.

Bedford, T., Cobey, S., Beerli, P., and Pascual, M. (2010). Global migration dynamics underlie evolution and persistence of human influenza A (H3N2). PLoS Pathogens *6,* e1000918.

Bedford, T., Rambaut, A., and Pascual, M. (2012). Canalization of the evolutionary trajectory of the human influenza virus. BMC Biology *10*, 38.

Bedford, T., Suchard, M.A., Lemey, P., Dudas, G., Gregory, V., Hay, A.J., McCauley, J.W., Russell, C.A., Smith, D.J., and Rambaut, A. (2014). Integrating influenza antigenic dynamics with molecular evolution. eLife *3*, e01914.

Beerenwinkel, N., Sing, T., Lengauer, T., Rahnenfuhrer, J., Roomp, K., Savenkov, I., Fischer, R., Hoffmann, D., Selbig, J., Korn, K., et al. (2005). Computational methods for the design of effective therapies against drug resistant HIV strains. Bioinformatics *21*, 3943–3950.

Belser, J.A., Katz, J.M., and Tumpey, T.M. (2011). The ferret as a model organism to study influenza A virus infection. Disease Models and Mechanisms *4*, 575–579.

Belshaw, R., Sanjuán, R., and Pybus, O.G. (2011). Viral mutation and substitution: units and levels. Current Opinion in Virology *1*, 430–435.

Bennett, S.N., Holmes, E.C., Chirivella, M., Rodriguez, D.M., Beltran, M., Vorndam, V., Gubler, D.J., and McMillan, W.O. (2003). Selection-driven evolution of emergent dengue virus. Molecular Biology and Evolution *20*, 1650–1658.

Bhatt, S., Holmes, E.C., and Pybus, O.G. (2011). The genomic rate of molecular adaptation of the human influenza A virus. Molecular Biology and Evolution *28,* 2443–2451.

Bhatt, S., Lam, T.T., Lycett, S.J., Leigh Brown, A.J., Bowden, T.A., Holmes, E.C., Guan, Y., Wood, J.L.N., Brown, I.H., Kellam, P., et al. (2013). The evolutionary dynamics of influenza A virus adaptation to mammalian hosts. Philosophical Transactions of the Royal Society B: Biological Sciences *368*, 20120382–20120382.

Bhatt, S., Katzourakis, A., and Pybus, O.G. (2010). Detecting natural selection in RNA virus populations using sequence summary statistics. Infection, Genetics and Evolution *10,* 421–430.

Bogner, P., Capua, I., Lipman, D.J., and Cox, N.J. (2006). A global initiative on sharing avian flu data. Nature *442*, 981–981.
Bollback, J.P., York, T.L., and Nielsen, R. (2008). Estimation of 2Nes from temporal allele frequency data. Genetics *179,* 497–502.
Bonhoeffer, S., Chappey, C., Parkin, N.T., Whitcomb, J.M., and Petropoulos, C.J. (2004). Evidence for positive epistasis in HIV-1. Science *306*, 1547–1550.
Boni, M.F., de Jong, M.D., van Doorn, H.R., and Holmes, E.C. (2010). Guidelines for identifying homologous recombination events in influenza A virus. PLoS One *5,* e10434.
Bouckaert, R., Heled, J., Kühnert, D., Vaughan, T., Wu, C.-H., Xie, D., Suchard, M.A., Rambaut, A., and Drummond, A.J. (2014). BEAST 2: a software platform for Bayesian evolutionary analysis. PLoS Computational Biology 10, e1003537.
Breban R, Riou J, Fontante, A (2013) Interhuman transmissibility of Middle East respiratory syndrome coronavirus: estimation of pandemic risk. The Lancet *382*, 694–699.
Breman, J.G., and Arita, I. (1980). The confirmation and maintenance of smallpox eradication. New England Journal of Medecine *303*, 1263–1273.
Brown, A.J. (1997). Analysis of HIV-1 env gene sequences reveals evidence for a low effective number in the viral population. Proceedings of the National Academy of Sciences *94*, 1862–1865.
Bush, R.M. (1999). Predicting the evolution of human influenza A. Science *286,* 1921–1925.
Campitelli, L., Mogavero, E., De Marco, M.A., Delogu, M., Puzelli, S., Frezza, F., Facchini, M., Chiapponi, C., Foni, E., Cordioli, P., et al. (2004). Interspecies transmission of an H7N3 influenza virus from wild birds to intensively reared domestic poultry in Italy. Virology *323*, 24–36.
Chao, L. (1990). Fitness of RNA virus decreased by Muller's ratchet. Nature *348*, 454–455.
Chare, E.R., Gould, E.A., and Holmes, E.C. (2003). Phylogenetic analysis reveals a low rate of homologous recombination in negative-sense RNA viruses. Journal of General Virology *84*, 2691–2703.
Cobey, S., and Koelle, K. (2008). Capturing escape in infectious disease dynamics. Trends in Ecology and Evolution *23*, 572–577.
Conzelmann, K.K. (1998). Nonsegmented negative-strand RNA viruses: genetics and manipulation of viral genomes. Annual Review of Genetics *32*, 123–162.
Cotten M, Watson SJ, Kellam P, Al-Rabeeah, AA, Makhdoom HQ, Assiri A, Al-Tawfiq, JA, Alhakeem RF, Madani H, AlRabiah FA, et al., (2013). Transmission and evolution of the Middle East respiratory syndrome coronavirus in Saudi Arabia: a descriptive genomic study. The Lancet *382,* 1993–2002.
Dawood, F.S., Iuliano, A.D., Reed, C., Meltzer, M.I., Shay, D.K., Cheng, P.-Y., Bandaranayake, D., Breiman, R.F., Brooks, W.A., Buchy, P., et al. (2012). Estimated global mortality associated with the first 12 months of 2009 pandemic influenza A H1N1 virus circulation: a modelling study. The Lancet Infectious Diseases *12*, 687–695.
de Visser, J.A.G.M., and Krug, J. (2014). Empirical fitness landscapes and the predictability of evolution. Nature Reviews Genetics *15*, 480–490.
Depledge, D.P., Kundu, S., Jensen, N.J., Gray, E.R., Jones, M., Steinberg, S., Gershon, A., Kinchington, P.R., Schmid, D.S., Balloux, F., et al. (2014). Deep sequencing of viral genomes provides insight into the evolution and pathogenesis of varicella zoster virus and its vaccine in humans. Molecular Biology and Evolution *31*, 397–409.
Desai, M.M., and Fisher, D.S. (2007). Beneficial mutation selection balance and the effect of linkage on positive selection. Genetics *176,* 1759–1798.
Dobzhansky, T. (1937). Genetics and the origin of species, Columbia University Press.
Dowdle, W.R. (1999). Influenza A virus recycling revisited. Bulletin World Health Organisation *77,* 820–828.

Drake, J.W. (1991). A constant rate of spontaneous mutation in DNA-based microbes. Proceedings of the National Academy of Sciences *88*, 7160–7164.

Drummond, A., Pybus, O.G., and Rambaut, A. (2003). Inference of viral evolutionary rates from molecular sequences. Advances in Parasitology. *54,* 331–358.

Duffy, S., Shackelton, L.A., and Holmes, E.C. (2008). Rates of evolutionary change in viruses: patterns and determinants. Nature Reviews Genetics *9*, 267–276.

Elena, S.F., and Sanjuán, R. (2007). Virus Evolution: Insights from an experimental approach. Annual Review of Ecology, Evolution, and Systematics *38*, 27–52.

Felsenstein, J. (1974). The evolutionary advantage of recombination. Genetics *78*, 737–756.

Ferguson, A.L., Mann, J.K., Omarjee, S., Ndung'u, T., Walker, B.D., and Chakraborty, A.K. (2013). Translating HIV sequences into quantitative fitness landscapes predicts viral vulnerabilities for rational immunogen design. Immunity *38*, 606–617.

Ferguson, N.M., Galvani, A.P., and Bush, R.M. (2003). Ecological and immunological determinants of influenza evolution. Nature *422*, 428–433.

Fisher, R. A. (1930). The genetical theory of natural selection, Clarendon Press.

Fitch, W.M., Leiter, J.M., Li, X.Q., and Palese, P. (1991). Positive Darwinian evolution in human influenza A viruses. Proceedings of the National Academy of Sciences *88*, 4270–4274.

Fitch, W.M., Bush, R.M., Bender, C.A., and Cox, N.J. (1997). Long term trends in the evolution of H(3) HA1 human influenza type A. Proceedings of the National Academy of Sciences *94,* 7712–7718.

Fraser, C., Donnelly, C.A., Cauchemez, S., Hanage, W.P., Van Kerkhove, M.D., Hollingsworth, T.D., Griffin, J., Baggaley, R.F., Jenkins, H.E., Lyons, E.J., et al. (2009). Pandemic potential of a strain of influenza A (H1N1): Early findings. Science *324,* 1557–1561.

Ganusov, V.V., Goonetilleke, N., Liu, M.K.P., Ferrari, G., Shaw, G.M., McMichael, A.J., Borrow, P., Korber, B.T., and Perelson, A.S. (2011). Fitness costs and diversity of the cytotoxic T lymphocyte (CTL) response determine the rate of CTL escape during acute and chronic phases of HIV Infection. Journal of Virology *85*, 10518–10528.

GGao, R., Cao, B., Hu, Y., Feng, Z., Wang, D., Hu, W., Chen, J., Jie, Z., Qiu, H., Xu, K., et al. (2013). Human infection with a novel avian-origin influenza A (H7N9) Virus. New England Journal of Medecine *368,* 1888–1897.

Gaschen, B., Taylor, J., Yusim, K., Foley, B., Gao, F., Lang, D., Novitsky, V., Haynes, B., Hahn, B.H., Bhattacharya, T., et al. (2002). Diversity considerations in HIV-1 vaccine selection. Science *296*, 2354–2360.

Gerrish, P. (2001). The rhythm of microbial adaptation. Nature *413*, 299-302.

Gerrish, P.J., and Lenski, R.E. (1998). The fate of competing beneficial mutations in an asexual population. Genetica *102-103*, 127–144.

Gire, S.K., Goba, A., Andersen, K.G., Sealfon, R., Park, D.J., Kanneh, L., Jalloh, S., Momoh, M., Dudas, G., Wohl, S., et al. (2014). Genomic surveillance elucidates Ebola virus origin and transmission during the 2014 outbreak. (2014). Science *345*, 1369-1372.

Gong, L.I., Suchard, M.A., and Bloom, J.D. (2013). Stability-mediated epistasis constrains the evolution of an influenza protein. eLife *2*, e00631.

Good, B.H., Rouzine, I.M., Balick, D.J., Hallatschek, O., and Desai, M.M. (2012). Distribution of fixed beneficial mutations and the rate of adaptation in asexual populations. Proceedings of the National Academy of Sciences *109*, 4950–4955.

Gould S. (1990). Wonderful Life: The burgess shale and the nature of history. New York: W. W. Norton and Company.

Grenfell, B.T. Pybus, O.G., Gog, J.R., Wood, J.L.N., Daly, J.M., Mumford, J.A., Holmes, E.C. (2004). Unifying the rpidemiological and rvolutionary dynamics of pathogens. Science *303,* 327–332.

Hay, A.J., Gregory, V., Douglas, A.R., and Lin, Y.P. (2001). The evolution of human influenza viruses. Philosophical Transactions of the Royal Society B: Biological Sciences *356*, 1861–1870.

Herfst, S., Schrauwen, E.J.A., Linster, M., Chutinimitkul, S., de Wit, E., Munster, V.J., Sorrell, E.M., Bestebroer, T.M., Burke, D.F., Smith, D.J., et al. (2012). Airborne transmission of influenza A/H5N1 virus between ferrets. Science *336,* 1534–1541.
Hill, W.G., and Robertson, A. (1966). The effect of linkage on limits to artificial selection. Genetical Research *8*, 269–294.
Hirst, G.K. (1943). Studies of antigenic differences among strains of influenza A by means of red cell agglutination. Journal of Experimental Medecine *78*, 407–423.
Holmes, E.C., Ghedin, E., Miller, N., Taylor, J., Bao, Y., St George, K., Grenfell, B.T., Salzberg, S.L., Fraser, C.M., Lipman, D.J., et al. (2005). Whole-genome analysis of human influenza A virus reveals multiple persistent lineages and reassortment among recent H3N2 viruses. PLoS Biology *3,* e300.
Huxley, Julian S. (1942). Evolution: the modern synthesis, John Wiley and Sons, 1942
Illingworth, C.J., and Mustonen, V. (2011). Distinguishing driver and passenger mutations in an evolutionary history categorized by interference. Genetics *189,* 989–1000.
Illingworth, C.J., and Mustonen, V. (2012). Components of selection in the evolution of the influenza virus: linkage effects beat inherent selection. PLoS Pathogens *8,* e1003091.
Illingworth, C.J., Fischer, A., and Mustonen, V. (2014). Identifying selection in the within-host evolution of influenza using viral sequence data. PLoS Computational Biology *10*, e1003755.
Illingworth, C.J., Parts, L., Schiffels, S., Liti, G., and Mustonen, V. (2012). Quantifying selection acting on a complex trait using allele frequency time series data. Molecular Biology and Evolution *29,* 1187–1197.
Illingworth, C.J., Parts, L., Bergström, A., Liti, G., and Mustonen, V. (2013). Inferring genome-wide recombination landscapes from advanced intercross lines: application to yeast crosses. PLoS One *8*, e62266.
Imai, M., Watanabe, T., Hatta, M., Das, S.C., Ozawa, M., Shinya, K., Zhong, G., Hanson, A., Katsura, H., Watanabe, S., et al. (2012). Experimental adaptation of an influenza H5 HA confers respiratory droplet transmission to a reassortant H5 HA/H1N1 virus in ferrets. Nature *486*, 420-428.
Kida, H., Ito, T., Yasuda, J., Shimizu, Y., Itakura, C., Shortridge, K.F., Kawaoka, Y., and Webster, R.G. (1994). Potential for transmission of avian influenza viruses to pigs. Journal of General Virology *75*, 2183–2188.
Koel, B.F., Burke, D.F., Bestebroer, T.M., van der Vliet, S., Zondag, G.C.M., Vervaet, G., Skepner, E., Lewis, N.S., Spronken, M.I.J., Russell, C.A., et al. (2013). Substitutions near the receptor binding site determine major antigenic change during influenza virus evolution. Science *342*, 976–979.
Koelle, K., Cobey, S., Grenfell, B., and Pascual, M. (2006). Epochal evolution shapes the phylodynamics of interpandemic influenza A (H3N2) in humans. Science *314*, 1898–1903.
Kryazhimskiy, S., Bazykin, G.A., and Dushoff, J. (2008a). Natural selection for nucleotide usage at synonymous and nonsynonymous sites in influenza A virus genes. Journal of Virology *82,* 4938–4945.
Kryazhimskiy, S., Bazykin, G.A., Plotkin, J., and Dushoff, J. (2008b). Directionality in the evolution of influenza A haemagglutinin. Proceedings of the Royal Society B: Biological Sciences *275*, 2455–2464.
Kryazhimskiy, S., Draghi, J.A., and Plotkin, J.B. (2011a). In evolution, the sum is less than its parts. Science *332,* 1160–1161.
Kryazhimskiy, S., Dushoff, J., Bazykin, G.A., and Plotkin, J.B. (2011b). Prevalence of epistasis in the evolution of influenza A surface proteins. PLoS Genetics *7,* e1001301.
Lam, T.T.-Y., Chong, Y.L., Shi, M., Hon, C.-C., Li, J., Martin, D.P., Tang, J.W.-T., Mok, C.-K., Shih, S.-R., Yip, C.-W., et al. (2013). Systematic phylogenetic analysis of influenza A virus reveals many novel mosaic genome segments. Infection, Genetics and Evolution *18*, 367–378.

Lander, E.S., Linton, L.M., Birren, B., Nusbaum, C., Zody, M.C., Baldwin, J., Devon, K., Dewar, K., Doyle, M., FitzHugh, W., et al. (2001). Initial sequencing and analysis of the human genome. Nature *409*, 860–921.
Lemey, P., Salemi, M., and Vandamme, A. (2009). The phylogenetic handbook, 2nd Edition, Cambridge University Press.
Lemey, P., Rambaut, A., Bedford, T., Faria, N., Bielejec, F., Baele, G., Russell, C.A., Smith, D.J., Pybus, O.G., Brockmann, D., et al. (2014). Unifying viral genetics and human transportation data to predict the global transmission dynamics of human influenza H3N2. PLoS Pathogens *10*, e1003932.
Lengauer, T., and Sing, T. (2006). Bioinformatics-assisted anti-HIV therapy. Nature *4*, 790–797.
Linz, B., Windsor, H.M., McGraw, J.J., Hansen, L.M., Gajewski, J.P., Tomsho, L.P., Hake, C.M., Solnick, J.V., Schuster, S.C., and Marshall, B.J. (2014). A mutation burst during the acute phase of Helicobacter pylori infection in humans and rhesus macaques. Nature Communications *5*, 4165.
Lipsitch, M., and Galvani, A.P. (2014). Ethical alternatives to experiments with novel potential pandemic pathogens. PLoS Medecine *11*, e1001646.
Lloyd, L.E., Jonczyk, M., Jervis, C.M., Flack, D.J., Lyall, J., Foote, A., Mumford, J.A., Brown, I.H., Wood, J.L., and Elton, D.M. (2011). Experimental transmission of avian-like swine H1N1 influenza virus between immunologically naïve and vaccinated pigs. Influenza and Other Respiratory Viruses *5*, 357–364.
Lozovsky, E.R., Chookajorn, T., Brown, K.M., Imwong, M., Shaw, P.J., Kamchonwongpaisan, S., Neafsey, D.E., Weinreich, D.M., and Harti, D.L. (2009). Stepwise acquisition of pyrimethamine resistance in the malaria parasite. Proceedings of National Academy of Sciences *106,* 12025–12030.
Luis, A.D., Hayman, D.T.S., O'Shea, T.J., Cryan, P.M., Gilbert, A.T., Pulliam, J.R.C., Mills, J.N., Timonin, M.E., Willis, C.K.R., Cunningham, A.A., et al. (2013). A comparison of bats and rodents as reservoirs of zoonotic viruses: are bats special? Proceedings of the Royal Society B: Biological Sciences *280*, 20122753–20122753.
Łuksza, M., and Lässig, M. (2014). A predictive fitness model for influenza. Nature *507*, 57-61.
Mancera, E., Bourgon, R., Brozzi, A., Huber, W., and Steinmetz, L.M. (2008). High-resolution mapping of meiotic crossovers and non-crossovers in yeast. Nature *454*, 479–485.
Marcus, P.I., Ngunjiri, J.M., and Sekellick, M.J. (2009). Dynamics of biologically active subpopulations of influenza virus: plaque-forming, noninfectious cell-killing, and defective interfering particles. Journal of Virology *83*, 8122–8130.
McDonald, J.H., and Kreitman, M. (1991). Adaptive protein evolution at the Adh locus in Drosophila. Nature *351*, 652–654.
Medina, R.A., Manicassamy, B., Stertz, S., Seibert, C.W., Hai, R., Belshe, R.B., Frey, S.E., Basler, C.F., Palese, P., and a-Sastre, A.G.I. (2010). Pandemic 2009 H1N1 vaccine protects against 1918 Spanish influenza virus. Nature Communications *1*, 1–6.
Mehle, A., and Doudna, J.A. (2009). Adaptive strategies of the influenza virus polymerase for replication in humans. Proceedings of the National Academy of Sciences *106*, 21312–21316.
Meyer, A.G., and Wilke, C.O. (2012). Integrating sequence variation and protein structure to identify sites under selection. Molecular Biology and Evolution *30,* 36–44.
Meyer, J.R., Dobias, D.T., Weitz, J.S., Barrick, J.E., Quick, R.T., and Lenski, R.E. (2012). Repeatability and contingency in the evolution of a key innovation in phage lambda. Science *335*, 428–432.
Mugal, C.F., Wolf, J.B.W., and Kaj, I. (2013). Why time matters: codon evolution and the temporal dynamics of dN/dS. Molecular Biology and Evolution *31,* 212–231.
Murcia, P.R., Hughes, J., Battista, P., Lloyd, L., Baillie, G.J., Ramirez-Gonzalez, R.H., Ormond, D., Oliver, K., Elton, D., Mumford, J.A., et al. (2012). Evolution of an

eurasian avian-like influenza virus in naïve and vaccinated pigs. PLoS Pathogens *8,* e1002730.
Mustonen, V., and Lässig, M. (2009). From fitness landscapes to seascapes: non-equilibrium dynamics of selection and adaptation. Trends in Genetics *25*, 111–119.
Neher, R.A., and Hallatschek, O. (2013). Genealogies of rapidly adapting populations. Proceedings of the National Academy of Sciences *110*, 437–442.
Neher, R.A., and Leitner, T. (2010). Recombination rate and selection strength in HIV intra-patient evolution. PLoS Computational Biology *6*, e1000660.
Neher, R.A., and Shraiman, B.I. (2009). Competition between recombination and epistasis can cause a transition from allele to genotype selection. Proceedings of the National Academy of Sciences *106*, 6866–6871.
Neher, R.A., Russell, C.A., and Shraiman, B.I. (2014). Predicting evolution from the shape of genealogical trees. eLife *3,* e03568
Nelson, M.I., Edelman, L., Spiro, D.J., Boyne, A.R., Bera, J., Halpin, R., Ghedin, E., Miller, M.A., Simonsen, L., Viboud, C., et al. (2008). Molecular epidemiology of A/H3N2 and A/H1N1 influenza virus during a single epidemic season in the United States. PLoS Pathogens *4,* e1000133.
Nelson, M.I., Simonsen, L., Viboud, C., Miller, M.A., and Holmes, E.C. (2007). Phylogenetic analysis reveals the global migration of seasonal influenza A viruses. PLoS Pathogens *3,* e131.
Nelson, M.I., Simonsen, L., Viboud, C., Miller, M.A., Taylor, J., George, K.S., Griesemer, S.B., Ghedin, E., Sengamalay, N.A., Spiro, D.J., et al. (2006). Stochastic Processes Are Key Determinants of Short-Term Evolution in Influenza A Virus. PLoS Pathogens *2,* e125.
Neverov, A.D., Lezhnina, K.V., Kondrashov, A.S., and Bazykin, G.A. (2014). Intrasubtype reassortments cause adaptive amino acid replacements in H3N2 influenza genes. PLoS Genetics *10,* e1004037.
Palese, P. (2004). Influenza: old and new threats. Nature Medicine *10*, S82–S87.
Palese, P., and Wang, T.T. (2011). Why do influenza virus subtypes die out? A hypothesis. mBio *2*, e00150-11.
Parera, M., and Martínez, M.A. (2014). Strong epistatic interactions within a single protein. Molecular Biology and Evolution *31,* 1546–1553.
Pennings, P.S., Kryazhimskiy, S., and Wakeley, J. (2014). Loss and recovery of genetic diversity in adapting populations of HIV. PLoS Genetics *10,* e1004000.
Plotkin, J.B., Dushoff, J., and Levin, S.A. (2002). Hemagglutinin sequence clusters and the antigenic evolution of influenza A virus. Proceedings of National Academy of Sciences *99,* 6263–6268.
Pybus, O.G., and Rambaut, A. (2009). Evolutionary analysis of the dynamics of viral infectious disease. Nature Reviews Genetics *10*, 540–550.
Pybus, O.G., Rambaut, A., Belshaw, R., Freckleton, R.P., Drummond, A.J., and Holmes, E.C. (2007). Phylogenetic evidence for deleterious mutation load in RNA viruses and its contribution to viral evolution. Molecular Biology and Evolution *24*, 845–852.
Pybus OG, Shapiro B (2009) Natural selection and adaptation of molecular sequences. In The Phylogenetic Handbook, 2nd Edition (P Lemey, M Salemi, A Vandamme, eds) pp.407-418. Cambridge University Press
Rambaut, A., Pybus, O.G., Nelson, M.I., Viboud, C., Taubenberger, J.K., and Holmes, E.C. (2008). The genomic and epidemiological dynamics of human influenza A virus. Nature *453*, 615–619.
Recker, M., Pybus, O.G., Nee, S., and Gupta, S. (2007). The Generation of influenza outbreaks by a network of host immune responses against a limited set of antigenic types. Proceedings of National Academy of Sciences *104,* 7711–7716.
Rhee, S.Y. (2003). Human immunodeficiency virus reverse transcriptase and protease sequence database. Nucleic Acids Research *31*, 298–303.
Richard, M., Schrauwen, E.J.A., de Graaf, M., Bestebroer, T.M., Spronken, M.I.J., van Boheemen, S., de Meulder, D., Lexmond, P., Linster, M., Herfst, S., et al. (2013).

Limited airborne transmission of H7N9 influenza A virus between ferrets. Nature *501*, 560–563.
Rocha, E.P.C., Smith, J.M., Hurst, L.D., Holden, M.T.G., Cooper, J.E., Smith, N.H., and Feil, E.J. (2006). Comparisons of dN/dS are time dependent for closely related bacterial genomes. Journal of Theoretical Biology *239*, 226–235.
Rouzine, I.M., Rodrigo, A., and Coffin, J.M. (2001). Transition between stochastic evolution and deterministic evolution in the presence of selection: general theory and application to virology. Microbiology and Molecular Biology Reviews *65,* 151–185.
Rouzine, I.M., Wakeley, J., and Coffin, J.M. (2003). The solitary wave of asexual evolution. Proceedings of National Academy of Sciences *100*, 587–592.
Russell, C.A., Jones, T.C., Barr, I.G., Cox, N.J., Garten, R.J., Gregory, V., Gust, I.D., Hampson, A.W., Hay, A.J., Hurt, A.C., et al. (2008). The global circulation of seasonal influenza A (H3N2) viruses. Science *320*, 340–346.
Russell, C.A., Fonville, J.M., Brown, A.E.X., Burke, D.F., Smith, D.L., James, S.L., Herfst, S., van Boheemen, S., Linster, M., Schrauwen, E.J., et al. (2012). The potential for respiratory droplet-transmissible A/H5N1 influenza virus to evolve in a mammalian host. Science *336*, 1541–1547.
Sanjuán, R., Nebot, M.R., Chirico, N., Mansky, L.M., and Belshaw, R. (2010). Viral mutation rates. Journal of Virology *84,* 9733–9748.
Sanjuán, R. (2006). Quantifying antagonistic epistasis in a multifunctional RNA secondary structure of the Rous sarcoma virus. Journal of General Virology *87*, 1595–1602.
Scholtissek, C. (1987). Molecular aspects of the epidemiology of virus disease. Experientia *43*, 1197–1201.
Shapiro, B., Rambaut, A., Pybus, O.G., and Holmes, E.C. (2006). A phylogenetic method for detecting positive epistasis in gene sequences and its application to RNA virus evolution. Molecular Biology and Evolution *23*, 1724–1730.
Sheridan, I., Pybus, O.G., Holmes, E.C., and Klenerman, P. (2004). High-resolution phylogenetic analysis of hepatitis C virus adaptation and its relationship to disease progression. Journal of Virology *78*, 3447–3454.
Shih, A.C.-C., Hsiao, T.-C., Ho, M.-S., and Li, W.-H. (2007). Simultaneous amino acid substitutions at antigenic sites drive influenza A hemagglutinin evolution. Proceedings of the National Academy of Sciences *104,* 6283–6288.
Shriner, D., Shankarappa, R., Jensen, M.A., Nickle, D.C., Mittler, J.E., Margolick, J.B., and Mullins, J.I. (2004). Influence of random genetic drift on human immunodeficiency virus type 1 env evolution during chronic infection. Genetics *166*, 1155–1164.
Sidorenko, Y., and Reichl, U. (2004). Structured model of influenza virus replication in MDCK cells. Biotechnology and Bioengineering *88*, 1–14.
Simmonds, P., Tuplin, A., and Evans, D.J. (2004). Detection of genome-scale ordered RNA structure (GORS) in genomes of positive-stranded RNA viruses: Implications for virus evolution and host persistence. RNA *10*, 1337–1351.
Smith, D.J., Lapedes, A.S., de Jong, J.C., Bestebroer, T.M., Rimmelzwaan, G.F., Osterhaus, A.D.M.E., and Fouchier, R.A.M. (2004). Mapping the antigenic and genetic evolution of influenza virus. Science *305*, 371–376.
Smith, G.J.D., Vijaykrishna, D., Bahl, J., Lycett, S.J., Worobey, M., Pybus, O.G., Ma, S.K., Cheung, C.L., Raghwani, J., Bhatt, S., et al. (2009). Origins and evolutionary genomics of the 2009 swine-origin H1N1 influenza A epidemic. Nature *459*, 1122–1125.
Smith, N.G.C., and Eyre-Walker, A. (2002). Adaptive protein evolution in Drosophila. Nature *415*, 1022–1024.
Sniegowski, P.D., and Gerrish, P.J. (2010). Beneficial mutations and the dynamics of adaptation in asexual populations. Philosophical Transactions of the Royal Society B: Biological Sciences *365*, 1255–1263.
Steinbruck, L., and McHardy, A.C. (2011). Allele dynamics plots for the study of evolutionary dynamics in viral populations. Nucleic Acids Research *39*, e4–e4.

Steinbrück, L., and McHardy, A.C. (2012). Inference of genotype–phenotype relationships in the antigenic evolution of human influenza A (H3N2) viruses. PLoS Computational Biology *8,* e1002492.

Strelkowa, N., and Lässig, M. (2012). Clonal interference in the evolution of influenza. Genetics *192*, 671–682.

Taubenberger, J.K., and Kash, J.C. (2010). Influenza virus evolution, host adaptation, and pandemic formation. Cell Host and Microbe *7*, 440–451.

Tusche, C., Steinbruck, L., and McHardy, A.C. (2012). Detecting patches of protein sites of influenza A viruses under positive selection. Molecular Biology and Evolution *29,* 2063–2071.

Venter, J.C., Adams, M.D., Myers, E.W., Li, P.W., Mural, R.J., Sutton, G.G., Smith, H.O., Yandell, M., Evans, C.A., Holt, R.A., et al. (2001). The sequence of the human genome. Science *291*, 1304–1351.

Waddington, C.H. (1942). Canalization of development and the inheritance of acquired characters. Nature *150*, 563-565.

Wain, L.V., Bailes, E., Bibollet-Ruche, F., Decker, J.M., Keele, B.F., Van Heuverswyn, F., Li, Y., Takehisa, J., Ngole, E.M., Shaw, G.M., et al. (2007). Adaptation of HIV-1 to its human host. Molecular Biology and Evolution *24*, 1853–1860.

Webster, R.G., Bean, W.J., Gorman, O.T., Chambers, T.M., and Kawaoka, Y. (1992). Evolution and ecology of influenza A viruses. Microbiology Reviews *56*, 152–179.

Wilke, C.O. (2004). The speed of adaptation in large asexual populations. Genetics *167*, 2045–2053.

Wilke, C.O., Lenski, R.E., and Adami, C. (2003). Compensatory mutations cause excess of antagonistic epistasis in RNA secondary structure folding. BMC Evolutionary Biology *3*, 3.

Williamson, S. (2003). Adaptation in the env gene of HIV-1 and evolutionary theories of disease progression. Molecular Biology and Evolution *20,* 1318–1325.

Wolf, Y.I., Viboud, C., Holmes, E.C., Koonin, E.V., and Lipman, D.J. (2006). Long intervals of stasis punctuated by bursts of positive selection in the seasonal evolution of influenza A virus. Biology Direct *1*, 34.

Wolfe, N.D., Dunavan, C.P., and Diamond, J. (2007). Origins of major human infectious diseases. Nature *447*, 279–283.

Worobey, M., and Holmes, E.C. (1999). Evolutionary aspects of recombination in RNA viruses. Journal of General Virology *80*, 2535–2543.

Worobey, M., Han, G.Z., and Rambaut, A. (2014). Genesis and pathogenesis of the 1918 pandemic H1N1 influenza A virus. Proceedings of the National Academy of Sciences *111*, 8107-8112..

Wright, S. (1931). Evolution in mendelian populations. Genetics *16,* 97–159.

Wylie, C.S., and Shakhnovich, E.I. (2011). A biophysical protein folding model accounts for most mutational fitness effects in viruses. Proceedings of the National Academy of Sciences *108*, 9916–9921.

Wynne, J.W., and Wang, L.-F. (2013). Bats and viruses: friend or foe? PLoS Pathogens *9,* e1003651.

Xu, R., Ekiert, D.C., Krause, J.C., Hai, R., Crowe, J.E., and Wilson, I.A. (2010). Structural basis of preexisting immunity to the 2009 H1N1 pandemic influenza virus. Science *328*, 357–360.

Yang, Z., and Bielawski, J. (2000). Statistical methods for detecting molecular adaptation. Trends in Ecology and Evolution *15*, 496–503.

Yang Z., and Nielsen, R. (2000) Estimating synonymous and nonsynonymous substitution rates under realistic evolutionary models. (2000). Molecular Biology and Evolution *17,* 32-43.

Zaki, A.M., van Boheemen, S., Bestebroer, T.M., Osterhaus, A.D.M.E., and Fouchier, R.A.M. (2012). Isolation of a novel coronavirus from a man with pneumonia in Saudi Arabia. New England Journal of Medecine *367*, 1814–1820.

Zanini, F., and Neher, R.A. (2013). Quantifying selection against synonymous mutations in HIV-1 env evolution. Journal of Virology *87*, 11843–11850..

Zhu, H., Wang, D., Kelvin, D.J., Li, L., Zheng, Z., Yoon, S.W., Wong, S.S., Farooqui, A., Wang, J., Banner, D., *et al.* (2013). Infectivity, transmission, and pathology of human-isolated H7N9 influenza virus in ferrets and pigs. Science *341*, 183–186.

12

Emerging Viral Infections

Michelle M. Becker, Everett Clinton Smith and Mark R. Denison*

Division of Pediatric Infectious Diseases, Vanderbilt University School of Medicine, Vanderbilt University School Medicine, Nashville, TN 37232-2581, USA

**Corresponding author, mark.denison@vanderbilt.edu*

Abstract

Viruses are small strands of nucleic acid housed within a shell of protein and sometimes lipid. For such minute particles, they are capable of causing devastating effects on the health of humans and other inhabitants of this planet. Due to large population sizes and short replication cycles, viruses evolve at an accelerated pace compared to larger organisms with longer life spans. This constant state of change allows them to continually sample sequence space and generate diverse populations containing viruses capable of transitioning into and adapting to a new host. When this happens, virus may emerge in a naïve population and cause disease, in which case, we call the virus and the resultant disease emerging. Emerging infectious diseases (EIDs) appear to be increasing, likely the result of human intrusions into ecological niches, changes in vector distribution, altered climate patterns and other factors. In this chapter, we will highlight select animal and plant emergent viruses and review characteristics that have allowed these viruses to utilize opportunities created by the changes listed above. We start with two, ebolavirus and coronavirus, that are currently epidemic. We also examine chikungunya, a virus with an expanding vector host and host range, and finish with parvovirus and coffee ringspot virus, pathogens of domestic animals and an economically important crop, respectively.

Introduction

Since the beginning of the 21st century, humans have experienced devastating health, social, and economic impacts from emerging infectious diseases (EIDs). Many EIDs have been caused by viruses, such as severe acute respiratory syndrome coronavirus (SARS-CoV), influenza (pandemic [p]H1N1, H5N1 and other highly pathogenic avian strains), chikungunya, and most recently, Middle East respiratory syndrome (MERS)-CoV and ebolavirus. These are striking examples of infectious diseases that have spread through multiple countries, resulting in high morbidity and mortality with significant social and economic disruption. These viruses and the diseases they cause can be classified as emerging, according to the definition presented in 2001: an infectious disease whose incidence is increasing following its first introduction into a new host population or whose incidence is increasing in an existing host population as a result of long-term changes in its underlying epidemiology (Woolhouse and Dye, 2001).

There are many factors that influence the emergence or re-emergence of an infectious disease, including microbial adaptation, changes in human ecology and behavior, and widespread international travel, among others. Anthropogenic factors, such as habitat destruction, have been described as the main driving forces leading to new EID events (Morse, 1995; Woolhouse *et al.,* 2005; Jones *et al.,* 2013) and it is likely that these human-driven changes will only increase over time. Because of the continued likelihood of new EIDs, it is important to identify adaptive changes necessary for viruses to move between and among human, animal and plant populations. Modern biological techniques, particularly high-throughput sequencing, have increased the amount of data collected during outbreaks. These data can be used to understand changes that occur within viral populations preceding, during and following emergence and could lead to better predictive capacity in the future.

Studies of EIDs have identified a higher prevalence of these emergences originating from viruses in animals, or zoonoses, than would be expected by chance (Taylor *et al.,* 2001; Woolhouse and Gowtage-Sequeria, 2005; Jones *et al.,* 2008). The original definition of a zoonosis was generated by the WHO and modified in 1959, but is still applicable: ‘Those diseases and infections which are naturally transmitted between vertebrate animals and man’ (WHO/FAO, 1959). In 2004 an emerging zoonosis was defined as “a zoonosis that is newly recognized or newly evolved, or that has occurred previously but shows an increase in incidence or expansion in geographical, host or vector range” (WHO *et al.,* 2004). RNA viruses in particular, have been touted as a major source of future EIDs, due to their high rates of mutation (Sanjuán *et al.,* 2010), which is thought to confer an increased capacity to move between and adapt to new hosts (Jones *et al.,* 2008). Viruses have large population sizes and short replication cycles, making them ideal candidates for studies of evolution in the laboratory. Viruses are constantly undergoing evolution, sampling the surrounding sequence space in each niche in which they find themselves and adapting through selection. Thus, this review concentrates on emerging diseases from a viro-centric perspective, focusing on viral evolution as a factor in emergence.

Emergence or re-emergence can occur through multiple pathways, including 1) evolution of existing viruses leading to transmission to and adaptation in new hosts, such as SARS-CoV, MERS-CoV, and parvovirus; 2) known pathogens spreading to new geographic areas or populations, such as chikungunya virus; 3) previously unrecognized viruses appearing in areas undergoing ecologic transformation, such as ebolaviruses and coffee ringspot virus, or 4) known viruses reemerging as a result of antimicrobial resistance in known agents or breakdowns in public health measures, also applicable to ebolaviruses.

Figure 12.1. Schematic of the negative-sense ebolavirus genome. The genome is approximately 19 kb and includes the nucleoprotein (NP, blue), virion protein 35 (VP35, brown), VP40 (red), glycoprotein (GP, green), VP30 (cyan), VP24 (purple), and the RNA-dependent RNA polymerase (L, orange).

Ebola

Introduction

Ebolaviruses (order *Mononegavirales*, family *Filoviridae*) have enveloped, nonsegmented, negative-sense RNA genomes and include three genera, *Marburgvirus*, *Ebolavirus*, and *Cuevavirus*. There is only one species each within the *Marbrugvirus* and *Cuevavirus* genera, represented by *Marburg virus* (MARV) and *Lloviu virus* (LLOV), respectively. The *Ebolavirus* genus includes five species, *Zaire ebolavirus* (EBOV), *Sudan ebolavirus* (SUDV), *Reston ebolavirus* (RESTV), *Tai Forrest ebolavirus* (TAFV), and *Bundibugyo ebolavirus* (BDBV) (Kuhn *et al.*, 2010; Feldmann and Geisbert, 2011). All five species within the *Ebolavirus* genus are capable of infecting humans, but only four cause fatal disease. RESTV has been associated with disease in nonhuman primates and pigs, but not humans (Jahrling *et al.*, 1990; Miranda *et al.*, 1999; Barrette *et al.*, 2009). Ebolavirus particles are long and filamentous, enclosing a ~19 kb genome (Figure 12.1), which encodes seven proteins consistently found in the same order from the 3' to the 5' end of the genome: nucleoprotein (NP), virion protein 35 (VP35), VP40, glycoprotein (GP), VP30, VP24, polymerase (L) (Kuhn *et al.*, 2010; Feldmann *et al.*, 2013).

Outbreak history

The first cases of viral hemorrhagic fever caused by a filovirus were recorded in 1967 in Frankfurt and Marburg, Germany and in Belgrade, former Yugoslavia. Laboratory personnel handling African green monkeys originating from Uganda became infected and the causative agent was named Marburg virus (Siegert *et al.*, 1967; Feldmann *et al.*, 2013). Almost 10 years later, two overlapping outbreaks were investigated in Zaire, now Democratic Republic of Congo (DRC) and Sudan (Team, 1978; Commission, 1978) (Figure 12.2). Though researchers initially tried to link the two outbreaks, they were unable to do so. Careful analyses with more modern techniques indicated that two different, but related viruses, EBOV and SUDV, were involved (Richman *et al.*, 1983). Identification of the index case in both outbreaks indicated that two different filoviruses independently entered the human population within just a couple months of each other. In addition, not only were there differences in mortality rates identified as a consequence of route of infection, but the attack rate for successive transmission generations appeared to decrease overall, suggesting that human-to-human transmission, when it occurred, may have resulted in viruses less capable of causing severe disease in humans (Team, 1978; Commission, 1978).

The most recent ebolavirus outbreak began in West Africa at the end of 2013 and is still ongoing in 2015. Due to the rapid onset of symptoms and often death, ebolavirus is of grave concern to the global community. The index case of the current EBOV outbreak in West Africa appears to have been a 2 year-old boy who

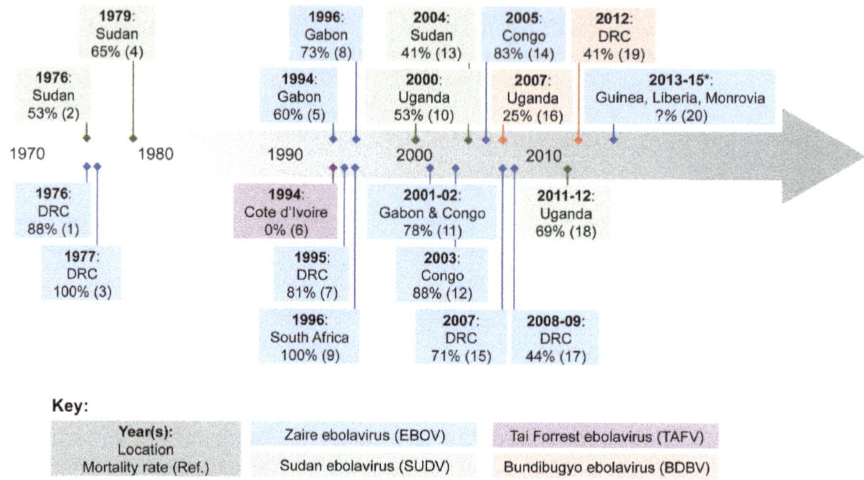

Figure 12.2. Timeline of ebolavirus disease outbreaks in humans from 1976 to 2015. Documented ebolavirus outbreaks are depicted and are colored according to the ebolavirus species responsible for each outbreak (see key). The date and location of the outbreak, as well as the case-fatality rate and reference, are also shown. DRC, Democratic Republic of Congo. References: (1) Commission, 1978; (2) Team, 1978; (3) Heymann *et al.,* 1980; (4) Baron *et al.*, 1983; (5) Georges *et al.*, 1999; (6) Le Guenno *et al.*, 1995; (7) Khan *et al.*, 1999; (8) Georges *et al.*, 1999; (9) Georges *et al.*, 1999; (10) CDC, 2001; (11) Leroy *et al.*, 2002b; (12) Formenty *et al.*, 2003; (13) WHO, 2005; (14) Nkoghe *et al.*, 2011; (15) WHO, 2007; (16) Towner *et al.*, 2008; (17) Muyembe Tamfum *et al.*, 2012; (18) Shoemaker *et al.*, 2012; (19) Albariño *et al.*, 2013; (20) Gire *et al.*, 2014. *Outbreak is still ongoing.

died at the end of 2013. Close relatives were the next to succumb to the disease (Baize *et al.,* 2014). Previously there were suggestions that ebolaviruses might be circulating in West Africa (Schoepp *et al.,* 2014), but this is the first time that EBOV infection in humans has been confirmed in this region of the continent (Dudas and Rambaut, 2014). This outbreak is remarkable for a number of reasons, especially its size, both in terms of the countries involved and the number of people affected, and the duration, which is unprecedented. Even more striking is that this outbreak appears to have originated with a single zoonotic transmission preceding sustained human-to-human transmission (Gire *et al.,* 2014) on an unprecedented scale for ebolaviruses. Within this outbreak, humans with EBOV infections have been diagnosed in Guinea, Sierra Leone, Liberia, Nigeria, Senegal, USA, Spain, and Mali (WHO, 2015b). This outbreak is a sobering reminder that emerging viruses can appear rapidly, potentially affecting the human population globally.

Description of disease

The incubation period for EBOV ranges from 2-21 days, with 4-10 days most commonly observed. Initial signs and symptoms are nonspecific and usually include fever, headache, malaise and myalgia, and sometimes vomiting and diarrhea. The lack of specificity of early symptoms frequently precludes accurate diagnosis of filovirus infection (Feldmann and Geisbert, 2011; Goeijenbier *et al.,* 2014). The onset of symptoms is usually sudden and patients either deteriorate rapidly and progress toward death, or have fever for several days and then begin to improve (Siegert *et al.*, 1967; Feldmann and Geisbert, 2011; Baize *et al.*, 2014). In severe disease, known as hemorrhagic fever (HF), multiple organ failure, lymphopenia, coagulative disorders, hypovolemic shock, and hemorrhage are often seen (Team, 1978; Commission, 1978; Feldmann and Geisbert, 2011).

Studies with nonhuman primates indicate that EBOV initially infects monocytes, macrophages, and dendritic cells, which then are disseminated through lymph nodes and the lymphatic system, resulting in systemic infection and viremia (Richman *et al.*, 1983; Geisbert *et al.*, 2003a). However, EBOV infection of dendritic cells occurs without activation of these professional antigen-presenting cells (Bosio *et al.*, 2003), thus delaying adaptive immune system involvement. Following dissemination, EBOV targets the liver and adrenal gland (Team, 1978; Geisbert *et al.*, 2003a; Feldmann and Geisbert, 2011), leading to organ failure. Infection and necrosis of the adrenal gland results in lack of enzymes needed to synthesize steroids involved in blood pressure homeostasis, contributing to hypotension and hypovolemia (Feldmann and Geisbert, 2011).The hypovolemic shock seen in EBOV patients has similarities to septic shock, including lymphocyte apoptosis, proinflammatory cytokine production and dysregulation of coagulation (Bray and Mahanty, 2003; Geisbert *et al.*, 2003b). Interestingly, there is also evidence that EBOV has superantigen activity (Leroy *et al.*, 2011), which nonspecifically activates the immune system and can lead to septic shock.

Viral load is negatively correlated with survival (Baize *et al.*, 1999; Sanchez *et al.*, 2004), but activation of the humoral and adaptive immune system early in infection is a positive predictor of survival. In patients who ultimately recover, improvement in symptoms is concomitant with a detectable antibody response, indicating that the adaptive immune system is capable of controlling EBOV infection, if enough survives the initial onslaught to be able to mount a defense (Rowe *et al.*, 1999; Sanchez *et al.*, 2004). In several outbreaks, individuals were identified who had close contact to patients with EBOV disease but did not become ill. The possibility that EBOV infection can result in asymptomatic or mild infection (Rowe *et al.*, 1999; Leroy *et al.*, 2000; Sanchez *et al.*, 2004; Gire *et al.*, 2012) raises the question of whether the mortality rates observed during outbreaks, 30-90% depending on the filovirus species (Baize *et al.*, 2014), tell the complete story. A more recent surveillance of individuals in Gabon, where several outbreaks have occurred, suggests that mild or asymptomatic EBOV infections are not only possible but maybe more frequent than previously appreciated (Becquart *et al.*, 2010). A previous study, significant in light of the current EBOV epidemic in West Africa, detected EBOV-specific antibodies in inhabitants of a Liberian rainforest (Van der Waals *et al.*, 1986).

Reservoir species

The sporadic and dramatic nature of the outbreaks attributable to EBOV raise the question of where the virus resides between these outbreaks as well as what is the inciting factor(s) that initiates an epidemic. Filoviruses can also infect and cause disease and death in nonhuman primates, specifically great apes, including gorillas and chimpanzees. Several epidemics in human populations have been correlated with outbreaks in nearby primate populations (Leroy *et al.*, 2004; Lahm *et al.*, 2007), suggesting that the same virus affecting humans can also harm nonhuman primates. The fact that nonhuman primates also are susceptible to disease outcomes similar to those in humans suggests that these species do not harbor the viruses over time as a reservoir species, but are subject to disease and death when infected with filoviruses. A reservoir species is hypothesized to be a species that can be infected, either transiently or persistently, but not exhibit morbidity or mortality as a result of infection (Swanepoel *et al.*, 1996). In this case, virus and host are well adapted and the virus has evolved to survive and be transferred to another host, continuing its existence. On the other side of the equation, hosts, even with their longer lifespans, can evolve to become to resistant to disease caused by a virus. Indeed, a prevailing theory suggests that the longer a virus and host co-exist the less disease that is caused by a viral infection (Weiss, 2001). The corollary of this is that emerging viruses may cause severe disease, which is seen with ebolaviruses.

Many species that come into contact with humans in and around villages where outbreaks occurred have been tested for EBOV antigen and antibodies, including plants, rodents, insects, and bats (Gonzalez *et al.*, 2007). In addition, attempts to experimentally infect rodents (Swanepoel *et al.*, 1996; Gire *et al.*, 2014) or arthropods (Turell *et al.*, 1996) were not successful, differentiating EBOV from hantaviruses and arboviruses, respectively. More recently, surveillance of rodents, birds, and bats in regions near EBOV outbreaks found evidence of infection in three species of African fruit and insectivorous bats *Hypsignathus monstrosus*, *Epomops franqueti*, and *Myonycteris torquata* (Leroy *et al.*, 2005). Interestingly, bats were RT-PCR positive or seropositive, but not both, likely indicating active infection or cleared infection and seroconversion, respectively (Leroy *et al.*, 2005). A study reported in 2007 found evidence of EBOV infection in fruit bats captured near regions where human outbreaks had taken place and in areas where endemic virus was not suspected (Pourrut *et al.*, 2007) suggesting that EBOV is circulating in a larger area than previously thought. A survey of bats in Ghana, West Africa indicated the presence of EBOV infection in bats outside the area of EBOV human and nonhuman primate outbreaks (Hayman *et al.*, 2012). A migratory fruit bat, *Eidolon helvum*, was found to be EBOV-seropositive and followed for 13 months with a radiotransmitter, further supporting the hypothesis that EBOV infection is not lethal for some species of bats (Hayman *et al.*, 2010). In addition, evidence of filovirus infection in bats has been seen in Asia (Olival *et al.*, 2013) and Europe (Negredo *et al.*, 2011).

Although mounting evidence implicates several species of bats as reservoirs of EBOV and other filoviruses, foreshadowing of this conclusion can be found in the history of outbreaks (Pourrut *et al.*, 2005). Specifically, index cases of outbreaks attributable to MARV were miners (Bausch *et al.*, 2006) or had visited caves populated by bats prior to becoming ill (Smith *et al.*, 1982; Johnson *et al.*, 1996). In two cases, workers from a cotton factory where bats were known to roost in the rafters were linked to SUDV outbreaks in 1976 and 1979 (Team, 1978; Baron *et al.*, 1983). The most conclusive evidence came from the 2007 DRC EBOV epidemic where the index case was a person who had purchased freshly killed bats from hunters during a massive annual migration of fruit bats (Leroy *et al.*, 2009). In the current West Africa outbreak, the index case was known to play in a hollow tree inhabited by a colony of insectivorous free-tailed bats (*Mops condylurus*) (Mari Saez *et al.*, 2015). This expands the potential reservoir species of bats to include insectivorous bats.

Viral diversity and evolution

There are five ebolavirus genera that exhibit a range of lethality in humans, from EBOV with a mortality rate of up to 90% to RESTV which has not been shown to cause any discernable disease in humans. There is about 50% nucleotide and amino acid difference between EBOV and MARV, and the four different ebolavirus genera are 30-40% divergent from each other (Towner *et al.,* 2008). However, viruses within each genus are remarkably similar. In one example, GP from a virus isolated from the 1976 EBOV outbreak was found to be almost identical to that from an EBOV isolate from 1995, even though almost 1000 km and 20 years separate the two outbreaks (Sanchez *et al.,* 1996). Examination of sequences isolated from different patients within the same epidemic and especially within the same transmission chain are remarkably similar, indicating unexpected genetic stability of these viruses during an outbreak (Pourrut *et al.,* 2005). Thus, genetic analysis of isolates can determine whether an outbreak is the result of multiple zoonotic transmissions or a single introduction into the human population. Sequences of GP of EBOV isolated between 2001-03 in Gabon and RC representing five outbreaks (Formenty *et al.,* 2003) revealed little to no variation within an epidemic chain (tracked human-to-human transmissions) and eight different viral lineages were suggested to each be the result of a separate introduction from an animal source (Leroy *et al.,* 2004). Intra-epidemic stability has been seen in outbreaks in DRC and Gabon as well (Rodriguez *et al.,* 1999; Leroy *et al.,* 2002a). Patterns from examining genetic relatedness of outbreaks in 2007 and 2008 suggest movement of the virus across Africa (Walsh *et al.,* 2005). However, previously unidentified viral lineages within these outbreaks (Grard *et al.,* 2011) indicate potential viral recombination or perhaps maintenance in a different reservoir species and transmission to humans.

Since EBOV and other filoviruses are not maintained within the human population, outbreaks are zoonotic in origin. There are similarities with viruses that rely on vectors for transmission. Genetic principles seen with vector-borne viruses may be applicable here as well, particularly the observation that the requirement to replicate in alternating hosts allows for genetic variation within either host, but restricts variation over time (Holmes, 2003). Less genetic variation could be seen as a lower level of nonsynonymous changes between outbreaks when compared to during outbreaks or within hosts (Gire *et al.,* 2014) possibly indicating that the virus is well-adapted to its natural host compared to humans as hosts.

In a study published in 1997, phylogenetic examination of filovirus sequences indicated that MARV and ebolaviruses likely share a common ancestor, and that the two lineages diverged within the last several thousand years (Suzuki and Gojobori, 1997). In addition, the calculated rate of nonsynonymous substitutions of GP, the protein with the most available sequences, was estimated to be 3.6×10^5 per site per year, which is the midrange of substitution rates calculated for RNA viruses (Sanjuán *et al.,* 2010). Another explanation for the lower genetic diversity seen among isolates is that a bottleneck event occurred recently within the reservoir species, resetting the molecular clock (Biek *et al.,* 2006). However, other studies suggest that EBOV has been around for a very long time and that viral genetic fragments persist in the human genome (Taylor *et al.,* 2010; Belyi *et al.,* 2010). Some of these issues are currently unresolved, but with time and more complete information about these pathogens, these questions can be answered.

Coronaviruses

Introduction

Coronaviruses (CoVs) had been studied as causes of human colds, models of demyelinating encephalitis, and pathogens of domestic animals until an unexpected discovery in 2003 that the etiological agent of the severe acute respiratory syndrome (SARS) described in China and Hong Kong was a novel CoV (Peiris *et al.,* 2003). Then, in 2012 cases of severe respiratory disease were described in the Middle East and the cause was determined to be another novel CoV. This virus was initially named EMC2012 (Zaki *et al.,* 2012), but was renamed Middle East Respiratory Syndrome (MERS)-CoV (de Groot *et al.,* 2013). The outbreak caused by MERS-CoV has been continuing for over two years and is still ongoing, with over 1000 confirmed infections and an overall mortality rate of greater than 30% (WHO, 2015a). Increased surveillance led to identification of two additional human CoVs, HKU1 (Woo *et al.,* 2005) and NL63 (van der Hoek *et al.,* 2004), capable of causing severe lower respiratory tract disease in humans. Since their identification, these viruses have been shown to be present worldwide, and have likely been circulating in humans for decades (Vabret *et al.,* 2005; Pyrc *et al.,* 2006; Woo *et al.,* 2006; Pyrc *et al.,* 2007). Thus to date, six human CoVs have been identified, and all are likely to be zoonotic.

CoVs (order *Nidovirales*, family *Coronaviridae*) are enveloped viruses with positive-sense RNA genomes ranging in size from 27 to 32 kb in length, making these the largest replicating RNA molecules known (Figure 12.3). CoV genomes contain up to nine open reading frames (ORFs), with the first encoding the replicase polyprotein (pp) which, when translated and proteolytically processed, results in up to 16 nonstructural proteins (nsps). The downstream ORFs encode structural and accessory proteins, of which the four structural proteins are always present in the same order: spike (S), envelope (E), membrane (M), and nucleocapsid (N). There are a number of other proteins encoded among the invariant structural proteins (Masters and Perlman, 2013). In the case of SARS-CoV, these can be deleted and the virus remains replication-competent in cell

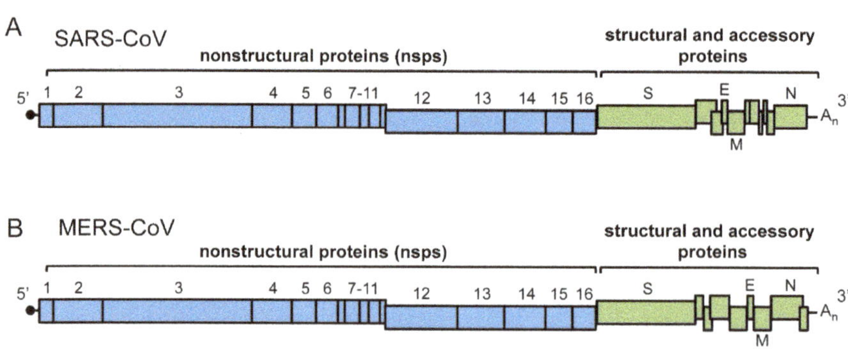

Figure 12.3. Schematic of SARS-CoV and MERS-CoV genomic organization. Coronaviruses have a single-stranded positive-sense RNA genome between ~27-32 kb in length. The 5' RNA cap and 3' poly-A tail are denoted. Nonstructural (nsp; blue) and structural protein (green) organization for both SARS-CoV (A) and MERS-CoV (B) is shown. Individual nsps are denoted numerically (1-16) and structural proteins are abbreviated: S, spike; E, envelope; M, matrix; N, nucleocapsid.

culture (Yount *et al.,* 2005), suggesting that these proteins carry out functions only required for virus replication *in vivo.* The S protein is present in the virion as trimers that project from the surface, resulting in the characteristic corona visible by EM that gives this family of viruses its name. The S protein is a class I fusion protein and functions as the viral attachment protein, and thus is the main viral determinant of tropism [reviewed in (Graham and Baric, 2010)] (Figure 12.4).

CoVs encode an RNA-dependent RNA polymerase (RdRp) with low inherent fidelity. However, CoVs also encode additional RNA-modifying enzymes unique among RNA viruses, including a 3'-5' exoribonuclease (ExoN) in nsp14 necessary for high fidelity viral replication and which appears to perform a proofreading function during replication (Minskaia *et al.,* 2006; Gorbalenya *et al.,* 2006; Eckerle *et al.,* 2010; Smith and Denison, 2013). This results in greater replication fidelity compared to other RNA viruses (Sanjuán *et al.,* 2010; Smith and Denison, 2013). Thus, conclusions about viral evolution generated from RNA viruses that do not proofread may not be generally applicable to CoVs. More recent studies have suggested that CoVs have been co-evolving with humans and animals for a much longer time than previously suggested (Wertheim *et al.,* 2013). However, the capacity of CoVs to proofread may have been more important for establishing and maintaining their large genomes than for preventing generation of mutations that

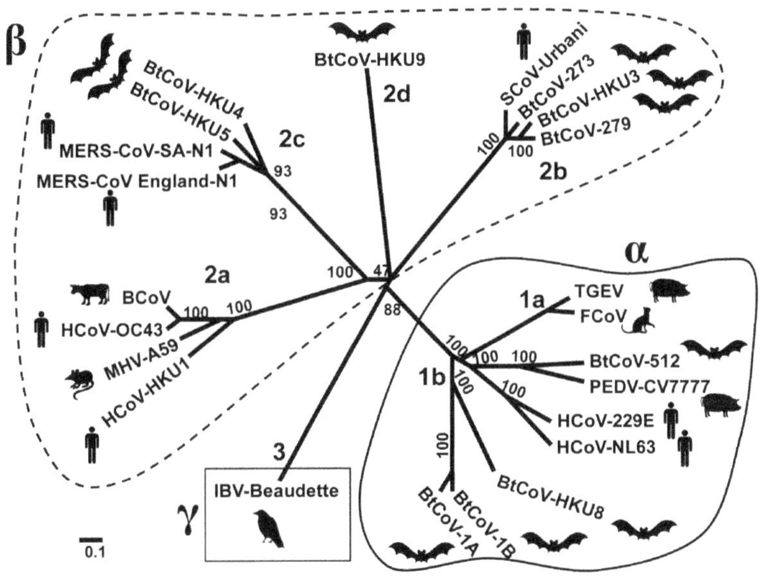

Figure 12.4. Phylogenetic tree of coronaviruses based on spike sequence. Full-length spike sequences of representative CoVs were downloaded from Genbank or PATRIC, aligned with ClustalX, and phylogenetically compared by Maximum Likelihood using 100 bootstraps. The tree shows that CoVs are divided into three distinct phylogenetic groups defined as α, β, and γ. This taxonomical nomenclature replaced the former group 1, 2 and 3 designation, respectively. Classical subgroup clusters are marked as 2a-2d for the β-CoVs and 1a and 1b for the α-CoVs. The tree was generated using Maximum Likelihood with the PhyML package. The scale bar represents nucleotide substitutions. Only nodes with bootstrap support above 70% are labeled. (Dr. Eric Donaldson)

result in a pool of viral variants capable of adapting to new environments. Even with RNA proofreading activity, CoVs still are quite capable of transmitting to new hosts and causing disease, as has been seen most recently with SARS-CoV and MERS-CoV.

Outbreak history

In February 2003, hospitals in Hong Kong began reporting cases of atypical severe pneumonia in patients, with fever followed by compromised respiratory function, and termed severe acute respiratory syndrome (SARS) (Tsang *et al.*, 2003). It was determined that a novel coronavirus, named SARS-CoV, was the cause of the disease (Peiris *et al.*, 2003). The outbreak was traced back to similar cases in Guangdong Province in China starting in November 2002, in which a majority were seen in health care workers or food handlers (WHO, 2003a). The spread of the virus from Guangdong Province to Hong Kong was linked to a physician who had treated patients in Guangdong, then traveled to Hong Kong. In only a couple of days, at least 16 people staying at the same hotel were infected and then traveled internationally (WHO, 2003a). In mainland China, the regions most severely affected were cities in Guangdong Province and Beijing, with a total of 5327 cases and 348 deaths reported (WHO, 2003a; Zhao, 2007). The epidemic spread to North America where the greatest disease burden was seen in Toronto, Canada, again mostly among health care workers and other patients (Poutanen *et al.*, 2003; Dwosh *et al.*, 2003). Isolation and quarantine were successful in limiting the epidemic, but a second wave of infections occurred following the declaration that the virus had ceased being transmitted. Overall, there was a total of 246 confirmed infections and 43 deaths attributable to SARS-CoV in Toronto (WHO, 2003a; b). Twenty-nine cases were reported in the United States, but 28 were attributable to importation with only one case where local transmission was suspected (WHO, 2003b; Isakbaeva *et al.*, 2004). The virus spread to a number of other Asian countries, including Singapore, Taiwan, Vietnam, Malaysia, South Korea, and Thailand (WHO, 2003b). Other cases were also reported in Europe, most of which were imported with no evidence of sustained human-to-human transmission (WHO, 2003b). It was determined that transmission required symptomatic illness, allowing for control through quarantine measures (Lipsitch *et al.*, 2003; Fraser *et al.*, 2004). The last confirmed cases of the pandemic were seen in July 2003. Other sporadic cases were seen in late 2003 and 2004 (Liang *et al.*, 2004), as well as several laboratory exposures (Lim *et al.*, 2004; Cheng *et al.*, 2007). SARS-CoV has not been seen in humans since this time.

In June 2012, a critically ill man with severe pneumonia and renal failure was seen in a hospital in Saudi Arabia. Virus was isolated from samples taken from the patient and the viral genome was sequenced by a group in the Netherlands. The virus was found to be a novel CoV, named EMC/2012 (Zaki *et al.*, 2012). Several months later, in September, a critically ill man was transferred to from Qatar to a hospital in the UK. The second patient had traveled to Saudi Arabia in late July and so when samples tested negative for known viral causes, they were tested for the novel CoV and were positive (Bermingham *et al.*, 2012). These two cases appeared to be linked since they were infected with similar viruses, but the patients lived in different cities and became infected several months apart. Following identification of the novel CoV in these two cases, retrospective analysis of a March 2012 outbreak in a Jordanian hospital indicated positive results for

MERS-CoV infection in two of the cases (Hijawi *et al.,* 2013). The virus has continued to emerge and spread in the Middle East, primarily in Saudi Arabia, and the outbreak is still ongoing.

Description of disease

Human CoVs, such as OC43 and 229E, first identified in the 1960s (Hamre and Procknow, 1966; McIntosh *et al.,* 1967) have been shown to be a cause of illnesses attributable to the common cold syndrome, second only to rhinoviruses (Myint, 1994). Human CoVs have also been linked to more severe lower respiratory tract disease often with other complicating health issues in the background (McIntosh *et al.,* 1974; Myint, 1994). SARS was initially described as 'atypical pneumonia' due to rapid disease progression (Zhao *et al.,* 2003) frequently accompanied by a prodromal phase followed by rapid onset of fever, often with cough, myalgia and malaise, and occasionally diarrhea. The cough was usually nonproductive and respiratory symptoms could deteriorate to the point of requiring ventilation. The WHO case definition for probable SARS included fever, radiological evidence of pneumonia, chills, cough, malaise, myalgia, or known exposure and positive test for SARS-CoV (WHO, 2003a). The pathology of SARS-CoV included the presence of neutrophils and hyaline membranes in the lung (Gu and Korteweg, 2007). MERS-CoV has similar prodromal and respiratory symptoms, however there does appear to be kidney involvement and failure that was not seen during the SARS-CoV epidemic (Assiri *et al.,* 2013a; van den Brand *et al.,* 2015).

Reservoir species

Serology of asymptomatic controls indicated the absence of antibodies specific for SARS-CoV (Zhong *et al.,* 2003; WHO, 2003a), suggesting that the virus was recently introduced into the human population. Samples were taken from animals housed in a wet market in Shenzen, Guangdong Province and evidence of viral infection was found in multiple Himalayan palm civets (*Paguma larvata*), a raccoon dog, (*Nyctereutes procyonoides*), and a Chinese ferret-badger (*Melogale moschata*) and virus was isolated from palm civet samples (Guan *et al.,* 2003). Palm civets were culled and bans on their trade and sale instituted. Following re-introduction of palm civets in a wet market in Xinyuan, 100% of those sampled had evidence of infection with SARS-CoV-like viruses (Kan *et al.,* 2005). However, none of the animals on the farms where these animals originated were positive for SARS-CoV-like viruses or SARS-CoV-specific antibodies (Tu *et al.,* 2004; Kan *et al.,* 2005). This suggested that palm civets and other animals in the wet markets could function as amplification hosts, but were not likely to be the reservoir species. This was attributed to bats in 2005, when SARS-CoV-like viruses were found in wild Chinese horseshoe bats (*Rhinolophus sp.*) (Lau *et al.,* 2005). Since then, CoVs have been found in bats in China, Africa, Europe, and North America (Drexler *et al.,* 2014). There are currently more bat CoV sequences deposited to Genbank than CoVs from all other species combined. However, the bat SARS-CoV-like viruses identified were not considered to be the direct precursors of the epidemic virus because of their genetic distance and lack of capacity to utilize angiotensin converting enzyme 2 (ACE2), the human receptor for SARS-CoV, to enter cells. Recently, two novel bat CoVs were identified that share over 95% sequence identity with SARS-CoV at the nucleotide level and were demonstrated to utilize ACE2 molecules from bats, palm civets, and humans as receptors to enter cells (Ge *et al.,* 2013). These data indicate that bat CoVs were the precursors to human SARS-CoV.

The hypothesis that CoVs have been passing into the human population either directly from bats or through an intermediate animal species gained support following determination of the genomic sequence of MERS-CoV and its location on the CoV phylogenetic tree between two previously reported bat CoVs, HKU4 and HKU5 (Woo *et al.,* 2007; Zaki *et al.,* 2012). Small clusters of human-to-human transmission of MERS-CoV have been documented (Assiri *et al.,* 2013b; HPA-UK, 2013); however the fact that the epidemic has continued for over two years with limited human-to-human spread suggests repeated introductions from an animal reservoir. HKU4 and HKU5 were identified in bats from China and MERS-CoV is circulating in the Middle East, so questions remain about whether the virus has moved across continents or similar viruses exist worldwide. Ongoing surveillance has identified a CoV from bats in Mexico that clustered near MERS-CoV based on sequence analysis of a portion of the RdRp (Anthony *et al.,* 2013). Additionally, CoVs have been found in bats in Saudi Arabia, with one similar to human MERS-CoV (Memish *et al.,* 2013), and CoVs have been seen in bats in Ghana and Europe, Thailand, and Africa (Annan *et al.,* 2013; Wacharapluesadee *et al.,* 2013; Corman *et al.,* 2014a).

Investigations of domesticated animals in the Middle East have demonstrated that dromedary camels are frequently seropositive for MERS-CoV, including those in Egypt, Jordan, Qatar, and Saudi Arabia (Perera *et al.,* 2013; Reusken *et al.,* 2013a; Hemida *et al.,* 2013; Reusken *et al.,* 2014a). Camels from Oman showed a much higher incidence of seropositivity than those from Spain (Reusken *et al.,* 2013b), and samples from both 2003 and 2013 from camels in the United Arab Emirates showed evidence of infection with MERS-CoV or a similar virus (Meyer *et al.,* 2014). These data indicate that MERS-CoV or MERS-CoV-like viruses may have been circulating in dromedary camels for an extended period of time. Many camels in the Middle East are imported from Africa, and examination of African camels indicates the presence of MERS-CoV-specific antibodies (Chu *et al.,* 2014; Corman *et al.,* 2014b; Reusken *et al.,* 2014b). Interestingly, a recent report of the complete genomic sequence of a CoV from a *Neoromicia capensis* bat in South Africa indicates the presence of a virus on that continent that is closely related to MERS-CoV (Corman *et al.,* 2014a). The placement of this virus on the phylogenetic tree, along with viral sequences obtained from a camel likely imported from Africa (Chu *et al.,* 2014) indicates that the origin of MERS-CoV could be from African bats, through African camels, to camels and humans in the Middle East (Corman *et al.,* 2014a). Evidence that camels infected with MERS-CoV show signs of disease (Memish *et al.,* 2014) indicates they may be an intermediate or amplifying host. These data suggest, but do not prove, the hypothesis that bats and camels together make up a reservoir community for MERS-CoV, from which the virus can then be transmitted to humans (Nishiura *et al.,* 2014).

Viral diversity and evolution

Soon after the identification of the etiologic agent of SARS as a CoV, the full genome sequence of several isolates was determined. The first two complete genome sequences were published back-to-back in 2003, Tor2 (Marra *et al.,*

2003) and Urbani (Rota *et al.,* 2003). These two strains differed at only 8 nucleotides (nt), resulting in 5 amino acid differences. Additional isolates were sequenced around this time, including HK-39 (Zeng *et al.,* 2003), Su-10, and CUHK-W1 (Tsui *et al.,* 2003). Su-10 and CUHK-W1 showed sequence variations, suggesting the possibility of two different viral strains co-circulating in Hong Kong (Tsui *et al.,* 2003). Additional CoV-positive samples were taken from humans and identified as early-, mid-, late-epidemic samples (He *et al.,* 2004; Zhao, 2007), and genome sequences compared with samples taken from animals in the wet markets and captured in the wild (Guan *et al.,* 2003). Viral sequences from animals in the wet markets in Shenzen, near the outbreak in Guangdong Province, China, are longer by 29 nt than other viruses during the epidemic. Strikingly, viruses isolated from humans from Guangdong Province early in the epidemic also have this sequence (Zhao, 2007), whereas other viruses isolated from humans are missing this 29 nt sequence, indicating that viruses found in humans early in the epidemic are most closely related to those seen in the animals in the wet markets. Other deletions were identified during the course of the epidemic as well, mainly located in the ORF8 region of the genome, located between the coding sequence for M and N structural proteins (He *et al.,* 2004; Zhao, 2007). Viruses identified in wild Chinese horseshoe bats (*Rhinolophus sinicus*) contain the 29 nt sequence seen in viruses from palm civets and humans early in the 2002-03 epidemic, again suggesting that these animals are a reservoir host for SARS-CoV precursor viruses (Lau *et al.,* 2005).

Many factors influence viral emergence and one of the main obstacles for CoVs is the interaction between the viral attachment protein, spike, and its cognate cellular receptor. CoVs bind different cellular molecules and the receptor-binding domain within spike, the minimum region competent for this interaction, is not located in exactly the same position in all CoVs (Graham and Baric, 2010). Examination of spike sequences from animal and human SARS-CoV strains from early-, mid-, and late-epidemic samples indicated that amino acid changes seen over time were linked to the increased binding capacity for human ACE2 that occurred over the course of the epidemic (Graham and Baric, 2010). In addition, examination of the evolution of bat ACE2 suggests that SARS-CoV or another virus that utilizes the same interface with bACE2 has been exerting evolutionary pressure on bACE2 prior to SARS-CoV emergence in humans (Demogines *et al.,* 2012). This was confirmed when a bat SARS-CoV-like virus was isolated and shown to be competent for utilizing ACE2 molecules from bat, civet, and human for cell entry (Ge *et al.,* 2013). Together these studies indicate the importance of spike-receptor interactions in CoV emergence into novel host species.

Similar analyses and evaluations of MERS-CoV sequences are currently ongoing. Initial reports of full genome sequences of MERS-CoV indicated that the virus is a betacoronavirus, but in lineage C, distinct from SARS-CoV and similar viruses in lineage B (van Boheemen *et al.,* 2012). There are at least 65 partial or complete MERS-CoV genomic sequences available (van Boheemen *et al.,* 2012; Assiri *et al.,* 2013a; Drosten *et al.,* 2013; Cotten *et al.,* 2013). Phylogenetic analysis indicates that viruses isolated from the same region group into clades, including Hafir-Al-Batin, Al-Hasa, Buraida, Riyadh, and clade A that includes viruses from the first two patients identified (Cotten *et al.,* 2014). These analyses also suggest that MERS-CoV originated in or near Riyadh with a most common recent ancestor around March 2012 (Cotten *et al.,* 2014). Examination of viral sequences taken from human and animals, specifically bats and camels, is a prominent component of many publications concerning MERS-CoV and phylogenetic analyses lead to a reconstruction of the potential movement of the virus over time and across countries. However, examination of viral genetic sequences alone does not reveal the function of viral proteins involved in virulence and pathogenesis. Further studies of viral evolution, not only of genomic sequences but also how changes alter viral protein function affecting disease outcomes, are needed.

Chikungunya

Introduction

Chikungunya virus (CHIKV) is an Old World alphavirus in the Semliki Forest antigenic complex in the family *Togaviridae*, with a plus-stranded, nonsegmented RNA genome (Solignat *et al.,* 2009). The ~11.8kb genome is capped and polyadenylated. The 5' ORF is translated to yield a polyprotein, cleaved by the viral nonstructural protein (nsP)2 protease, resulting in nsPs 1-4. Three major structural genes are translated as a polyprotein from subgenomic RNAs originating from the 3' one-third of the genome, resulting in capsid and two glycoproteins, E1 and E2 (Figure 12.5), which combine to form the 70nm icosahedral virion. CHIKV is enveloped and sensitive to ether and detergents (Griffin, 2013).

Outbreak history

The first epidemic attributed to CHIKV occurred in Tanganyika (now Tanzania) in East Africa in 1952-1953 (Robinson, 1955). Although retrospective case reports suggest that CHIKV has been circulating and causing disease in India and Southeast Asia over the last 200 years (Carey, 1971; Griffin, 2013; Halstead, 2015), the first virus isolation, and subsequent characterization, from a CHIKV-infected patient was from this epidemic (Ross, 1956). Since the identification of a CHIKV as a novel virus causing human disease in 1952, there have been sporadic outbreaks in Africa and Asia, with a periodicity ranging from 2-20 years, potentially linked to immunity in primate hosts in sylvatic cycles (Tsetsarkin *et al.,* 2011b). Following absences of CHIKV for 20 years in the Democratic Republic of Congo (formerly Zaire) and 39 years in Indonesia, outbreaks occurred in both countries, in 2000 and 2001-03, respectively (Burt *et al.,* 2012). In 2004, CHIKV was identified in Kenya and had spread to the Comoros Islands by January 2005 (Burt *et al.,* 2012). The virus was then detected on the neighboring islands of Seychelles and Mauritius (Pialoux *et al.,* 2007) and then on La Reunion, where the epidemic continued until 2007 (Renault *et al.,* 2007; Burt *et al.,* 2012). The epidemic on La Reunion Island was more severe than outbreaks reported on other islands, with an

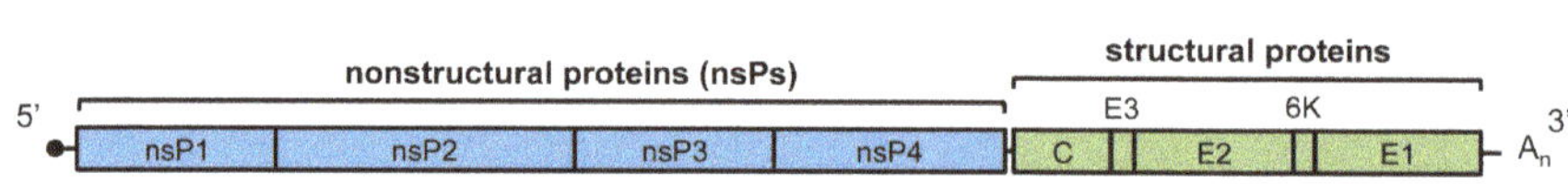

Figure 12.5. Schematic of the chikungunya virus genome. The single-stranded, positive-sense chikungunya virus genome is around ~11.8 kb in length. The 5' RNA cap structure and 3' poly-A tail are shown. Nonstructural proteins (nsP, blue) and structural proteins (green) are shown. Abbreviations: C, capsid protein; E1-E3, envelope protein; 6K, small hydrophobic protein.

initial attack rate reported to be around 35% (Renault *et al.*, 2007). However, serologic studies completed later to investigate the extent of asymptomatic infections led to estimates that 60% or more of the population may have been infected during the epidemic (Sergon *et al.*, 2007; Nakkhara *et al.*, 2013). In addition, CHIKV was thought to have disappeared from India and Southeast Asia due to the over 30-year long absence of virus detection and lack of serological evidence of infection (Lahariya and Pradhan, 2006; Arankalle *et al.*, 2007). However, massive CHIKV epidemics occurred in these regions between 2004 and 2006 (Tsetsarkin *et al.*, 2011b; Burt *et al.*, 2012). Interestingly, these new outbreaks have been attributed to viruses within both the East/Central/South African (ECSA) and the Asian genotype, indicating that both clades may be circulating in these regions of the world (Yergolkar *et al.*, 2006; Arankalle *et al.*, 2007).

Description of disease

Chikungunya means ‘that which bends up’ in the language of the Makonde people, who inhabit the area where the first recorded epidemic took place, and refers to the contorted position adopted by those infected due to the severe joint pain associated with this disease (Robinson, 1955; Lumsden, 1955). CHIKV causes chikungunya fever in humans, with acute infection characterized by fever, rash, myalgia and incapacitating polyarthralgia. The virus is usually cleared within 5 to 7 days, along with the acute symptoms, but in some cases, the polyarthralgia can last for months to years, causing significant morbidity (Sissoko *et al.*, 2009; Hoarau *et al.*, 2010; Schwartz and Albert, 2010; Schilte *et al.*, 2013; Thiberville *et al.*, 2013).

Vectors

Alphaviruses are transmitted between their mammalian hosts by an arthropod vector, and thus are arthropod-borne viruses, or arboviruses. CHIKV is endemic in Africa and exists in both sylvatic and urban transmission cycles. In the sylvatic cycle, CHIKV alternates primarily between a variety of forest-dwelling *Aedes* mosquito species, namely *Ae. furcifer*, *Ae. taylori*, *Ae. luteocephalus*, *Ae. africanus*, and *Ae. neoafricanus*, and nonhuman primates (Diallo *et al.*, 1999; Her *et al.*, 2009). Spillover events can initiate a non-sylvatic or urban cycle, where humans are the vertebrate host, causing an outbreak which tends to be self-limiting in rural areas inhabited by smaller populations (Her *et al.*, 2009). Within the urban cycle, CHIKV has been historically transmitted by *Ae. aegypti* mosquitoes. Until recently, CHIKV in Asia has been primarily detected in *Ae. aegypti* (Rao, 1964), suggesting that it was maintained in urban cycles only and dependent on immunologically naïve human populations for continued presence (Her *et al.*, 2009; Tsetsarkin *et al.*, 2011b). However, isolation of CHIKV from nonhuman primates in Malaysia (Apandi *et al.*, 2009) indicates that sylvatic cycles are possible in Asia as well as Africa (Powers and Logue, 2007; Tsetsarkin *et al.*, 2011b), but the role of nonhuman primates as possible reservoirs for the virus in this region remains unclear.

Viral evolution and diversity

To complete their transmission cycle, arboviruses must be competent to replicate in their disparate vectors and hosts. It has been hypothesized that the requirement to replicate in both invertebrate and vertebrate hosts imposes restrictions on virus adaptation and evolution so to maintain replication competence in both hosts. It is thought that adaptation for replication or transmission in one host may lead to a fitness deficit in the other resulting in less genetic variability and less rapid evolution of alphaviruses compared to non-vectored RNA viruses (Ciota and Kramer, 2010; Arias-Goeta *et al.,* 2014). This theory has been tested *in vitro* with arboviruses such as eastern equine encephalitis virus, Sindbis virus, dengue virus (DENV) and CHIKV (Weaver *et al.,* 1999; Greene *et al.,* 2005; Vasilakis *et al.,* 2009; Coffey *et al.,* 2011). Viruses were serially passaged in cell lines established from the vertebrate or invertebrate host or by alternating between cell lines derived from the two different hosts. In general, viruses passaged in a single cell line generated more mutations and replicated more efficiently in the cell line in which they were passaged, but lost replicative capacity in the other cell line, whereas those passaged in alternating cell lines exhibited fewer mutations, and either maintained or gained fitness in both cell lines. Some of these results are difficult to compare, as there are differences in how fitness was measured and questions remain as to the link between genetic diversity generated through passage and viral fitness in the two cells lines and how this relates to viral fitness in the actual host. Similar experiments were carried out with Venezuelan equine encephalitis virus and St. Louis encephalitis virus *in vivo* (Coffey *et al.,* 2008; Ciota *et al.,* 2009), passaging viruses in vertebrates or mosquitoes or alternating between the two hosts, with somewhat different results, highlighting the distinctions between cell lines and living organisms and the interactions of viruses with their animal hosts. For WNV, it was suggested that the fitness constraints were lower in the mosquito compared to the avian host, so viruses generated more diverse populations while replicating in the mosquito and then the populations were (Deardorff *et al.,* 2011). This hypothesis also was investigated by sampling infected birds and pools of mosquitoes on Long Island, NY during the 2003 peak transmission season of WNV, where evidence of genetic diversity in both hosts and purifying selection were found (Jerzak *et al.,* 2005). Fitness may not be the only parameter important in alternating host passage. In Rift Valley fever virus, viruses lost important genomic components necessary for virulence when passaged in a single cell line compared to when alternatively passaged in cell lines from two different hosts (Moutailler *et al.,* 2011).

Examining phylogenetic analyses of the CHIKV E1 gene initially identified two lineages, the ECSA and West African clades (Powers *et al.,* 2000). Viruses isolated in Asia were genetically distinct from these two lineages, but most similar to the East/Central African clade, indicating that CHIKV likely originated in Africa, diverged within the continent and was subsequently introduced into Asia (Powers *et al.,* 2000; Her *et al.,* 2009). Currently, CHIKV is grouped into three genotypes based on phylogenetic analyses, the Asian, ECSA, and West African clades (Powers *et al.,* 2000; Schuffenecker *et al.,* 2006). These three lineages correlated with their endemic locations until about 2000, when CHIKV emergence events increased throughout endemic areas and into new areas. Since then, the genetics of the different lineages have been used to track the movement of these viruses globally during epidemics.

A significant change in CHIKV re-emergence has been observed since 2004, with more frequent outbreaks over an expanded territory, and increased detection of more severe disease manifestations, including neurological symptoms and death

(Rajapakse *et al.*, 2010; Das *et al.*, 2010). *Ae. aegypti* have long been associated with urban transmission of CHIKV and have been the source of virtually all virus isolated from mosquitoes in Asia (Powers and Logue, 2007). Adaptation to a new vector may be a more critical step in virus emergence, similar to plant viruses, than adaptation to new hosts (Chen *et al.*, 2013). In the case of CHIKV, disease emergence over the last decade has been linked to enhanced virus competence in the new *Ae. albopictus* vector. In the naïve, more temperate regions where CHIKV began circulating during the 2000s, there was a lack of *Ae. aegypti* (Burt *et al.*, 2012), but an abundance of *Ae. albopictus* mosquitoes. For DENV, when epidemics occurred in areas where *Ae. aegypti* were absent, it was assumed that the vector was the most abundant *Aedes* species present, *Ae. albopictus* (Gratz, 2004). The distribution of *Ae. albopictus* has expanded over the last several decades, spreading to regions in Africa, the Middle East, Europe, and the Americas from their origins in Southeast Asia, the western Pacific islands and the Indian Ocean region (Gratz, 2004). In some areas *Ae. albopictus* are displacing *Ae. aegypti* (Gratz, 2004). Although both species are anthropophilic (Gratz, 2004), *Ae. albopictus* exhibit a higher level of anthropophily compared to *Ae. aegypti*, but both will also take a blood meal from other vertebrates such as birds. *Ae. aegypti* are found more often indoors and *Ae. albopictus* are more exophilic (Delatte *et al.*, 2010), potentially coming into contact with multiple species and thus capable of transmitting viruses between humans and animals.

The mechanism of CHIKV adaptation to *Ae. albopictus* has been examined by several studies (Stapleford *et al.*, 2014). During the outbreak in La Reunion, a mutation in E1 (A226V) was noted consistently in viral isolates that correlated with the capacity of the virus to replicate more efficiently in *Ae.albopictus* as well as *Ae.aegypti* (Tsetsarkin *et al.*, 2007; Vazeille *et al.*, 2007). These data begged the question of why this single amino acid change had not been seen previously. Addressing this question, another study found that the E1:A226V substitution was linked epistatically to another residue, E1:T98A and that the presence of E1:T98 in strains of the Asian genotype prevented the emergence of the E1:A226V substitution due to a fitness defect in these strains. This restriction was not present in the ECSA strain that was the founder strain of the 2004 epidemic that encoded the E1:98A substitution, allowing emergence of the E1:A226V substitution (Tsetsarkin *et al.*, 2011a). Another residue, E2:T211 in the ECSA lineage was found to enhance replication in *Ae. albopictus*, but only when E1:V226 was present. However, prior to 2005, most ECSA strains had E2:I211 at this residue (Tsetsarkin *et al.*, 2009). These data indicated that the emergence of E1:A226V is important for the capacity of the virus to be efficiently vectored by *Ae. albopictus*, but the process of adaptation is complicated and involves more than a single residue change. Viruses are constantly sampling sequence space, and it may require several steps away from the consensus sequence to elicit a change in vector or host that results in emergence or re-emergence of the virus (Antia *et al.*, 2003).

Recently, two studies of CHIKV evolution have been performed to better understand mechanisms of CHIKV adaptation to its vector and/or mammalian hosts. Stapleford *et al.* evaluated viral populations in both mosquitoes and mice in a setting that recapitulated a natural transmission cycle (Stapleford *et al.*, 2014). They identified important bottlenecks within the mosquito, including the midgut, salivary gland, and saliva, the latter of which contained particular virus variants specifically enhanced in saliva (Stapleford *et al.*, 2014). Similar bottlenecks have been identified for WNV and VEEV (Ciota *et al.*, 2012; Forrester *et al.*, 2012). In contrast to the lack of mutations seen when *Ae. aegypti* mosquitoes were infected, the E1:A226V mutation was seen at high frequency using deep sequencing when *Ae. albopictus* were infected experimentally, suggesting this mutation occurred as an adaptation to *Ae. albopictus*, recapitulating the emergence event seen in nature (Stapleford *et al.*, 2014). Following passage, two additional mutations in E1 (V80I and A129V) were selected at high frequency in the saliva of multiple mosquitoes. To test the capacity of this experimental methodology to predict further adaptation to the vector or greater virulence in the mammalian host, Stapleford *et al.* also infected both *Ae. aegypti* and *Ae. albopictus* with a virus containing all three of these changes and found that this triple mutant virus was capable of replicating better in both mosquito strains and was more virulent in a neonatal mouse model (Stapleford *et al.*, 2014). In another study with a similar goal, Tsetsarkin *et al.* looked at sequence variations that had arisen naturally in different viral lineages with the E1:A226V adaptation and introduced them into isogenic viruses and examine viral replication in cell lines, mosquitoes, and neonatal mice (Tsetsarkin *et al.*, 2011c). They determined that E2:L210Q and E2:K252Q substitutions enhanced replication in the presence of the E1:A226V mutation (Niyas *et al.*, 2010). These two substitutions in E2 arose in different geographical areas and have not occurred in any naturally occurring isolates to date; however, when combined experimentally, the resulting virus replicated more efficiently than either alone (Tsetsarkin *et al.*, 2014). These data remind us that viruses, even after taking an adaptive step, continue to sample sequence space and second-step adaptive mutations may yield a more fit virus that could emerge, disseminate and potentially replace its predecessor.

Parvovirus

Introduction

A striking example of virus evolution leading to emergence is found in parvoviruses. Parvoviruses (genus *Protoparvovirus*, family *Parvoviridae*) are non-segmented single-stranded DNA viruses with a genome of ~5 kb encapsidated in non-enveloped virions ~25nm in diameter, making them one of the smallest DNA viruses, both in genome length and virion size. Parvoviruses are hardy in nature and resistant to many disinfectants, and thus capable of persisting outside of host cells for a protracted period of time. The genome consists of two ORFs; the first encodes two nonstructural proteins (NS1 and NS2) and the second encodes two capsid proteins (VP1 and VP2). These viruses require host machinery to replicate and productively infect dividing cells, although the virus does not code for proteins to initiate cellular DNA replication, as do larger DNA viruses such as papillomaviruses. Parvovirus infection in mammals tends to impact rapidly dividing cells most severely, such as those found in the gastrointestinal tract and lymphatic systems, often resulting in diarrhea and leukopenia (Berns and Parrish, 2013).

Outbreak description

In 1978, reports from veterinarians identified a severe and sometimes fatal infection in dogs characterized by myocarditis in puppies and enteritis in young and old dogs (Tratschin *et al.*, 1982; Parrish *et al.*, 1985). Reports were published in the US, the UK, Australia, and Canada describing this new disease, which resembled feline panleukopenia virus (FPV). Although the first published reports were in the US, the virus appears to have originated in Europe, then rapidly spread around the world (Parrish, 1990; 1995). This spread was attributed to movement of infected dogs or contaminated materials, such as shoes worn where infected dogs were present, or other fomites (Carmichael, 2005; Parrish and Kawaoka, 2005). This virus was named canine parvovirus 2 (CPV-2) to distinguish from another canine parvovirus (CPV-1) also known as minute virus of canines (MVC) (Binn *et al.,* 1970; Parrish *et al.,* 1985). Following on the heels of this outbreak, another virus, which was antigenically and genetically distinct from CPV-2, emerged. This virus, known as CPV-2a, completely replaced CPV-2 worldwide between 1979 and 1981. Since then, CPV-2a has been supplanted by CPV-2b (Parrish *et al.,* 1985; 1988; 1991; Parrish, 1995) and more recently, CPV-2c has been identified and is circulating among canids (Parrish, 1995; Truyen, 2006).One of the changing characteristics seen as different variants have emerged is host range. CPV-2 was capable of infecting canid and felid cells *in vitro*, in contrast with *in vivo* experiments and naturally occurring infections, where dogs were competent hosts, but cats were not (Parrish, 1991; Truyen and Parrish, 1992). However, subsequent variants CPV-2a, -2b, -2c have all regained the capacity to bind and infect felid cells in addition to canid cells (Truyen *et al.,* 1996; Mochizuki *et al.,* 1996). In an attempt to determine whether CPV-2 was a newly emerged virus or a newly recognized virus, archived dog sera collected prior to 1976 were studied, but no antibodies specific for CPV-2 were found (Schwers *et al.,* 1979; Carmichael, 2005). This gave credence to the hypothesis that CPV-2 was a novel emerging virus, but questions still remain as to the exact origin or the mechanism of its rapid and efficient global spread of CPV-2 (Figure 12.6).

Disease description

In the early 1930s it was determined that a filterable agent, a virus, was a causative agent of feline distemper (Hindle and Findlay, 1932), and FPV was first identified in the late 1930s (Hammon and Enders, 1939; Lawrence *et al.,* 1940). FPV infects domestic cats and other felids, causing diarrhea, lack of appetite and leukopenia (Parrish, 1995; Mochizuki *et al.,* 1996). If infection occurs in young kittens, involvement of the brain and retina has also been observed (Kilham *et al.,* 1967; Parrish, 1995). This disease was well known by veterinarians and was most often seen at cat shows and in boarding or other crowded conditions (Hindle and Findlay, 1932).

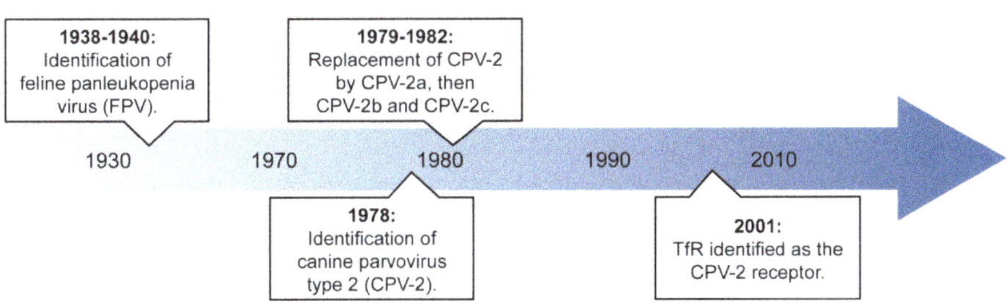

Figure 12.6. Timeline of parvovirus identification and emergence. Identification of FPV and then emergences of CPV-2 and variants are shown. CPV-2b and -2c are currently co-circulating. The FPV and CPV-2 receptor was identified as transferrin receptor in 2001.

Viral evolution and diversity

Antigenically and genetically, CPV-2 and its variants are most closely related to FPV, and it was initially thought that CPV-2 emerged from FPV following a small number of mutations that allowed it to infect cells from a different species (Parrish *et al.,* 1985; Truyen *et al.,* 1996; Shackelton *et al.,* 2005; Parrish *et al.,* 2008). Also, with the rapid replacement of CPV-2 with CPV-2a, a concern was raised that one of these viruses could have originated from FPV live virus vaccines, but sequence analysis of these strains demonstrated no evidence for this (Parrish, 1995; Truyen *et al.,* 1998). Another possibility is that CPV-2 originated from previously unrecognized viral strains, potentially circulating in wild animal populations not normally subject to human scrutiny. A recent surveillance study found a much larger variety of parvoviruses circulating in non-domesticated animals than previously known and demonstrated that these viruses move between domesticated and wild animals with high frequency (Allison *et al.,* 2014). This adds to the mounting evidence that both CPV-2 and FPV are descendants of parvoviruses circulating in the wild, such as viruses affecting related species, such as mink, raccoons, Asian raccoon dogs or foxes (Truyen and Parrish, 1995; Allison *et al.,* 2012; 2013; 2014).

Examination of genetic changes over time indicates that while FPV strains have remained fairly constant, CPV strains have changed dramatically over the same time period (Shackelton *et al.,* 2005). This observation reflects the unexpected determination that the mutation rate for CPV, a DNA virus, is more similar to those seen for RNA viruses than other DNA viruses (Shackelton *et al.,* 2005; Hoelzer *et al.,* 2008). CPV is also undergoing positive selection, specifically in the capsid gene, suggesting ongoing adaptation to its host or response to immune pressure. FPV has not exhibited positive selection, suggesting that it exists in equilibrium with its host, highlighting the differences between these two viruses as emerging or endemic pathogens (Hoelzer *et al.,* 2008).

The receptor for CPV-2 was identified in 2001 as the transferrin receptor type-1 (TfR) (Parker *et al.,* 2001) and differences between canine and feline TfR molecules help explain host range differences seen with FPV and CPV (Allison *et al.,* 2014). Domestic dog TfRs have a glycosylation site in the TfR apical domain, where the virus binds, which is not seen in felid TfRs and interferes with the capacity of FPV-like viruses to bind and infect dogs and related canids (Allison *et al.,* 2014). This is not the only difference between domestic cat and dog TfRs, which are 88% identical at the amino acid level, but the presence or absence of this glycosylation site is the major contributor to the differences in parvovirus binding to TfR. Recently, another theory about the origin of CPV-2 and its subsequent variants has emerged following examination not only of viral evolution, but evolution of the host as well (Kaelber *et al.,* 2012). It is clear that viruses have short replication cycles and large population sizes and thus are capable of rapid evolution, likely in response to a changing environment, such as pressure from a host immune response (Meyerson and Sawyer, 2011). What is not as clear is that the host may also undergo evolution as a response to pressure from an infectious agent. When this theory is probed in relationship to parvoviruses and TfRs from different carnivore species, there is evidence of a high degree of positive selection of the canid TfR compared to the felid TfR, which is surprising given the fact that CPV-2 has been infecting canids for only about 40 years. An alternative hypothesis is that CPV precursors infected canids for an extensive portion of their

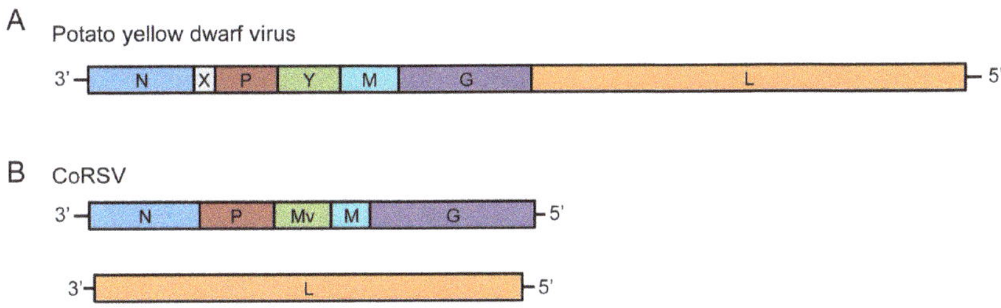

Figure 12.7. Genomic organization of Potato yellow dwarf virus (PYDV), a plant nucleorhabdovirus, compared to CoRSV. Shown are the negative-sense RNA genomes of PYDV (A) and CoRSV (B). PYDV has a single-stranded genome, while CoRSV encodes the same proteins within a bipartite genome. Abbreviations: N, nucleoproteins (blue); X, protein of unknown function (grey); P, phosphoprotein (red); (PYDV) Y or (CoRSV) Mv, movement protein (green); M, matrix protein (cyan); G, glycoprotein (purple); L, RNA-dependent RNA polymerase (orange). Intergenic junction sequences are not shown.

evolutionary history and exhibited evolutionary pressure that resulted in the selection of a glycosylation site mutant in the TfR which significantly decreased the capacity of CPV precursors to infect dog and coyotes. It then took several million years for parvoviruses to evolve a combination that could overcome this block to binding, and thus infect dogs and related species once again (Kaelber *et al.,* 2012).

Coffee ringspot virus

Introduction

In addition to the emerging viruses described in this chapter that impact human and animal health, emergence of plant viruses represent a continuing threat to the flora that contribute to human life and health (Scholthof *et al.,* 2011; Miller and Whitham, 2013). There are more viruses capable of infecting plants than all other animals (King *et al.,* 2012). Many discoveries were seen first in studies with plant viruses before being confirmed or recapitulated in animal viruses, starting with Tobacco mosaic virus as the first virus isolated and shown to be infectious (Miller and Whitham, 2013). One plant virus currently emerging in South America, specifically Brazil and surrounding countries, is coffee ringspot virus (CoRSV), a nonenveloped virus with a single-stranded negative-sense RNA genome divided into two segments, each around 6kb in length (Ramalho *et al.,* 2014) (Figure 12.7). CoRSV was initially thought to be most similar to viruses within the family *Rhabdoviridae* in the order *Mononegavirales*, which include enveloped viruses with monoparite genomes, (Miller and Whitham, 2013; Dietzgen *et al.,* 2014). Due to these differences, CoRSV and two other viruses, orchid fleck virus (OFV) and citrus leprosis virus (CiLV), have been proposed to comprise a new viral family, *Dichorhaviridae*. These viruses have a similar genetic organization to rhabdoviruses, but have a bipartite genome and lack a lipid envelope (Dietzgen *et al.,* 2014). The first RNA segment of CoRSV has 5 ORFs, while RNA2 encodes one large ORF, with sequence similarity to other RdRps. The 5 ORFs in RNA1 are proposed to be nucleocapsid, phosphoprotein, movement protein, matrix protein, and glycoprotein, respectively (Ramalho *et al.,* 2014).

Figure 12.8. CoRSV lesions. Characteristic ringspot lesions caused by CoRSV on coffee plant leaves (A) and cherries (fruit) (B) are shown. On the leaf (A) mites feeding on the center of the ringspots (small brown dot) can be seen. Minas Gerais, Brazil. (Profa. Antonia dos Reis Figueira, Dr. Michael Goodin).

Outbreak history

CoRSV, previously of minor economic importance since it was only found in a small region in Brazil and *Coffea arabica* was the only identified host (Kitajima *et al.*, 2011), now appears to be emerging and increasing the damage it causes. CoRSV was initially identified in São Paulo State (Bitancourt, 1938), but has now been seen in several regions of Brazil and also in Costa Rica (Chagas *et al.*, 2003; Kitajima *et al.*, 2011; Ramalho *et al.*, 2014). The expansion of CoRSV range and possibly plant hosts is likely due to a combination of factors, of which genetic changes within the virus itself may be one. Research into this virus using modern molecular biology is in its infancy, so this remains to be tested (Ramalho *et al.*, 2014).

Description of disease

Plant virus names are based on the plant which they were first found to infect and a description of the damage inflicted (Miller and Whitham, 2013). The lesions caused by CoRSV were first described in 1938 (Bitancourt, 1938; Scholthof *et al.*, 2011) and over the course of the infection ringspots on the leaves grow in size, leading to leaf drop. The virus also causes lesions on the coffee berries, which decreases their size and quality, leading to reduction of the value of the crop (Bertrand *et al.*, 2012) (Figure 12.8).

Vector

Plants are sessile, thus plant viruses often rely on vectors, usually insects, to be transmitted between hosts (Gray and Banerjee, 1999; Power, 2000; Miller and Whitham, 2013). CoRSV is transmitted by the false spider mite *Brevipalpus phoenicis* (Geijskes) (Acari: Tenuipalpidae) (Chagas *et al.*, 2003). Although it is not known whether CoRSV replicates in the mite, CoRSV can be propagated in cultured insect cells suggesting that viral replication occurs in the mite as part its role as vector. Both OFV and CiLV are transmitted by mites in the same genus, further supporting their classification together (Dietzgen *et al.*, 2014). However, *B. phoenicis* mites infected with CiLV could not transmit CiLV to coffee plants, and mites infected with CoRSV could not transmit CoRSV to citrus. These data indicate a specificity between virus, host plant and mite populations that is not well understood (Chagas *et al.*, 2003). In addition, the interaction between virus and vectors may be the limiting factor in virus disease emergence in that it is possible for viruses to have few vectors but large host ranges, but the reverse is not seen (Power, 2000). This is likely due to the same limitations seen in viruses of vertebrates that require a vector (see section on CHIKV) through constraints on the virus to maintain replication in diverse hosts. Interestingly, *B. phoenicis* mites exist predominantly as a haploid female population, reproducing through parthenogenesis due to the feminizing effect of a symbiotic bacteria from the genus *Cardinium* (Weeks *et al.*, 2001; Groot and Breeuwer, 2006). Being vectored by an insect population that does not vary genetically through sexual reproduction likely restricts the genetic variability of CoRSV more significantly compared to viruses vectored by genetically variable insects (Power, 2000).

Viral diversity and evolution

Factors affecting the emergence of new plant viruses are still being elucidated, but include climate change and human intervention, in the form of harvesting techniques, for example (Chagas *et al.*, 2003). Coffee plants have been shown to be exquisitely sensitive to climate change (Davis *et al.*, 2012). In addition, increasing temperatures cause plants to be more susceptible to disease (Király *et al.*, 2008; Prasch and Sonnewald, 2013) while at the same time increasing the population of insects that carry viruses capable of infecting plants (Melo de Almeida *et al.*, 2012) resulting in multiple hits, decreasing plant survival and plant food production. Because coffee is such an important export crop in Brazil and other countries, destruction of this resource would be devastating to the farmers and their livelihood as well as the economies of these countries (McCook, 2006). CoRSV is a newly emerging virus of coffee and research into its diversity and evolution is just beginning, but this virus has the potential to significantly impact coffee production in many areas. For many subsistence farmers, there exists a firmly established link between plant production and human health. When the plant production decreases, more land must be farmed and there are fewer resources for pest control, leading to poor land management practices. These cannot be maintained over time, resulting in short- or long-term decreases in the capacity of these farmers to be self-sufficient. A lack of resources also leads to poor nutrition and health, increasing the human health burden caused by viruses, albeit indirectly in this case. Therefore, investigating mechanisms by which plant viruses infect their hosts, focusing on viral change over time leading to emergence and more significant disease in their hosts will generate greater understanding of these processes. This will open up avenues for intervention, which may include finding or developing plants resistant to these virus and the diseases they cause (Sera *et al.*, 2013). Ideally, this will break these destructive cycles that indirectly, but significantly impact human health, due to the intricate connections between plant and food production, climate change, poverty and health.

Acknowledgements

We would like to thank Alison Ashbrook, Laurie Silva, and Michael Goodin for critical reading of the manuscript. We would also like to thank Eric Donaldson for generating the phylogenetic tree shown in Figure 12.4, and Antonia dos Reis Figueira and Michael Goodin for the images shown in Figure 12.8. The authors are supported by AI108197 and AI109680. Space limitations did not allow for citing the important work of all of our colleagues, for which we apologize.

References

Albariño, C.G., Shoemaker, T., Khristova, M.L., Wamala, J.F., Muyembe, J.J., Balinandi, S., et al. 2013. Genomic analysis of filoviruses associated with four viral hemorrhagic fever outbreaks in Uganda and the Democratic Republic of the Congo in 2012. Virology *442*: 97–100.

Allison, A.B., Harbison, C.E., Pagán, I., Stucker, K.M., Kaelber, J.T., Brown, J.D., et al. 2012. Role of multiple hosts in the cross-species transmission and emergence of a pandemic parvovirus. J. Virol. *86*: 865–872.

Allison, A.B., Kohler, D.J., Fox, K.A., Brown, J.D., Gerhold, R.W., Shearn-Bochsler, V.I., et al. 2013. Frequent cross-species transmission of parvoviruses among diverse carnivore hosts. J. Virol. *87*: 2342–2347.

Allison, A.B., Kohler, D.J., Ortega, A., Hoover, E.A., Grove, D.M., Holmes, E.C., et al. 2014. Host-specific parvovirus evolution in nature is recapitulated by *in vitro* adaptation to different carnivore species. PLoS Pathog. *10*: e1004475.

Annan, A., Baldwin, H.J., Corman, V.M., Klose, S.M., Owusu, M., Nkrumah, E.E., et al. 2013. Human betacoronavirus 2c EMC/2012-related viruses in bats, Ghana and Europe. Emerg. Infect. Dis. *19*: 456–459.

Anthony, S.J., Ojeda-Flores, R., Rico-Chávez, O., Navarrete-Macias, I., Zambrana-Torrelio, C.M., Rostal, M.K., et al. 2013. Coronaviruses in bats from Mexico. J Gen Vir *94*: 1028–1038.

Antia, R., Regoes, R.R., Koella, J.C. and Bergstrom, C.T. 2003. The role of evolution in the emergence of infectious diseases. Nature *426*: 658–661.

Apandi, Y., Nazni, W.A., Azleen, Z.A.N., Vythilingam, I., Noorazian, M.Y., Azahari, A.H., et al. 2009. The first isolation of chikungunya virus from non-human primates in Malaysia. JGMV *1*: 035–039.

Arankalle, V.A., Shrivastava, S., Cherian, S., Gunjikar, R.S., Walimbe, A.M., Jadhav, S.M., et al. 2007. Genetic divergence of Chikungunya viruses in India (1963-2006) with special reference to the 2005-2006 explosive epidemic. J. Gen. Virol. *88*: 1967–1976.

Arias-Goeta, C., Moutailler, S., Mousson, L., Zouache, K., Thiberge, J.-M., Caro, V., et al. 2014. Chikungunya virus adaptation to a mosquito vector correlates with only few point mutations in the viral envelope glycoprotein. Infect. Genet. Evol. *24*: 116–126.

Assiri, A., Al-Tawfiq, J.A., Rabeeah, Al, A.A., Al-Rabiah, F.A., Hajjar, Al, S., Al-Barrak, A., et al. 2013a. Epidemiological, demographic, and clinical characteristics of 47 cases of Middle East respiratory syndrome coronavirus disease from Saudi Arabia: a descriptive study. Lancet Infect Dis *13*: 752–761.

Assiri, A., McGeer, A., Perl, T.M., Price, C.S., Rabeeah, Al, A.A., Cummings, D.A.T., et al. 2013b. Hospital outbreak of Middle East respiratory syndrome coronavirus. N. Engl. J. Med. *369*: 407–416.

Baize, S., Leroy, E.M., Georges-Courbot, M.C., Capron, M., Lansoud-Soukate, J., Debré, P., et al. 1999. Defective humoral responses and extensive intravascular apoptosis are associated with fatal outcome in Ebola virus-infected patients. Nat. Med. *5*: 423–426.

Baize, S., Pannetier, D., Oestereich, L., Rieger, T., Koivogui, L., Magassouba, N., et al. 2014. Emergence of Zaire Ebola Virus Disease in Guinea. N. Engl. J. Med. *371*: 1418–1425.

Baron, R.C., McCormick, J.B. and Zubeir, O.A. 1983. Ebola virus disease in southern Sudan: hospital dissemination and intrafamilial spread. B World Health Organ *61*: 997–1003.

Barrette, R.W., Metwally, S.A., Rowland, J.M., Xu, L., Zaki, S.R., Nichol, S.T., et al. 2009. Discovery of swine as a host for the Reston ebolavirus. Science *325*: 204–206.

Bausch, D.G., Nichol, S.T., Muyembe-Tamfum, J.-J., Borchert, M., Rollin, P.E., Sleurs, H., et al. 2006. Marburg hemorrhagic fever associated with multiple genetic lineages of virus. N. Engl. J. Med. *355*: 909–919.

Becquart, P., Wauquier, N., Mahlakõiv, T., Nkoghe, D., Padilla, C., Souris, M., et al. 2010. High prevalence of both humoral and cellular immunity to Zaire ebolavirus among rural populations in Gabon. PLoS ONE *5*: e9126.
Belyi, V.A., Levine, A.J. and Skalka, A.M. 2010. Unexpected inheritance: multiple integrations of ancient bornavirus and ebolavirus/marburgvirus sequences in vertebrate genomes. PLoS Pathog. *6*: e1001030.
Bermingham, A., Chand, M.A., Brown, C.S., Aarons, E., Tong, C., Langrish, C., et al. 2012. Severe respiratory illness caused by a novel coronavirus, in a patient transferred to the United Kingdom from the Middle East, September 2012. Euro Surveill. *17*: 20290.
Berns, K.I. and Parrish, C.R. 2013. Parvoviridae. In: Fields Virology (D. M. Knipe and P. M. Howley, eds), pp. 1768–1791. Wolters Kluwer Health/Lippincott Williams and Wilkins, Philadelphia.
Bertrand, B., Boulanger, R., Dussert, S., Ribeyre, F., Berthiot, L., Descroix, F., et al. 2012. Climatic factors directly impact the volatile organic compound fingerprint in green Arabica coffee bean as well as coffee beverage quality. Food Chem *135*: 2575–2583.
Biek, R., Walsh, P.D., Leroy, E.M. and Real, L.A. 2006. Recent common ancestry of Ebola Zaire virus found in a bat reservoir. PLoS Pathog. *2*: e90.
Binn, L.N., Lazar, E.C., Eddy, G.A. and Kajima, M. 1970. Recovery and characterization of a minute virus of canines. Infect Immun *1*: 503–508.
Bitancourt, A.A. 1938. A mancha anular, uma nova ameaça do cafeeiro. Biológico *4*: 404–405.
Bosio, C.M., Aman, M.J., Grogan, C., Hogan, R., Ruthel, G., Negley, D., et al. 2003. Ebola and Marburg viruses replicate in monocyte-derived dendritic cells without inducing the production of cytokines and full maturation. J. Infect. Dis. *188*: 1630–1638.
Bray, M. and Mahanty, S. 2003. Ebola hemorrhagic fever and septic shock. J. Infect. Dis. *188*: 1613–1617.
Burt, F.J., Rolph, M.S., Rulli, N.E., Mahalingam, S. and Heise, M.T. 2012. Chikungunya: a re-emerging virus. Lancet *379*: 662–671.
Carey, D.E. 1971. Chikungunya and dengue: a case of mistaken identity? J Hist Med Allied Sci *26*: 243–262.
Carmichael, L.E. 2005. An annotated historical account of canine parvovirus. J. Vet. Med. B Infect. Dis. Vet. Public Health *52*: 303–311.
CDC (Centers for Disease Control and Prevention). 2001. Outbreak of Ebola hemorrhagic fever Uganda, August 2000-January 2001. Morb Mortal Wkly Rep *50:* 73–77.
Chagas, C.M., Kitajima, E.W. and Rodrigues, J. 2003. Coffee Ringspot Virus Vectored by Brevipalpus phoenicis (Acari: Tenuipalpidae) in Coffee. Exp. Appl. Acarol. *30*: 203–213.
Chen, R., Wang, E., Tsetsarkin, K.A. and Weaver, S.C. 2013. Chikungunya virus 3' untranslated region: adaptation to mosquitoes and a population bottleneck as major evolutionary forces. PLoS Pathog. *9*: e1003591.
Cheng, V.C.C., Lau, S.K.P., Woo, P.C.Y. and Yuen, K.-Y. 2007. Severe acute respiratory syndrome coronavirus as an agent of emerging and reemerging infection. Clin Microbiol Rev *20*: 660–694.
Chu, D.K.W., Poon, L.L.M., Gomaa, M.M., Shehata, M.M., Perera, R.A.P.M., Abu Zeid, D., et al. 2014. MERS coronaviruses in dromedary camels, Egypt. Emerg. Infect. Dis. *20*: 1049–1053.
Ciota, A.T. and Kramer, L.D. 2010. Insights into arbovirus evolution and adaptation from experimental studies. Viruses *2*: 2594–2617.
Ciota, A.T., Ehrbar, D.J., Van Slyke, G.A., Payne, A.F., Willsey, G.G., Viscio, R.E., et al. 2012. Quantification of intrahost bottlenecks of West Nile virus in *Culex pipiens* mosquitoes using an artificial mutant swarm. Infect. Genet. Evol. *12*: 557–564.

Ciota, A.T., Jia, Y., Payne, A.F., Jerzak, G., Davis, L.J., Young, D.S., et al. 2009. Experimental passage of St. Louis encephalitis virus *in vivo* in mosquitoes and chickens reveals evolutionarily significant virus characteristics. PLoS ONE *4*: e7876.

Coffey, L.L., Beeharry, Y., Bordería, A.V., Blanc, H. and Vignuzzi, M. 2011. Arbovirus high fidelity variant loses fitness in mosquitoes and mice. P Natl Acad Sci USA *108*: 16038–16043.

Coffey, L.L., Vasilakis, N., Brault, A.C., Powers, A.M., Tripet, F. and Weaver, S.C. 2008. Arbovirus evolution *in vivo* is constrained by host alternation. P Natl Acad Sci USA *105*: 6970–6975.

Commission, R.O.A.I. 1978. Ebola haemorrhagic fever in Zaire, 1976. B World Health Organ *56*: 271–293.

Corman, V.M., Ithete, N.L., Richards, L.R., Schoeman, M.C., Preiser, W., Drosten, C., et al. 2014. Rooting the phylogenetic tree of middle East respiratory syndrome coronavirus by characterization of a conspecific virus from an african bat. J. Virol. *88*: 11297–11303.

Corman, V.M., Jores, J., Meyer, B., Younan, M., Liljander, A., Said, M.Y., et al. 2014. Antibodies against MERS coronavirus in dromedary camels, Kenya, 1992-2013. Emerg. Infect. Dis. *20*: 1319–1322.

Cotten, M., Watson, S.J., Kellam, P., Rabeeah, Al, A.A., Makhdoom, H.Q., Assiri, A., et al. 2013. Transmission and evolution of the Middle East respiratory syndrome coronavirus in Saudi Arabia: a descriptive genomic study. Lancet *382*: 1993–2002.

Cotten, M., Watson, S.J., Zumla, A.I., Makhdoom, H.Q., Palser, A.L., Ong, S.H., et al. 2014. Spread, circulation, and evolution of the Middle East respiratory syndrome coronavirus. MBio *5 (1)*: e01062-13.

Das, T., Jaffar-Bandjee, M.-C., Hoarau, J.-J., Krejbich Trotot, P., Denizot, M., Lee-Pat-Yuen, G., et al. 2010. Chikungunya fever: CNS infection and pathologies of a re-emerging arbovirus. Prog. Neurobiol. *91*: 121–129.

Davis, A.P., Gole, T.W., Baena, S. and Moat, J. 2012. The impact of climate change on indigenous Arabica coffee (Coffea arabica): predicting future trends and identifying priorities. PLoS ONE *7*: e47981.

de Groot, R.J., Baker, S.C., Baric, R.S., Brown, C.S., Drosten, C., Enjuanes, L., et al. 2013. Middle East respiratory syndrome coronavirus (MERS-CoV): announcement of the Coronavirus Study Group. J. Virol. *87*: 7790–7792.

Deardorff, E.R., Fitzpatrick, K.A., Jerzak, G.V.S., Shi, P.-Y., Kramer, L.D. and Ebel, G.D. 2011. West Nile virus experimental evolution *in vivo* and the trade-off hypothesis. PLoS Pathog. *7*: e1002335.

Delatte, H., Desvars, A., Bouétard, A., Bord, S., Gimonneau, G., Vourc'h, G., et al. 2010. Blood-feeding behavior of Aedes albopictus, a vector of Chikungunya on La Réunion. Vector-Borne Zoonot *10*: 249–258.

Demogines, A., Farzan, M. and Sawyer, S.L. 2012. Evidence for ACE2-utilizing coronaviruses (CoVs) related to severe acute respiratory syndrome CoV in bats. J. Virol. *86*: 6350–6353.

Diallo, M., Thonnon, J., Traore-Lamizana, M. and Fontenille, D. 1999. Vectors of Chikungunya virus in Senegal: current data and transmission cycles. Am. J. Trop. Med. Hyg. *60*: 281–286.

Dietzgen, R.G., Kuhn, J.H., Clawson, A.N., Freitas-Astúa, J., Goodin, M.M., Kitajima, E.W., et al. 2014. Dichorhavirus: a proposed new genus for Brevipalpus mite-transmitted, nuclear, bacilliform, bipartite, negative-strand RNA plant viruses. Arch Virol *159*: 607–619.

Drexler, J.F., Corman, V.M. and Drosten, C. 2014. Ecology, evolution and classification of bat coronaviruses in the aftermath of SARS. Antiviral Res. *101*: 45–56.

Drosten, C., Seilmaier, M., Corman, V.M., Hartmann, W., Scheible, G., Sack, S., et al. 2013. Clinical features and virological analysis of a case of Middle East respiratory syndrome coronavirus infection. Lancet Infect Dis *13*: 745–751.

Dudas, G. and Rambaut, A. 2014. Phylogenetic Analysis of Guinea 2014 EBOV Ebolavirus Outbreak. PLoS Curr, doi: 10.1371/currents.outbreaks. 84eefe5ce43ec9dc0bf0670f7b8b417d.
Dwosh, H.A., Hong, H.H.L., Austgarden, D., Herman, S. and Schabas, R. 2003. Identification and containment of an outbreak of SARS in a community hospital. Can Med Assoc J *168*: 1415–1420.
Eckerle, L.D., Becker, M.M., Halpin, R.A., Li, K., Venter, E., Lu, X., et al. 2010. Infidelity of SARS-CoV Nsp14-exonuclease mutant virus replication is revealed by complete genome sequencing. PLoS Pathog. *6*: e1000896.
Feldmann, H. and Geisbert, T.W. 2011. Ebola haemorrhagic fever. Lancet *377*: 849–862.
Feldmann, H., Sanchez, A. and Geisbert, T.W. 2013. Filoviridae. In: Fields Virology (D. M. Knipe and P. M. Howley, eds), pp. 923–956. Wolters Kluwer Health/Lippincott Williams and Wilkins, Philadelphia.
Formenty, P., Libama, F., Epelboin, A., Allarangar, Y., Leroy, E., Moudzeo, H., et al. 2003. [Outbreak of Ebola hemorrhagic fever in the Republic of the Congo, 2003: a new strategy?]. Med Trop (Mars) *63*: 291–295.
Forrester, N.L., Guerbois, M., Seymour, R.L., Spratt, H. and Weaver, S.C. 2012. Vector-borne transmission imposes a severe bottleneck on an RNA virus population. PLoS Pathog. *8*: e1002897.
Fraser, C., Riley, S., Anderson, R.M. and Ferguson, N.M. 2004. Factors that make an infectious disease outbreak controllable. P Natl Acad Sci USA *101*: 6146–6151.
Ge, X.-Y., Li, J.-L., Yang, X.-L., Chmura, A.A., Zhu, G., Epstein, J.H., et al. 2013. Isolation and characterization of a bat SARS-like coronavirus that uses the ACE2 receptor. Nature *503*: 535–538.
Geisbert, T.W., Hensley, L.E., Larsen, T., Young, H.A., Reed, D.S., Geisbert, J.B., et al. 2003a. Pathogenesis of Ebola hemorrhagic fever in cynomolgus macaques: evidence that dendritic cells are early and sustained targets of infection. Am. J. Pathol. *163*: 2347–2370.
Geisbert, T.W., Young, H.A., Jahrling, P.B., Davis, K.J., Kagan, E. and Hensley, L.E. 2003b. Mechanisms underlying coagulation abnormalities in ebola hemorrhagic fever: overexpression of tissue factor in primate monocytes/macrophages is a key event. J. Infect. Dis. *188*: 1618–1629.
Georges, A.J., Leroy, E.M., Renaut, A.A., Benissan, C.T., Nabias, R.J., Ngoc, M.T., et al. 1999. Ebola hemorrhagic fever outbreaks in Gabon, 1994-1997: epidemiologic and health control issues. J. Infect. Dis. *179* Suppl 1: S65–75.
Gire, S.K., Goba, A., Andersen, K.G., Sealfon, R.S.G., Park, D.J., Kanneh, L., et al. 2014. Genomic surveillance elucidates Ebola virus origin and transmission during the 2014 outbreak. Science *345*: 1369–1372.
Gire, S.K., Stremlau, M., Andersen, K.G., Schaffner, S.F., Bjornson, Z., Rubins, K., et al. 2012. Epidemiology. Emerging disease or diagnosis? Science *338*: 750–752.
Goeijenbier, M., van Kampen, J.J.A., Reusken, C.B.E.M., Koopmans, M.P.G. and van Gorp, E.C.M. 2014. Ebola virus disease: a review on epidemiology, symptoms, treatment and pathogenesis. Neth J Med *72*: 442–448.
Gonzalez, J.-P., Pourrut, X. and Leroy, E. 2007. Ebolavirus and other filoviruses. Curr. Top. Microbiol. Immunol. *315*: 363–387.
Gorbalenya, A.E., Enjuanes, L., Ziebuhr, J. and Snijder, E.J. 2006. Nidovirales: Evolving the largest RNA virus genome. Virus Res. *117*: 17–37.
Graham, R.L. and Baric, R.S. 2010. Recombination, reservoirs, and the modular spike: mechanisms of coronavirus cross-species transmission. J. Virol. *84*: 3134–3146.
Grard, G., Biek, R., Muyembe Tamfum, J.J., Fair, J., Wolfe, N., Formenty, P., et al. 2011. Emergence of divergent Zaire ebola virus strains in Democratic Republic of the Congo in 2007 and 2008. J. Infect. Dis. *204*: S776–S784.
Gratz, N.G. 2004. Critical review of the vector status of Aedes albopictus. Med Vet Entomol *18*: 215–227.

Gray, S.M. and Banerjee, N. 1999. Mechanisms of arthropod transmission of plant and animal viruses. Microbiol. Mol. Biol. Rev. *63*: 128–148.

Greene, I.P., Wang, E., Deardorff, E.R., Milleron, R., Domingo, E. and Weaver, S.C. 2005. Effect of alternating passage on adaptation of sindbis virus to vertebrate and invertebrate cells. J. Virol. *79*: 14253–14260.

Griffin, D.E. 2013. Alphaviruses. In: Fields Virology (D. M. Knipe and P. M. Howley, eds), pp. 651–686. Wolters Kluwer Health/Lippincott Williams and Wilkins, Philadelphia.

Groot, T.V.M. and Breeuwer, J.A.J. 2006. Cardinium symbionts induce haploid thelytoky in most clones of three closely related Brevipalpus species. Exp. Appl. Acarol. *39*: 257–271.

Gu, J. and Korteweg, C. 2007. Pathology and pathogenesis of severe acute respiratory syndrome. Am. J. Pathol. *170*: 1136–1147.

Guan, Y., Zheng, B.J., He, Y.Q., Liu, X.L., Zhuang, Z.X., Cheung, C.L., et al. 2003. Isolation and characterization of viruses related to the SARS coronavirus from animals in southern China. Science *302*: 276–278.

Halstead, S.B. 2015. Reappearance of chikungunya, formerly called dengue, in the Americas. Emerg. Infect. Dis. *21 (4):* 557-561.

Hammon, W.D. and Enders, J.F. 1939. A virus disease of cats, principally characterized by aleucocytosis, enteric lesions and the presence of intranuclear inclusion bodies. J. Exp. Med. *69*: 327–352.

Hamre, D. and Procknow, J.J. 1966. A new virus isolated from the human respiratory tract. Proc. Soc. Exp. Biol. Med. *121*: 190–193.

Hayman, D.T.S., Emmerich, P., Yu, M., Wang, L.-F., Suu-Ire, R., Fooks, A.R., et al. 2010. Long-term survival of an urban fruit bat seropositive for Ebola and Lagos bat viruses. PLoS ONE *5*: e11978.

Hayman, D.T.S., Yu, M., Crameri, G., Wang, L.-F., Suu-Ire, R., Wood, J.L.N., et al. 2012. Ebola virus antibodies in fruit bats, Ghana, West Africa. Emerg. Infect. Dis. *18*: 1207–1209.

He, J., Peng, G., Min, J., Yu, D., Liang, W., Zhang, S., et al. 2004. Molecular evolution of the SARS coronavirus during the course of the SARS epidemic in China. Science *303*: 1666–1669.

Hemida, M.G., Perera, R.A., Wang, P., Alhammadi, M.A., Siu, L.Y., Li, M., et al. 2013. Middle East Respiratory Syndrome (MERS) coronavirus seroprevalence in domestic livestock in Saudi Arabia, 2010 to 2013. Euro Surveill. *18*: 20659.

Her, Z., Kam, Y.-W., Lin, R.T.P. and Ng, L.F.P. 2009. Chikungunya: a bending reality. Microbes Infect. *11*: 1165–1176.

Heymann, D.L., Weisfeld, J.S., Webb, P.A., Johnson, K.M., Cairns, T. & Berquist, H. 1980. Ebola hemorrhagic fever: Tandala, Zaire, 1977-1978. J. Infect. Dis. *142:* 372–376.

Hijawi, B., Abdallat, M., Sayaydeh, A., Alqasrawi, S., Haddadin, A., Jaarour, N., et al. 2013. Novel coronavirus infections in Jordan, April 2012: epidemiological findings from a retrospective investigation. East. Mediterr. Health J. *19 Suppl 1*: S12–8.

Hindle, E. and Findlay, G.M. 1932. Studies on feline distemper. J Comp Pathol Therap *45*: 11–26.

Hoarau, J.-J., Jaffar-Bandjee, M.-C., Krejbich Trotot, P., Das, T., Li-Pat-Yuen, G., Dassa, B., et al. 2010. Persistent chronic inflammation and infection by Chikungunya arthritogenic alphavirus in spite of a robust host immune response. J. Immunol. *184*: 5914–5927.

Hoelzer, K., Shackelton, L.A., Parrish, C.R. and Holmes, E.C. 2008. Phylogenetic analysis reveals the emergence, evolution and dispersal of carnivore parvoviruses. J. Gen. Virol. *89*: 2280–2289.

Holmes, E.C. 2003. Patterns of intra- and interhost nonsynonymous variation reveal strong purifying selection in dengue virus. J. Virol. *77*: 11296–11298.

HPA-UK. 2013. Evidence of person-to-person transmission within a family cluster of novel coronavirus infections, United Kingdom, February 2013. Euro Surveill. *18*: 20427.
Isakbaeva, E.T., Khetsuriani, N., Beard, R.S., Peck, A., Erdman, D., Monroe, S.S., et al. 2004. SARS-associated coronavirus transmission, United States. Emerg. Infect. Dis. *10*: 225–231.
Jahrling, P.B., Geisbert, T.W., Dalgard, D.W., Johnson, E.D., Ksiazek, T.G., Hall, W.C., et al. 1990. Preliminary report: isolation of Ebola virus from monkeys imported to USA. Lancet *335*: 502–505.
Jerzak, G., Bernard, K.A., Kramer, L.D. and Ebel, G.D. 2005. Genetic variation in West Nile virus from naturally infected mosquitoes and birds suggests quasispecies structure and strong purifying selection. J. Gen. Virol. *86*: 2175–2183.
Johnson, E.D., Johnson, B.K., Silverstein, D., Tukei, P., Geisbert, T.W., Sanchez, A.N., et al. 1996. Characterization of a new Marburg virus isolated from a 1987 fatal case in Kenya. Arch. Virol. Suppl. *11*: 101–114.
Jones, B.A., Grace, D., Kock, R., Alonso, S., Rushton, J., Said, M.Y., et al. 2013. Zoonosis emergence linked to agricultural intensification and environmental change. P Natl Acad Sci USA *110*: 8399–8404.
Jones, K.E., Patel, N.G., Levy, M.A., Storeygard, A., Balk, D., Gittleman, J.L., et al. 2008. Global trends in emerging infectious diseases. Nature *451*: 990–993.
Kaelber, J.T., Demogines, A., Harbison, C.E., Allison, A.B., Goodman, L.B., Ortega, A.N., et al. 2012. Evolutionary reconstructions of the transferrin receptor of Caniforms supports canine parvovirus being a re-emerged and not a novel pathogen in dogs. PLoS Pathog. *8*: e1002666.
Kan, B., Wang, M., Jing, H., Xu, H., Jiang, X., Yan, M., et al. 2005. Molecular evolution analysis and geographic investigation of severe acute respiratory syndrome coronavirus-like virus in palm civets at an animal market and on farms. J. Virol. *79*: 11892–11900.
Khan, A.S., Tshioko, F.K., Heymann, D.L., Le Guenno, B., Nabeth, P., Kerstiëns, B., et al. 1999. The reemergence of Ebola hemorrhagic fever, Democratic Republic of the Congo, 1995. J. Infect. Dis. *179* Suppl 1: S76–86.
Kilham, L., Margolis, G. and Colby, E.D. 1967. Congenital infections of cats and ferrets by feline panleukopenia virus manifested by cerebellar hypoplasia. Lab. Invest. *17*: 465–480.
King, A., Adams, M.J., Lefkowitz, E.J. and Carstens, E.B. 2012. Virus Taxonomy. Ninth Report of the International Committee on Taxonomy of Viruses. Elsevier, Philadelphia.
Király, L., Hafez, Y.M., Fodor, J. and Király, Z. 2008. Suppression of tobacco mosaic virus-induced hypersensitive-type necrotization in tobacco at high temperature is associated with downregulation of NADPH oxidase and superoxide and stimulation of dehydroascorbate reductase. J. Gen. Virol. *89*: 799–808.
Kitajima, E.W., Chagas, C.M., Braghini, M.T., Fazuoli, L.C., Locali-Fabris, E.C. and Salaroli, R.B. 2011. Natural infection of several Coffea species and hybrids and Psilanthus ebracteolatus by the coffee ringspot virus (CoRSV). Sci. agric. (Piracicaba, Braz.) *68*: 503–507.
Kuhn, J.H., Becker, S., Ebihara, H., Geisbert, T.W., Johnson, K.M., Kawaoka, Y., et al. 2010. Proposal for a revised taxonomy of the family Filoviridae: classification, names of taxa and viruses, and virus abbreviations. Arch Virol *155*: 2083–2103.
Lahariya, C. and Pradhan, S.K. 2006. Emergence of chikungunya virus in Indian subcontinent after 32 years: A review. J Vector Borne Dis *43*: 151–160.
Lahm, S.A., Kombila, M., Swanepoel, R. and Barnes, R.F.W. 2007. Morbidity and mortality of wild animals in relation to outbreaks of Ebola haemorrhagic fever in Gabon, 1994–2003. Trans. R. Soc. Trop. Med. Hyg. *101*: 64–78.
Lau, S.K.P., Woo, P.C.Y., Li, K.S.M., Huang, Y., Tsoi, H.-W., Wong, B.H.L., et al. 2005. Severe acute respiratory syndrome coronavirus-like virus in Chinese horseshoe bats. P Natl Acad Sci USA *102*: 14040–14045.

Lawrence, J.S., Syverton, J.T., Shaw, J.S. and Smith, F.P. 1940. Infectious feline agranulocytosis. Am. J. Pathol. *16*: 333–354.9.

Le Guenno, B., Formenty, P., Formentry, P., Wyers, M., Gounon, P., Walker, F., et al. 1995. Isolation and partial characterisation of a new strain of Ebola virus. Lancet *345*: 1271–1274.

Leroy, E.M., Baize, S., Mavoungou, E. and Apetrei, C. 2002a. Sequence analysis of the GP, NP, VP40 and VP24 genes of Ebola virus isolated from deceased, surviving and asymptomatically infected individuals during the 1996 outbreak in Gabon: comparative studies and phylogenetic characterization. J. Gen. Virol. *83*: 67–73.

Leroy, E.M., Baize, S., Volchkov, V.E., Fisher-Hoch, S.P., Georges-Courbot, M.C., Lansoud-Soukate, J., et al. 2000. Human asymptomatic Ebola infection and strong inflammatory response. Lancet *355*: 2210–2215.

Leroy, E.M., Becquart, P., Wauquier, N. and Baize, S. 2011. Evidence for Ebola virus superantigen activity. J. Virol. *85*: 4041–4042.

Leroy, E.M., Epelboin, A., Mondonge, V., Pourrut, X., Gonzalez, J.-P., Muyembe-Tamfum, J.-J., et al. 2009. Human Ebola outbreak resulting from direct exposure to fruit bats in Luebo, Democratic Republic of Congo, 2007. Vector-Borne Zoonot *9*: 723–728.

Leroy, E.M., Kumulungui, B., Pourrut, X., Rouquet, P., Hassanin, A., Yaba, P., et al. 2005. Fruit bats as reservoirs of Ebola virus. Nature *438*: 575–576.

Leroy, E.M., Rouquet, P., Formenty, P., Souquière, S., Kilbourne, A., Froment, J.-M., et al. 2004. Multiple Ebola virus transmission events and rapid decline of central African wildlife. Science *303*: 387–390.

Leroy, E.M., Souquière, S., Rouquet, P. and Drevet, D. 2002b. Re-emergence of ebola haemorrhagic fever in Gabon. Lancet 359: 712.

Liang, G., Chen, Q., Xu, J., Liu, Y., Lim, W., Peiris, J.S.M., et al. 2004. Laboratory diagnosis of four recent sporadic cases of community-acquired SARS, Guangdong Province, China. Emerg. Infect. Dis. *10*: 1774–1781.

Lim, P.L., Kurup, A., Gopalakrishna, G., Chan, K.P., Wong, C.W., Ng, L.-C., et al. 2004. Laboratory-acquired severe acute respiratory syndrome. N. Engl. J. Med. *350*: 1740–1745.

Lipsitch, M., Cohen, T., Ben Cooper, Robins, J.M., Ma, S., James, L., et al. 2003. Transmission dynamics and control of severe acute respiratory syndrome. Science *300*: 1966–1970.

Lumsden, W.H. 1955. An epidemic of virus disease in Southern Province, Tanganyika Territory, in 1952-53. II. General description and epidemiology. Trans. R. Soc. Trop. Med. Hyg. *49*: 33–57.

Mari Saez, A., Weiss, S., Nowak, K., Lapeyre, V., Zimmermann, F., Dux, A., et al. 2015. Investigating the zoonotic origin of the West African Ebola epidemic. EMBO Mol Med *7*: 17–23.

Marra, M.A., Jones, S.J.M., Astell, C.R., Holt, R.A., Brooks-Wilson, A., Butterfield, Y.S.N., et al. 2003. The Genome sequence of the SARS-associated coronavirus. Science *300*: 1399–1404.

Masters, P.S. and Perlman, S. 2013. Coronaviridae. In: Fields Virology (D. M. Knipe and P. M. Howley, eds), pp. 825–858. Wolters Kluwer Health/Lippincott Williams and Wilkins, Philadelphia.

McCook, S. 2006. Global rust belt: Hemileia vastatrix and the ecological integration of world coffee production since 1850. JGH *1*: 177.

McIntosh, K., Chao, R.K., Krause, H.E., Wasil, R., Mocega, H.E. and Mufson, M.A. 1974. Coronavirus infection in acute lower respiratory tract disease of infants. J. Infect. Dis. *130*: 502–507.

McIntosh, K., Dees, J.H., Becker, W.B., Kapikian, A.Z. and Chanock, R.M. 1967. Recovery in tracheal organ cultures of novel viruses from patients with respiratory disease. P Natl Acad Sci USA *57*: 933–940.

Melo de Almeida, J.E., Mori, A.E., Pozza, E.A., Reis, P.R. and Figueira, A.D.R. 2012. Temporal analysis and control of the ringspot disease and mite vector of Coffee ringspot virus. Pesquisa Agropecuaria Brasileira *47*: 913–919.
Memish, Z.A., Cotten, M., Meyer, B., Watson, S.J., Alsahafi, A.J., Rabeeah, Al, A.A., et al. 2014. Human infection with MERS coronavirus after exposure to infected camels, Saudi Arabia, 2013. Emerg. Infect. Dis. *20*: 1012–1015.
Memish, Z.A., Mishra, N., Olival, K.J., Fagbo, S.F., Kapoor, V., Epstein, J.H., et al. 2013. Middle East respiratory syndrome coronavirus in bats, Saudi Arabia. Emerg. Infect. Dis. *19 (11)*: 1819-1823.
Meyer, B., Müller, M.A., Corman, V.M., Reusken, C.B.E.M., Ritz, D., Godeke, G.-J., et al. 2014. Antibodies against MERS coronavirus in dromedary camels, United Arab Emirates, 2003 and 2013. Emerg. Infect. Dis. *20*: 552–559.
Meyerson, N.R. and Sawyer, S.L. 2011. Two-stepping through time: mammals and viruses. Trends Microbiol *19*: 286–294.
Miller, W.A. and Whitham, S.A. 2013. Plant Viruses. In: Fields Virology (D. M. Knipe and P. M. Howley, eds), pp. 2289–2325. Wolters Kluwer Health/Lippincott Williams and Wilkins, Philadelphia.
Minskaia, E., Hertzig, T., Gorbalenya, A.E., Campanacci, V., Cambillau, C., Canard, B., et al. 2006. Discovery of an RNA virus 3′→ 5′ exoribonuclease that is critically involved in coronavirus RNA synthesis. P Natl Acad Sci USA *103*: 5108–5113.
Miranda, M.E., Ksiazek, T.G., Retuya, T.J., Khan, A.S., Sanchez, A., Fulhorst, C.F., et al. 1999. Epidemiology of Ebola (subtype Reston) virus in the Philippines, 1996. J. Infect. Dis. *179 Suppl 1*: S115–9.
Mochizuki, M., Horiuchi, M., Hiragi, H., San Gabriel, M.C., Yasuda, N. and Uno, T. 1996. Isolation of canine parvovirus from a cat manifesting clinical signs of feline panleukopenia. J Clin Microbiol *34*: 2101–2105.
Morse, S.S. 1995. Factors in the emergence of infectious diseases. Emerg. Infect. Dis. *1*: 7–15.
Moutailler, S., Roche, B., Thiberge, J.-M., Caro, V., Rougeon, F. and Failloux, A.-B. 2011. Host alternation is necessary to maintain the genome stability of rift valley fever virus. PLoS Negl Trop Dis *5*: e1156.
Muyembe Tamfum, J.J., Mulangu, S., Masumu, J., Kayembe, J.M., Kemp, A. and Paweska, J.T. 2012. Ebola virus outbreaks in Africa: Past and present. Onderstepoort J Vet Res 79(2), Art. #451, 8 p.
Myint, S.H. 1994. Human Coronaviruses: A Brief Review. Rev. Med. Virol. *4*: 35–46.
Nakkhara, P., Chongsuvivatwong, V. and Thammapalo, S. 2013. Risk factors for symptomatic and asymptomatic chikungunya infection. Trans. R. Soc. Trop. Med. Hyg. *107*: 789–796.
Negredo, A., Palacios, G., Vázquez-Morón, S., González, F., Dopazo, H., Molero, F., et al. 2011. Discovery of an ebolavirus-like filovirus in Europe. PLoS Pathog. *7*: e1002304.
Nishiura, H., Ejima, K. and Mizumoto, K. 2014. Missing information in animal surveillance of MERS-CoV. Lancet Infect Dis *14*: 100–100.
Niyas, K.P., Abraham, R., Unnikrishnan, R.N., Mathew, T., Nair, S., Manakkadan, A., et al. 2010. Molecular characterization of Chikungunya virus isolates from clinical samples and adult Aedes albopictus mosquitoes emerged from larvae from Kerala, South India. Virol. J. *7*: 189.
Nkoghe, D., Kone, M.L., Yada, A. & Leroy, E. 2011. A limited outbreak of Ebola haemorrhagic fever in Etoumbi, Republic of Congo, 2005. Trans. R. Soc. Trop. Med. Hyg. *105*: 466–472.
Olival, K.J., Islam, A., Yu, M., Anthony, S.J., Epstein, J.H., Khan, S.A., et al. 2013. Ebola virus antibodies in fruit bats, bangladesh. Emerg. Infect. Dis. *19*: 270–273.
Parker, J.S., Murphy, W.J., Wang, D., O'Brien, S.J. and Parrish, C.R. 2001. Canine and feline parvoviruses can use human or feline transferrin receptors to bind, enter, and infect cells. J. Virol. *75*: 3896–3902.

Parrish, C.R. 1990. Emergence, natural history, and variation of canine, mink, and feline parvoviruses. Adv. Virus Res. *38*: 403–450.

Parrish, C.R. 1991. Mapping specific functions in the capsid structure of canine parvovirus and feline panleukopenia virus using infectious plasmid clones. Virology *183*: 195–205.

Parrish, C.R. 1995. Pathogenesis of feline panleukopenia virus and canine parvovirus. Baillieres Clin. Haematol. *8*: 57–71.

Parrish, C.R. and Kawaoka, Y. 2005. The origins of new pandemic viruses: the acquisition of new host ranges by canine parvovirus and influenza A viruses. Annu. Rev. Microbiol. *59*: 553–586.

Parrish, C.R., Aquadro, C.F., Strassheim, M.L., Evermann, J.F., Sgro, J.Y. and Mohammed, H.O. 1991. Rapid antigenic-type replacement and DNA sequence evolution of canine parvovirus. J. Virol. *65*: 6544–6552.

Parrish, C.R., Have, P., Foreyt, W.J., Evermann, J.F., Senda, M. and Carmichael, L.E. 1988. The global spread and replacement of canine parvovirus strains. J. Gen. Virol. *69 (Pt 5)*: 1111–1116.

Parrish, C.R., Holmes, E.C., Morens, D.M., Park, E.-C., Burke, D.S., Calisher, C.H., et al. 2008. Cross-species virus transmission and the emergence of new epidemic diseases. Microbiol. Mol. Biol. Rev. *72*: 457–470.

Parrish, C.R., O'Connell, P.H., Evermann, J.F. and Carmichael, L.E. 1985. Natural variation of canine parvovirus. Science *230*: 1046–1048.

Peiris, J.S.M., Lai, S.T., Poon, L.L.M., Guan, Y., Yam, L.Y.C., Lim, W., et al. 2003. Coronavirus as a possible cause of severe acute respiratory syndrome. Lancet *361*: 1319–1325.

Perera, R.A., Wang, P., Gomaa, M.R., El-Shesheny, R., Kandeil, A., Bagato, O., et al. 2013. Seroepidemiology for MERS coronavirus using microneutralisation and pseudoparticle virus neutralisation assays reveal a high prevalence of antibody in dromedary camels in Egypt, June 2013. Euro Surveill. *18*: pii=20574.

Pialoux, G., Gaüzère, B.-A., Jauréguiberry, S. and Strobel, M. 2007. Chikungunya, an epidemic arbovirosis. Lancet Infect Dis *7*: 319–327.

Pourrut, X., Délicat, A., Rollin, P.E., Ksiazek, T.G., Gonzalez, J.-P. and Leroy, E.M. 2007. Spatial and temporal patterns of Zaire ebolavirus antibody prevalence in the possible reservoir bat species. J. Infect. Dis. *196 Suppl 2*: S176–83.

Pourrut, X., Kumulungui, B., Wittmann, T., Moussavou, G., Délicat, A., Yaba, P., et al. 2005. The natural history of Ebola virus in Africa. Microbes Infect. *7*: 1005–1014.

Poutanen, S.M., Low, D.E., Henry, B., Finkelstein, S., Rose, D., Green, K., et al. 2003. Identification of severe acute respiratory syndrome in Canada. N. Engl. J. Med. *348*: 1995–2005.

Power, A.G. 2000. Insect transmission of plant viruses: a constraint on virus variability. Curr. Opin. Plant Biol. *3*: 336–340.

Powers, A.M. and Logue, C.H. 2007. Changing patterns of chikungunya virus: re-emergence of a zoonotic arbovirus. J. Gen. Virol. *88*: 2363–2377.

Powers, A.M., Brault, A.C., Tesh, R.B. and Weaver, S.C. 2000. Re-emergence of Chikungunya and O'nyong-nyong viruses: evidence for distinct geographical lineages and distant evolutionary relationships. J. Gen. Virol. *81*: 471–479.

Prasch, C.M. and Sonnewald, U. 2013. Simultaneous application of heat, drought, and virus to Arabidopsis plants reveals significant shifts in signaling networks. Plant Physiol. *162*: 1849–1866.

Pyrc, K., Berkhout, B. and van der Hoek, L. 2007. The novel human coronaviruses NL63 and HKU1. J. Virol. *81*: 3051–3057.

Pyrc, K., Dijkman, R., Deng, L., Jebbink, M.F., Ross, H.A., Berkhout, B., et al. 2006. Mosaic structure of human coronavirus NL63, one thousand years of evolution. J. Mol. Biol. *364*: 964–973.

Rajapakse, S., Rodrigo, C. and Rajapakse, A. 2010. Atypical manifestations of chikungunya infection. Trans. R. Soc. Trop. Med. Hyg. *104*: 89–96.

Ramalho, T.O., Figueira, A.R., Sotero, A.J., Wang, R., Geraldino Duarte, P.S., Farman, M., et al. 2014. Characterization of Coffee ringspot virus-Lavras: a model for an emerging threat to coffee production and quality. Virology *464-465*: 385–396.
Rao, T.R. 1964. Vectors of dengue and chikungunya viruses: a brief review. Indian J. Med. Res. *52*: 719–726.
Renault, P., Solet, J.-L., Sissoko, D., Balleydier, E., Larrieu, S., Filleul, L., et al. 2007. A major epidemic of chikungunya virus infection on Reunion Island, France, 2005-2006. Am. J. Trop. Med. Hyg. *77*: 727–731.
Reusken, C.B., Ababneh, M., Raj, V.S., Meyer, B., Eljarah, A., Abutarbush, S., et al. 2013. Middle East Respiratory Syndrome coronavirus (MERS-CoV) serology in major livestock species in an affected region in Jordan, June to September 2013. Euro Surveill. *18*: 20662.
Reusken, C.B., Farag, E.A., Jonges, M., Godeke, G.J., El-Sayed, A.M., Pas, S.D., et al. 2014. Middle East respiratory syndrome coronavirus (MERS-CoV) RNA and neutralising antibodies in milk collected according to local customs from dromedary camels, Qatar, April 2014. Euro Surveill. *19*.
Reusken, C.B.E.M., Haagmans, B.L., Müller, M.A., Gutierrez, C., Godeke, G.-J., Meyer, B., et al. 2013. Middle East respiratory syndrome coronavirus neutralising serum antibodies in dromedary camels: a comparative serological study. Lancet Infect Dis *13*: 859–866.
Reusken, C.B.E.M., Messadi, L., Feyisa, A., Ularamu, H., Godeke, G.-J., Danmarwa, A., et al. 2014. Geographic distribution of MERS coronavirus among dromedary camels, Africa. Emerg. Infect. Dis. *20*: 1370–1374.
Richman, D.D., Cleveland, P.H., McCormick, J.B. and Johnson, K.M. 1983. Antigenic analysis of strains of Ebola virus: identification of two Ebola virus serotypes. J. Infect. Dis. *147*: 268–271.
Robinson, M.C. 1955. An epidemic of virus disease in Southern Province, Tanganyika Territory, in 1952-53. I. Clinical features. Trans. R. Soc. Trop. Med. Hyg. *49*: 28–32.
Rodriguez, L.L., De Roo, A., Guimard, Y., Trappier, S.G., Sanchez, A., Bressler, D., et al. 1999. Persistence and genetic stability of Ebola virus during the outbreak in Kikwit, Democratic Republic of the Congo, 1995. J. Infect. Dis. *179 Suppl 1*: S170–6.
Ross, R.W. 1956. The Newala epidemic. III. The virus: isolation, pathogenic properties and relationship to the epidemic. J Hyg (Lond) *54*: 177–191.
Rota, P.A., Oberste, M.S., Monroe, S.S., Nix, W.A., Campagnoli, R., Icenogle, J.P., et al. 2003. Characterization of a novel coronavirus associated with severe acute respiratory syndrome. Science *300*: 1394–1399.
Rowe, A.K., Bertolli, J., Khan, A.S., Mukunu, R., Muyembe Tamfum, J.J., Bressler, D., et al. 1999. Clinical, virologic, and immunologic follow-up of convalescent Ebola hemorrhagic fever patients and their household contacts, Kikwit, Democratic Republic of the Congo. J. Infect. Dis. *179 Suppl 1*: S28–35.
Sanchez, A., Lukwiya, M., Bausch, D., Mahanty, S., Sanchez, A.J., Wagoner, K.D., et al. 2004. Analysis of human peripheral blood samples from fatal and nonfatal cases of Ebola (Sudan) hemorrhagic fever: cellular responses, virus load, and nitric oxide levels. J. Virol. *78*: 10370–10377.
Sanchez, A., Trappier, S.G., Mahy, B.W., Peters, C.J. and Nichol, S.T. 1996. The virion glycoproteins of Ebola viruses are encoded in two reading frames and are expressed through transcriptional editing. P Natl Acad Sci USA *93*: 3602–3607.
Sanjuán, R., Nebot, M.R., Chirico, N., Mansky, L.M. and Belshaw, R. 2010. Viral mutation rates. J. Virol. *84*: 9733–9748.
Schilte, C., Staikowsky, F., Staikovsky, F., Couderc, T., Madec, Y., Carpentier, F., et al. 2013. Chikungunya virus-associated long-term arthralgia: a 36-month prospective longitudinal study. PLoS Negl Trop Dis *7*: e2137.
Schoepp, R.J., Rossi, C.A., Khan, S.H., Goba, A. and Fair, J.N. 2014. Undiagnosed acute viral febrile illnesses, Sierra Leone. Emerg. Infect. Dis. *20*: 1176–1182.

Scholthof, K.-B.G., Adkins, S., Czosnek, H., Palukaitis, P., Jacquot, E., Hohn, T., et al. 2011. Top 10 plant viruses in molecular plant pathology. Mol. Plant Pathol. *12*: 938–954.

Schuffenecker, I., Iteman, I., Michault, A., Murri, S., Frangeul, L., Vaney, M.-C., et al. 2006. Genome microevolution of chikungunya viruses causing the Indian Ocean outbreak. PLoS Med. *3*: e263.

Schwartz, O. and Albert, M.L. 2010. Biology and pathogenesis of chikungunya virus. Nat. Rev. Microbiol. *8*: 491–500.

Schwers, A., Pastoret, P.P., Burtonboy, G. and Thiry, E. 1979. Frequence en Belgique de l'infection a Parvovirus chez le chien, avant et apres l'observation des premiers cas cliniques. Revue de Medecine Veterinaire.

Sera, T., Sera, G.H., Fazuoli, L.C. and Bettencourt, A.J. 2013. IPR 99-Dwarf arabica coffee cultivar resistant to coffee ringspot virus. Crop Breeding and Applied Biotechnology *13*: 91–94.

Sergon, K., Yahaya, A.A., Brown, J., Bedja, S.A., Mlindasse, M., Agata, N., et al. 2007. Seroprevalence of Chikungunya virus infection on Grande Comore Island, union of the Comoros, 2005. Am. J. Trop. Med. Hyg. *76*: 1189–1193.

Shackelton, L.A., Parrish, C.R., Truyen, U. and Holmes, E.C. 2005. High rate of viral evolution associated with the emergence of carnivore parvovirus. P Natl Acad Sci USA *102*: 379–384.

Shoemaker, T., MacNeil, A., Balinandi, S., Campbell, S., Wamala, J.F., McMullan, L.K., et al. 2012. Reemerging Sudan Ebola virus disease in Uganda, 2011. Emerg. Infect. Dis. 18: 1480–1483.

Siegert, R., Shu, H.L., Slenczka, W., Peters, D. and Müller, G. 1967. [On the etiology of an unknown human infection originating from monkeys]. Dtsch. Med. Wochenschr. *92*: 2341–2343.

Sissoko, D., Malvy, D., Ezzedine, K., Renault, P., Moscetti, F., Ledrans, M., et al. 2009. Post-epidemic Chikungunya disease on Reunion Island: course of rheumatic manifestations and associated factors over a 15-month period. PLoS Negl Trop Dis *3*: e389.

Smith, D.H., Johnson, B.K., Isaacson, M., Swanapoel, R., Johnson, K.M., Killey, M., et al. 1982. Marburg-virus disease in Kenya. Lancet *1*: 816–820.

Smith, E.C. and Denison, M.R. 2013. Coronaviruses as DNA wannabes: a new model for the regulation of RNA virus replication fidelity. PLoS Pathog. *9*: e1003760.

Solignat, M., Gay, B., Higgs, S., Briant, L. and Devaux, C. 2009. Replication cycle of chikungunya: a re-emerging arbovirus. Virology *393*: 183–197.

Stapleford, K.A., Coffey, L.L., Lay, S., Bordería, A.V., Duong, V., Isakov, O., et al. 2014. Emergence and transmission of arbovirus evolutionary intermediates with epidemic potential. Cell Host Microbe *15*: 706–716.

Suzuki, Y. and Gojobori, T. 1997. The origin and evolution of Ebola and Marburg viruses. Mol. Biol. Evol. *14*: 800–806.

Swanepoel, R., Leman, P.A., Burt, F.J., Zachariades, N.A., Braack, L.E., Ksiazek, T.G., et al. 1996. Experimental inoculation of plants and animals with Ebola virus. Emerg. Infect. Dis. *2*: 321–325.

Taylor, D.J., Leach, R.W. and Bruenn, J. 2010. Filoviruses are ancient and integrated into mammalian genomes. BMC Evol Biol *10*: 193.

Taylor, L.H., Latham, S.M. and Woolhouse, M.E. 2001. Risk factors for human disease emergence. Philos. Trans. R. Soc. Lond., B, Biol. Sci. *356*: 983–989.

Team, R.O.A.W.I.S. 1978. Ebola haemorrhagic fever in Sudan, 1976. B World Health Organ *56*: 247–270.

Thiberville, S.-D., Boisson, V., Gaudart, J., Simon, F., Flahault, A. and de Lamballerie, X. 2013. CChikungunya fever: A clinical and virological investigation of outpatients on Reunion Island, south-west Indian Ocean. PLoS Negl Trop Dis *7*: e2004.

Towner, J.S., Sealy, T.K., Khristova, M.L., Albariño, C.G., Conlan, S., Reeder, S.A., et al. 2008. Newly discovered ebola virus associated with hemorrhagic fever outbreak in Uganda. PLoS Pathog. *4*: e1000212.

Tratschin, J.D., McMaster, G.K., Kronauer, G. and Siegl, G. 1982. Canine parvovirus: relationship to wild-type and vaccine strains of feline panleukopenia virus and mink enteritis virus. J. Gen. Virol. *61 (Pt I)*: 33–41.
Truyen, U. 2006. Evolution of canine parvovirus--a need for new vaccines? Vet Microbiol *117*: 9–13.
Truyen, U. and Parrish, C.R. 1992. Canine and feline host ranges of canine parvovirus and feline panleukopenia virus: distinct host cell tropisms of each virus *in vitro* and *in vivo*. J. Virol. *66*: 5399–5408.
Truyen, U. and Parrish, C.R. 1995. The evolution and control of parvovirus host ranges. Semin Virol *6*: 311–317.
Truyen, U., Evermann, J.F., Vieler, E. and Parrish, C.R. 1996. Evolution of canine parvovirus involved loss and gain of feline host range. Virology *215*: 186–189.
Truyen, U., Geissler, K., Parrish, C.R., Hermanns, W. and Siegl, G. 1998. No evidence for a role of modified live virus vaccines in the emergence of canine parvovirus. J. Gen. Virol. *79 (Pt 5)*: 1153–1158.
Tsang, K.W., Ho, P.L., Ooi, G.C., Yee, W.K., Wang, T., Chan-Yeung, M., et al. 2003. A Cluster of Cases of Severe Acute Respiratory Syndrome in Hong Kong. N. Engl. J. Med. *348*: 1977–1985.
Tsetsarkin, K.A., Chen, R., Leal, G., Forrester, N., Higgs, S., Huang, J., et al. 2011a. Chikungunya virus emergence is constrained in Asia by lineage-specific adaptive landscapes. P Natl Acad Sci USA *108*: 7872–7877.
Tsetsarkin, K.A., Chen, R., Sherman, M.B. and Weaver, S.C. 2011b. Chikungunya virus: evolution and genetic determinants of emergence. Curr Opin Virol *1*: 310–317.
Tsetsarkin, K.A., Chen, R., Yun, R., Rossi, S.L., Plante, K.S., Guerbois, M., et al. 2014. Multi-peaked adaptive landscape for chikungunya virus evolution predicts continued fitness optimization in Aedes albopictus mosquitoes. Nat Comms *5*: 4084.
Tsetsarkin, K.A., McGee, C.E. and Higgs, S. 2011c. Chikungunya virus adaptation to Aedes albopictus mosquitoes does not correlate with acquisition of cholesterol dependence or decreased pH threshold for fusion reaction. Virol. J. *8*: 376.
Tsetsarkin, K.A., McGee, C.E., Volk, S.M., Vanlandingham, D.L., Weaver, S.C. and Higgs, S. 2009. Epistatic roles of E2 glycoprotein mutations in adaption of chikungunya virus to Aedes albopictus and Ae. aegypti mosquitoes. PLoS ONE *4*: e6835.
Tsetsarkin, K.A., Vanlandingham, D.L., McGee, C.E. and Higgs, S. 2007. A single mutation in chikungunya virus affects vector specificity and epidemic potential. PLoS Pathog. *3*: e201.
Tsui, S.K.W., Chim, S.S.C., Lo, Y.M.D.Chinese University of Hong Kong Molecular SARS Research Group. 2003. Coronavirus genomic-sequence variations and the epidemiology of the severe acute respiratory syndrome. N. Engl. J. Med. *349*: 187–188.
Tu, C., Crameri, G., Kong, X., Chen, J., Sun, Y., Yu, M., et al. 2004. Antibodies to SARS coronavirus in civets. Emerg. Infect. Dis. *10*: 2244–2248.
Turell, M.J., Bressler, D.S. and Rossi, C.A. 1996. Short report: lack of virus replication in arthropods after intrathoracic inoculation of Ebola Reston virus. Am. J. Trop. Med. Hyg. *55*: 89–90.
Vabret, A., Mourez, T., Dina, J., van der Hoek, L., Gouarin, S., Petitjean, J., et al. 2005. Human coronavirus NL63, France. Emerg. Infect. Dis. *11*: 1225–1229.
van Boheemen, S., de Graaf, M., Lauber, C., Bestebroer, T.M., Raj, V.S., Zaki, A.M., et al. 2012. Genomic characterization of a newly discovered coronavirus associated with acute respiratory distress syndrome in humans. MBio *3*.
van den Brand, J.M.A., Smits, S.L. and Haagmans, B.L. 2015. Pathogenesis of Middle East respiratory syndrome coronavirus. J. Pathol. *235*: 175–184.
van der Hoek, L., Pyrc, K., Jebbink, M.F., Vermeulen-Oost, W., Berkhout, R.J.M., Wolthers, K.C., et al. 2004. Identification of a new human coronavirus. Nat. Med. *10*: 368–373.

Van der Waals, F.W., Pomeroy, K.L., Goudsmit, J., Asher, D.M. and Gajdusek, D.C. 1986. Hemorrhagic fever virus infections in an isolated rainforest area of central Liberia. Limitations of the indirect immunofluorescence slide test for antibody screening in Africa. Trop Geogr Med *38*: 209–214.

Vasilakis, N., Deardorff, E.R., Kenney, J.L., Rossi, S.L., Hanley, K.A. and Weaver, S.C. 2009. Mosquitoes put the brake on arbovirus evolution: experimental evolution reveals slower mutation accumulation in mosquito than vertebrate cells. PLoS Pathog. *5*: e1000467.

Vazeille, M., Moutailler, S., Coudrier, D., Rousseaux, C., Khun, H., Huerre, M., et al. 2007. Two Chikungunya isolates from the outbreak of La Reunion (Indian Ocean) exhibit different patterns of infection in the mosquito, Aedes albopictus. PLoS ONE *2*: e1168.

Wacharapluesadee, S., Sintunawa, C., Kaewpom, T., Khongnomnan, K., Olival, K.J., Epstein, J.H., et al. 2013. Group C betacoronavirus in bat guano fertilizer, Thailand. Emerg. Infect. Dis. *19*: 1349–1351.

Walsh, P.D., Biek, R. and Real, L.A. 2005. Wave-like spread of Ebola Zaire. PLoS Biol. *3*: e371.

Weaver, S.C., Brault, A.C., Kang, W. and Holland, J.J. 1999. Genetic and Fitness Changes Accompanying Adaptation of an Arbovirus to Vertebrate and Invertebrate Cells. J. Virol. *73*: 4316–4326.

Weeks, A.R., Marec, F. and Breeuwer, J. 2001. A mite species that consists entirely of haploid females. Science *292*: 2479–2482.

Weiss, R.A. 2001. The Leeuwenhoek Lecture 2001. Animal origins of human infectious disease. Philos. Trans. R. Soc. Lond., B, Biol. Sci. *356*: 957–977.

Wertheim, J.O., Chu, D.K.W., Peiris, J.S.M., Kosakovsky Pond, S.L. and Poon, L.L.M. 2013. A case for the ancient origin of coronaviruses. J. Virol. *87*: 7039–7045.

WHO (World Health Organization). 2003a. Consensus document on the epidemiology of severe acute respiratory syndrome (SARS). Rep. WHO/CDS/CSR/GAR/ 2003. 11. Geneva, Switzerland. http://www.who.int/csr/sars/en/WHOconsensus.pdf

WHO (World Health Organization). 2003b. Summary of probable SARS cases with onset of illness from 1 November 2002 to 31 July 2003. http://www.who.int/csr/sars/country/table2004_04_21/en/

WHO/FAO/OIE. 2004. Report of the WHO/FAO/OIE joint consultation on joint consultation on emerging zoonotic diseases. In: (World Health Organization (WHO), Food and Agriculture Organization of the United Nations (FAO), World Organization for Animal Health (OIE), eds), pp. 1–72. Geneva, Switzerland. http://apps.who.int/iris/bitstream/10665/68899/1/WHO_CDS_CPE_ZFK_2004.9.pdf

WHO (World Health Organization). 2005. Outbreak of Ebola haemorrhagic fever in Yambio, south Sudan, April – June 2004. Wkly. Epidemiol. Rec. *80:* 369–376.

WHO (World Health Organization). 2007. Outbreak news. Ebola virus haemorrhagic fever, Democratic Republic of the Congo--update. Wkly. Epidemiol. Rec. *82:* 345–346.

WHO (World Health Organization). 2015a. Middle East respiratory syndrome coronavirus (MERS-CoV) – Saudi Arabia. http://www.who.int/csr/don/18-august-2015-mers-saudi-arabia/en/

WHO (World Health Organization). 2015b. Ebola. http://www.who.int/csr/disease/ebola/en/

WHO/FAO. 1959. JOINT WHO/FAO expert committee on zoonoses. In: World Health Organization (WHO)/Food and Agriculture Organization of the United Nations (FAO), eds, pp. 1–84. Geneva, Switzerland.

Woo, P.C.Y., Lau, S.K.P., Chu, C.-M., Chan, K.-H., Tsoi, H.-W., Huang, Y., et al. 2005. Characterization and complete genome sequence of a novel coronavirus, coronavirus HKU1, from patients with pneumonia. J. Virol. *79*: 884–895.

Woo, P.C.Y., Lau, S.K.P., Yip, C.C.Y., Huang, Y., Tsoi, H.-W., Chan, K.-H., et al. 2006. Comparative analysis of 22 coronavirus HKU1 genomes reveals a novel genotype and evidence of natural recombination in coronavirus HKU1. J. Virol. *80*: 7136–7145.

Woo, P.C.Y., Wang, M., Lau, S.K.P., Xu, H., Poon, R.W.S., Guo, R., et al. 2007. Comparative analysis of twelve genomes of three novel group 2c and group 2d coronaviruses reveals unique group and subgroup features. J. Virol. *81*: 1574–1585.
Woolhouse, M.E.J. and Dye, C. 2001. Preface. Philos. Trans. R. Soc. Lond., B, Biol. Sci. *356*: 981–982.
Woolhouse, M.E.J. and Gowtage-Sequeria, S. 2005. Host range and emerging and reemerging pathogens. Emerg. Infect. Dis. *11*: 1842–1847.
Woolhouse, M.E.J., Haydon, D.T. and Antia, R. 2005. Emerging pathogens: the epidemiology and evolution of species jumps. Trends Ecol. Evol. *20*: 238–244.
Yergolkar, P.N., Tandale, B.V., Arankalle, V.A., Sathe, P.S., Sudeep, A.B., Gandhe, S.S., et al. 2006. Chikungunya outbreaks caused by African genotype, India. Emerg. Infect. Dis. *12*: 1580–1583.
Yount, B., Roberts, R.S., Sims, A.C., Deming, D., Frieman, M.B., Sparks, J., et al. 2005. Severe acute respiratory syndrome coronavirus group-specific open reading frames encode nonessential functions for replication in cell cultures and mice. J. Virol. *79*: 14909–14922.
Zaki, A.M., van Boheemen, S., Bestebroer, T.M., Osterhaus, A.D.M.E. and Fouchier, R.A.M. 2012. solation of a novel coronavirus from a man with pneumonia in Saudi Arabia. N. Engl. J. Med. *367*: 1814–1820.
Zeng, F.Y., Chan, C.W.M., Chan, M.N., Chen, J.D., Chow, K.Y.C., Hon, C.C., et al. 2003. The complete genome sequence of severe acute respiratory syndrome coronavirus strain HKU-39849 (HK-39). Exp Biol Med *228*: 866–873.
Zhao, G.-P. 2007. SARS molecular epidemiology: a Chinese fairy tale of controlling an emerging zoonotic disease in the genomics era. Philos. Trans. R. Soc. Lond., B, Biol. Sci. *362*: 1063–1081.
Zhao, Z., Zhang, F., Xu, M., Huang, K., Zhong, W., Cai, W., et al. 2003. Description and clinical treatment of an early outbreak of severe acute respiratory syndrome (SARS) in Guangzhou, PR China. J. Med. Microbiol. *52*: 715–720.
Zhong, N.S., Zheng, B.J., Li, Y.M., Poon, Xie, Z.H., Chan, K.H., et al. 2003. Epidemiology and cause of severe acute respiratory syndrome (SARS) in Guangdong, People's Republic of China, in February, 2003. Lancet *362*: 1353–1358.

www.ingramcontent.com/pod-product-compliance
Ingram Content Group UK Ltd.
Pitfield, Milton Keynes, MK11 3LW, UK
UKHW062006071026
15591UKWH00023B/858

9 781910 190234